quaternary dating methods

FURTHER TITLES IN THIS SERIES

1. A.J. Boucot
   EVOLUTION AND EXTINCTION RATE CONTROLS

2. W.A. Berggren and J.A. van Couvering
   THE LATE NEOGENE — BIOSTRATIGRAPHY, GEOCHRONOLOGY AND PALEOCLIMATOLOGY OF THE LAST 15 MILLION YEARS IN MARINE AND CONTINENTAL SEQUENCES

3. L.J. Salop
   PRECAMBRIAN OF THE NORTHERN HEMISPHERE

4. J.L. Wray
   CALCAREOUS ALGAE

5. A. Hallam (Editor)
   PATTERNS OF EVOLUTION, AS ILLUSTRATED BY THE FOSSIL RECORD

6. F.M. Swain (Editor)
   STRATIGRAPHIC MICROPALEONTOLOGY OF ATLANTIC BASIN AND BORDERLANDS

Developments in Palaeontology and Stratigraphy, 7

quaternary dating methods

Edited by

WILLIAM C. MAHANEY

Department of Geography
York University
Downsview, Ont. M3J 2R7 (Canada)

ELSEVIER
Amsterdam – Oxford – New York – Tokyo 1984

ELSEVIER SCIENCE PUBLISHERS B.V.
1, Molenwerf,
P.O. Box 211, 1000 AE Amsterdam, The Netherlands

Distribution for the United States and Canada:

ELSEVIER SCIENCE PUBLISHING COMPANY INC.
52, Vanderbilt Avenue
New York, N.Y. 10017, U.S.A.

ISBN 0-444-42392-3 (Vol.7)
ISBN 0-444-41142-9 (Series)

Printed in The Netherlands

Professor Kolb was a native of Vicksburg, Mississippi where he spent most of his professional life associated with the geology and rock mechanics division of the Waterways Experiment Station, the chief research facility of the U.S. Army Corp of Engineers. He was made chief of that division in 1956 and continued in that capacity until he retired in 1973. His only break in service with the Corps of Engineers was two years which he spent as head of the Army Research Office in Fairbanks, Alaska, during the period 1963-64.

Much of his work and his publications involved the geomorphology and engineering geology of the lower Mississippi Valley, other gulf coast rivers, and the Gulf coastal and offshore areas. However, his division was also active in diverse geologic projects on a national and worldwide basis. Since his retirement he has continued to work as consulting geologist for various engineering firms and for various state agencies, and agencies of the Federal Government.

THIS VOLUME IS DEDICATED TO THE MEMORY OF DR. CHARLES R. KOLB.

Institute for Environmental Studies
Faculty and Staff
Louisiana State University
Baton Rouge, La.

. . .to determine the duration of such an ice age as that of the Pleistocene is not easy and there is much difference of opinion as to the number of thousands of years required for the advances and retreats of the ice and the interglacial times between. The time since the last ice sheet departed has been estimated in several ways, such as the rate of recession of waterfalls, the cutting of cliffs by waves, the building of bars by waves, the deposit of materials in the deltas of rivers, the weathering of glacial deposits, and the counting of annual layers of clay (varves) in glacial lakes.

---from A.P. Coleman, 1926, Ice Ages Recent and Ancient, London, Macmillan & Co., p. 67-68.

PREFACE

Dating of geological materials is of universal concern to earth scientists working on various problems within the Quaternary. If we are to adequately determine the ages of diverse materials in numerous Quaternary sequences all over the world, we must bring together specialists from several disciplines who are working, in some cases, with the newest dating techniques, and in others, with the older "classical" methods. In many research situations the use of a combination of dating methods provides the greatest return for time and labor spent in the field and laboratory. Over one hundred research scientists, from many parts of North America, convened at York University in May, 1981, to discuss dating methods of interest to Quaternary scientists. The publication of this proceedings volume should be viewed as a statement of where we are at present and where we intend to go in the future. Papers are organized into groups dealing with absolute, relative, and multiple dating methods.

The first group of papers on absolute dating (AD) methods opens with discussions of radiocarbon by J. Terasmae and J. Rucklidge. They report on some of the limitations and problems using gas counting methods in radiocarbon dating, and a new particle accelerator under construction at the University of Toronto. Both authors stress the importance of detecting very small concentrations of ^{14}C to permit dating of materials older than 40,000 years. Uranium series dating is discussed by H.P. Schwarcz, M. Gascoyne and C.E. Stearns. They summarize the problems of dating different materials, such as corals, mollusca, and speleothems, as well as the importance of using U-series dates in reconstructing sea level and climatic fluctuations. The application of $^{40}Ar/^{39}Ar$ to dating young volcanic rocks provides a topic of discussion for C. Hall and D. York. They summarize numerous problems, including mineral separations and enrichment of radiogenic ^{40}Ar. They stress that ^{40}Ar can be used for very young samples down to zero age, and further, that results are often superior to K/Ar determinations. In a much younger time frame of approximately the last hundred years, ^{210}Pb dating of lake sediments is examined by R.W. Durham and S.R. Joshi.

N.D. Naeser and C.W. Naeser describe fission track dating where ages are determined from tracks counted in separate grains, and tephra characterization as it applies to dating continental sequences. R.W. Barendregt discusses paleomagnetism as a technique for relative and absolute age dating of a variety of deposits. Short-lived excursions in the continental paleomagnetic record will likely provide a useful correlative tool as more work is carried out on terrestrial sequences. A second contribution outlining the problems of matching polarity transitions in undated sequences of deposits with the radiometrically dated magnetic polarity time scale is given by M. Stupavsky and C.P. Gravenor. They investigate the problems of using small amplitude time variable characteristics (excursions) in successful dating and correlation of sediments. Improved sampling techniques, magnetic cleaning, screening and smoothing of sediment magnetic remanence help to refine the magnetic calendar. Dating obsidian by measuring hydration thicknesses on clasts and determination of rates of hydration is discussed by I. Friedman and F. Trembour. They stress the immense time frame over which the method operates - several million years - as well as the precision of measuring hydration thickness, and factors controlling the rate of hydration. Thermoluminescence dating is reported on by M. Lamothe, A. Dreimanis, M. Morency and A. Raukas. They explore the use of quartz particles in the 100-140 μm and 40-70 μm sizes and polymineralic fine silts 4-11 μm to determine TL ages over most of late-Pleistocene time.

Amino acid dating of mollusks which involves the racemization reaction where the L-form (left-handed) develops into an equilibrium mixture (50-50) of D (right-handed) and L amino acids provides the subject of discussion for J. Wehmiller. He evaluates the method utilizing samples within radiometrically dated sections and stresses that several genera give increasing D/L values with older ages, and that each genera achieves racemic equilibrium in samples ranging from early Pleistocene to Miocene age. The utility of using amino acid D/L ratios in dating wood samples is reviewed by N.W. Rutter and R.J. Crawford. They assess ratios based on aspartic acid, but caution that correlation of ratios in sedimentary sequences must be based on assumptions of similar climatic and environmental histories. The use of amino acid ratios in wood does not appear to be applicable to units older than Sangamonian age. The principle, limitations and applications of tree-ring dating are discussed by M.L. Parker, L.A. Jozsa, S.G. Johnson and P.A. Bramhall. They review the method and stress the importance of X-ray densitometry in measuring ring width and density as well as computer cross-dating employed to match samples that are difficult to date using other matching techniques. P.E. Calkin and J.M. Ellis examine the successful use of lichen measurements to date Holocene geologic events in the Brooks Range, Alaska. The problems of determining a time-size curve for slow growing *Rhizocarpon geographicum* form the central part of their discussion.

In the second group of papers contributors discuss various individual relative dating (RD) methods including landform characteristics, till sheet characteristics, surface and buried paleosols, pollen, and vertebrate fossils as they are used in age assignment. The use of landform characteristics in determining the relative ages of landscapes is reviewed by D.R. Coates. He synthesizes the use of various criteria such as drainage density changes, hypsometric integrals, bifurcation ratios, stream junction angles, spur morphology and hillslope geometry in arriving at relative age relationships. The use of paleosols as a relative age dating tool is examined by W.J. Vreeken. He reviews several principles of age relationships between soils and surfaces, and summarizes the use of soil properties as time indicators. A. MacS. Stalker discusses till sheet characteristics from several sections in the southwestern Canadian Prairies. Deposits are differentiated on postglacial chemical alteration, compaction, structure, jointing, style of breakage and oxidation features. C.W. Finkl investigates the chronological ordering of pedological episodes vital to the reconstruction of paleoenvironments. He stresses the importance of micromorphological data as a tool in establishing chronological sequences of soils, especially in deeply weathered tropical terrain. Using pollen assemblages to check the accuracy of radiocarbon and for relative dating beyond the range of 50,000 yrs., A.M. Davis explores numerous problems resulting from variations in pollen spectra, redeposition of pollen from older sediments, size of the pollen sum, and use of numerical clustering methods. Changes in vertebrate faunal groups from Blancan, through Irvingtonian to Rancholabrean Land Mammal ages is discussed by C.R. Harington. He reviews the evidence for the first appearance of *Bison* in North America at the end of the Irvingtonian Age as well as conflicting absolute age determinations on the same materials.

The third group of papers centers on the use of multiple criteria used in age assignment. Here contributors stress the importance of using several relative dating methods to establish geological sequences. An evaluation of multiple dating methods used to differentiate Quaternary deposits in the Wind River and Teton Ranges, Wyoming, by W.C. Mahaney, D. Halvorson, J. Piegat and K. Sanmugadas centers on the importance of lichenometry, weathering characteristics and soils as age indicators. They pay particular attention to the use of Fe ratios (oxalate extractable/dithionite extractable iron) and quartz/feldspar ratios, clay mineral composition, and organic/chemical properties in soils for age differentiation. Relative and absolute dating methods applied to dating late glacial sediments in the Lake Agassiz Basin provide the basis for discussion by R.W. Klassen. He assesses the discrepancy between absolute and relative chronologies, and suggests the oldest ^{14}C dates may be contaminated with dead carbon. W.J. Wayne describes the use of several relative dating methods used in the Río Blanco Basin, Cordon del Plata, Mendoza Province, Argentina, to date glacial and periglacial deposits. These methods include zircon fission track, deposit morphology, loess thicknesses, soil profile development, vegetation characteristics and lichens.

The papers were followed by a panel discussion on Saturday evening. The panel was chaired by C.S. Churcher (Royal Ontario Museum, Toronto), D.R. Coates, H.B.S. Cooke (Vancouver, B.C.) and J. Terasmae. They reviewed and assessed the information presented by various speakers. This ended with a short but lively discussion led by C. Kolb (Louisiana State Univ., Baton Rouge, La.) on the suitability of salt domes in Louisiana for nuclear waste disposal. The importance of using various dating methods to determine the age of stream terraces cut in late Tertiary sediments overlying the salt domes formed the major thrust of discussion.

On the last day of the Symposium, I.P. Martini, M.E. Brookfield and Q.H.J. Gwyn led a field trip to the Bowmanville Bluffs, along the north shore of Lake Ontario. In the field trip guide they discuss the sequence of Wisconsinan tills and interlayered lacustrine and fluvial sediments.

The late Harry S. Crowe, Dean of Atkinson College and the York University Ad-Hoc Fund assisted with financial support for the Symposium and publication of this volume. The technical and organization details for the Symposium were worked out by A. Cote and his facilities staff. E. Yates typed some of the papers and handled numerous pre-registration problems. G. Wahab supervised registration. Only minimal editorial changes were made to the manuscripts to create a uniform design throughout the volume. Evelyn Cassalman diligently typed and proofread the abstracts-with-program and field guide as well as this proceedings volume. Linda Mahaney proofread the entire volume in draft and camera-ready copy stages. Milo Dowden provided encouragement in the preparation of the proceedings. I am most grateful to all of the above-named individuals for their valuable contribution and support toward the publication of this volume.

Finally, a number of students assisted with the planning, organization and logistics for the Symposium. In particular I would like to thank G. Berssenbrugge, V. Elchuk, R. Foxall, L.J. Gowland, P. Julig, J. Kolodiz, D. McWilliams, R. Prukner and G. Yamada. Sessions were taped by J. Briggs, D.I.A.R., York University. Some illustrations were prepared by G. Berssenbrugge. All illustrations were photographically reproduced by J. Dawson and J. Nolty, D.I.A.R., York University.

WILLIAM C. MAHANEY
Downsview Ont.

CONTENTS

CONTRIBUTING AUTHORS

R.W. Barendregt — Lethbridge University, Department of Geography, Lethbridge, Alberta

P.A. Bramhall — Forintek Canada Ltd., Vancouver, British Columbia

M.E. Brookfield — University of Guelph, Department of Land Resource Science, Guelph, Ontario

P. Calkin — State University of New York, Department of Geological Sciences, Buffalo, New York 14226

D.R. Coates — State University of New York, Department of Geological Sciences, Binghamton, New York 13901

R.J. Crawford — University of Alberta, Department of Chemistry, Edmonton, Alberta

A.M. Davis — University of Toronto, Department of Geography, Toronto, Ontario M5S 1A1

A. Dreimanis — University of Western Ontario, Department of Geology, London, Ontario N6A 5B7

R.W. Durham — National Water Research Institute, Canada Centre for Inland Waters, Environment Canada, Burlington, Ontario L7R 4A6

J.M. Ellis — State University of New York, Department of Geological Sciences, Buffalo, New York 14226

C.W. Finkl, Jr. — Nova University, Ocean Sciences Center, Institute of Coastal Studies, 8000 North Ocean Drive, Dania, Florida 33004

I. Friedman — U.S. Geological Survey, Denver Federal Center, Denver, Colorado

M. Gascoyne — McMaster University, Department of Geology, Hamilton, Ontario

C.P. Gravenor — University of Windsor, Department of Geology, Windsor, Ontario N9B 3P4

Q.H.J. Gwyn — Département de géographie, Université de Sherbrooke, Sherbrooke, Québec

C.M. Hall
University of Toronto
Department of Physics
Toronto, Ontario

D. Halvorson
University of North Dakota
Department of Geology
Grand Forks, North Dakota 58201

C.R. Harington
National Museums of Canada
Paleobiology Division
Ottawa, Ontario

S.G. Johnson
Forintek Canada Ltd.
Vancouver, British Columbia

S.R. Joshi
National Water Research Institute
Canada Centre for Inland Waters
Environment Canada
Burlington, Ontario L7R 4A6

L.A. Jozsa
Forintek Canada Ltd.
Vancouver, British Columbia

R.W. Klassen
Geological Survey of Canada
3303 - 33rd Street North West
Calgary, Alberta

M. Lamothe
University of Western Ontario
Department of Geology
London, Ontario N6A 5B7

W.C. Mahaney
York University
Atkinson College
Department of Geography
4700 Keele Street
Downsview, Ontario M3J 2R7

I.P. Martini
University of Guelph
Department of Land Resource Science
Guelph, Ontario

M. Morency
Université du Québec a Montréal
Département des Sciences de la Terre
Montréal, Québec H3C 3P8

C.W. Naeser
U.S. Geological Survey
Denver Federal Center
Denver, Colorado

N.D. Naeser
U.S. Geological Survey
Denver Federal Center
Denver, Colorado

M.L. Parker
M.L. Parker Co., Box 638
Point Roberts, Washington 98281

J. Piegat
Purdue University
Department of Geosciences
West Lafayette, Indiana 47907

A. Raukas
Academy of Sciences of Estonian SSR
Institute of Geology
200101 Tallin, Estonia, U.S.R.R.

J.C. Rucklidge
University of Toronto
Department of Geology
Toronto, Ontario M5S 1A1

N.W. Rutter	University of Alberta Department of Geology Edmonton, Alberta
K. Sanmugadas	York University Department of Geography 4700 Keele Street Downsview, Ontario M3J 1P3
H. Schwarcz	McMaster University Department of Geology Hamilton, Ontario
A. MacS. Stalker	Geological Survey of Canada 601 Booth Street Ottawa, Ontario K1A 0E8
C.E. Stearns	Tufts University Department of Geology Medford, Massachusetts 02155
M. Stupavsky	University of Windsor Department of Geology Windsor, Ontario N9B 3P4
J. Terasmae	Brock University Department of Geology St. Catharines, Ontario
F. Trembour	U.S. Geological Survey Denver Federal Center Denver, Colorado
W.J. Vreeken	Queens University Department of Geography Kingston, Ontario
W.J. Wayne	University of Nebraska Department of Geology Lincoln, Nebraska
J.F. Wehmiller	University of Delaware Department of Geology Newark, Delaware 19711
D. York	University of Toronto Department of Physics Toronto, Ontario

SESSIONS CHAIRMEN

H.B.S. Cooke — 2133, 154 Street
White Rock, British Columbia

B.D. Fahey — Guelph University
Department of Geography
Guelph, Ontario

D. Ford — McMaster University
Department of Geography
Hamilton, Ontario

D.R. Grant — Geological Survey of Canada
601 Booth Street
Ottawa, Ontario

V.K. Prest, Retired — Geological Survey of Canada
601 Booth Street
Ottawa, Ontario

A. MacS. Stalker — Geological Survey of Canada
601 Booth Street
Ottawa, Ontario

J. Terasmae — Brock University
Department of Geological Sciences
St. Catharines, Ontario

J. Welsted — Brandon University
Department of Geography
Brandon, Manitoba

PANEL DISCUSSION

C.S. Churcher — University of Toronto
Department of Zoology
Toronto, Ontario

D.R. Coates — State University of New York
Department of Geological Sciences
Binghamton, New York 13901

H.B.S. Cooke — 2133 154 Street
White Rock, British Columbia
V4A 4S5

J. Terasmae — Brock University
Department of Geological Sciences
St. Catharines, Ontario L2S 3A1

RADIOCARBON DATING: SOME PROBLEMS AND POTENTIAL DEVELOPMENTS

J. TERASMAE

ABSTRACT

Research on the radiocarbon dating method during the last 20 years has increased almost exponentially in terms of both volume and diversity, and there has been also an increase in the number of problems relating to various aspects of radiocarbon dating. The multidisciplinary scope of many radiocarbon dating problems has required involvement of expertise from several fields of research and the problems can be grouped into a few broad categories: secular variations of radiocarbon, laboratory techniques, reference standards, sample contamination, calibration and data reporting, radiocarbon in oceans, fresh water, and soils, correction and evaluation of radiocarbon dates, dating of various materials, *etc.*

This abundance of problems has led to some negative criticism of either the whole or certain aspects of the method. On the positive side, however, the recognition of problems has led to a better understanding of the method and significant improvements have been made with regard to both precision and accuracy of radiocarbon dating. Current active research on many problems by different laboratories, and especially the concerted efforts directed towards solving specific problems by groups of laboratories clearly indicate the healthy condition of the radiocarbon dating field.

Some expected developments are likely to include additional standardization of laboratory methods and procedures, improved standards of calibration and correction of radiocarbon dates, better understanding of secular variations of radiocarbon, and some extension of the range of the method although the common practical limit will probably remain at 30,000 to 50,000 years BP.

INTRODUCTION

Radiocarbon dating is the method of age determination that is probably considered first in most studies of late Quaternary events where a time scale or a chronological sequence needs to be established. This is not because it is the only method available for such studies, or even the most appropriate one in many cases, but rather because the potential users are more familiar with radiocarbon dating than with many of the other new dating methods.

Radiocarbon dating has been in use for about 30 years since Libby (1952) established the basic principles and techniques of the method. In more recent years the numbers of both radiocarbon dating laboratories and radiocarbon dates have increased almost exponentially, and according to an estimate in 1976 about 15,000 radiocarbon dates were produced annually (Libby, 1979).

As could be expected, the number of problems relating to radiocarbon dating also increased as experience accumulated from dating different kinds of samples from different environments and radiocarbon dates were checked against other dating methods and historical records. These problems have been the object of active research from the time of the 1st International Radiocarbon Conference in 1954. The 10th Radiocarbon Conference was held in 1979 and if the number of pages of conference proceedings is any indication of the amount of research (about 1000 pages in the 10th Conference Proceedings) then this research will certainly continue and even increase in the future years. Of course, a large volume of additional data has been published in various journals and reports, including Radiocarbon, that contains date lists from a number of laboratories and specific technical notes.

It would be quite impossible to cover fully all problems of radiocarbon dating in this brief paper and therefore I have intentionally chosen only a few problems for discussion while realizing that other people most likely would have made a different selection of problems. Radiocarbon dating is by no means unique with regard to problems of methodology but the range of this dating method covers the time of human activities and historical records and, therefore, some errors in accuracy are easier to detect than in the case of other methods. Furthermore, the overenthusiastic users have expected greater accuracy than the method can normally offer and this has led to some disappointment and sometimes rather unwarranted criticism.

There is still some confusion about the matter of what a radiocarbon date really represents. The user should remember that a date is based simply on a best estimate of radiocarbon content of the sample submitted to the laboratory. It does not include any of the possible sampling errors and normally none of the physical and biological error sources that can affect the date. The reported date is a mean value with a stated error figure, for example 11,180 ± 180 years BP (GSC-649) where the data in brackets identify the laboratory and sample number.

As pointed out by Ogden (1977) the statistics of nuclear disintegration appoximate a Poisson distribution which means that the sample events, recorded as counts in a radiation detector, are asymmetrically grouped about the mean. For ages of less than about 25,000 years the difference is small enough so that the total calculated error range (at the one sigma level) is simply divided by 2 and this value is reported as the one sigma error for the radiocarbon date. However, for greater ages the asymmetry effect of a Poisson distribution is significant and these dates are reported as, for example, 48,300 + 500, -400 years BP (GrN-6695). The reported error is based on the statistics of counting alone and normally includes uncertainties relating to measurement of modern, background, and sample activities.

Owing to the statistics of radiocarbon determination, the probability that the true age of the sample is exactly the reported mean is zero, and all that is implied is that the reported radiocarbon age of the submitted sample has two chances out of three of being within the quoted limits.

It is interesting to note that, according to Ogden (1977), fewer than 50% of the radiocarbon dates from geological and archaeological samples in northeastern North America have been adopted as "acceptable" by the users. To me this means that there is a need for discussion of the problems concerning the radiocarbon dating method and that the users should be encouraged to explain why some radiocarbon dates are rejected whereas others are considered acceptable. One might well wonder whether the saying "my mind is made up, please do not confuse me with facts" may have some relevance in this context.

THE SIMPLE MODEL AND SOME BASIC ASSUMPTIONS

In principle the radiocarbon dating method is rather straightforward. As described by Libby (1952), radiocarbon (^{14}C) is produced in the upper atmosphere by reaction between cosmic radiation and nitrogen. Radiocarbon combines with oxygen to form radioactive carbon dioxide that is uniformly mixed throughout the atmosphere and incorporated into the biosphere (primarily through photosynthesis) and exchanged with the hydrosphere resulting in a global equilibrium state of radiocarbon - the initial radiocarbon activity. When a subsystem (a tree, a sea shell, *etc.*) is isolated from the global system (*i.e.* a tree is cut down, or dies and is buried in sediment) then no more radiocarbon is added to it and the initial activity (amount of ^{14}C in the subsystem or sample) begins to decrease according to laws of radioactive decay. Provided that the rate of decay is known (reflected by the half-life of the isotope) the activity of radiocarbon in the sample is determined and the length of time (age) that the sample has been isolated from the global equilibrium state can be calculated. The above description can be considered as "the simple model" of radiocarbon dating.

The validity of this model depends on several basic assumptions which include:

1. The production of radiocarbon in the atmosphere has been constant for the last 50,000 - 100,000 years.

2. The mixing, uptake, and exchange of radiocarbon in the atmosphere - biosphere - hydrosphere system have been uniform and rapid on a global scale to provide the same initial activity for all samples used for radiocarbon dating.

3. The decay rate of radiocarbon has been constant.

4. No "young" or "old" carbon has been added to the sample since it was isolated from the global equilibrium state.

5. No isotopic fractionation has occurred to alter the standard $^{14}C : {}^{13}C : {}^{12}C$ ratios in the sample.

Radiocarbon analysts, including Libby, realized almost from the beginning that although the above assumptions were generally valid they were not exactly correct and subsequent research has been directed towards investigation of problems that can affect various aspects of the radiocarbon dating method. These problems can be grouped in a few general categories: (a) secular fluctuations of radiocarbon, (b) laboratory techniques, (c) reference standards, (d) sample contamination, (e) calibration and data reporting, (f) radiocarbon in oceans, fresh water, and soils, (g) correction and evaluation of radiocarbon dates, and (h) dating of various materials.

RADIOCARBON FLUCTUATIONS

Although probable temporal radiocarbon fluctuations in the atmosphere - biosphere - hydrosphere system were speculatively anticipated during the early years of radiocarbon dating, it required technological, theoretical, and empirical advancements in several disciplines (including astrophysics, oceanography, and geochemistry) before some of the hypotheses could be tested and at least partly quantified.

Recently Damon *et al.* (1978) summarized the possible causes of radiocarbon fluctuations.

1. Variations in the rate of radiocarbon production in the atmosphere.

   a. Variations in the cosmic-ray flux throughout the solar system.

      a. Cosmic-ray bursts from supernovae and other stellar phenomena.
      b. Interstellar modulation of the cosmic-ray flux.

   b. Modulation of the cosmic-ray flux by solar activity.

   c. Modulation of the cosmic-ray flux by changes in the geomagnetic field.

   d. Production by antimatter meteorite collisions with the earth.

e. Production by nuclear weapons testing and nuclear technology.

2. Variations in the rate of exchange of radiocarbon between various geochemical reservoirs and changes in the relative carbon dioxide content of the reservoirs.

a. Control of CO_2 solubility and dissolution as well as residence times by temperature variations.

b. Effects of sea-level variations on ocean circulation and capacity.

c. Assimilation of CO_2 by the terrestrial biosphere in proportion to biomass and CO_2 concentration, and dependence of CO_2 on temperature, humidity and human activity.

d. Dependence of CO_2 assimilation by the marine biosphere upon ocean temperature and salinity, availability of nutrients, upwelling of CO_2-rich water, and turbidity of the mixed layer of the ocean.

3. Variations in the total amount of carbon dioxide in the atmosphere, biosphere and hydrosphere.

a. Changes in the rate of introduction of CO_2 into the atmosphere by volcanism and other processes that result in CO_2 degassing of the lithosphere.

b. The various sedimentary reservoirs serving as a sink of CO_2 and ^{14}C. Tendency for changes in the rate of sedimentation to cause changes in the total CO_2 content of the atmosphere.

c. Combustion of fossil fuels by human industrial and domestic activity.

There is no longer any doubt that natural radiocarbon fluctuations have, indeed, occurred through time, and in addition to those caused by human activities (burning of fossil fuels and nuclear weapons testing). However, the causes, magnitude, and possible cycles of the natural ^{14}C fluctuations are still the object of active research, as are, for example, the probable linkages of these fluctuations with solar activity and climate change.

Dendrochronology and radiocarbon dating have had a long association because tree-ring series have provided a time scale against which radiocarbon dates could be checked, as well as providing samples of wood from the same series that could be used for radiocarbon activity measurements. For example, Suess (1955) used wood samples to demonstrate the decreasing atmospheric radiocarbon activity due to combustion of fossil fuels, and de Vries (1958) showed that atmospheric radiocarbon concentration had fluctuated due to natural causes during the last few hundred years, using samples from dendrochronologically dated tree-ring series. Subsequent research has confirmed the results of these early studies and perhaps the best known example is the study, about 7,500 years long, of the bristlecone pine tree ring series established in the southwestern United States by Ferguson (1970). Some 550 to 600 samples from this series have been analysed for radiocarbon by five different laboratories (Damon *et al.*, 1978, Suess, 1980). It appears that both short-term (decades and centuries) and long-term (millennia) fluctuations of natural radiocarbon concentration have occurred and range from about one or two percent for the short events and about 10% for the last 10,000 years. A 10% change in radiocarbon concentration would cause an approximately 1000 years change in a corresponding radiocarbon date.

Although these fluctuations comprise a problem in radiocarbon dating they do not invalidate the method because when the nature of the fluctuations is sufficiently well known they can be corrected for in the radiocarbon age calculation.

The nuclear weapons testing caused a sudden increase in the atmospheric radiocarbon concentration (almost 100%) in the early 1960-s and there has been a steady decline in more recent years (Nydal *et al.*, 1979, Stenhouse and Baxter, 1979).

This unintentional 'experiment' has been used to study the dispersal, residence time, uptake, and exchange of radiocarbon in the atmosphere - biosphere- hydrosphere system. On the other hand, human activities have altered the natural radiocarbon concentration to the extent that it is no longer possible to use the present ^{14}C concentration as "normal" for comparison with past concentrations and this certainly has created a problem for radiocarbon dating.

REFERENCE STANDARDS

Reference standards in radiocarbon dating are necessary for calibration and periodic checking of the proper operation of laboratory equipment, as well as for the calculation of radiocarbon dates that are reported to the user. The use of reference standards assures the uniformity of radiocarbon analysis in individual laboratories and provides a common base for comparing radiocarbon dates produced by different laboratories.

For many years the oxalic acid standard, provided by the U.S. National Bureau of Standards (first batch of 1000 pounds was produced in 1957) has been used by radiocarbon dating laboratories. The carbon-14 activity standard is defined as 95% of the carbon-14 activity in the NBS Oxalic Acid Standard. The rapid increase in the number of radiocarbon dating laboratories and the corresponding increase in user demand for the Oxalic Acid Standard had depleted the supply by about 1978. A new batch (1000 pounds) was produced for the NBS by fermentation of French beet molasses from the 1977 spring-summer-fall harvest. However, the ^{13}C abundance in the new standard was found to be higher by 1.51 ± 0.17%, and hence, the new standard is not identical with the depleted supply of the old standard.

A number of studies of the stable isotopes of carbon (^{13}C and ^{12}C), carried out since the 1950-s, have shown that the ratio of $^{13}C/^{12}C$ is not the same in all sample materials used for radiocarbon dating. Therefore, another standard was required for use as reference in respect to the carbon isotope ratios, and for correcting radiocarbon dates due to the variance of the carbon isotope ratios. It had been established, furthermore, that the fractionation of ^{14}C is very closely twice that for ^{13}C, and because of this relationship the measured $^{13}C/^{12}C$ ratio can be used for correcting the initial ^{14}C activity of the sample. For this purpose the PDB Carbonate Standard (or "Chicago Limestone Standard") was established (Craig, 1957). This standard is based on belemnites *(Belemnitella americana)* from the Peedee Formation (Cretaceous) in South Carolina. The $^{13}C/^{12}C$ ratio of a sample is compared with the same isotopic ratio in the PDB Carbonate Standard and expressed as $\delta^{13}C_{PDB}$ in per mille ($^{o}/_{oo}$).

In addition to the oxalic acid and belemnite standards, other reference standards have been introduced by different laboratories for different reasons (Polach, 1979), for example:

1. Arizona 1850 Wood, primary radiocarbon dating standard. 1.2 kilograms of ten rings (AD 1846-1855) from douglas fir growing near Tucson. $\delta^{13}C_{PDB}$ = 23.0 ± 1 $^{o}/_{oo}$ (range -22 to -25 $^{o}/_{oo}$).

2. ANU (Australian National University) Sucrose, secondary radiocarbon dating standard. 1000 kilograms of analytically pure sucrose produced from sugar cane (Sept. 1965 - June 1971). A 10 ton batch of raw sugar was processed. $\delta^{13}C_{PDB}$ = -11 ± 1.5 $^{o}/_{oo}$ (range -0 to -13 $^{o}/_{oo}$).

3. Scientists in New Zealand have introduced contemporary reference standards for marine shells and fresh-water shells, Antarctic specimens, terrestrial bone specimens and land snails

while standards for soils are being investigated (Rafter *et al.*, 1972). The $\delta^{13}C_{PDB}$ values have been determined for these standards that are based on collections made in the 1950-s which are not subject to the nuclear weapons testing effects.

4. Other additional reference standards have been introduced by some laboratories primarily for their own use.

The proliferation of reference standards, especially if it continues at an increasing rate, can certainly lead to problems for radiocarbon dating. Some standards are produced in small quantity that will be quickly depleted and new batches may or may not be identical to the previous ones. The main problem, however, will be the use of various standards for correcting radiocarbon dates. Although a dating laboratory normally provides adequate documentation on the reported radiocarbon dates, especially in cases when sample preparation or age calculation differs from the generally followed practice, such documentation is commonly omitted by many users who publish dates without the qualifying data provided by the laboratory. Without qualifying data the reader has no way of knowing which dates have been corrected (if they were corrected?) using which reference standards (or procedures). Subsequent use of such unqualified published dates will only compound the confusion.

CARBON ISOTOPE FRACTIONATION

Recent studies have demonstrated that isotopic fractionation of carbon isotopes occurs during the growth processes of all organic materials. Hundreds of samples of different species (especially plants) have been analysed for carbon isotope fractionation and it has been found that isotope ratios vary between species and also between different parts of the same organism.

Plants can be divided into two groups according to the extent of carbon isotope fractionation that is directly correlated with the pathway of carbon metabolism (Troughton, 1972). Most temperate-climate hardwoods and conifers operate on the Calvin or C_3 carbon-fixation cycle, whereas many arid-land plants operate on the Slack-Hatch or C_4 cycle. However, a third group of plants (the CAM plants, referring to Crassulacean Acid Metabolism) can operate either as C_3 or C_4 plants depending on environmental conditions. There is considerable variation in respect to carbon isotope composition within each of these groups of plants (Lerman, 1972) and the isotope fractionation clearly is another problem in radiocarbon dating.

It is recommended that radiocarbon activities of all standards and samples be normalized to a standard $^{13}C/^{12}C$ ratio, or $\delta^{13}C$ value on the PDB scale: -25 $^{o}/_{oo}$ for organic matter and -19 $^{o}/_{oo}$ for the oxalic acid standard (Damon *et al.*, 1978). This normalization will compensate for environmental, biological, and laboratory isotope fractionation. However, only about 20% of laboratories publish lists of corrected dates (Lerman, 1972). Therefore the user should be aware of this problem and be certain whether a date has or has not been corrected for isotope fractionation.

From a practical viewpoint, the corrections for fractionation are rather insignificant on a 30,000 year old sample because a correction of even 400 years is unlikely to be seen within the quoted plus or minus error for that sample. However, these corrections become increasingly significant for younger samples.

Figure 1 illustrates the magnitude of corrections for various kinds of samples that result from isotope fractionation. The corrections for apparent radiocarbon ages are shown relative to "average wood" that belongs in the C_3 group of plants and has a $\delta^{13}C_{PDB}$ value of -25 $^{o}/_{oo}$.

One should remember that it is important to identify the dated sample to the species of plant or animal if a correction for isotope

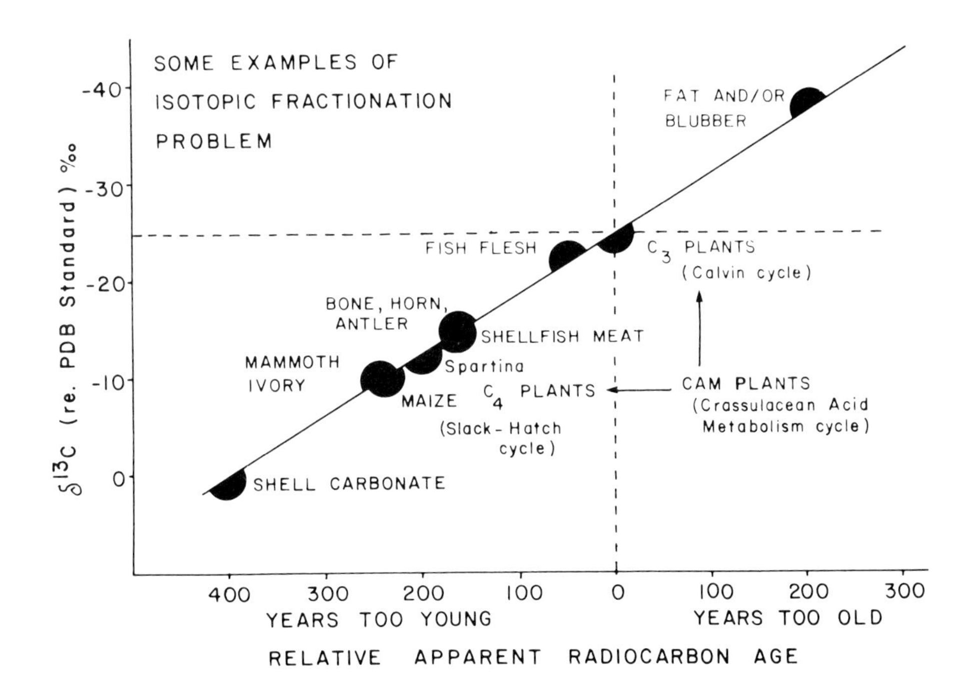

Figure 1 Variation in the apparent radiocarbon age of different kinds of samples due to carbon isotope fractionation, plotted relative to "average wood" (C_3 plants) that has a $\delta^{13}C_{PDB}$ value of -25 $^o/_{oo}$ (modified after Stuckenrath, 1977).

fractionation is to be made at a later date on the basis of available data on fractionation. Detrital samples (composed of fragments of many species of plants, for example) present an unresolved problem in respect to correction for isotope fractionation. Lake sediments, peat, soils, and alluvial plant-bearing deposits are all detrital and collectively comprise a large proportion of all radiocarbon dated samples.

CORRECTION AND REPORTING OF RADIOCARBON DATES

Technological advancements in the radiocarbon dating method since the 1950-s improved significantly the precision of the method and also confirmed the discrepancy between radiocarbon dates and other time scales such as the Egyptian chronology (based on historically dated samples) and dendrochronologically dated wood.

Research by Pearson *et al.* (1977) has demonstrated that it is possible to make accurate measurements of radiocarbon in dendrochronological samples with an overall precision of less than 25 years standard deviation. The general conclusion is that "radiocarbon years" do not correspond exactly to calendar years or the absolute time scale and, therefore, attempts have been made to establish methods for correcting the radiocarbon chronology.

The problem is that correction for natural atmospheric radiocarbon variation is only one of several corrections that have been necessary to relate the radiocarbon assay of organic sample materials to a calendar time scale. Carbon isotope fractionation has been mentioned above, and there are many reports that deal with the problems of sample contamination, including the growth processes of organisms and post-depositional addition of either "old" or "young" carbon, and the effects of such contamination on radiocarbon dates (for example, Blake, 1981;

Karrow and Anderson, 1975; Mangerud and Gulliksen, 1975; Ogden, 1977; Olson and Broecker, 1958; Olsson, 1970, 1974; Olsson and Blake, 1961-62; Olsson and Florin, 1980; Pardi and Marcus, 1977; Sissons, 1981; Stuckenrath, 1977; Stuckenrath *et al.*, 1979; Stuiver and Borns, 1975; Thurber, 1972).

The attempts of correcting radiocarbon dates in respect to the various sources of error have resulted in many proposed methods of calibration so that radiocarbon dates could be expressed in calendar years or as an absolute time scale or chronology. One of the best known methods of calibration is based on the dendrochronology of bristlecone pine (Suess, 1980). However, Damon *et al.* (1978) list references for no less than 13 radiocarbon calibration schemes (or "curves") that follow the same overall trend but differ in detail. At the moment there is no "official calibration curve" for correcting radiocarbon dates and Stuckenrath (1977) cautioned that publication of dates "corrected" by one scheme or another would merely add to the existing confusion because "eventually, those 'corrected' dates will have to be uncorrected in order to be recorrected in order to be correct".

The reporting of radiocarbon dates also presents some problems. By international agreement, all laboratories use a common set of standards for calculation and reporting of radiocarbon dates so that the published results will be comparable. The NBS Oxalic Acid Standard and the PDB Carbonate Standard were already mentioned above. In addition, the "Libby" half-life of 5568 ± 30 years is used instead of the more recently determined value of 5730 ± 40 years, and 1950 AD is used as the "zero year" (or "present") when radiocarbon dates are expressed as the number of years before present, or BP. Unfortunately, however, complete uniformity with respect to the use of standards has not been achieved and differences exist in sample preparation procedures, the age calculation statistics, and the reporting format of dates. For example, Figure 2 illustrates the significance of designations AD, BC and BP, especially with reference to reporting relatively young radiocarbon ages.

REPORTING OF RADIOCARBON DATES
AS **N** YEARS **AD, BC, BP.**

EXAMPLE: **The age of event E is "600 years"**
(relative to 1980)

600 years AD = 1980 - 600 = 1380 years old
600 years BC = 1980 + 600 = 2580 years old
600 years BP = (1980 - 1950) + 600 = 630 years old

Radiocarbon dates are commonly reported as ages in years before present (BP) where "PRESENT" means 1950 AD.

11 180 ± 180 yr BP (GSC - 649)

Figure 2 The effect of designations AD, BC, and BP in reporting relatively young radiocarbon dates.

The <u>9th International Radiocarbon Conference</u> (Berger and Suess, 1979) passed the following resolutions.

1. No change is recommended in the use of conventional ^{14}C years. A conventional ^{14}C year implies the use of the Libby half-life of 5568 years.

2. The reference standard remains 95% of the NBS oxalic acid activity, corrected for isotopic fractionation to a $\delta^{13}C$ value of -19.0 $^{o}/_{oo}$ with regard to PDB. The year AD 1950 continues to be the reference year for conventional ^{14}C ages in years BP. The use of lower case bp was rejected.

3. It is recommended that 1950 be no longer subtracted from conventional ^{14}C ages in order to arrive at a so-called AD/BC age. AD*/BC* nomenclature is to be used after application of one of the available age correction curves or tables. The asterisk indicates a tree-ring calibrated age (*e.g.*, 1250 BC*), whereas the text should specify the curve or table used.

4. It is recommended that a format for reporting radiocarbon dates compatible with computer-based retrieval systems be established.

Subsequent workshops have dealt with several areas of concern, for example the calibration of the radiocarbon time scale (Damon *et al.*, 1980), the use of the new NBS Oxalic Acid Standard (Stuiver, 1980), and the reporting of marine sample ages. It is expected that continued research and international cooperation will resolve some of the problems that can all too easily lead to a chaotic situation in radiocarbon dating.

The user of radiocarbon dates also has a responsibility in helping to avoid possible confusion in respect to published dates. As mentioned earlier, laboratories normally provide adequate background information (sample preparation, age calculation, corrections) for radiocarbon dates reported to the user, particularly when special procedures have been involved. However, many users commonly omit the rather essential supporting data, provided by the laboratory, when they publish radiocarbon dates and a "third generation" user of already published dates has no way of knowing what qualifying statements were made by the laboratory initially. This can certainly lead to problems.

It is important, therefore, that users publish radiocarbon dates properly, for example, 11,180 ± 180 years BP (GSC-649), including the laboratory sample identification data and any qualifying information that the dating laboratory provides. Correct identification of the sample material is also an important matter. All this information is essential if a radiocarbon date is to be corrected at a later time when the appropriate procedures have been established and accepted, or if intermediate corrections are desirable for some particular purpose. Without adequate background data no meaningful correction of radiocarbon dates is possible.

COMMENTS ON SOME OTHER PROBLEMS

Sample contamination has been an ever present problem in radiocarbon dating. It ranges from the obvious - modern roots, wood fragments with paint on them, and a piece of wood with a rusty nail in it - to the visually undetectable, such as an archaeological sample of charcoal pieces derived from a mixture of multiple age wood (for example, previous structures, driftwood, *etc.*). Usually the mixed sample problem can be resolved by dating a set of samples from the same stratigraphic level but frequently a single sample is collected that, furthermore, might be so small that only one age determination can be made. The cost of dating (about $180 per sample charged by commercial laboratories) may also be a problem due to limited funds for a project or the long time that careful dating of a set of samples can require.

A general survey of the influence of some possible allochthonous material of different origin on the apparent age of radiocarbon dated samples can be summarized as follows.

1. Apparent age too low.
   a. Dissolved humus products from stratigraphically higher levels.
   b. Modern roots.
   c. Sample containers (paper bags, cloth bags, cardboard boxes).
   d. Careless sampling and sample storage can contribute younger or modern carbon to the sample.
2. Apparent age too high.
   a. Dissolved humus products from older deposits.
   b. Dissolved groundwater carbonate.
   c. Older organic particles.
   d. Eroded or reworked deposits (especially carbonate rock particles).
   e. Coal, lignite, graphite particles.
3. "Mixed" or "average" apparent age.
   a. Soil organic matter (from soil profiles).
   b. Alluvial deposits (erosion and redeposition).
   c. Mixing of deposits by bioturbation, frost action, or human activities.

The sample contamination problems can be at least partly alleviated by user awareness of all aspects of this matter, including the recognition of potential contamination sources in the field. The value of careful observation and field notes cannot be overemphasized because the age determination in the laboratory can be only as good as the quality of the submitted sample, and supporting field data can be very helpful in dealing with contamination problems.

Sample size is another common problem. Normally about 5 to 10 grams of carbon is required for a radiocarbon age determination. To attain high accuracy in radiocarbon measurements Pearson *et al.* (1977) used 180 - 200 gram samples of wood. Unfortunately, many users submit samples for dating that are well below the minimum required weight. Insufficient sample size can affect a radiocarbon age determination in at least two different ways. It will not be possible to properly pretreat (or chemically clean) the sample, and the error limits of the date will be greater for a small sample (Figure 3) even at long counting times. This problem is aggravated by the commonly very low organic carbon content of many Pleistocene deposits (especially glacial deposits) and it is simply not feasible to obtain sufficiently large samples for radiocarbon dating within the practical limitations of time and cost in the field.

In archaeological research it can be a difficult decision whether to submit one or more valuable objects that comprise a sufficiently large sample for radiocarbon dating (that will destroy the sample) or submit a very small sample that cannot yield the desired accuracy of the date.

SOME POTENTIAL DEVELOPMENTS

One of the most exciting recent developments in radiocarbon dating

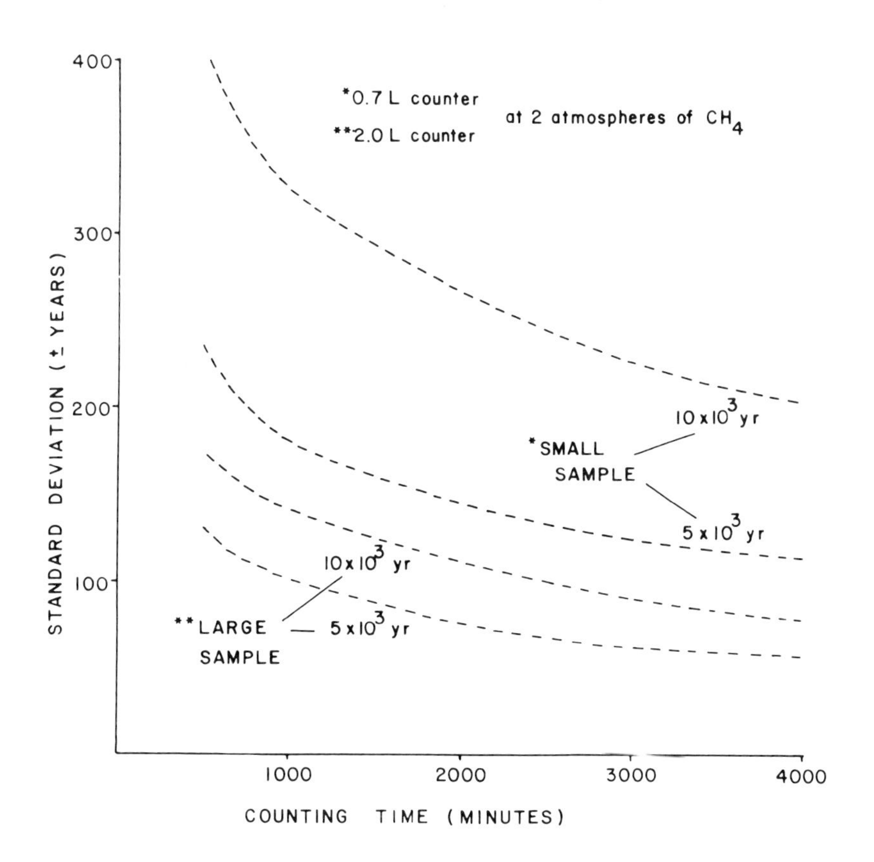

Figure 3 The difference between small and large samples with respect to age, error limits, and counting time (modified after Stuckenrath, 1977).

has been the use of techniques from nuclear physics for counting radiocarbon atoms directly in the sample, rather than the disintegration events (Muller, 1977; Litherland, 1979). Since the details of this method will be described in another paper presented at this conference, only a few potential applications will be mentioned here.

Perhaps the greatest advantage of this new method is its capability of using very small samples (only a few milligrams) for dating and thus resolve one of the difficult problems of conventional radiocarbon dating. Other advantages include the short counting time that, coupled with the small sample size capability, can make it feasible to obtain sets of dates (assuming that the cost is not prohibitive) from, for example, detrital samples and sort out some of the contamination problems, and the potential possibility of extending the range of radiocarbon dating.

Stuiver (1978) has compared the new atom counting method and the conventional beta counting method, and one should temper overenthusiasm with realism in respect to the new method because it does not instantly solve all problems of radiocarbon dating and make the conventional method obsolete.

Improvements have been made also in the techniques used in the conventional (or beta radiation counting) method of radiocarbon dating. For example, Harbottle *et al.* (1979) describe a method that makes it possible to extend the conventional radiocarbon dating by means of a specially designed small gas proportional counter to samples containing as little as 10 milligrams of carbon. The advantage of this method is that it is much less expensive than an atom counting facility which costs well in excess of $500,000 for the basic installation alone.

Another improvement is the isotopic enrichment of radiocarbon by thermal diffusion that extends the dating range to about 75,000 years BP (Grootes, 1978; Stuiver *et al.*, 1978). However, this technique is rather complicated, time consuming, costly, and certainly not without problems as illustrated by the following example. The Salmon Springs nonglacial interval (peat) near Sumner, Washington, was dated at 71,500 + 1700, -1400 years BP (QL-110) by Stuiver *et al.* (1978). On the other hand, Easterbrook *et al.* (1981) obtained fission track ages (coupled with paleomagnetic studies) of 0.66 ± 0.04 million years (m.y.), 0.84 ± 0.21 m.y. and 0.87 ± 0.27 m.y. from the same stratigraphic sequence, indicating that the Salmon Springs Glaciation is of middle Quaternary age. In other cases, the isotopic enrichment method seems to yield ages that are in agreement with independent stratigraphic and chronological evidence (Stuiver *et al.*, 1978).

The radiocarbon fluctuations caused by changes in the global carbon cycle probably will be the object of further research. A box-diffusion model for the carbon cycle, proposed by Siegenthaler *et al.* (1980), indicates that carbon - cycle - induced ^{14}C variations may have been significant during the transition episode from glacial to postglacial conditions when rather drastic environmental changes occurred within a short time. Significant variations of reservoir parameters can be expected during periods of major climatic changes, and Siegenthaler *et al.* (1980) believe that changes in the supposed oceanic circulation may be the most important factor (in addition to the atmospheric production rate of ^{14}C) affecting the global distribution of radiocarbon. They also postulate that ^{14}C variations due to changes of the atmospheric CO_2 level or the air - sea exchange probably did not exceed one to a few percent, and that fluctuations of the forest biomass which may have occurred between glacial and postglacial conditions hardly affected the ^{14}C concentration over a long term. These matters are clearly of importance to improvement of the accuracy of radiocarbon dating.

Geomagnetic studies will probably play a greater role in future radiocarbon research. For example, Barbetti (1980) postulated that long-term variations in the ^{14}C time-scale are readily explained by known geomagnetic changes, and Stuiver and Quay (1980) postulated that solar wind magnetic as well as geomagnetic forces modulate the incoming cosmic ray flux and explain the main features of the atmospheric ^{14}C record. They argue that climatic fluctuations are not a dominant cause of atmospheric ^{14}C changes.

A recent study by Eicher *et al.* (1981), concerning pollen and oxygen isotope analyses of late Quaternary lacustrine deposits in France, indicates some interesting trends in these independent paleoclimatic records. If such studies can be correlated with geomagnetic research and carbon isotope ratio measurements it is possible that some useful information can be obtained on the climate - radiocarbon - geomagnetic (solar activity) relationships.

The monitoring of the radiocarbon concentration (activity) in the multiple components of the global atmosphere - biosphere - hydrosphere system will probably be expanded significantly in the future and hopefully will improve our understanding of the global carbon cycle. Such monitoring is not without some "fringe benefits" as pointed out by Ogden (1977) because both whiskey and wine have been used as samples for tracing the atmospheric bomb carbon fluctuations in Scotland, Belgium and France.

I wish to conclude with the optimistic prediction that radiocarbon research will proceed vigorously on many fronts in future years and will continue to involve scientists from many disciplines for the benefit of all concerned. When we understand the causes of radiocarbon fluctuations it may well happen, for example, that the radiocarbon record from tree-rings can help the astrophysicist learn more about past solar activity. It seems certain that the more we know about the problems of radiocarbon dating the better chance we have to improve the method, make it more accurate, and more applicable to a greater variety of users and disciplines.

REFERENCES CITED

Barbetti, M., 1980, Geomagnetic strength over the last 50,000 years and changes in atmospheric ^{14}C concentration: emerging trends: Radiocarbon, v. 22, no. 2, p. 192-199.

Berger, R and Suess, H.E., eds., 1979, Radiocarbon Dating: Univ. California Press, Berkeley, 787 p.

Blake,W., Jr., 1981, Glacial history of Svalbard and the problem of the Barents Shelf ice sheet: comments: Boreas, v. 10, p. 125-131.

Craig, H., 1957, Isotopic standards for carbon and oxygen and correction factors for mass-spectrometric analysis of carbon dioxide: Geochimica et Cosmochimica Acta, v. 12, p. 133-149.

Damon, P.E., Lerman, J.C. and Long, A., 1978, Temporal fluctuations of atmospheric ^{14}C: causal factors and implications: Annual Review of Earth and Planetary Sciences, v. 6, p. 457-494.

Damon, P.E., Lerman, J.C., Long, A., Bannister, B., Klein, J. and Linick, T.W., 1980, Report on the workshop on the calibration of the radiocarbon time scale: Radiocarbon, v. 22, no. 3, p. 947-949.

De Vries, Hl., 1958, Variation in concentration of radiocarbon with time and location on earth: K. Ned. Akad. Wet., Proc. Ser. B, v. 61, p. 94-102.

Easterbrook, D.J., Briggs, N.D., Westgate, J.A. and Gorton, M.P., 1981, Age of the Salmon Springs Glaciation in Washington: Geology, v. 9, p. 87-93.

Eicher, U., Siegenthaler, U. and Wegmüller, S., 1981, Pollen and oxygen isotope analyses on late- and post-glacial sediments of the Tourbiere de Chirens (Dauphine, France): Quaternary Research, v. 15, p. 160-170.

Ferguson, C.W., 1970, Dendrochronology of bristlecone pine, *Pinus aristata*. Establishment of a 7484-year chronology in the White Mountains of eastern-central California: p. 237-259 in Olsson, I.U., Radiocarbon Variations and Absolute Chronology: Wiley-Interscience, 657 p.

Grootes, P.M., 1978, Carbon-14 time scale extended: comparison of chronologies: Science, v. 200, p. 11-15.

Harbottle, G., Sayre, E.V. and Stoenner, R.W., 1979, Carbon-14 dating of small samples by proportional counting: Science, v. 206, p. 683-685.

Karrow, P.F. and Anderson, T.W., 1975, Palynological study of lake sediment profiles from southwestern New Brunswick: discussion: Canadian Jour. Earth Sci., v. 12, p. 1808-1812.

Lerman, J.C., 1972, Carbon-14 dating: origin and correction of isotope fractionation errors in terrestrial living matter: p. 612-624 in Rafter, T.A. and Grant-Taylor, T., eds., 8th Intern. Radiocarbon Conference Proceedings, Royal Society of New Zealand, Wellington, N.Z.

Libby, W.F., 1952, Radiocarbon Dating: University of Chicago Press, Chicago, 124 p.

__________, 1979, Forward to Proceedings of the 9th Intern. Conference (1976): in Berger, R. and Suess, H.E., eds., Radiocarbon Dating, Univ. California Press, Berkeley, p. ix-x.

Litherland, A.E., 1979, Dating methods of Pleistocene deposits and their problems: the promise of atom counting: Geoscience Canada, v. 6, no. 2, p. 80-82.

Mangerud, J. and Gulliksen, S., 1975, Apparent radiocarbon ages of Recent marine shells from Norway, Spitsbergen, and arctic Canada: Quaternary Research, v. 5, p. 263-273.

Muller, R.A., 1977, Radioisotope dating with a cyclotron: Science, v. 196, p. 489-494.

Nydal, R., Lövseth, K. and Gulliksen, S., 1979, A survey of radiocarbon variation in nature since the test ban treaty: in Berger, R. and Suess, H.E., eds., Radiocarbon Dating, Univ. California Press, Berkeley, p. 313-323.

Ogden, J.G., III, 1977, The use and abuse of radiocarbon dating: Annals New York Acad. Sciences, v. 288, p. 167-173.

Olson, E.A. and Broecker, W.S., 1958, Sample contamination and reliability of radiocarbon dates: Trans. New York Acad. Sci., v. 20, p. 593-604.

Olsson, I.U., ed., 1970, Radiocarbon Variations and Absolute Chronology: Wiley-Interscience, 657 p.

__________, 1974, Some problems in connection with the evaluation of ^{14}C dates: Geol. Fören, Stockholm Förhandl., v. 96, p. 311-320.

Olsson, I.U. and Blake, W., Jr., 1961-62, Problems of radiocarbon dating of raised beaches, based on experience in Spitsbergen: Norsk Geografisk Tidsskrift, v. 18, p. 47-64.

Olsson, I.U. and Florin, M.B., 1980, Radiocarbon dating of dy and peat in the Getsjö area, Kolmården, Sweden, to determine the rational limit of Picea: Boreas, v. 9, p. 289-305.

Pardi, R. and Marcus, L., 1977, Non-counting errors in ^{14}C dating: Annals New York Acad. Sciences, v. 288, p. 174-180.

Pearson, G.W., Pilcher, J.R., Baillie, M.G.L. and Hillam, J., 1977, Absolute radiocarbon dating using a low altitude European tree-ring calibration: Nature, v. 270, no. 5632, p. 25-28.

Polach, H.A., 1979, Correlation of ^{14}C activity of NBS Oxalic Acid with Arizona 1850 Wood and ANU Sucrose Standards: in Berger, R. and Suess, H.E., eds., Radiocarbon Dating, Univ. California Press, Berkeley, p. 115-124.

Rafter, T.A., Jansen, H.S., Lockerbie, L. and Trotter, M.M., 1972, New Zealand radiocarbon reference standards: in Rafter, T.A. and Grant-Taylor, T., eds., 8th Intern. Radiocarbon Conf. Proc., Royal Society of New Zealand, Wellington, N.Z., p. 625-675.

Siegenthaler, U., Heimann, M. and Oeschger, H., 1980, ^{14}C variations caused by changes in the global carbon cycle: Radiocarbon, v. 22, no. 2, p. 177-191.

Sissons, J.B., 1981, The last Scottish ice-sheet: facts and speculative discussion: Boreas, v. 10, p. 1-17.

Stenhouse, M.J. and Baxter, M.S., 1979, The uptake of bomb ^{14}C in humans: in Berger, R. and Suess, H.E., eds., Radiocarbon Dating, Univ. California Press, Berkeley, p. 324-341.

Stuckenrath, R., 1977, Radiocarbon: some notes from Merlin's diary: Annals New York Acad. Sciences, v. 288, p. 181-188.

Stuckenrath, R., Miller, G.H. and Andrews, J.T., 1979, Problems of radiocarbon dating Holocene organic-bearing sediments, Cumberland Peninsula, Baffin Island, N.W.T., Canada: Arctic and Alpine Research, v. 11, no. 1, p. 109-120.

Stuiver, M., 1978, Carbon-14 dating: a comparison of beta and ion counting: Science, v. 202, p. 881-883.

__________, 1980, Workshop on ^{14}C data reporting: Radiocarbon, v. 22, no. 3, p. 964-966.

Stuiver, M. and Borns, H.W., Jr., 1975, Late Quaternary marine invasion in Maine: its chronology and associated crustal movements: Geol. Soc. Amer. Bull., v. 86, p. 99-104.

Stuiver, M., Heusser, C.J. and Yang, I.C., 1978, North American glacial history extended to 75,000 years ago: Science, v. 200, p. 16-21.

Stuiver, M. and Quay, P.D., 1980, Patterns of atmospheric ^{14}C changes: Radiocarbon, v. 22, no. 2, p. 166-176.

Suess, H.E., 1955, Radiocarbon concentration in modern wood: Science, v. 122, p. 415-417.

__________, 1980, The radiocarbon record in tree rings of the last 8000 years: Radiocarbon, v. 22, no. 2, p. 200-209.

Thurber, D.L., 1972, Problems of dating non-woody material from continental environments: in Bishop, W.W. and Miller, J.A., eds., Calibration of Hominid Evolution, Univ. Toronto Press, Toronto, p. 1-17.

Troughton, J.H., 1972, Carbon Isotope fractionation by plants: in Rafter, T.A. and Grant Taylor, T., eds., 8th Intern. Radiocarbon Conf. Proc., Royal Society of New Zealand, Wellington, N.Z., p. 420-438.

RADIOISOTOPE DETECTION AND DATING WITH PARTICLE ACCELERATORS

J.C. RUCKLIDGE

ABSTRACT

Radiocarbon dating is conventionally performed by the accurate measurement of the beta rays emitted during the decay of ^{14}C. The direct determination of ^{14}C atoms may also be accomplished by ultrasensitive mass spectrometry using a particle accelerator. Signal to noise ratios in excess of 10^{15}:1 are required for this task, and can only be attained if the background from the isobar ^{14}N and from interfering molecular ions is eliminated. By using a tandem Van de Graaff accelerator as a molecule disintegrator and accepting only negative ions, this condition may be attained and individual ^{14}C atoms may thus be counted directly. The higher counting rates inherent in this method, and the ability to use much smaller sample sizes offer significant advantages over the conventional method. In addition, other naturally occurring radioactive isotopes with half lives relevant to Quaternary dating may be measured using the same approach. At the present time the isotopes ^{10}Be, ^{14}C, ^{26}Al, ^{32}Si, ^{36}Cl and ^{129}I have been detected at natural levels without prior isotope enrichment by using negative ions, molecular dissociation with tandem accelerators and atom counting. Ratios of $^{14}C/^{12}C$ and $^{36}Cl/Cl$ near 10^{-15} have been reached during experiments being carried out to develop radiocarbon and ^{36}Cl dating of milligramme samples. With the ability to use such small samples, valuable geological and archaeological specimens will be datable without significant damage, and the costs involved may be reduced. New facilities dedicated to this type of measurement are being constructed at various centres, and are expected to begin producing routine radiocarbon dates in 1982.

INTRODUCTION

Several radioisotopes of light elements are produced through spallation processes caused by cosmic rays interacting with the common components of the atmosphere. Of these the better known are ^{14}C (half life 5730a) from ^{14}N, and ^{36}Cl (half life 3.01×10^5a) from ^{36}Ar. If equilibrium is assumed, the rate at which the radioactive isotope is produced is balanced by the rate of its loss due to beta decay. The equilibrium concentration of ^{14}C in living organic matter is about 1.2×10^{-12} atoms per atom of ^{12}C. After the organism's death the ^{14}C is no longer replenished, and it decays away with a half life of 5730 years. Conventional radiocarbon dating, as pioneered by Libby (Anderson *et al.*, 1947) is carried out by measuring the radioactivity of the sample due to the beta decay of the ^{14}C atoms in it. For contemporary carbon the specific activity is about 15 disintegrations per minute per gram, though each gram of material contains 6×10^{10} atoms of ^{14}C. Clearly, any technique which can identify these ^{14}C atoms directly, even with very modest collection efficiencies of, say, 10^{-3} would be so much more efficient than radioactive measurements (the fractional disintegration of ^{14}C is 1.4×10^{-8}/hour) that the size of the sample needed for direct atom counting would be dramatically less than that required by the conventional method. At the same time, higher accuracy is in principle possible, because many more counts should be recorded in much shorter counting times. Even samples as old as 100,000 years contain 3.35 x 10^5 atoms ^{14}C/gram which should provide enough counts to obtain satisfactory statistics on specimens more than twice as old as those which can be dated by the beta decay method. Contamination of samples would probably prove to be the factor limiting the age which could be measured reliably. To emphasize the problems of detecting one ^{14}C atom in an ocean of 10^{12} atoms of ^{12}C and ^{13}C, it is instructive to consider the analagous problem of identifying a single specific grain of sand, 0.25mm diameter, in a pile with a volume of 88 cubic metres, equivalent to the contents of five loaded dump trucks!

One of the major problems in performing this sorting by conventional mass spectrometry is the existence of the isobar of ^{14}C, ^{14}N which is ubiquitous in comparison, and which differs from ^{14}C in mass by only one part in 10^5, hence an extremely high resolution mass spectrometer would be needed to separate the few ^{14}C atoms from the many ^{14}N atoms. An additional problem is the existence of molecules such as $^{12}CH_2$ and ^{13}CH which have masses even closer to ^{14}C and hence contribute to an overwhelming background which prevents this simple mass spectrometric approach from being successful. One early attempt which came close to attaining the necessary sensitivity of 1 in 10^{12} to detect ^{14}C in contemporary carbon was that of Anbar *et al.* in 1974 (Anbar, 1978). The stable molecule $^{14}C^{15}N^-$ was used to discriminate against $^{14}N^{15}N^-$ ions which were expected to be unstable. The limitations were backgrounds from $^{12}C_2H_5^-$, $^{12}C^{16}OH^-$, $^{29}Si^-$ and $^{28}SiH^-$. By careful elimination of hydrogen from the system a sensitivity of 1 in 10^{11} was obtained, but the ultimate stumbling block proved to be $^{29}Si^-$. The breakthrough in this problem had to await the development of C^- ion sources and the realization that particle accelerators, as used in nuclear physics experiments, could be made to function as ultrasensitive mass spectrometers, and in this mode of operation the two major difficulties outlined above could be overcome. Reviews of the use of accelerators for ultrasensitive analysis may be found in articles by Litherland 1980, and Litherland *et al.* (1981).

ACCELERATOR MASS SPECTROMETRY

The problem of molecular interferences in the study of rare isotopes by mass spectrometry can be solved by the use of a tandem accelerator as a molecule disintegrator (Purser, 1977). Negative molecular ions dissociate rapidly and completely after losing three or more electrons during charge changing collisions in the central electrode of a tandem accelerator. The analysing and switching magnets that usually follow a tandem accelerator can then be used to remove most of the molecular fragments leaving the atomic species alone for study.

The success of this procedure in the detection of ^{14}C at natural levels was first demonstrated in 1977 at the University of Rochester (Purser *et al.*, 1977). The instability of $^{14}N^-$ and the stability of $^{14}C^-$ was also used to eliminate the interfering mass 14 isobar. Shortly afterwards the ^{14}C isotope was detected in graphite samples with $^{14}C/^{12}C$ less than 10^{-15} (Bennett *et al.*, 1977). The ability to make such measurements on carbon isotopes also required the development of high current negative ion sputter sources for nuclear physics. These ion sources involve the sputtering of solid materials and consequently have a very low memory effect which is essential for the measurement of such large and variable isotope ratios.

Almost simultaneously and independently, measurements on the tandem accelerator at McMaster University (Nelson *et al.*, 1977) showed that ^{14}C at near natural levels could be detected and at Berkeley a cyclotron was used, following a suggestion by Muller (1977), to demonstrate that ^{14}C at or near natural levels could be detected using positive ions and range separation of the ^{14}C from the intense interfering ^{14}N beams (Stephenson *et al.*, 1979).

The three pioneering experiments on ^{14}C have now been extended to many stable and radioisotopes, and the significance of the new principles for rare isotope analysis by mass spectrometry has been perceived rapidly. New specialized tandem accelerators and their peripheral equipment are being developed to exploit the discoveries (Purser *et al.*, 1980).

Tandem Accelerator

The layout of the ion beam transport system of a tandem accelerator is illustrated in Figure 1. In the negative ion sputter source the sample is bombarded by a beam of Cs^+ ions which erodes the sample surface, releasing atoms as both neutral and charged (both positive and negative) ions. A positive extraction electrode draws negative ions away from the sample and into the inflection magnet which selects ions with a particular M/q for injection into the accelerator. The high voltage electrode at the centre of the tandem is maintained at a high positive potential of the order of megavolts, 8MV being the voltage used in the initial ^{14}C experiments. Negative ions accelerated to this point pass through a gas canal containing Ar, where charge changing collisions take place. Electrons are stripped from the ions in the beam, the precise number of electrons stripped depending on the accelerating voltage and gas pressure. The negative ions are thus converted to positive, but because of their high energies and high rigidity remain essentiallly undeviated from their paths. In their new multiple positive change state they are accelerated further through the remainder of the accelerator, as the electric field gradient is now reversed. Emerging at the high energy end the positive ions pass through a 90° magnet which selects a specific mass energy product (ME/q^2). A 45° switching magnet directs the particles down the appropriate detection line, and a 10° electrostatic analyzer defines E/q so that ions entering the final detector have unique values of M/q, or what is equivalent, uniquely defined velocities. The final detector is a gas filled heavy ion counter developed by Shapira *et al.* (1975) which provides several dE/dx signals from separate electodes along the path of the ion, as well as the total energy deposited. When the velocity of the ions incident on the counter is defined, a measurement of dE/dx defines the ion's nuclear charge Z. This provides valuable information and sometimes conclusive identification of the isobar of interest.

Molecular Destruction

Molecules are bound by their outer electrons and if enough of these electrons are removed the positively charged components will repel each other and the molecule will fragment. It is easy to show that molecules with two or more electrons removed should be destroyed by the coulomb repulsion of the components. However, it was found (Litherland, 1978) that the molecule $^{12}CH_2{}^{+2}$ was stable enough to interfere with $^{14}C^{+2}$ detection. It is therefore necessary to select the charge state +3 as triply charged molecules are unknown except for some transition metal elements. For this reason in the carbon experiments the high energy analysis stages were tuned to accept $^{14}C^{+3}$ for which there are no interfering molecular species.

The molecular fragments can be selectively eliminated by a suitable combination of electric and magnetic fields (Purser *et al.*, 1979) leaving only the atoms to be analyzed. However, if low resolution ($M/\Delta M<400$) mass spectrometers are used, certain ambiguities are possible. The energy, E, of an ion after acceleration in a tandem accelerator divided by the charge, q, is given by

$$\frac{E}{q} = \frac{M}{m} . \;(v+V)\frac{e}{q} + Ve$$

where m is the mass of the negative ion injected, and M is the mass of the ion accelerated to energy E. V and v are the voltage of the central electrode of the tandem and the injection voltage. e is the electronic charge. Clearly if M and q have common factors then ions with the same values of M/q and E/q will be transmitted through all mass spectrometric elements of low resolution. An example might be the ions $^{195}Pt^{+5}$, $^{156}Dy^{+4}$, $^{117}Sn^{+3}$, $^{78}Se^{+2}$ and $^{39}K^{+1}$ all of which would be present following the injection into a tandem accelerator of negative ions of mass 195. In principle such ambiguities can be eliminated for stable isotopes by a simple energy measurement or by another charge changing collision followed by magnetic or electric spectrometry. However, for the much rarer radioisotopes a charge state q relatively prime to M should be chosen, for example +7 when measuring ^{36}Cl (Elmore *et al.*, 1979). The optimum energy for molecular destruction has been discussed by Litherland (1980) and corresponds simply to the tandem accelerator voltage needed to optimize the appropriate charge state of the positive ions.

Isobar Selection Methods

All radioisotopes eventually decay to stable isotopes and in the case of beta decay the masses of the radioisotope and stable isotope are so close that they are called isobars. ^{14}C and ^{14}N are members of a well known pair. At present there is no universal method for separating isobars other than extremely high resolution mass spectrometry and as this implies low atom efficiency it is unsuitable for rare isotopes. However, a number of procedures have been used with varying degrees of success.

1. The use of negative ions provides a spectacular solution to some isobar problems. In the case of ^{14}C and ^{14}N the apparent complete instability of N^- makes isobar separation possible at the ion source (Litherland, 1980). Isotope ratios below 10^{-15} for $^{14}C/^{12}C$ have been measured (Bennett *et al.*, 1977), and other examples are discussed below. Fortunately many studies of negative ions have been made in recent years (Middleton, 1977) primarily to discover how best to produce them for tandem acceleration. Cesium negative ion sputter sources have been developed which optimize the production of negative ions, and which are suitable for mass spectrometry with tandem accelerators.

2. Isobars can also be separated chemically, but this method is of limited use because of the very high degree of purity needed in the reagents. The partial chemical separation of sulphur and chlorine has so far proved essential for the distinguishing ^{36}S and ^{36}Cl which both form negative ions readily.

3. The range of the isobar with lower nuclear charge, Z, is longer in matter than that of the isobar with larger Z, and this makes their separation possible (Muller, 1977). The method is very useful for the separation of ^{10}Be and ^{10}B and is the only reliable method at present. However, for ^{14}C and ^{14}N separation it is more difficult (Stephenson *et al.*, 1979) and this method is probably limited to light isotopes because of range straggling.

4. Individual ions can be identified by dE/dx measurements in the heavy ion counter at multi-MeV energies, and this method is of great value provided the relative intensity of the interfering isobar is not too great. A relative intensity of up to 10^5 seems to be near the limit for discrimination by rate of energy loss (Elmore *et al.*, 1979).

5. If the isobar with the larger Z is the rare radioactive isobar then complete stripping of the electrons from the nucleus makes isobar selection possible. This method can be applied at intermediate energies (Kilius and Litherland, 1978) or at very high energies (Raisbeck *et al.*, 1979) but is probably of limited use because of the complexity and cost of the equipment. However, the separation of $^{7}Be^{+4}$ and $^{7}Li^{+3}$ can be accomplished at low (a few MeV) energies and has recently been done (Kutschera, pers. commun.). The separation of $^{26}Al^{+13}$ and $^{26}Mg^{+12}$ requires at least 100MeV ions (Raisbeck *et al.*, 1979). It is interesting to note that the rare isotope ^{3}He was first identified by this method in 1939 by Alvarez and Cornog.

Other methods of separating isobars have been discussed by Litherland (1980), but at present it appears that each isobaric pair has to be considered separately to find the best method.

RADIOCARBON DATING

The best known experiments in direct radioisotope measurements have been on ^{14}C (Bennett *et al.*, 1977, 1978; Gove *et al.*, 1980) and have been summarized in a more general article by Bennett (1979). The method of sample preparation was originally to prepare amorphous carbon mixed with an equal amount of KBr which served as a binder. This was pressed into an aluminium cylinder which was mounted on the negative ion source wheel. In more recent measurements, a mixture of amorphous carbon and finely powdered silver or copper was pressed into a 2mm diameter shallow re-entrant hole in the aluminium cylinder. The weight of carbon included in the sample ranged from less than 1mg up to 10mg, although in most cases the major part of the material still remained after the measurement was completed. Attempts are presently being made in several laboratories to crack carbon from methane or acetylene onto metal substrates which can be mounted on the source wheel. This will produce a graphite-like material which, in the cesium sputter source, should give substantially higher currents of C^-.

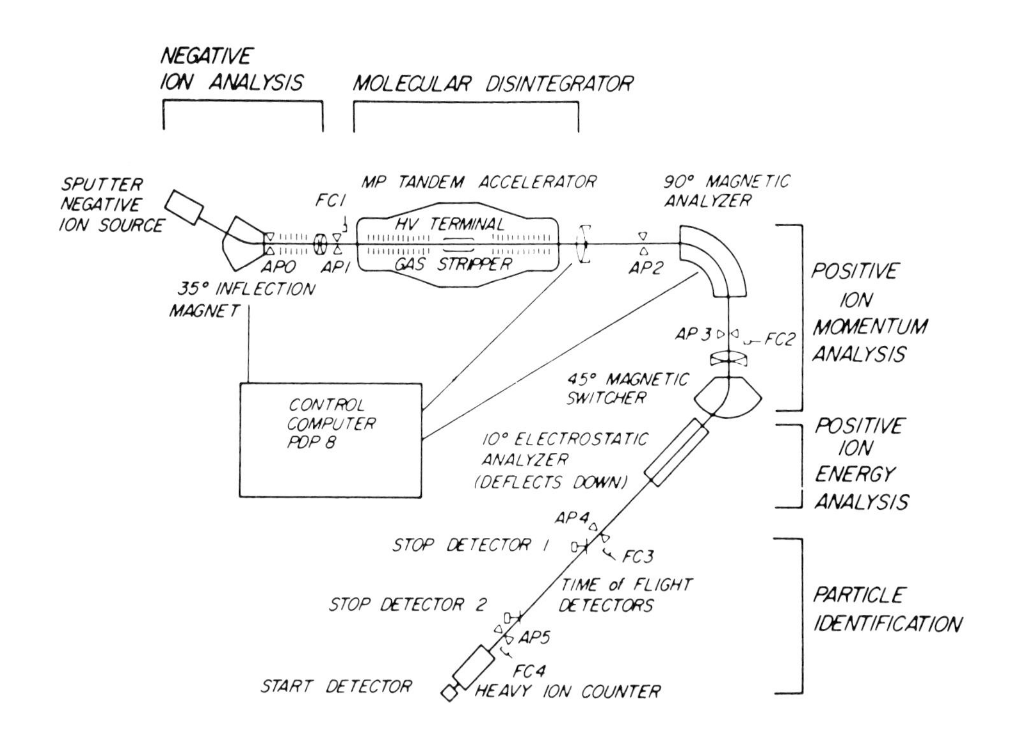

Figure 1 The ion beam transport system of the University of Rochester tandem accelerator used as an ultra-sensitive mass spectrometer is shown schematically. Ion-beam defining apertures are designated AP and Faraday cups are designated FC.

Figure 2 shows a typical mass spectrum of negative ions extracted from the sputter source, prior to injection into the accelerator. It can be seen how the peak at mass 14, which is made up almost entirely of the molecules $^{12}CH_2^-$ and $^{13}CH^-$ is four orders of magnitude lower than ^{12}C at mass 12. The subsequent molecular elimination removes all but one part in 10^8 of this mass 14 peak in order to reveal the miniscule component of ^{14}C. Measurement times of one to six hours have been found to be adequate to obtain the counts necessary to achieve accuracy comparable with the standard method. If higher currents can be produced by improved sample preparations, then these counting times may be substantially reduced.

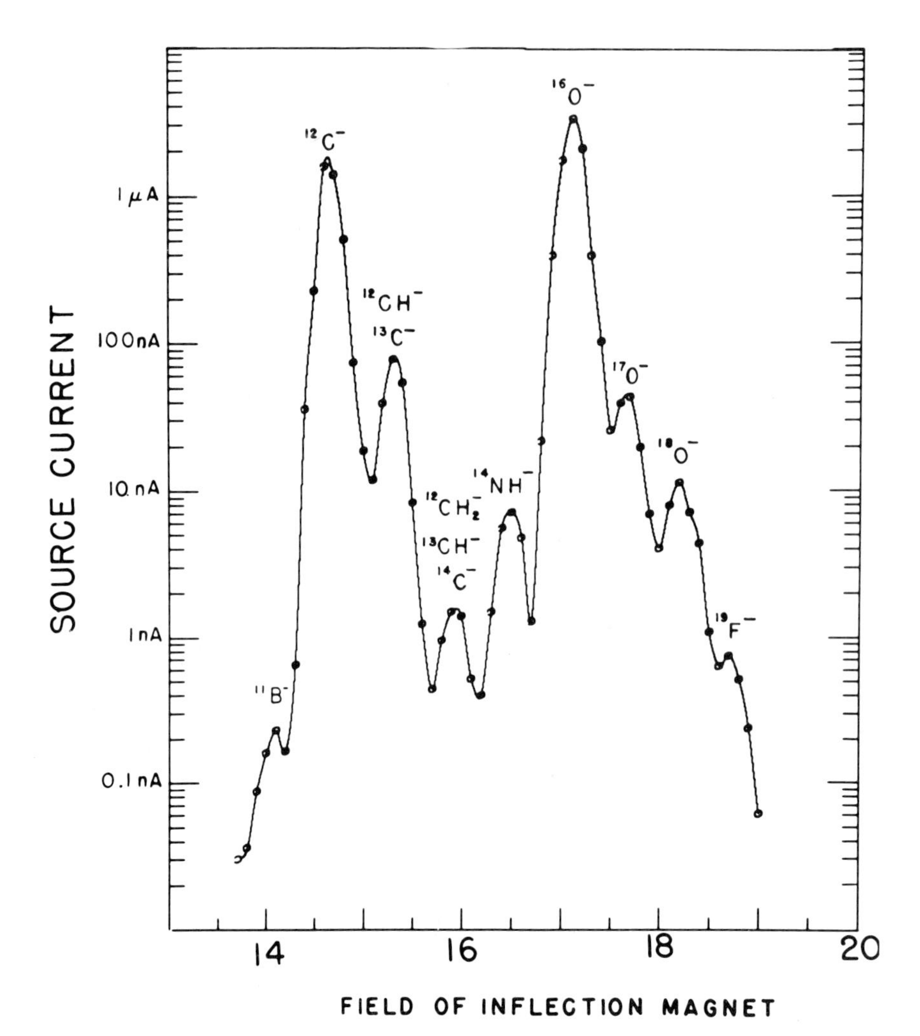

Figure 2 Output of the negative ion beams from the sputter source using a carbon sample, prior to injection into the accelerator. The ions have an energy of 20 keV and are magnetically analyzed.

Figure 3 shows a three dimensional plot of the output of the final heavy ion detector when a sample of carbon from Mt. Shasta, dated by the U.S. Geological Survey at 4590 ± 250 years old, was run in the ion source. The two parameters E_{TOTAL} and E_{FINAL} refer to the signals from the detector corresponding to the total energy deposited in the detector from the ions and the energy from some fraction of the track near the end of its range. The peaks are plotted on a logarithmic scale. The carbon ion energy was 40MeV. In Figure 4 the great sensitivity of the system is again illustrated in two pulse height spectra for carbon samples. The one from Mt. Hood, 220 years old, is almost contemporary, while the Hillsdale sample is 40,000 years old. The $^{14}C/^{12}C$ ratio for the Mt. Hood sample is about 10^{-12} while the Hillsdale sample is about 8×10^{-15}. The ^{12}C and ^{13}C peaks come from $^{12}CH_2^-$ and $^{13}CH^-$ molecules from the source, which are dissociated in the terminal and then undergo improbable charge changes during their acceleration in the second half of the tandem, so they have the same ME/q^2 as the ^{14}C ions. These peaks could be eliminated with an electrostatic deflector, but as they are clearly resolved from the ^{14}C peak there is no need to be concerned with them. The ^{14}N arises from negative ion molecular hydrides which also undergo low probability charge changes and slip through to the detector, but again the peak is well resolved, and offers no interference.

The abundant isotope of carbon, ^{12}C, is measured by its current in a Faraday cup after the 90° magnet, as the beam is far too intense for atom counting. The experimental procedure is to cycle repeatedly the tuning of the inflection magnet and accelerator voltage through the

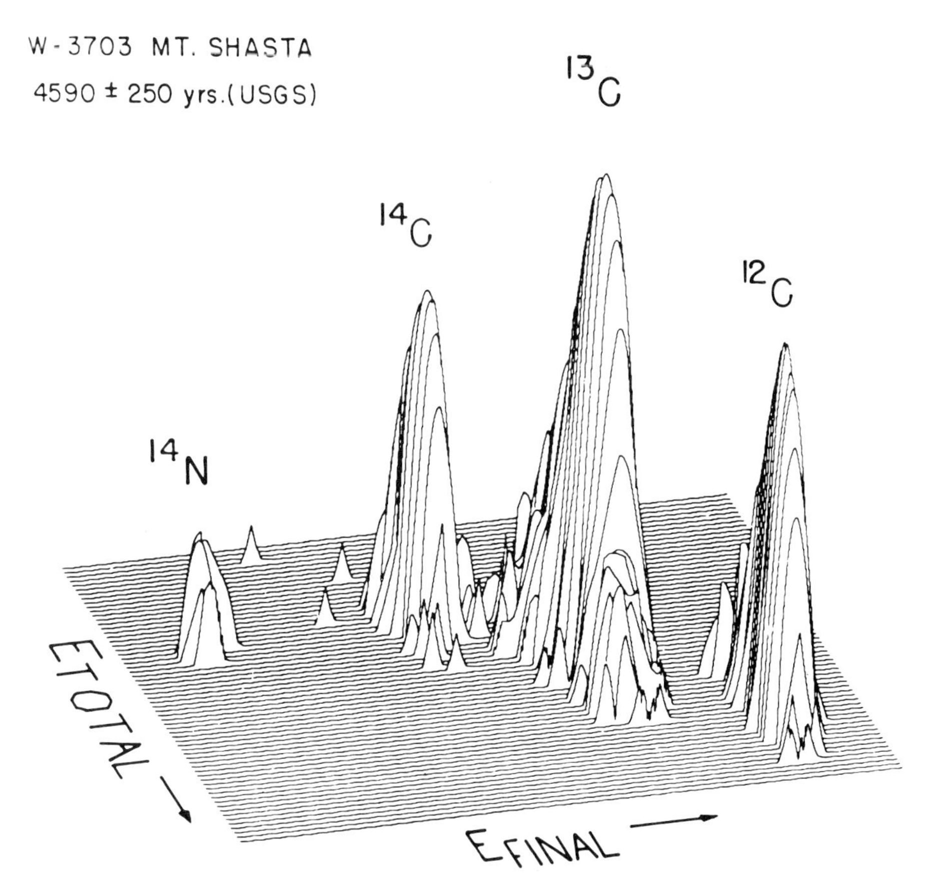

Figure 3 A three dimensional representation of data from the heavy ion counter used at Rochester. The vertical axis is the logarithm of the number of counts. E_{TOTAL} is the total ion energy measured in the counter and E_{FINAL} is the energy measured in the final section of the counter. E_{FINAL} is related to the ion range and the rate of energy loss dE/dX.

mass range 12 to 14. While tuned to masses 12 and 13 current measurements are made for short periods and then at mass 14 counts from ^{14}C are accumulated so that the ratio $^{14}C/^{12}C$ can be derived. Each microampere of ^{12}C yields about 450 ions of ^{14}C per minute for a sample of contemporary carbon. Up to 10 microamps of $^{12}C^-$ may be extracted from the ion source under favourable conditions, so with a transmission efficiency of 10% ^{14}C count rates in the order of hundreds per minute may be expected. This must be compared with the rate of 15 beta rays emitted per minute per gram for the conventional method.

Results from some of the early measurements made at Rochester are shown in Figure 5 and details listed in Table 1. The deviation of the first Mt. Shasta determination was due to inhomogeneities in the sample. When a larger fraction of the charcoal sample was taken and homogenized by pulverization, the second set of data was recorded. The repeat measurements were in excellent agreement with the expected age, and there is an important lesson to be learned here. While the new method is able to date milligram quantities of material, it must be established clearly that this small sample is truly representative of the sample as a whole. Obviously, for priceless archaeological specimens a few milligrams may be all that can be spared, but where more material is available it should be carefully homogenized, or the results interpreted accordingly.

In June 1977 an intact, completely frozen baby mammoth (named Dima) was uncovered by a bulldozer operator in the Magadan region of northeastern Siberia. A sample of 1.3g of muscle was made available to M.

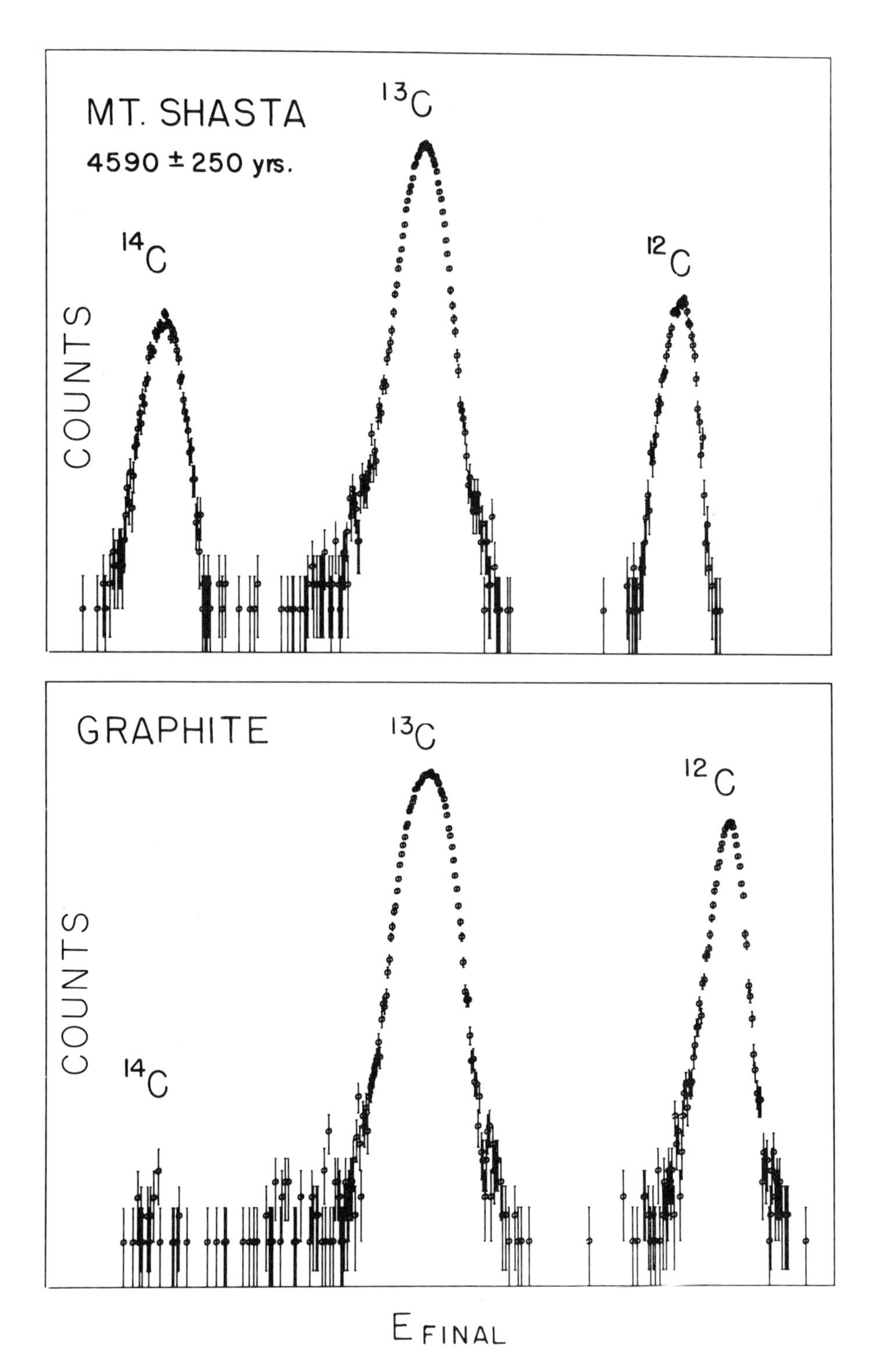

Figure 4 The pulse height spectra from the final section of the gas ionization chamber, with dE/dX selection for carbon, are shown for two carbon samples. The difference between the ^{14}C counts is significant.

Goodman of the Anatomy Department at Wayne State University. At his request a sample of about 0.9mg (obtained from about 4mg of muscle) was dated and the age shown in Table 1 was measured (Gove *et al.*, 1980).

The accuracy of the most recent measurements, based on reproducibility, approaches 1%. In order to compete with the highly developed beta ray counting method an accuracy of at least ±1% (± 80 years on recent material) must be attained consistently. This has not been easy to achieve with existing apparatus, and the problems seem to be associated with variations in transmission efficiency through the electron

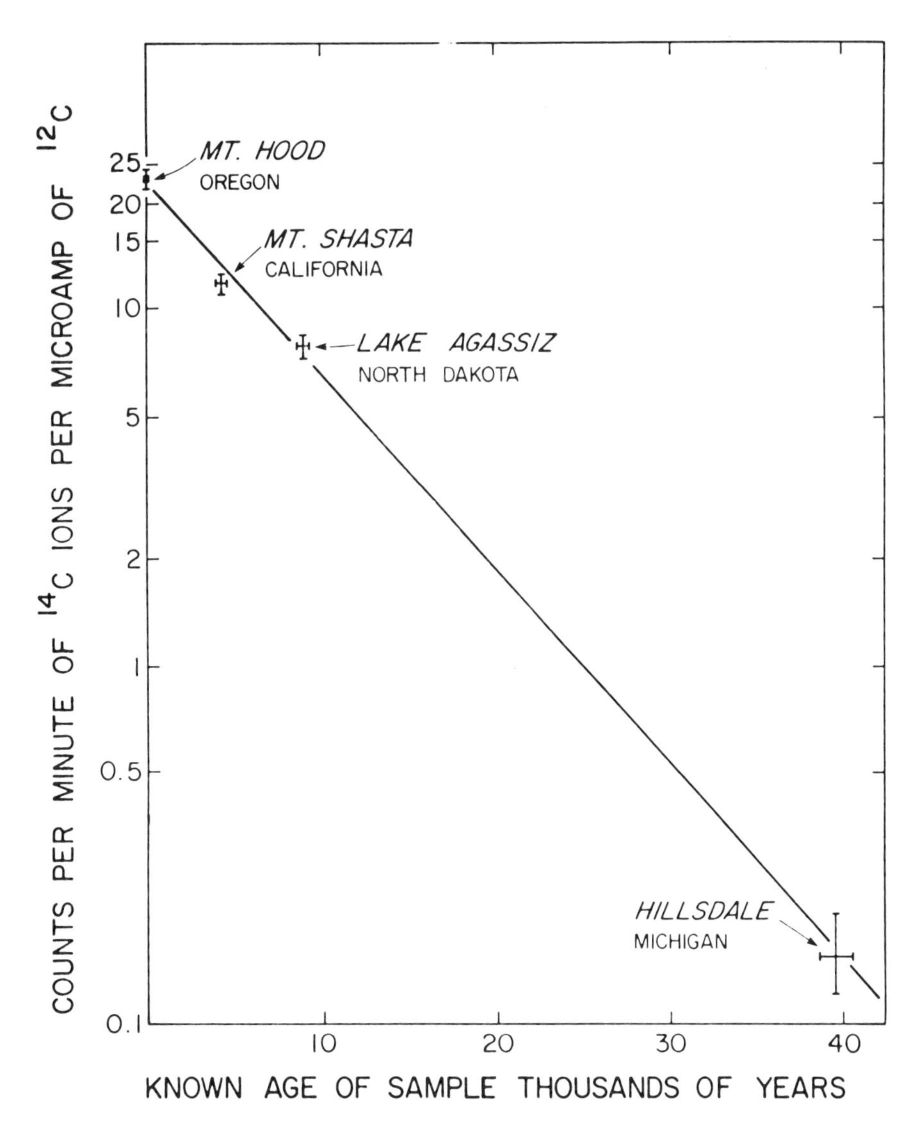

Figure 5 The results of dating milligram samples by atom counting compared to the ages measured by beta-decay counting. The ^{14}C is compared with the $^{12}C^-$ current.

stripping canal due to variations in ion source emittance during a measurement. This problem can be solved in principle by increasing the diameter of the stripping canal or by carefully controlling the emittance of the ion beams injected into the accelerator. However, the isotope ratios also appear to vary with time (Tombrello and Kellogg, 1979) due to the nature of the cesium sputtering process so that great care will have to be taken to ensure that all the various fractionation processes at all stages are under control. Fortunately measurements of isotope ratios are made with respect to some standard so that it is necessary only to ensure that the various fractionation processes during measurement are the same for the unknown sample as for the calibration sample. Careful control of contamination is also necessary.

Up to the present the use of a tandem for radiocarbon dating of archaeological samples has not been possible as the accuracy required (0±1%) has not been reached consistently. However the motivation for the tandem work, at least at Rochester, was to carry out research and development towards the design of machines specifically for radiocarbon dating. In this respect, the work has been quite successful as a dedicated 3MV machine, the Tandetron, is now constructed by General Ionex Corporation. Such instruments will be installed at the Universities of Oxford, Arizona, Toronto and Nagoya, and may be expected to begin producing routine dates in 1982. In the meantime, while several laboratories are preparing to use existing tandems for dating, relatively few results have been published.

Table 1. ^{14}C results from the Toronto-Rochester-General Ionex Group

Sample	Run Time (min)	Rochester Age (years BP)[a]	Expected Age[b]
Mt. Hood (Oregon)	90	220 ± 300	220 ± 150
Mt. Shasta (California)	180	5700 ± 400	4590 ± 250
L. Agassiz (N. Dakota)	90	8800 ± 600	9150 ± 250
Hillsdale (Michigan)	400	41,000 ± 1200	39,500 ± 1000
Graphite	65	48,000 ± 1400	
Mt. Shasta (repeat)		4580 ± 90	4590 ± 250
Bull Mummy cloth (Dahshur, Egypt)		2200 ± 150	2050 ± 200
Baby woolly mammoth (USSR)		27,000 ± 1000	Available sample too small for conventional dating

[a]BP = years before 1950. ^{14}C half-life assumed to be 5568 years

[b]As measured by Meyer Rubin of the U.S. Geological Survey

Tucker *et al.* (1981) reported recently on earthquake dating by ^{14}C atom counting on a tandem accelerator in Zurich. Trenches were cut across fracture lines associated with the Wasatch fault in Utah, and milligram sized specimens of detrital charcoal were collected from each side in the surface soil layers. The charcoal was assumed to be associated with prehistoric earthquakes along the fault, and $^{14}C/^{12}C$ were measured which correspond to ages of 7800, 8800 and 9000 years BP with uncertainties of ±600 years. These results were consistent with geological evidence and one conventional beta count from a larger sample. In this study the point was stressed that one of the major advantages of the atom counting method was that results could be obtained while collecting parties were still in the field, and so they could be directed to collect more material guided by the initial results, rather than having to wait for the next field season before following it up.

Other groups which have reported on the experimental measurement of ^{14}C using a tandem accelerator include Nelson *et al.* (1977) at McMaster University, who successfully observed ^{14}C at contemporary levels; Andrews *et al.* (1978) at Chalk River detected a background of ^{14}C corresponding to an age of 60,000 years; Barratt *et al.* (1978) at Oxford University detected ^{14}C in old and new carbon.

Another approach to atom counting with accelerators has been made using a cyclotron by a group at the University of California at Berkeley (Muller, 1977). The cyclotron acts as a multistage mass spectrometer, and because of the resonant nature of the acceleration process this has the effect of improving the resolution function by completely removing ions which are off resonance. The cyclotron itself is therefore superior to a single or double mass spectrometer with the same mass resolution. The fact that the ions must travel about 1km during acceleration unfortunately makes the transmission rather sensitive to vacuum pressure. In addition, the problems of getting the ions into and out of the cyclotron are more difficult than those of the tandem accelerator and this complicates accurate isotope ratio measurements. Up to the present, the hope that the cyclotron would be useful for radioisotope dating, especially with ^{14}C, has not been realized. However, the Berkeley group successfully determined a 'blind' measurement of a sample already dated by the conventional method at 6000 years (Muller *et al.*, 1978), though a second 'blind' measurement of an 8000 year old sample yielded an erroneous age of 18,000 years (Muller, 1978).

OTHER RADIOISOTOPES

^{14}C is the radioisotope which has found wide application in dating archaeological objects and Quaternary strata. The unique feature of carbon is its association with organic remains so that it has direct relevance in the dating of formerly living organisms, and secondly its half-life is suited to determining ages in the appropriate time frame, namely back to about 50,000 years BP. ^{14}C is produced by the action of cosmic rays on ^{14}N in the atmosphere, but there are several other spallations from cosmic rays which give rise to radioisotopes with different half-lives and different chemical affinities.

Beryllium-10 Analysis

^{10}Be is generated by the action of cosmic rays on nitrogen and oxygen in the atmosphere, and has a half-life of 1.6 x 10^6a. Because the advantage of counting atoms increases in direct proportion to the half-life of the radionuclide, the only effective way of measuring ^{10}Be is through ultrasensitive mass spectrometry (Raisbeck *et al.*, 1978a). After being precipitated out of the atmosphere, cosmogenic ^{10}Be is deposited in geological reservoirs, such as sedimentary rocks and polar ice sheets. Here is an isotope through which such strata a few million years old may be dated, if the geochemical behaviour of beryllium can be ascertained with sufficient precision.

Many of the measurements of ^{10}Be have been made using a cyclotron as part of a mass spectrometer, and the interfering isobar, ^{10}B, has been rejected by the range technique. The feasibility of the method was established with samples already determined by beta decay measurements (Raisbeck *et al.*, 1978b), and ratios of $^{10}Be/^{9}Be$ ranging from 10^{-8} to 10^{-10} were confirmed. Samples containing as few as 10^7 ^{10}Be atoms, which would only emit 5 beta rays per year, could be measured by this method. These early tests were followed by a series of exploratory measurements which included the detection of Be in Antarctic ice (Raisbeck, Yioux, Fruneau, Lieuvin and Louiseaux, 1979), ^{10}Be variations in marine sediments (Raisbeck, Yioux, Fruneau, Louiseaux, Lieuvin, Ravel and Hays, 1979) and a study of the residence time of ^{10}Be and its concentration in the ocean surface layer (Raisbeck, Yioux, Fruneau, Louiseaux and Lieuvin, 1979).

^{10}Be has subsequently been measured in tandem accelerators, but because the yield of $^{10}Be^-$ is very low the molecular ion $^{10}Be^{16}O^-$ was selected for acceleration. After the breakup of this molecule $^{10}Be^{+3}$ was accelerated further and counted, being separated from the interfering ^{10}B by the range technique (Kilius *et al.*, 1980). It appears that observation of $^{10}Be/^{9}Be$ ratios as low as 7 x 10^{-15} can be made without excessive difficulty, and it is to be expected that applications of the ^{10}Be method will become more common in the near future.

Chlorine-36 Analysis

^{36}Cl (half-life 3.1 x 10^5a) is created in the atmosphere by cosmic rays, from whence it joins the stable isotopes of chlorine, ^{35}Cl and ^{37}Cl, and for the most part enters the hydrosphere. The main use of ^{36}Cl seems to be in hydrogeological studies, and through it reliable methods of dating groundwaters may be developed. An important application may be in the evaluation of sites proposed for nuclear waste storage. Experiments with a tandem accelerator have resulted in ^{36}Cl being detected in groundwater, and some of the spectra obtained in this work are presented in Figure 6 (Elmore *et al.*, 1979).

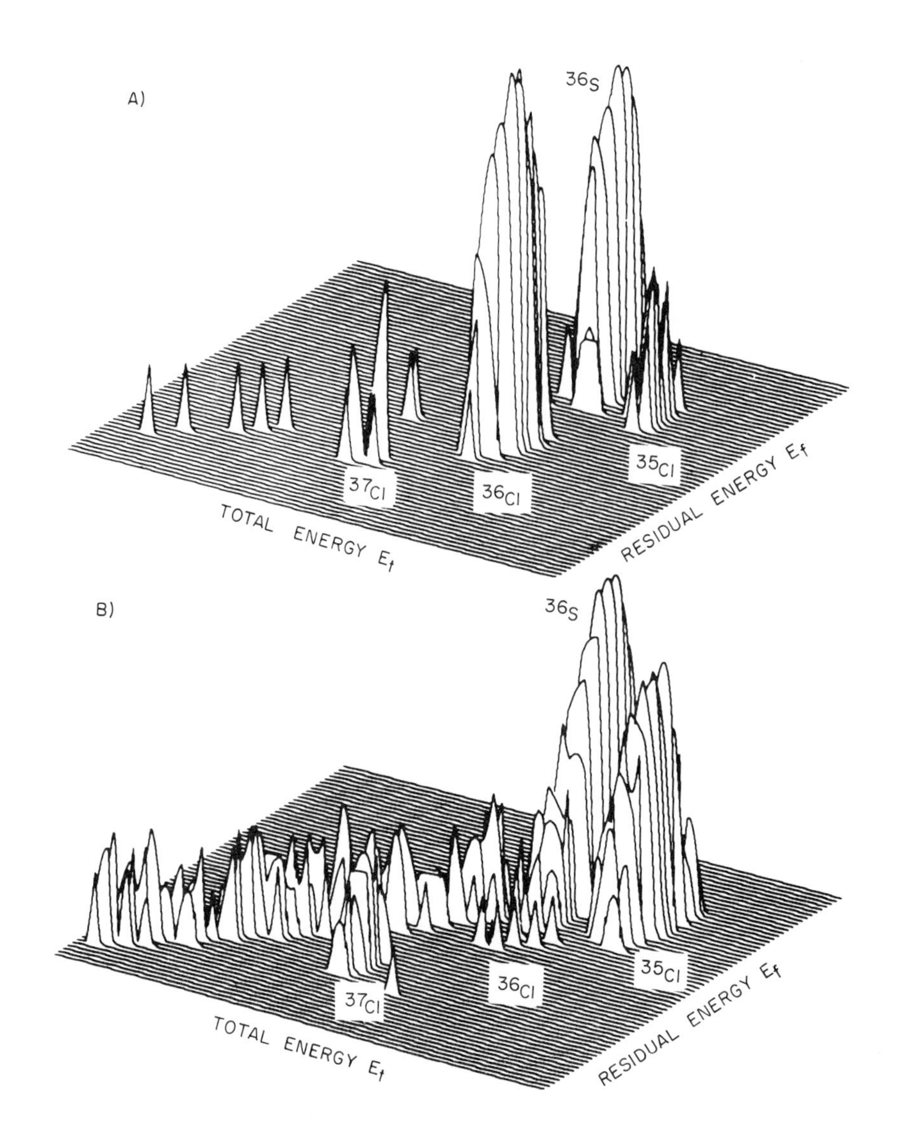

Figure 6 A) A three dimensional representation, similar to that shown in Figure 3, of data from the heavy ion counter at Rochester. The ions are observed from an AgCl sample from Lake Ontario when the apparatus was adjusted to transmit only mass-37 charge-7 ions. The ^{35}Cl and ^{37}Cl peaks are part of the background and the ^{36}S peak is due to the residual sulphur in the purified sample.
B) The $^{36}Cl^{+7}$ ions from a zone refined reagent-grade AgCl sample are shown together with the background ions. The few counts shown for ^{36}Cl represent a ^{36}Cl/Cl ratio of $(3\pm1) \times 10^{-15}$.

^{36}Cl has to be distinguished from the isobars ^{36}Ar and ^{36}S as well as from molecules such as $^{12}C_3$ and $^{18}O_2$. Ar^- is unstable and does not enter the accelerator, and molecules are of course broken up. ^{36}S, although a rare isotope (0.014% of sulphur) forms negative ions readily and is the major difficulty in the measurement. Samples have to be carefully purified to minimize sulphur contamination, and are prepared in the form of AgCl. In the final analysis the residual ^{36}S is distinguished from ^{36}Cl by the range technique. In the measurement of ^{36}Cl in Antarctic meteorites by Nishiizumi *et al.* (1979) the lower limit reached was 7×10^{-16} for ^{36}Cl/Cl, and this is the smallest radioisotope ratio yet measured by this technique.

Aluminium-26 Analysis

^{26}Al (half-life 7.2×10^5a) is also created by cosmic rays in the atmosphere by the spallation of argon, but because its production yield is about 100 times less than that of ^{10}Be it is much more difficult to study. The separation of ^{26}Al and its isobar ^{26}Mg can be performed satisfactorily by negative ion mass spectrometry, and Kilius *et al.* (1979) have measured $^{26}Al/Al$ ratios less than 10^{-14} in milligram samples. This is comparable to measurements made on large samples by counting the gamma rays following the beta decay to ^{26}Mg. The areas where applications have been attempted include studies of the growth rates of manganese nodules on the ocean floor (Guichard *et al.* (1978), and determination of the constancy of cosmic ray intensity in the past (Hampel and Schaeffer, 1979). It seems possible that measuring $^{26}Al/^{10}Be$ on small samples in the same apparatus may provide a several million year chronology for ocean sediments which would be independent of fluctuations in cosmic ray flux (Lal, 1962; Reyss *et al.*, 1976).

Iodine-129 Analysis

^{129}I (half-life 1.6×10^7a) occurs in meteorites through the action of cosmic rays. It can also be produced in the spontaneous fission of ^{238}U and in the neutron induced fission of ^{235}U. Neutron activation analysis provides a satisfactory method for studying ^{129}I, but recent experiments by Elmore *et al.* (1980) have shown that negative ion tandem mass spectrometry with a time-of-flight detection system can determine $^{129}I/^{127}I$ ratios less than 10^{-12} in measurement times less than half an hour. The detection of ^{129}I in two meteorite samples has been accomplished this way.

CONCLUSIONS

The experiments described in this article have been mainly of an exploratory nature. The feasibility of measuring the various radioisotopes by accelerator mass spectrometry has been established, but routine production of results for application to archaeology and Quaternary geology must await the imminent establishment of laboratories dedicated to these tasks. In this second development stage much attention will be devoted to improving the precision with which the measurements can be made, so that present uncertainties will be eliminated. There is every indication that with improved design of ion sources, ion beam optics and detectors, precisions better than ±1% may be obtained. In addition it may be possible to extend the range of radioisotopes studied to some not mentioned here. One which holds promise for archaeologists and Quaternary geologists is ^{41}Ca. This isotope, which has a half-life of 1.3×10^5a, presents problems in its identification because of the difficulty in discriminating against the isobar ^{41}K. However, there may be the possibility of using the molecule $^{41}CaF^+$ to discriminate against the unstable $^{41}KF^+$ (Ravn, 1979), in which case the implications for the dating of calciferous organisms are obvious.

A further aspect of the advent of this method of ultrasensitive analysis is in the assay of stable isotopes at ultra-low levels. The recently publicised iridium anomaly at the Cretaceous-Tertiary boundary (Alvarez *et al.*, 1980) may represent a global event which has left its mark in contemporaneous strata around the world. Correlation of such strata by matching trace element patterns in different locations may offer another approach to geochronology. However the method and its applications evolve, there can be little doubt that an important bridge has been crossed, and the path on the other side may be expected to lead to exciting new fields.

ACKNOWLEDGEMENTS

I would like to thank my colleagues at the Universities of Toronto and Rochester for the opportunity to take part in these exciting developments. In particular I thank A.E. Litherland for many discussions and advice in the preparation of this article. I also acknowledge the receipt of grants from the Natural Sciences and Engineering Research Council, Ottawa.

REFERENCES CITED

Alvarez, L.W. and Cornog, R., 1939, '^{3}He in Helium' and 'Helium and Hydrogen of Mass 3': Physical Review, v. 56, p. 379 and 613.

Alvarez, L.W., Alvarez, W., Asaro, F. and Michel, H.V., 1980, Extra-terrestrial cause for the Cretaceous-Tertiary Extinction: Science, v. 208, p. 1095-1108.

Anbar, M., 1978, The limitations of mass spectrometric radiocarbon dating using CN^- ions, in Gove, H.E., ed., Proceedings of the First Conference on Radiocarbon Dating with Accelerators: University of Rochester, p. 152-155.

Anderson, E.C., Libby, W.F., Weinhouse, S., Reid, A.F., Kirshenbau, A.D. and Grosse, A.V., 1947, Natural Radiocarbon from Cosmic Radiation: Physical Review, v. 72, p. 931-936.

Andrews, H.R., Ball, G.C., Brown, R.M., Burn, N., Davies, W.G., Imahori, Y. and Milton, J.C.D., 1978, Radiocarbon dating with the Chalk River MP tandem accelerator, in Gove, H.E., ed., Proceedings of the First Conference on Radiocarbon Dating with Accelerators: University of Rochester, p. 114-126.

Barratt, P.J.S.B., Doucas, G., Garman, E.F., Hyder, H.R.McK., Sinclair, D., Hedges, R.E.M. and White, N.R., 1978, Experiments to establish the design of a dedicated tandem accelerator for ^{14}C dating at Oxford, in Gove, H.E., ed., Proceedings of the First Conference on Radiocarbon Dating with Accelerators: University of Rochester, p. 127-151.

Bennett, C.L., 1979, Radiocarbon Dating with Accelerators: American Scientist, v. 67, p. 450-457.

Bennett, C.L., Beukens, R.P., Clover, M.R., Gove, H.E., Liebert, R.P., Litherland, A.E., Purser, K.H. and Sondheim, W.E., 1977, Radiocarbon dating using electrostatic accelerators: Negative ions provide the key: Science, v. 198, p. 508-510.

Bennett, C.L., Beukens, R.P., Clover, M.R., Elmore, D., Gove, H.E., Kilius, L.R., Litherland, A.E. and Purser, K.H., 1978, Radiocarbon dating with electrostatic accelerators: Dating milligram samples: Science, v. 201, p. 345-347.

Elmore, D., Fulton, B.R., Clover, M.R., Marsden, J.R., Gove, H.E., Naylor, H., Purser, K.H., Kilius, L.R., Beukens, R.P. and Litherland, A.E., 1979, Analysis of ^{36}Cl in environmental water samples using an electrostatic accelerator: Nature, v. 277, p. 22-25, and *errata* p. 246.

Elmore, D., Gove, H.E., Ferraro, R., Kilius, L.R., Lee, H.W., Chang, K.H., Beukens, R.P., Litherland, A.E., Russo, C.J., Purser, K.H., Murrell, M.T. and Finkel, R.C., 1980, Determination of ^{129}I using tandem accelerator mass spectrometry: Nature, v. 286, p. 138-140.

Gove, H.E., Elmore, D., Ferraro, R.D., Beukens, R.P., Chang, K.H., Kilius, L.R., Lee, H.W., Litherland, A.E., Purser, K.H. and Rubin, M., 1980, Radiocarbon dating with tandem electrostatic accelerators: Radiocarbon, v. 22, p. 785-793.

Guichard, R., Reyss, J-L. and Yokoyama, Y., 1978, Growth rate of manganese nodules: Nature, v. 272, p. 155-156.

Hampel, W. and Schaeffer, O.A., 1979, ^{26}Al in iron meteorites and the constancy of cosmic ray intensity in the past: Earth and Planetary Science Letters, v. 42, p. 348-358.

Kilius, L.R. and Litherland, A.E., 1978, On the use of negative ions and electrostatic accelerators for the study of low concentrations of long lived radionuclei, in Davis, S.N., ed., Rep. Workshop on Dating Old Ground Water: University of Arizona, Tucson, p. 62-79.

Kilius, L.R., Beukens, R.P., Chang, K.H., Lee, H.W., Litherland, A.E., Elmore, D., Ferraro, R. and Gove, H.E., 1979, Separation of ^{26}Al and ^{26}Mg isobars by negative ion mass spectrometry: Nature, v. 282, p. 488-489.

Kilius, L.R., Beukens, R.P., Chang, K.H., Lee, H.W., Litherland, A.E., Elmore, D., Ferraro, R., Gove, H.E. and Purser, K.H., 1980, Measurement of $^{10}Be/^{9}Be$ Ratios Using an Electrostatic Tandem Accelerator: Nuclear Instruments and Methods, v. 171, p. 355-360.

Lal, D., 1962, Cosmic ray produced radio nuclides in the sea: Journal of the Oceanographical Society of Japan, 20th Anniversary Issue, p. 600-614.

Litherland, A.E., 1978, Radiocarbon dating with accelerators, in Gove, H.E., ed., Proceedings of the First Conference on Radiocarbon Dating with Accelerators: University of Rochester, p. 70-113.

__________, 1980, Ultrasensitive Mass Spectrometry with Accelerators: Annual Review of Nuclear and Particle Science, v. 30, p. 437-473.

Litherland, A.E., Beukens, R.P., Kilius, L.R., Rucklidge, J.C., Gove, H.E., Elmore, D. and Purser, K.H., 1981, Ultrasensitive mass spectrometry with tandem accelerators: Nuclear Instruments and Methods, v. 186, p. 463-477.

Middleton, R., 1977, A Survey of Negative Ions from a Cesium Sputter Source: Nuclear Instruments and Methods, v. 144, p. 373-399.

Muller, R.A., 1977, Radioisotope dating with a cylcotron: Science, v. 196, p. 489-494.

__________, 1978, Radioisotope dating with the LBL 88" cyclotron, in Gove, H.E., ed., Proceedings of the First Conference on Radiocarbon Dating with Accelerators: University of Rochester, p. 33-37.

Muller, R.A., Stephenson, E.J. and Mast, T.S., 1978, Radioisotope measurement with an accelerator: A blind measurement: Science, v. 201, p. 347.

Nelson, D.E., Korteling, R.G. and Stott, W.R., 1977, Carbon-14 Direct Detection at Natural Concentrations: Science, v. 198, p. 507-508.

Nishiizumi, K., Arnold, J.R., Elmore, D., Ferraro, R.D., Gove, H.E., Finkel, R.C., Beukens, R.P., Chang. K.H. and Kilius, L.R., 1979, Measurements of ^{36}Cl in Antarctic meteorites and Antarctic ice using a van de Graaff accelerator: Earth Planetary Sci. Lett., v. 45, p. 285-292.

Purser, K.H., 1977, U.S. Patent No. 4037 100.

Purser, K.H., Liebert, R.B., Litherland, A.E., Beukens, R.P., Gove, H.E., Bennett, C.L., Clover, M.R. and Sondheim, W.E., 1977, An attempt to detect stable N from a sputter ion source and some implications of the results for the design of tandems for ultra-sensitive carbon analysis: Revue de Physique Appliquee, v. 12, p. 1487-1492.

Purser, K.H., Litherland, A.E. and Rucklidge, J.C., 1979, Secondary Ion Mass Spectrometry at Close to Single-atom Concentration Using DC Accelerators: Surface and Interface Analysis, v. 1, p. 12-19.

Purser, K.H., Liebert, R.B. and Russo, C.J., 1980, MACS: An accelerator based radioisotope measuring system: Radiocarbon, v. 22, p. 794-806.

Raisbeck, G.M., Yiou, F., Fruneau, Lieuvin, M. and Loiseaux, J.M., 1978a, ^{10}Ve Detection Using a Cyclotron Equipped with an External Ion Source, in Gove, H.E., ed., Proceedings of the First Conference on Radiocarbon Dating with Accelerators: University of Rochester, p. 38-46.

__________, 1978b, Beryllium-10 Mass Spectrometry with a Cyclotron: Science, v. 202, p. 215-217.

Raisbeck, G.M., Yiou, F. and Stephan, C., 1979, ^{26}Al measurement with a cyclotron: Journal de Physique, Lettres, v. 40, p. 241-244.

Raisbeck. G.M., Yiou, F., Fruneau, Lieuvin, M. and Loiseaux, J.M., 1979, Measurement of ^{10}Be in 1000- and 5000-year-old Antarctic ice: Nature, v. 275, p. 731-733.

Raisbeck, G.M., Yiou, F., Fruneau, Loiseaux, J.M., Lieuvin, M., Ravel, J.C. and Hays, J.D., 1979, A search in a marine sediment core for %10&Be concentration variations during a geomagnetic field reversal: Geophys. Res. Lett., v. 6, p. 717-719.

Raisbeck, G.M., Yiou, F., Fruneau, Loiseaux, J.M. and Lieuvin, M., 1979, ^{10}Be concentration and residence time in the ocean surface layer: Earth and Planetary Science Letters, v. 43, p. 237-240.

Ravn, H.L., 1979, Experiments with intense secondary beams of radioactive ions: Physics Reports, v. 54, p. 201-259.

Reyss, J.L., Yokayama, Y. and Tanaka, S., 1976, Aluminum-26 in deep-sea sediments: Science, v. 193, p. 1119-1121.

Shapira, D., DeVries, R.M., Fulbright, H.W., Toke, J. and Clover, M.R., 1975, The Rochester heavy-ion detector: Nuclear Instruments and Methods, v. 129, p. 123-130.

Stephenson, E.J., Mast, T.S. and Muller, R.A., 1979, Radiocarbon dating with a cyclotron: Nuclear Instruments and Methods, v. 158, p. 571-577.

Tombrello, T.A. and Kellogg, W.K., 1979, Simulation experiments and planetary sputtering phenomena: Tenth Lunar and Planetary Science Conference, Abstracts, p. 1233-1235.

Tucker, A.B., Bonani, G., Suter, M. and Woefli, W., 1981, Earthquake dating by ^{14}C atom counting, Proc. Symp. on Accelerator Mass Spectrometry: Argonne National Laboratory Publication ANL/PHY-81-1, p. 285-292.

URANIUM-SERIES DATING OF QUATERNARY DEPOSITS

H. SCHWARCZ and M. GASCOYNE

ABSTRACT

A wide variety of Quaternary deposits can be dated by U-series methods, over the time range from 400 ky to 1 ky. In order to be datable, the deposits must contain material which:

a) crystallized at the time of deposition;

b) initially contained some U (0.1 ppm or more) but no Th or Pa;

c) has remained a chemically-closed system since the time of deposition.

In continental regimes the best such materials are travertines formed either in caves (speleothems) or as spring deposits. The latter are often intercalated with detrital clastic deposits containing vertebrate fossils, or relatable to glacial stratigraphy. The travertines may have been porous at the time of deposition, and subsequent filling of pores by later carbonate can result in spuriously young ages. Calcrete (caliche) formed in soil profiles can be dated; however. calcreted horizons grow continuously over long periods, and material must be selected to be representative of discrete depositional events. Speleothem is often intercalated with fluvial sediments in the deep interior of caves, marking periods of flooding. In cave-mouth deposits, wind-blown detritus is also present, as well as cultural materials to which ages can be assigned by dating of speleothem. Lacustrine marls are datable if the carbonate component is dominantly authigenic. Molluscan carbonate, present as snail or clam shells in lacustrine sediments, appears to give less reliable ages. Bone, although containing the highest concentrations of U (up to 1000 ppm) has given questionable results in some cases, evidently due to "open system", continued uptake of U from groundwater. U-series dates of marine, biogenic carbonates have been used to establish high sea-stands related to eustatic fluctuations. Low sea-stands can be established by dating speleothems found in caves that have been drowned by rise of sea level.

These dating methods can be related to climatic fluctuations in various ways. Speleothem, dated by U-series methods contains an isotopic record of paleotemperature and variation in isotopic composition of rainfall. Dated travertines in arid regions mark ancient periods of higher rainfall and lake levels. Calcretes deposited intermittently in soil sequences or at varying depths may reflect changes in past rainfall intensity. Glacial outwash deposits in some localities in the U.S. Rockies have been buried by or interstratified with travertines formed by warm-water springs. Speleothem formed near cave mouths can entrap pollen which can be used to define the climatic conditions at the time of deposition. This would permit the construction of dated pollen profiles beyond the range of ^{14}C.

INTRODUCTION

Before the discovery of absolute radiometric dating methods, time scales for the evolution of geomorphic features had to be estimated on the basis of observation of modern erosion rates and extrapolation to earlier times. This method suffered, at least in temperate and arctic regimes, from the problem that erosion rates have altered drastically during the past few hundreds of thousands of years due to intense changes in climate that have accompanied global, continental glaciation. If man had evolved during a period of persistently stable climate, then he would have been able to use this method more effectively. In fact, it is now apparent that one of the chief tasks of the geomorphologist is not only to estimate the rates of evolution of landforms, but also to note the fluctuations of these rates through time, and from them, to evolve a picture of climatic change. This record can be compared in turn with the record of faunal and floral change and with the now well-calibrated record of global ice volume obtained from the isotopic study of deep sea cores (Shackleton and Opdyke, 1976). In view of the interest in this problem of time scales in geomorphology it is fortunate that we now have at our disposal a number of geochronometers applicable to the measurement of the age of such young deposits. These are summarized in Figure 1, which compares the time ranges over which each is applicable. This paper is specifically devoted to those which arise from the disequilibrium between the daughter isotopes of U-238 and U-235, and their respective parents.

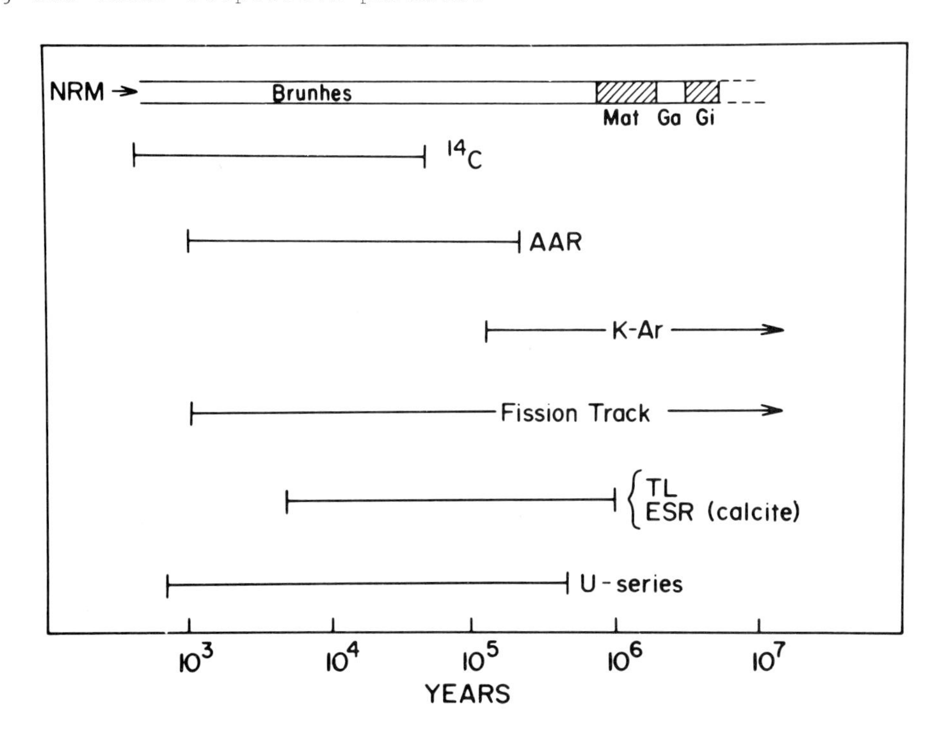

Figure 1 Dating methods applicable to the Quaternary. NRM: natural remanent magnetism; normal (blank) and reversed (shaded) chrons are shown; Mat = Matuyama; Ga = Gauss; Gi = Gilbert. AAR: amino acid racemization; TL: thermoluminescence; ESR: electron spin resonance; U-series range is shown for $^{230}Th/^{234}U$ dating.

Uranium series dating is based on the decay of the daughters of U isotopes by emission of alpha particles. In each method ($^{230}Th/^{234}U$, $^{234}U/^{238}U$, $^{231}Pa/^{235}U$ *etc.*) the following assumptions are made: 1) at the time of formation of the dated sample the parent isotope(s) was present but the daughter in question was either absent or was present at a known activity level: 2) after a time which is long with respect to the half-life of the daughter, the activity of the daughter becomes equal to that of the parent (that is, the sample achieves secular radioactive equilibrium); 3) the sample has not been chemically disturbed since the time of its formation. The half life of the daughters of the uranium isotopes range from a few milliseconds to 2.5 x 10^5y, and allow us to estimate the age of samples over the mid- and upper-Pleistocene though with varying degrees of precision.

The applicability of the method rests on the observation that the daughter isotopes behave chemically quite differently from their respective parents. Therefore, in low-temperature depositional processes such as growth of calcite from solution, the parent uranium may be trapped as a trace element in a crystal while the daughters may be excluded or at least greatly depleted relative to their equilibrium value. The various decay series in question are shown in Figure 2. Thorium-230, the daughter of ^{234}U (and in turn, ^{238}U is very insoluble in ground and surface-water). Therefore calcite precipitated in continental or marine environments is found to contain up to several ppm of U but negligible amounts of Th. Therefore its age of crystallization can be determined by measurement of the amount of ^{230}Th that has grown into equilibrium with the U parent. Similar relations are observed between ^{231}Pa (also insoluble in surface waters) and its parent ^{235}U. The ages of samples are given by the following equations:

$$\left[\frac{^{230}Th}{^{234}U}\right]_t = \frac{1 - e^{-\lambda_{230}t}}{(^{234}U/^{238}U)_t} + \left[\frac{\lambda_{230}}{\lambda_{230} - \lambda_{234}}\right] \bullet \left[1 - \frac{1}{(^{234}U/^{238}U)_t}\right] \bullet \left(1 - e^{-(\lambda_{230} - \lambda_{234})t}\right) \quad (1)$$

$$\left[\frac{^{231}Pa}{^{235}U}\right]_t = \left[1 - e^{-\lambda_{231}t}\right] \quad (2)$$

where ^{234}U, *etc.*, are the respective activities in disintegrations per minute per gram of sample, λ_{230}, *etc.* are the decay constants of the isotopes, and t is the time since crystallization. Note that equation (1) is transcendental and must be solved by iterative approximations. Figure 3 is a graph of solutions of this equation which shows that the limit of the age determinable by this method is about 400 ky depending on the $^{234}U/^{238}U$ ratio. The precision of measurement of the activity ratios depends on the concentration of U in the sample, and the chemical yield of the extraction method. Generally $^{231}Pa/^{235}U$ ages can only be obtained on samples containing more than 1 ppm U, due to the much lower abundance of the parent ^{235}U($^{235}U/^{238}U$ = 1/137.8 on atom basis).

The ratio $^{234}U/^{238}U$ should be equal to unity in rocks and waters, since these are two isotopes of the same heavy element and thus not subject to isotope fractionation like light elements. However, it has been found that ^{234}U if preferentially leached from rocks during the weathering cycle, so that the $^{234}U/^{238}U$ ratio in ground waters is normally somewhat greater than unity (and ranges up to values of about 20 in rare instances) while the ratio in sediments is generally somewhat less than unity. The value of this ratio in a Quaternary deposit which has behaved as a closed system since its formation, could be used to estimate the age of the sediment. The ratio changes with time according to the relation:

$$\left(\frac{^{234}U}{^{238}U}\right)_t - 1 = \left(\frac{^{234}U}{^{238}U}\right)_o - 1 \bullet e^{-\lambda_{234}t} \quad (3)$$

However, the initial ratio, $(^{234}U/^{238}U)_o$, cannot be independently known, in most cases. For marine samples (corals, *etc.*) it can be assumed to have been near the present-day value of 1.14, but the ratio of

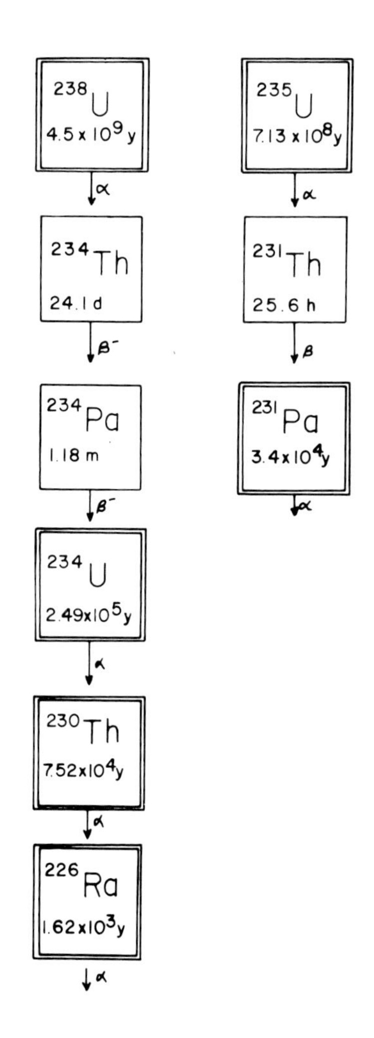

Figure 2 Decay chains for ^{238}U and ^{235}U. Heavy borders are around longer-lived isotopes, useful for dating Quaternary sediments.

marine samples approaches unity within about 300 ky, which is within the range of the ^{230}Th/^{234}U method. For older continental deposits one can sometimes assume that the initial ratio has remained constant if younger deposits, datable by ^{230}Th, demonstrate this fact. However, many examples of sites at which the initial ratio has varied drastically with time are now known, *e.g.*, for stalagmite deposition in caves (see below).

The ideal sample for U-series dating is one which was chemically precipitated from a low-termperature solution. A large variety of poetentially-datable materials satisfy this criterion. Principally, these are various forms of calcium carbonate, either calcite, aragonite or some other polymorph or hydrated species. We shall refer to all of them here as calcite, though the same methods are applicable to other polymorphs or states of hydration. The calcite may occur as travertine, that is, massive, chemically-precipitated limestone formed either in a cave (speleothem) or at a lime-rich spring. If the travertine is initially dense, non-porous, and free of detrital contaminants, then it generally satisfies the assumptions of the method. However, many surficial deposits of this sort (tufa) are porous, and contain detritus. They may be recrystallized or leached by water permeating through them long after they were deposited; they may also contain non-radiogenic

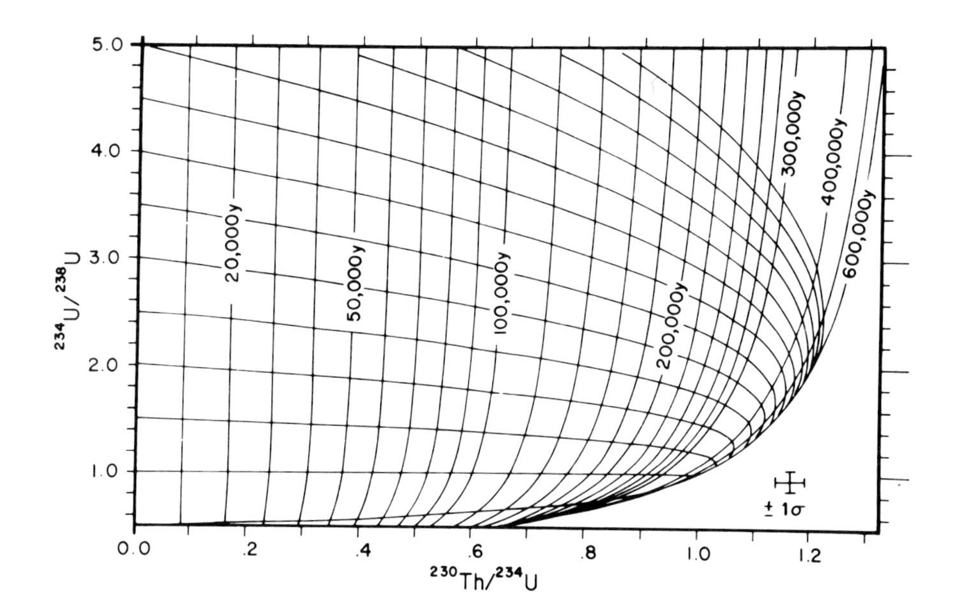

Figure 3 Graph of activity ratios $^{234}U/^{238}U$ *vs* $^{230}Th/^{234}U$ for closed systems of varying initial $^{234}U/^{238}U$. Isochrons are near-vertical curves; sub-horizontal curves are loci of samples of constant initial $^{234}U/^{238}U$ ratio. Approximate error cross is shown for a sample containing 0.5 ppm U. Effective limit of dating method is 350,000 years.

^{230}Th (common Th) trapped on the detrital particles, as well as particles of limestone which behave chemically like the travertine, but which contained uranium in secular equilibrium. Recrystallization will make the sample appear to be anomalously young, while detrital contamination has the opposite effect.

Other types of potentially datable materials are: calcite deposited in the soil as calcrete (caliche); shells of molluscs or other invertebrate organisms; bones of vertebrates; egg-shells of large birds. All of these deposits can initially contain significant amounts of uranium or acquire them soon after deposition from groundwater. Peat deposits, while initially lacking in uranium, can also adsorb this element from groundwater some time after deposition, and if they are subsequently dried out, and compacted they may behave as closed systems. Other types of chemical deposits such as phosphatic concretions, gypsum, evaporites in general, and even deposits of ferric or manganese hydroxides (*e.g.* concretions) can contain significant amount of co-precipitated uranium although in many instances there will also be sizable amounts of common thorium either in the same crystalline material, or adsorbed on detritus.

The analysis of isotope ratios for age determination is usually carried out by dissolving a sample of the material in question, in nitric acid or some other oxidizing medium. This ensures that uranium remains (or is converted to) the 6+ oxidation state, so it can be extracted by ion exchange from the solution. In the case of carbonates (travertine, *etc.*) the insoluble residue may be simply discarded, or it may be separately dissolved in hydrofluoric acid, and added to the remaining solution. The HNO_3 insoluble portion commonly contains the detrital component responsible for the presence of common Th or Pa, and in some cases it can be separately analysed to permit a correction for the component (Ku *et al.*, 1979). The solution is then processed to separate U and Th by ion exchange; these elements are plated onto metal discs and the activity of the various alpha-emitting isotopes is determined by alpha spectrometry. Usually a tracer containing artificial nuclides is added (*e.g.* ^{228}Th in equilibrium with ^{232}U) to determine the chemical yields of U and Th, and thus allow such ratios as ^{230}Th/^{234}U to be determined. The age is obtained by solution of eq. (1), *etc.*, using a computer. Error estimates are based on the number of alpha disintegrations counted, assuming Poisson statistics.

If the sample contains much insoluble detritus, or if it displays the presence of ^{232}Th in the alpha spectrum, then the spectrum is likely to contain some common ^{230}Th, and possibly some common U leached from the detritus. If the abundance of common Th is high, as shown by a low ^{230}Th/^{232}Th activity ratio (typically less than 20), then it is necessary to correct for this component. Various schemes for making this correction have been devised, and are summarized elsewhere (Schwarcz, 1980).

In the following sections we shall describe some of the applications of these dating methods to Quaternary continental deposits, with special emphasis to those applications which reveal information about climatic change. We shall omit the dating of coral reefs, which will be discussed by Stearns (this volume).

TRAVERTINE

Speleothems

Caves are formed in limestone due to dissolution of the rock by waters containing carbonic acid from the soil. Subsequently, the caves may be partly filled by various forms of chemically-precipitated carbonate: stalagmites, stalactites, flowstone, *etc.*, which are collectively called speleothems. When deposited in the deep interior of caves, and above the level where streams can transport detritus, these deposits can be exceptionally pure and free of detrital contaminants. Such speleothems are typically quite impermeable, and are composed of aggregates of a few very large crystals, up to several cm in smallest dimension. They are therefore ideally suited to U-series dating, as well as to other types of geochemical analyses (Gascoyne *et al.*, 1978). Normally the content of U in speleothem is below 1 ppm. Therefore ages can generally only be obtained using the ^{230}Th/^{234}U method. However, if there are layers of shale or other detrital sedimentary rocks hydrologically situated so as to contribute U to infiltrating groundwaters, then speleothems precipitated from these waters may contain up to 100 ppm U. The ^{234}U/^{238}U ratio of water dripping from a single source in a cave is relatively constant although some short term variation has been observed (P. Thompson *et al.*, 1975). The possible use of this ratio in speleothem as a method of age determination has been investigated by P. Thompson *et al.* (1975), G. Thompson *et al.* (1975), and Harmon and Schwarcz (1982). The latter authors found that at two discrete sites within a single cave a few hundred metres apart the behaviour of the ^{234}U/^{238}U ratio varied markedly. At one site the initial ratio remained essentially constant over a period of 250,000 years, while at the other site the ratio varied over the same time range, and was significantly different from that at the first site. G. Thompson *et al.* (1975) found that the measured ^{234}U/^{238}U ratio decreased regularly from the top to the base of a single speleothem. This allowed them to determine its growth history but the ages which they obtained were much older than the apparent ^{230}Th/^{234}U ages for the same sample (see Harmon *et al.*, 1978a for a discussion of these results).

Absolute ages of speleothems can be used in various ways to help define the Quaternary history of a region. First, they provide a basic chronology for the caves in which they are formed. If they are intercalated between clastic detritus carried into the caves by meltwater streams from glaciers, then they give potential information about the timing of glaciation (Ford, 1973). If they are from caves that have been developed at progressively deeper levels in a limestone formation due to lowering of local base level, then the age of a given deposit is a minimum estimate of the time the base level stood at the elevation of the speleothem. This has been used by Ford *et al.* (1981) to determine the average rate of downcutting of the carbonate-based part of the Canadian Rockies. Here a number of phreatic caves have been left high and dry by lowering of valley floors. Many of the caves are represented today only by remanent tunnels truncated at both ends by steep mountain cliffs (Figure 4). Within some of these caves speleothems are found ranging in age from a few thousand years to beyond the range of $^{230}Th/^{234}U$ dating. One sample has in fact been shown by paleomagnetic measurements to be older than 700 ky, since it preserves a reversed natural remanent magnetisation (Latham *et al.*, 1979). By relating the elevations and ages of these speleothems, it has been possible to construct a model for the timing of the uplift and downcutting of the mountain range, and thus to estimate the age of local relief in the eastern part of the Canadian Rocky Mountains. A summary of these data is given in Table 1. Using the heights of the cave mouths above the adjacent valley floors, and taking into account the hydraulic gradients that must have existed between the mouths of the caves and the inner portions represented by the preserved remanents, the maximum rate of downcutting was between 0.13 and 2.1 m per 1000 y while the minimum rate ranged between 0.04 and 0.07 m/1000 y. The mean relief of 1.3 km that exists today in this region would have developed in a time of between 6 and 12 million years.

Figure 4 View from inside of erosional cave remanent, near crest of Rocky Mountains, Alberta.

The existence of speleothem deposits in a cave implies that there was liquid water in the soil above the cave and that organic activity to produce CO_2 was occurring in the soil. Therefore, the rate of deposition of speleothem is modulated by changes in climate, and the frequency of occurrence of speleothem ages in a region gives an indication of the intensity of the climate through time. Both aridity and intense prolonged cold could have the same effect on growth rates but

Table 1. Some ages for the Extant Relief of the Canadian Rockies calculated for a mean local relief of 1.34 ± 0.40 km

Selected rates of valley deepening	Years (Ma)
At a greatest maximum rate of 2.07 m/1000 years	0.65 ± .19
At a mean maximum rate of 1.04 m/1000 years	1.30 ± .36
At a lowest maximum rate of 0.13 m/1000 years	10.30 ± 3.1
At the Bearjaw Cave minimum rate of 0.07 m/1000 years	19.10 ± 5.7
At the Eagle Cave minimum rate of 0.04 m/1000 years	33.50 ± 10
Solutional limiting age	12-14
from Ford *et al.*, 1981	

in a given region it will generally be apparent which of these factors was responsible. In alpine caves of western North America we have obtained a frequency distribution of speleothem ages over the last 300 ky (Harmon *et al.*, 1977) that parallels climate variations as inferred from the deep-sea isotope record. In England a similar pattern slightly displaced in time, has been obtained (Gascoyne *et al.*, 1982) (Figure 5).

As noted, U-series dates on speleothems can also be used as a time base for various studies which treat speleothems as repository records of the past. For example, their stable isotopic composition can be used to infer past climate (Harmon *et al.*, 1978b; Gascoyne *et al.*, 1981), their paleomagnetic record can be determined and compared with records from other types of samples (Latham *et al.*, 1979), and they have been shown to contain pollen which may also preserve a paleoclimatic record (Bastin, 1978). Speleothem from the mouths of caves can be interstratified with detrital sediments containing archaeological remains, as well as fauna and flora (including pollen), all of which can be set in a chronological framework by use of the dated speleothem layers (Schwarcz, 1980; Blackwell *et al.*, 1982).

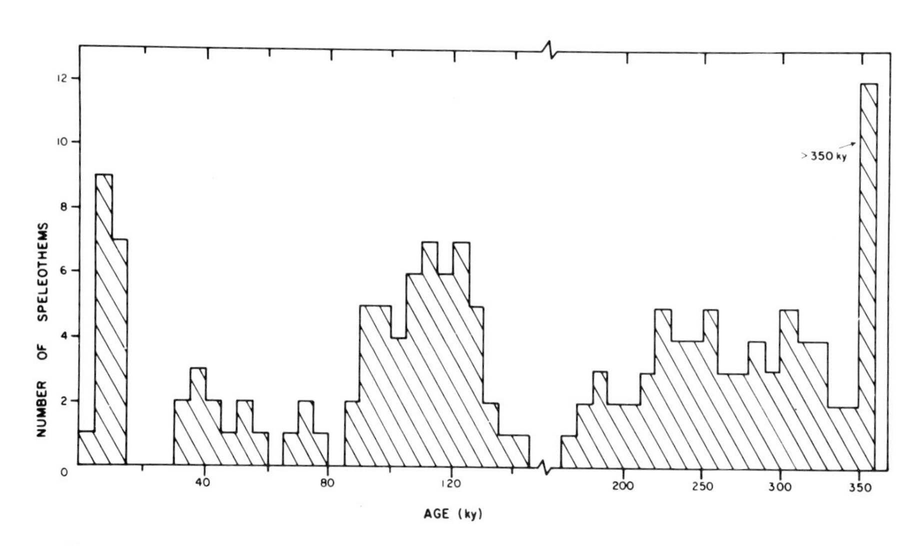

Figure 5 Histogram of dates of speleothems from caves in Lancashire, England. Note absence of speleothem deposits during glacial maxima and peaks in frequency during interglacials and interstadials. From Gascoyne *et al.*, 1982.

Spring-Deposited Travertines

Lime-rich waters emerging from springs in areas of carbonate bedrock, or in volcanic terrain, can deposit terrace-like accumulations of calcium carbonate, which are commonly found to be intercalated with other types of sediment. The carbonates, commonly described as tufa, are typically quite porous, and include large amounts of vegetable matter derived from the growth of reeds, mosses and other plants, as well as algal or bacterial mats that may be partly responsible for carbonate precipitation. The organic material is largely lost by oxidation and decay within a few thousand years after deposition, adding to the porosity of the deposit. Finally, further precipitation of calcite may occur due to passage of water through the porous mass, filling primary pores with secondary calcite whose absolute age is much younger than the time of deposition of the terrace. Nevertheless, in ideal circumstances, it is possible to obtain meaningful U-series dates on such travertines. Optimally the sample material should be collected from layers which appear initially to have been as dense (non-porous) as possible, and lacking in detrital contaminant. Study of some modern carbonate-depositing springs suggests that the detritus problem is alleviated by the action of the flowing spring water carrying away allochthonous sediment that is blown in or carried in by animals.

Dates of such travertines can be significant in studies of geomorphic evolution. Where the travertine forms a broad, sheet-like deposit, it serves as a chronostratigraphic marker which may be deposited on, or intercalated with, other Quaternary sediments whose ages are sought. For example, terraces of tufa may be superimposed on fluvial terraces, and may thus provide an age for the cutting of the latter. This situation is found in northwestern Hungary where terraces have been cut at various stages of development of the Danube River and its tributaries. Warm-spring-deposited tufa mounds are built up on these platforms at many points west of Budapest, and some have been dated by the $^{230}Th/^{234}U$ method. The first studies were by Cherdyntsev *et al.* (1965) who were concerned with the archaeological sites at Tata and Vértessöllös. Later, Pécsi (1973) pointed out the relation between these tufa deposits and the fluvial terraces, and correlated them with discrete glacial stages (Table 2); dates on the tufas obtained by J.K. Osmond (quoted in Pécsi, 1973), confirm the greater antiquity of the higher tufa mounds, showing that in general they appear to have been deposited not long after the terrace cutting. More recent data on two of these terraces by ourselves are also shown, indicating that the tufa surface at Vértesszöllös cannot be as old as was suggested by Pécsi. Note however that the age of such a tufa gives a minimum age for cutting of the fluvial terrace and the true age can be much older, since there is no special tendency for the terrace cutting to stimulate formation of warm springs.

Another type of setting in which travertine ages are applicable, less commonly however, is where glacial, glaciofluvial or galciolacustrine deposits are capped by spring deposits. In the western United States the glacial outwash deposits of some of the alpine glaciers of the Rocky Mountains are locally capped by warm-spring deposited travertines which are widespread in this region (Feth and Barnes, 1979). One example of this phenomenon which we have studied is on the north slope of the Wind River Range of Wyoming where Richmond (1976) has mapped a series of lobes of drift which have been pushed northward down the canyons of this range by successive tongues of ice emerging from alpine ice caps on the range, during each of the past glacial stages. The drift lobes can be traced in turn to sheets of fluvial outwash deposits, which partly fill the valley of the Wind River, north of the range. These outwash deposits are themselves cut into a series of fluvial terraces, and are locally capped with sheets of warm-spring-fed travertine. We have dated some of these travertines in order to establish a chronology for the glacial outwash events. The Circle Terrace is believed to be equivalent in age to the emplacement of the Bull Lake moraine. Travertine resting on the Circle Terrace, dated by $^{230}Th/^{234}U$, yields ages ranging from 42 ky at the base to 30 ky at the

Table 2. Terraces of the Tata River, Hungary.

The Tata is a small tributary of the Danube; its flood plain is below the flood level of the Danube. The following terraces are attributed to erosional stages of the Pleistocene (heights in metres, measured in vicinity of Vértesszöllös).

Terrace No.	Height above flood plain	Proposed correlative glacial-stage	Absolute age of travertine (Ky)
I	0	modern	--
IIa	4-5	Würm	--
IIb	8-10	Riss II-Riss/Würm	120-78(a)
III	20-25	Riss I	190 ± 45(b)
IV	c. 40	Mindel	--
V	55-65	?	210 ± 10(a)
			>270 (b)

Sources of data:

a) H.P. Schwarcz and A. Latham, unpublished data, 1981

b) Pécsi, 1973.

top. These ages lie within the middle-Wisconsin glacial stage and therefore cannot be very close to the ages of the outwash events, although they put a minimum age on the event. Outwash correlated with the younger Pinedale moraine is capped by a porous travertine that yields an age of 12 ± 3 ky, consistent with its formation during the melting of the ice cap present on the crest of the range during the Wisconsin.

While these methods have so far only been applied to spring-deposited travertines, there is another class of carbonate deposits which should be investigated. Many rivers fed by Ca-rich spring waters will deposit $CaCO_3$ where the water becomes sufficiently aerated to lose CO_2 and become supersaturated in calcite (or aragonite). Tufa mounds accumulate along the thalweg and their presence serves as a turbulence-producing structure that accelerates the aeration process. They can thus grow into sizeable structures which mark a stage in the evolution of the stream. Generally these mounds of stream-deposited travertine are themselves highly susceptible to erosion during further lowering of base level. For example, in the Wadi Paran, in the Negev Desert of Israel, large mounds of travertine are built up on the sides and crests of intermittently active waterfalls. These accumulations probably developed during the Wisconsin glacial stage which was a period of high rainfall in this region (Horowitz 1979). No U-series dates have yet been obtained on these deposits. Similar tests of pluvial/interpluvial climate could be obtained from speleothems obtained from caves in such regions.

LACUSTRINE LIMESTONES

Calcareous sediments which were deposited in pluvial lakes are known from several continents. Both calcareous bottom sediments, formed partly by chemical precipitation from the lake waters, and tufa accumulations along the lake shores, are known from such environments, and have in some instances been dated by U-series methods, as well as by the ^{14}C method.

The Pleistocene lakes Lahontan and Bonneville in Utah and Nevada accumulated a considerable thickness of marl, in which is contained an abundant fauna of ostracods and gastropods; tufa mounds dating from former high lake-stands occur along the margin. These deposits have been analysed by ^{14}C and U-series methods by Kaufman and Broecker (1965). There was agreement between the two methods for those samples lacking in excess ^{226}Ra relative to uranium; the dates lay in the interval 25 to 10 ky, as would be expected for the last pluvial period in this region. Some samples possibly dating to older pluvials were also recognized. Evaporitic lake deposits in the Searles Lake basin of California were dated by Peng *et al.* (1978), also using both U-series and ^{14}C, and yielding ages in the last part of the Wisconsin glacial stage (33 to 9 ky). Interestingly, there was satisfactory concordance between ^{14}C and ^{230}Th ages, although some correction for detrital matter had to be made to the latter.

In Israel, Kaufman (1971) was able to use $^{230}Th/^{234}U$ ratios to determine the time of accumulation of marly sediments of the Lisan Formation which were deposited in the Jordan graben, where the smaller Dead Sea now remains. The Lisan sediments (Figure 6) yielded ages in the interval from 62 to 20 ky, again showing that lake levels were high during the midst of the last, Wisconsin, Glacial stage in this region. Significant corrections had to be made for common ^{230}Th in the analysis of these sediments, which consist of finely interlaminated, chemically-precipitated calcite, and detrital silt brought into the lake by spring floods. The U-series ages for the younger parts of the sequence were in excellent agreement with ^{14}C ages. Agreement in the absolute ages of stratigraphically correlative parts of the Lisan Formation was also observed, over distances of tens of kilometers.

Taylor Dry Valley in Antarctica, at present a polar desert, contains a series of proglacial lake sediments including authigenic carbonates and sulphates. Absolute dates of these by $^{230}Th/^{234}U$ show that they were deposited during at least the last three interglacial stages, possibly at the very end of the interglacials during a time of thickening and expansion of the adjacent ice sheets (Hendy *et al.*, 1979). Hollin (1980) has used these dates in support of his proposal that a surge of the East Antarctic ice sheet brought about rapid cooling of the global ocean and the onset of the following glacial stage.

Dating of pluvial lake sequences in arid regions remains one of the most powerful techniques for defining climatic change in these regions. In the North African desert many such basins exist, some containing extensive faunal and archaeological remains as well as carbonate sediments. One basin which has begun to yield such data is the Shati Cardium Lake in Libya, from which ages between 200 and 80 ky have been obtained (Gaven *et al.*, 1981). In this and other pluvial lake sequences special care must be taken to sort out the isotopic effects attributable to admixture of detrital sediment and to possible recrystallization of primary aragonite to calcite, or fine-grained magnesian calcite to coarser, sparry low-Mg calcite. Such recrystallization effects were inferred by Szabo and Butzer (1979) in order to account for an apparent inversion in the ages of successive carbonate deposits in a South African calc-pan.

CALCRETE, AND OTHER PEDOGENIC CARBONATES

Authigenic calcite can form in soils of arid or semi-arid environments. If the carbonate is sufficiently abundant, it may occur as a discrete layer in the soil (calcrete, caliche) or as discrete concretionary masses. The time of formation of any particular part of such a layer or concretion can be determined roughly by U-series dating. Microscopic analysis of such deposits generally shows them to be extensively interspersed with detrital particles which originally comprised the soil. These will invariably contribute both U and Th to any analysis of the carbonate, and must be extensively corrected for. Ku *et al.* (1979) have used multiple analyses of coeval samples to determine the time of formation of calcretes from alluvial soils in the Mojave Desert, California. Multivariate regression techniques allowed

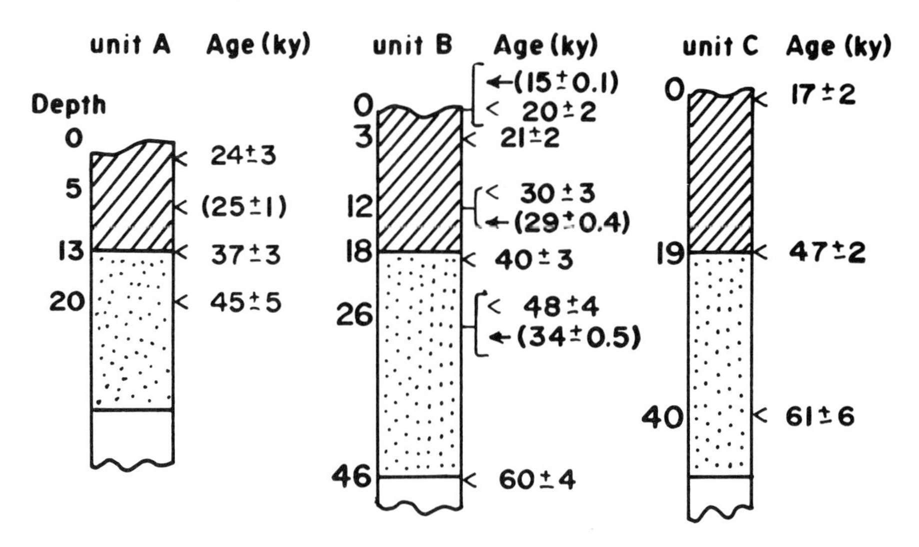

Upper Lisan sediments

Lower Lisan sediments

unexposed

< 60±1 ky $^{230}Th/^{234}U$ age

←(20±1) ky ^{14}C age

Figure 6 Chronostratigraphy of lacustrine sediments of the Lisan formation, showing comparison between ^{14}C and $^{230}Th/^{234}U$ ages. Data from Kaufman (1971).

them to infer the $^{230}Th/^{234}U$ and $^{234}U/^{238}U$ ratios of the authigenic component of the soil.

The main problem in using such data may ultimately be the question of geomorphological significance of the deposit. Soils, and calcrete layers in particular, are the product of slow dvelopment over long time periods. Therefore it is somewhat dubious to speak of the "age" of a calcrete, unless the process of pedogenesis was initiated and interrupted at discrete points in time. This situation can be observed, for example, where soils have formed on top of volcanogenic sediment (such as ash-fall tuffs). Here pedogenesis can be assumed to have begun at the time of eruption, and in many cases will have been terminated at the time of the next eruption. This situation is described, for example, by Williams *et al.* (1979) in the Gadeb region of Ethiopia, where successive ash beds are each capped by soils with well-developed Ca-horizons.

Another serious problem in the dating of calcretes is the requirement that the sediment from which the soil developed must have been lacking in calcareous detrital particles, since these will contribute an apparently infinite-age component to the analysis of the calcrete, one that cannot be compensated for by analytical methods such as those used by Ku *et al.* This is not a problem where the calcrete has developed from a volcanic ash.

ORGANIC DEPOSITS

Unquestionably one of the most widespread and abundant sources of primary carbonate precipitates associated with Quaternary sediments, both fluvial and lacustrine, is molluscan shell material. This material is even more abundant in marine settings where it occurs, for example, on raised beaches marking old high-sea-stands. Many examples of U-series dating of molluscs occur in the literature. However, the validity of such dates has been severely criticized by Kaufman *et al.* (1971) who give clear evidence for post-mortem migration of radionuclides into and out of mollusc tests. This is shown for example by the observation that the shells of living clams contain less than 0.2 ppm U while fossil clam shells a few tens of thousands of years old contain 2 ppm or more U. While Szabo and Rosholt (1969) have attempted to correct for this migration by use of an open system model, Kaufman *et al.* (1971) conclude that this model was not valid for most cases of nuclide migration, and that even concordance between $^{230}Th/^{234}U$ and $^{231}Pa/^{235}U$ ages did not ensure that the ages were correct.

Nevertheless, some of the most important estimates of climatic events of the Quaternary have been made through dates on molluscs. Stearns and Thurber (1967) used $^{230}Th/^{234}U$ ages of molluscs from ancient strandlines around the Mediterranean to obtain dates of high sea-stands at greater than 250, 200, 120 and about 80 ky. Andrews *et al.* (1975) have used dates of molluscs from arctic shorelines to define the age of last interglacial high sea-stands. However, extreme caution should be taken in using such data as definitive estimates of Quaternary chronology. Further work ought to be undertaken to review the methods of dating shells and the selection of samples for dating.

Similar problems seem to plague the equally promising field of dating ancient bones. Like mollusc shells, these organic deposits of calcium phosphate are initially free of all but minute traces of uranium. After a few thousand years of burial in the soil they may acquire from a few ppm to as much as several thousand ppm U. The uptake of U is poorly understood and may include both the replacement of Ca by U in the apatite crystal structure, and the reduction of soluble U^{6+} to insoluble U^{4+} by organic matter (collagen) of the bone. Szabo (1980) has used comparison between ^{14}C and U-series ages to determine that U-uptake ceases after about 2700 years in bones in some continental sediments. However, it is questionable that this interval is universally applicable to all settings in which fossil bones occur. Szabo has used the dates on bones from continental, fluvial deposits to place limits on their time of deposition (Table 3). Szabo *et al.*(1973) give references to several other U-series studies of fossil bones.

Finally we should note that some possibility exists for dating the organic matter found in some Quaternary sedimentary deposits. Cherdyntsev (1971) found that, for peat deposits with sufficiently high $^{230}Th/^{232}Th$ ratios (indicating low contamination with detritus) there was a good correlation between ^{14}C and $^{230}Th/^{234}U$ ages. It is presumed that U is adsorbed by the organic matter soon after deposition, from groundwater or the swamp water from which the peat was deposited after compaction, the migration of U into the peat ceases. Vogel and Kronfeld (1980) have recently described a successful attempt at dating peat by the $^{230}Th/^{234}U$ method. Samples from Emmen and Peelo, the Netherlands; Tenagi Philippon, Macedonia; and Zell am Inn and Grossweil in S. Bavaria, were dated by first ashing and then leaching with dilute hydrochloric acid. All samples, except those from Emmen, contained high U concentrations (7 to 21.5 ppm) and $^{230}Th/^{232}Th$ ratios ranging from 5 to 36. Close agreement was found between ^{14}C and $^{230}Th/^{234}U$ ages for these samples, and a clearer age determination was obtained for the S. Bavarian peats (85 to 90 ky) than obtained by ^{14}C. In a separate attempt at dating Sangamon peat deposits from Baffin Island and Toronto, Canada, Gascoyne (unpub. results) found high ^{232}Th contents in all samples, accompanied by low $^{230}Th/^{232}Th$ ratios, ranging between 0.7 and 1.9. No attempt was made to correct for this large detrital component and the samples were regarded as undatable by the method used.

Table 3. Comparison of U-series and ^{14}C ages on bones from different archaeological sites

Location	U-series age (Ky)	^{14}C age (Ky)	^{14}C age-U series age (Ky)
Lindenmeier Site, Colorado	4250± 500[a]	10780± 375[b]	6530
Lindenmeier Site, Colorado	5500± 500	10780± 375[b]	5280
Dent Site, Colorado	7700± 500	11200± 500[c]	3500
Lehner Site, Arizona	7700±1500[d]	11115± 500	3415
Domebo Mammoth Site, Oklahoma	11500±2000[d]	11045± 647[c]	-455
Murray Springs Site, Arizona	10250±2000[a,d]	11230± 340[e]	980
Caulapan, Mexico	18500±1500[a,f,j]	30600±1000[g]	12100
Caulapan, Mexico	21000±1500[a,f]	21850± 850[g]	850
Medicine Hat, Canada	9500±1000[a,h]	11200± 200[i]	1700

[a]Average concordant $^{230}Th/^{234}U$ and $^{231}Pa/^{235}U$ age; [b]Haynes and Agogino 1960, Folsom level; [c]Haynes 1967, Clovis levels; [d]unpublished results of B.J. Szabo; [e]Haynes 1968, Clovis level; [f]Szabo *et al.*, 1969; [g]Kelley *et al.*, 1978; [h]Szabo *et al.*, 1973; [i]Lowdon and Blake 1968; [j]date obtained on dentine, now considered unreliable for dating.

From Szabo (1980) with permission

DETRITAL SEDIMENTS

In all of the above examples of dating of chemically-precipitated natural substances (calcite, apatite, *etc.*) we have considered that the main source of contamination is the largely siliceous detrital, clastic sediment that is present as a contaminant in samples collected for U-series analysis. It is possible, in principle at least, to turn the argument about and use surficial, detrital sediments as the sample. Although such materials are not likely to form a closed system, it is possible that some sort of regular behaviour could be detected through comparison of several coeval aliquots which have some common chemical characteristics revealing the time they were laid down.

Rosholt (1980) had derived such a method, which he calls "uranium trend analysis" where plots of U and Th isotope ratios with respect to one another generate lines whose slopes can then be interpreted in terms of the time since the system came into existence (the sediment was deposited). Rosholt assumes that the system has been continuously modified after deposition, by a combination of processes of leaching and uptake (by adsorption) of radio-isotopes. A parameter "$\underline{F}$" is introduced which represents the flux of U through the system but which is itself described as a decaying function of time, implying that the intensity of U transport through the system decreases from an initial maximum at or near the time of deposition. He has obtained ages as great as 900 ky by this method, by which point it is only changes in the $^{234}U/^{238}U$ ratio which determine the age of the system. It is necessary to calibrate the method for each sedimentary sequence, in order to evaluate the $\underline{F}$ value for that sequence. Also, several analyses of varying composition are necessary from a site in order to obtain a single age.

Ultimately we may suppose that where any primary, authigenic phase in a soil or sediment has remained a closed system since formation, it might be datable, as long as some control on its initial isotopic composition is possible. This could, for example, be used on clay minerals or ferric hydroxide minerals in a soil, if they could be separated cleanly from the soil, and if they could be shown to have had a well-defined initial isotopic composition. The method is, in principle, analogous to the dating of deep-sea sediments or manganese nodules by study of the decay of excess ^{230}Th or ^{231}Pa (Ku, 1976), but has yet to be tried out on soils.

VOLCANIC ROCKS

Volcanic rocks less than 10^5 y old can in some cases be dated by U-series methods more precisely than by the more conventional K/Ar method. Condomines (1978) has used the ^{230}Th/^{238}U method to date the lavas in which were recorded the Laschamp magnetic reversal at 39 ky. We have applied the method to several other volcanic sequences and generally obtain considerably better precision than obtainable by K/Ar over the same time interval. The assumption in the method is that at the time of extrusion and crystallization of the volcanic rock, all minerals in the rock have the same ^{230}Th/^{232}Th ratio, but different ^{234}U/^{232}Th contents. Then, with passage of time, ^{230}Th grows into each mineral phase of the rock in proportion to the amount of U in that phase. Minerals and whole rocks crystallized from the same magma reservoir at the same time will then lie along an isochron (Figure 7). This method is applicable to rocks as old as 350 ky, and can be used to make a cross-check on K/Ar analysis of rocks suspected of contamination with small amounts of old rocks. This contamination would have a large effect on K/Ar dates, but would have a smaller, or negligible effect on the U-series methods, depending on the U-content of the contaminant.

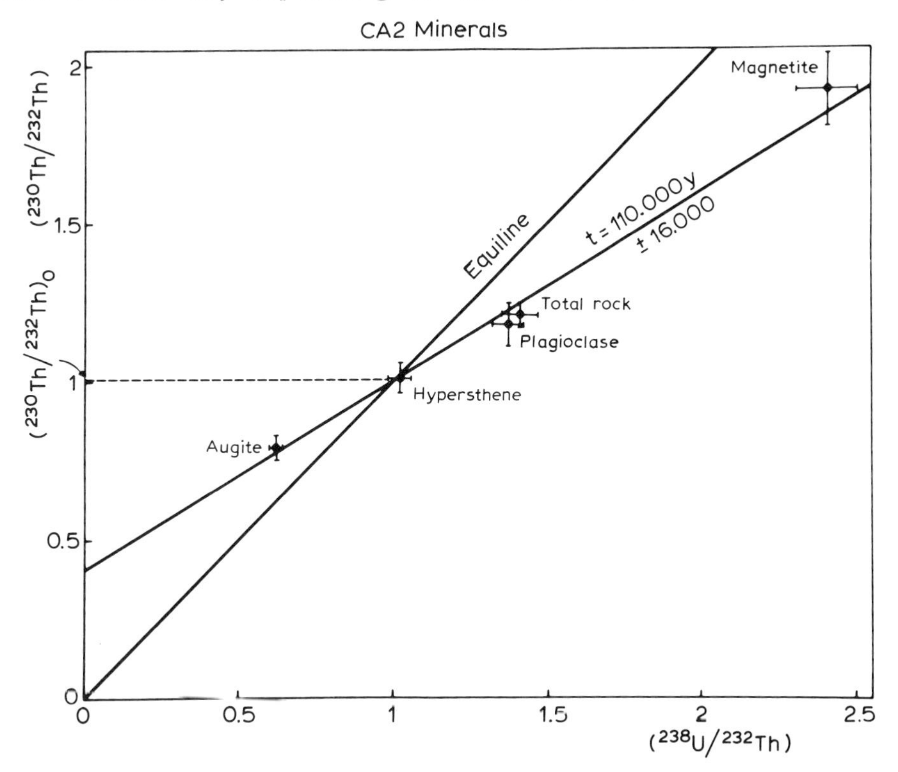

Figure 7 Plot of ^{230}Th/^{232}Th *vs* ^{238}U/^{232}Th for mineral separates and total rock for an andesite volcanic flow from Tierra Blanca, Costa Rica. Equiline would be locus of points if sample were at secular equilibrium. Slope of best-fit line to data points gives ^{230}Th/^{238}U ratio from which age has been calculated. Data from Allegre and Condomines, 1976.

CONCLUSIONS

We have seen that the U-series methods are well-suited to dating a wide variety of continental deposits formed during the last 350,000 years of the Quaternary Period. The methods described here can be applied to problems of geomorphic evolution, paleoclimate, dating of archaeological, faunal and floral assemblages, and to questions of ancient sea level. In regard to the latter, it should be noted that the dating of speleothems from caves on oceanic islands such as Bermuda or the Bahamas allows the possibility of determining not only the level of high sea-stands as are conventionally obtained from U-series dates on coral reefs. In addition, it is possible, by dating of "drowned" stalagmites and stalactites, to place limits on the time of low sea-stands and to thereby follow the curve of changing sea level (and continental ice volume) through the past few glacial stages (Gascoyne *et al.*, 1979; Harmon *et al.*, 1981).

The future of U-series methods would appear to lie in the more reliable application of these methods to sample materials which have heretofore caused the greatest problems: animal bones and molluscan shells. The former have been shown to give results which are concordant with ^{14}C ages on one end of the scale, and with U-series dates on corals, on the other. However, a significant number of cases of gross disequilibrium have also been identified, as indicated by discordance between ^{230}Th and ^{231}Pa dates for the same sample, and failure of the dates to agree with stratigraphic sequence (Cherdyntsev, 1971). The other field for improvement in the methods, is in the possible application of $^{234}U/^{238}U$ ratios to age determination. This ratio can in principle be used to measure ages up to about 1000 ky, if some control on the initial ratio is available. While the method of Rosholt (1980) shows some promise along these lines another approach which might prove to be more valuable is to take advantage of the observation that ^{234}U atoms in a crystal can be divided into two categories: "hot" atoms which grew *in situ* from parent ^{238}U atoms; and "cold" atoms that were present when the crystal formed. With the passage of time, "cold" atoms disappear, and "hot" atoms grow, while the total ratio of hot + cold to total ^{238}U atoms (measured, as usual, in activity ratios) approaches unity. By selective leaching of the sample, it may be possible to identify the hot and cold atoms separately and thus to determine the extent of approach to equilibrium (Schwarcz and Wintle, 1980)

ACKNOWLEDGEMENTS

We acknowledge the financial assistance of the N.S.E.R.C. and the S.S.H.R.C. Much of the research described in this paper was carried out by us in collaboration with Dr. D.C. Ford, whose assistance we gratefully acknowledge.

REFERENCES CITED

Allegre, C.J. and Condomines, M., 1976, Fine chronology of volcanic processes using $^{238}U-^{230}Th$ systematics: Earth Planet. Sci. Lett., v. 28, p. 395-404.

Andrews, J.T., Szabo, B.J. and Isherwood, W., 1975, Multiple tills, radiometric ages, and assessment of the Wisconsin Glaciation in eastern Baffin Island, N.W.T., Canada: A progress report: Arctic Alpine Res., v. 7, p. 39-59.

Bastin, B., 1978, L'analyse pollinique des stalagmites: Une nouvelle possibilité d'approche des fluctuations climatiques du Quaternaire: Annales de la Société Géologique de Belique, T. 101, p. 13-19.

Blackwell, B., Schwarcz, H.P. and Debénath, A., 1982, Absolute dating of hominids and Paleolithic artifacts of the cave of La Chaise-de-Vouthon (Charente), France: In preparation.

Cherdyntsev, V.V., 1971, Uranium-234. Israel Program for Scientific

Translations, Monson, Jerusalem.

Cherdyntsev, V.V., Kazachevskii, I.V. and Kuz'mina, E.A., 1965, Dating of Pleistocene carbonate formations by the thorium and uranium isotopes: Geochem. Int., v. 2, p. 794-801.

Condomines, M., 1978, Datation de la coulée à aimantation inverse d'Olby (Chaine des Puys) par le déséquilibre radioactif $^{230}Th-^{238}U$: Comptes R. Géol. Soc. Fr., v. 5, p. 254-257.

Feth, J. and Barnes, I., 1979, Spring deposited travertine in eleven western states: U.S. Geological Surv. Water Res. Inv., 79-35 (Open File Report).

Ford, D.C., 1973, Developments of the canyons of the South Nahanni River, N.W.T.: Canadian Jour. Earth Sci., v. 10, no. 3, p. 366-378.

Ford, D.C., Schwarcz, H.P., Drake, J.J., Gascoyne, M., Harmon, R.S. and Latham, A.G., 1981, Estimates of the age of existing relief within the southern Rocky Mountains of Canada: Arctic and Alpine Research, v. 13, no. 1, p. 1-10.

Gascoyne, M., Schwarcz, H.P. and Ford, D.C., 1978, Uranium series dating and stable isotope studies of speleothems: Part I. Theory and techniques, 1978: Trans. British Cave Res. Assoc., v. 5, no. 2, p. 91-111.

Gascoyne, M., Benjamin, G.J., Schwarcz, H.P. and Ford, D.C., 1979, Sea-level lowering during the Illinoian glaciation: Evidence from a Bahama "Blue Hole": Science, v. 205, p. 806-808.

Gascoyne, M., Ford, D.C. and Schwarcz, H.P., 1981, Late Pleistocene chronology and paleoclimate of Vancouver Island determined from cave deposits: Canadian Jour. Earth Sci.. (in press).

Gascoyne, M., Schwarcz, H.P. and Ford, D.C., 1982, Uranium-series ages of speleothems from north-west England: Correlation with Quaternary climate: Submitted to Jour. Geol. Soc. London.

Gaven, C., Hillaire-Marcel, C. and Petit-Maire, N., 1981, A Pleistocene lacustrine episode in southeastern Libya: Nature, v. 290, p. 131-133.

Harmon, R.S., Ford, D.C. and Schwarcz, H.P., 1977, Interglacial chronology of the Rocky and MacKenzie Mountains based on $^{230}Th/^{234}U$ dating of calcite speleothems: Canadian Jour. Earth Sci., v. 14, p. 2543-2552.

Harmon, R.S., Schwarcz, H.P., Thompson, P. and Ford, D.C., 1978a, Critical comment on 'Uranium series dating of stalagmites from Blanchard Springs Cavern, Arkansas, U.S.A.': Geochim. Cosmochim. Acta, v. 42, p. 433-439.

__________ 1978b, Late Pleistocene paleoclimates of North America as inferred from stable isotope studies of speleothems: Quaternary Research, v. 9, p. 54-70.

Harmon, R.S., Land, L.S., Mitterer, R.M., Garrett, P., Schwarcz, H.P. and Larson, G.J., 1981, Bermuda sea level during the last interglacial: Nature, v. 289, p. 481-483.

Harmon, R.S. and Schwarcz, H.P., 1982, Isotopic studies of speleothems from a cave in southern Missouri: Submitted to Bull. Geol. Soc. America.

Hendy, C.H., Healy, T.R., Rayner, E.M., Shaw, J. and Wilson, A.T., 1979, Late Pleistocene glacial chronology of the Taylor Valley, Antarctica, and the global climate: Quaternary Research, v. 11, p. 172-

184.

Hollin, J., 1980, Climate and sea level in isotope stage 5: an East Antarctic ice surge at ~ 95,000 BP?: Nature, v. 283, p. 629-633.

Horowitz, A., 1979, The Quaternary of Israel: N.Y., Academic Press.

Kaufman, A., 1971, U-series dating of Dead Sea Basin carbonates: Geochim. Cosmochim. Acta, v. 35, p. 1269-1281.

Kaufman, A. and Broecker, W.S., 1965, Comparison of Th^{230} and C^{14} ages for carbonate materials from Lakes Lahontan and Bonneville: Jour. Geophys. Res., v. 70, p. 4039-4054.

Kaufman, A., Broecker, W.S., Ku, T.-L. and Thurber, D.L., 1971, The status of U-series methods of mollusk dating: Geochim. Cosmochim. Acta, v. 35, p. 1155-1183.

Ku, T.-L., 1976, The uranium series methods of age determination: Ann. Rev. Earth Plan. Sci., v. 4, p. 347-380.

Ku, T.-L., Bull, W.G., Freeman, S.T. and Knauss, K.G., 1979, Th^{230}/U^{234} dating of pedogenic carbonates in gravelly desert soils of Vidal Valley, Southeastern California: Geol. Soc. Amer. Bull., v. 980, p. 1063-1073.

Latham, A.G., Schwarcz, H.P., Ford, D.C. and Pearce, G.W., 1979, Palaeomagnetism of stalagmite deposits: Nature, v. 280, p. 383-385.

Pécsi, M., 1973, Geomorphological position and absolute age of the lower Paleolithic site at Vértesszöllös, Hungary: Magyar Tud. Akad., Földrajztud. Kutato Intézet, Földrajzi Köz., v. 2, p. 109-119.

Peng, T.-H., Goddard, J.G. and Broecker, W.S., 1978, A direct comparison of ^{14}C and ^{230}Th ages at Searles Lake, California: Quaternary Research, v. 9, p. 319-329.

Richmond, G.M., 1976, Pleistocene stratigraphy and chronology in the mountains of western Wyoming, in Mahaney, W.C., ed., Quaternary Stratigraphy of North America: Stroudsburg, Dowden, Hutchinson and Ross, p. 353-379.

Rosholt, J.N., 1980, Uranium-Trend Dating of Quaternary sediments, U.S. Geol. Surv., Open-File Rep. 80-1087.

Schwarcz, H.P., 1980, Absolute age determination of archaeological sites by uranium series dating of travertines: Archaeometry, v. 22, no. 1, p. 3-24.

Schwarcz, H.P. and Wintle, A.G., 1980, $^{234}U/^{238}U$ dating of speleothem by hot-atom method. Tech. Memo 80-1, Dept. of Geology, McMaster Univ., Hamilton, Ont., Canada.

Shackleton, N.J. and Opdyke, N.D., 1976, Oxygen isotope and paleomagnetic stratigraphy of Pacific core V28-239, Late Pliocene to latest Pleistocene, in Cline, R.M. and Hays, J.D., eds., Investigation of Late Quaternary Paleoceanography and Paleoclimatology, Geol. Soc. Amer. Mem. no. 145, p. 449-464.

Stearns, C.E. and Thurber, D.L., 1967, Th^{230}/U^{234} dates of late Pleistocene marine fossils from the Mediterranean and Moroccan littorals: Progress in Oceanography, v. 4, p. 293-305.

Szabo, B.J., 1980, Results and assessment of uranium-series dating of vertebrate fossils from Quaternary alluviums in Colorado: Arctic Alpine Res., v. 12, p. 95-100.

Szabo, B.J., Stalker, A. MacS. and Churcher, C.S., 1973, Uranium-series ages of some Quaternary deposits near Medicine Hat, Alberta, Canada: Can. Jour. Earth Sci., v. 10, p. 1464-1469.

Szabo, B.J. and Rosholt, J.N., 1969, Uranium series dating of Pleistocene molluscan shells from Southern California -- An open system model: Jour. of Geophys. Res., v. 74, no. 12, p. 3253-3260.

Szabo, B.J. and Butzer, K.W., 1979, Uranium-series dating of lacustrine limestones from pan deposits with final Acheulian assemblage at Rooidam, Kimberley District, South Africa: Quaternary Research, v. 11, p. 257-260.

Thompson, G.M., Lumsden, D.N., Walker, R.L. and Carter, J.A., 1975, Uranium series dating of stalagmites from Blanchard Springs Caverns, U.S.A.: Geochim. Cosmochim. Acta, v. 39, p. 1211-1218.

Thompson, P., Ford, D.C. and Schwarcz, H.P., 1975, $^{234}U/^{238}U$ ratios in limestone cave seepage waters and speleothems from West Virginia: Geochim. Cosmochim. Acta, v. 39, p. 661-669.

Vogel, J.C. and Kronfeld, J., 1980, A new method for dating peat: South African Jour. of Sci., v. 76, p. 557-558.

Williams, M.A.J., Williams, F.M., Gasse, F., Curtis, G.H. and Adamson, D.A., 1979, Plio-Pleistocene environments at Gadeb prehistoric site, Ethiopia: Nature, v. 282, p. 29-33.

URANIUM-SERIES DATING AND THE HISTORY OF SEA LEVEL

CHARLES E. STEARNS

ABSTRACT

U-series dating of coral associated with former shorelines has provided the abscissa of time for diagrams of fluctuation in sea level during, especially, the last 250,000 years. Ordinates of amplitude have been provided by morphostratigraphic detail, by assumptions of rates of uplift, or by ^{18}O values in either littoral deposits or deep-sea sediments. The latter case provides mutual reinforcement of ^{18}O and sea level chronologies.

The most complete reference sections come from the Huon peninsula of New Guinea. Twelve separate reef crests have been distinguished and dated: four in the interval 250 ka to 180 ka (core stage 7), seven in the interval 140 ka to 28 ka (core stages 5-4-3), and one in the interval 9 ka to 5 ka. Were $^{230}Th/^{234}U$ dating sufficiently accurate, sea level histories from other parts of the world might be compared to the New Guinea record, even with the identification of individual sea level events. Such, alas, is not the case. At least the following sources of error plague us.

Statistical errors in $^{230}Th/^{234}U$ values are commonly quoted as ± 0.02, but an interlaboratory comparison suggests that ± 0.04 may be a fairer index of precision. This is ± 15 ka at 125 ka, nearly equal to the 17 ka interval between sea level maxima on New Guinea. Thus, individual samples from other areas cannot be assigned to a specific episode by "closed-system" ages alone. Close correlation must depend upon position in a morphostratigraphic sequence as well.

Corals are not ideal closed systems. ^{230}Th ages may be either too old or too young. Concordancy of ^{230}Th and ^{231}Pa ages is the only useful check, but $^{231}Pa/^{235}U$ is seldom measured.

Molluscan data are notoriously unreliable and would not be studied, were we not curious about the extra-tropical seas. Coexisting coral and mollusca can yield identical ^{230}Th ages. More commonly, molluscan ages are significantly younger. Concordant ^{230}Th and ^{231}Pa ages can be as little as 50% of those from coexisting coral. The molluscan systems have been "closed" to loss of daughter products, but "open" to addition of parent U. We are measuring the average age of the parent, which enters molluscan shell after death.

Both coral and mollusca may also be "open systems" to which ^{234}U, ^{230}Th, and/or ^{231}Pa have been added in excess of that produced within the shell. In some cases, addition may have been systematic. The S-R model, which allows correction to "concordant" ages, is not demonstrably useful, but alternative models might be developed from sufficient data.

There can be no doubt that "dating" the maximum sea level of the last interglacial at 125,000 ± 10,000 years, and confirmation of its wide-spread preservation are achievements of U-series dating. Detailed sea level records of the "last interglacial" (core stage 5), based on local morphostratigraphic evidence, still show conflicts (differences?) which cannot be resolved by U-series dating.

INTRODUCTION

U-series dating of coral associated with former shorelines has provided the abscissa of time for diagrams of fluctuation in sea level during, especially, the last 250,000 years. Ordinates of amplitude have been provided by morphostratigraphic detail, by assumptions of rates of uplift, or by ^{18}O values in either littoral deposits or deep-sea sediments. The latter case provides mutual reinforcement of ^{18}O and sea level chronologies.

Were $^{230}Th/^{234}U$ dating sufficiently accurate, sea level histories from various parts of the world might be compared, even to the identification of individual sea level events. Such, alas, is not the case. ^{230}Th ages are subject to errors arising from (1) failure of calcareous fossils to be ideal closed systems, and (2) laboratory precision less than that implied by commonly quoted statistical errors of measurement. In the use of ^{230}Th ages, we must hold realistic limits of error constantly in mind.

THE CLOSED SYSTEM MODEL

A living organism may incorporate U but neither ^{230}Th nor ^{231}Pa in its calcareous skeleton. U, commonly with ^{234}U in excess of equilibrium with ^{238}U (activity ratio 1.14 ± 0.03), is present in sea water. ^{230}Th and ^{231}Pa are not. After death, the fossil skeleton may remain closed to further addition of U, but retain daughter products of U already incorporated. This is the model conventionally used in connection with unrecrystallized coral: the closed-system model. Ages (time elapsed since death) may be estimated from the growth of daughter products toward equilibrium with their parents. Activity ratios record this growth and are simpler to determine than individual quantities.

Usefulness of the model with unrecrystallized coral has been confirmed by (a) concordancy of $^{230}Th/^{234}U$, $^{231}Pa/^{235}U$, and $^{234}U/^{238}U$ ages, (b) consistency with morphostratigraphic sequence, and (c) consistency with independent estimates of age (Ku, 1976). Reasonable $^{230}Th/^{234}U$ ratios alone, however, do not prove the closed-system assumption. Common checks on reliability are largely negative: absence of recrystallization, plausible U and Th concentrations, $^{234}U/^{238}U_o$ = 1.14 ± 0.03, high $^{230}Th/^{232}Th$ ratios. If any of these be suspect, the closed-system model is suspect. If they be reasonable, however, they still do not guarantee reliability.

At present, the best check of reliability is concordancy between $^{230}Th/^{234}U$ and $^{231}Pa/^{235}U$ ages (Ku, 1968, 1976). These ratios relate to different decay series. If they yield the same age, they strongly support the inference that the two daughter products have been produced by a common source (containing both parents) and that they have been quantitatively retained in the sample. It is still important to remember that we are measuring the age of the uranium source, but it may also be the age of the fossil. The check is seldom made, because analysis for ^{231}Pa is tedious.

Oahu data (Ku *et al.*, 1974) suggest that ^{231}Pa may leak somewhat. In eleven samples, mean $^{231}Pa/^{235}U$ age is less than mean $^{230}Th/^{234}U$ age, and the spread in $^{231}Pa/^{235}U$ ages is greater. Thus, although $^{231}Pa/^{235}U$ ratios provide the best confirmation of "closed system" ages, they may themselves tend to be too low. $^{230}Th/^{234}U$ is the more precise measure of age.

Coral from Barbados has yielded concordant ages for five discrete reef-terraces (Ku, 1968, James *et al.*, 1971). In a diagram of $^{230}Th/^{234}U$ *vs* $^{231}Pa/^{235}U$ (Figure 1A), they provide us with five reference points of "known" ages. They also define a concordia (for $^{234}U/^{238}U$ = 1.11 ± 0.02), or locus of points representing equal ^{230}Th and ^{231}Pa ages.

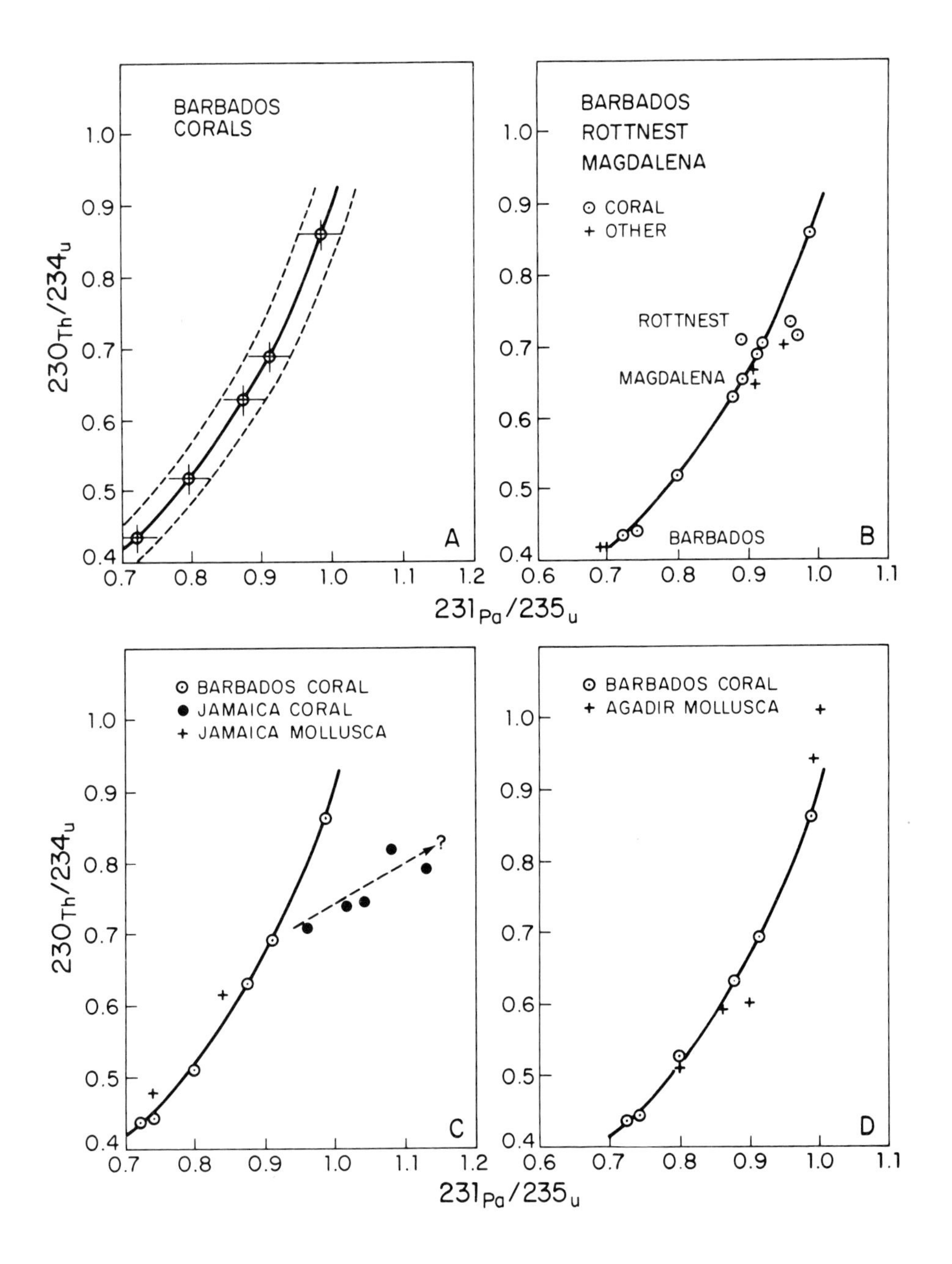

Figure 1 $^{230}Th/^{234}U$ *vs* $^{231}Pa/^{235}U$ (activity ratios) For sources of data, see text.

The plot has been helpful to me as a base-line against which to compare and evaluate other data. Oahu data, for example, plots mostly to the left of concordia.

Szabo (1979a) has measured both $^{230}Th/^{234}U$ and $^{231}Pa/^{235}U$ ratios in four coral and one mollusc from Rottnest Island, Australia (Figure 1B). Within reasonable limits of error all, including the mollusc, yield concordant and indistinguishable closed-system ages. One coral and two echinoderms from Magdalena, Baja California (Omura *et al.*, 1979) are more closely concordant. In at least these cases, $^{230}Th/^{234}U$ is a "reliable" basis for estimation of age. Note, however, that Magdalena falls midway between two Barbados reference points and could not be specifically correlated with one or the other by isotopic ratios alone.

Bender *et al.* (1979) have shown that Barbados coral older than 150 ka have demonstrably excess ^{230}Th and that those older than 250 ka have demonstrably excess ^{234}U. ^{231}Pa was not studied. Total U does not increase perceptibly with age. These coral are not ideal closed systems but, rather, accept at least daughter products from the external environment after death.

Szabo (1979b) has measured both $^{230}Th/^{234}U$ and $^{231}Pa/^{235}U$ in five coral and three mollusca from a Jamaican deposit (Figure 1C). Only one coral yields a nearly concordant "closed-system" age. In the others, $^{230}Th/^{234}U$ and $^{231}Pa/^{235}U$ are both too high: individual samples drift to the right and upward, away from concordia. These young coral have accepted daughter products from the external environment, additional to those produced with the shell. It is important to note that these samples meet all ordinary criteria of "reliability": calcite <3%, $^{230}Th/^{232}Th$ 100, U >2.25 <3.90 ppm, $^{234}U/^{238}U$ >1.08 <1.18. $^{231}Pa/^{235}U$ should be regularly checked for concordancy in coral as well as mollusca.

Szabo and Rosholt (1969) proposed that discordant ages, in which ^{231}Pa age > ^{230}Th age (or $^{231}Pa/^{235}U$ >1), might be "corrected" by a model in which: (a) U, once incorporated, is retained in the shell, (b) daughter products produced within the shell are retained, and (c) daughter products of U in an "external" environment (interstitial fluid or solid matrix) are also fixed by implosion and/or absorption. The system is "open" to addition of daughter products, but "closed" to their loss. In my reference diagram, data points lie to the right of concordia (Figure 1C, Jamaica coral). With appropriate assumptions, it might be possible to segregate the two components of daughter products added through time, substract those components added from the external environment, and recover the concordant ^{230}Th and ^{231}Pa ages of U fixed in the shell.

Szabo and Rosholt (1969) proposed a specific model, but in the absence of independent evidence of age of the California mollusca from which data were then available, were unable to evaluate its usefulness. As pointed out by Kaufman *et al.* (1971), it is simply not possible to tell whether apparent ages derived from the model are preferable to those based on other models of migration. Returning the Jamaican data (Figure 1C) to concordia by a best-fitting straight line does as well.

Fossil marine mollusca typically contain more U than do living mollusca (Kaufman *et al.*, 1971), which in turn contain less than living coral. Thus, most of the U in fossil mollusca is incorporated after death. Kaufman *et al.* (1971) concluded that U concentrations might increase by a factor of 25 within 10 ka and by another 50 percent in the next 100 ka. It is of paramount importance to remember that, if anything, one may be able to measure the age of the uranium in a fossil mollusc. Even for this limited objective, concordancy of ^{230}Th and ^{231}Pa ages is the <u>only</u> check on "reliability".

Limited data from coexisting coral and mollusca (James *et al.*, 1971, Szabo, 1979a; Figure 1B) show that the latter can yield concordant ^{230}Th and ^{231}Pa ages indistinguishable from coral. ^{230}Th ages of mollusca are more commonly less than those of coexisting coral (Veeh and Chappell, 1970, Hoang *et al.*, 1974, 1978). ^{230}Th ages of mollusca in a Virginia deposit, believed to be at least 500 ka old ($^{230}Th/^{234}U$ should have reached transient equilibrium), range from 5 to 160 ka (Wehmiller *et al.*, 1980). None have reached equilibrium.

Szabo (1979b; Figure 1C) has shown that mollusca can yield concordant ^{230}Th and ^{231}Pa ages which differ among themselves and may be as little as fifty percent of coexisting coral ages. This single study makes the important suggestion that molluscan shell may have been "open" to addition of U, but "closed" in the retention of daughter products produced within the shell and refusal of daughter products from the external environment. Hoang (personal communication) has now measured both $^{230}Th/^{234}U$ and $^{231}Pa/^{235}U$ in three molluscan samples from Ouljian deposits near Agadir, Morocco (Figure 1D). Estimated ages for each are concordant within reasonable limits of error, and not significantly different from those of Szabo's "young" mollusca from

Jamaica. Furthermore, two samples from an older terrace at Agadir yield activity ratios near equilibrium - not in excess - suggesting that daughters have not been added from the external environment, while daughters produced in the shell have been retained. In both sets of samples, we are measuring the average age of uranium addition and not the full age since death of the organism.

In summary, for ^{230}Th "ages" to be meaningful, some reassurance must be provided that isotopic accumulation in fossil shell has been systematic. The negative reassurance that recrystallization has not occurred is insufficient. At present, concordancy of ^{230}Th and ^{231}Pa ages is the best check on the assumption that a system has remained "closed"; it implies that daughter products have been produced by U fixed in the shell.

In coral, if U be >2 <4 ppm and $^{234}U/^{238}U_0$ be 1.14 ± 0.03, "closed-system" ages may be true ages. In mollusca, concordant ages are always minimum ages. They may be indistinguishable from true ages (Rottnest, Figure 1B), but they may also be as little as fifty percent of true age (Jamaica, Figure 1C). Furthermore, coexisting mollusca need not yield the same age. Presumably, the highest concordant value in a series of samples from one locality is closest to true age, but U-series data provide no basis for estimating the discrepancy.

Mild discordancy, with ^{231}Pa age < ^{230}Th age, may reflect incomplete retention of ^{231}Pa. In such samples, ^{230}Th age is probably the better estimate of true (coral) or apparent (mollusca) age. The Szabo-Rosholt (1969) model correction does not apply to such samples.

Discordancy with ^{231}Pa age > ^{230}Th age (or $^{231}Pa/^{235}U$ >1) implies that daughter products from sources external to the shell have been added to those produced by U fixed within the shell. ^{230}Th age will be high. If the gain could be shown to be systematic, it might be possible to reduce the data to discordancy. Usefulness of the Szabo-Rosholt (1969) model has not been demonstrated (Kaufman *et al.*, 1971) and at best needs serious reevaluation with an enlarged data base of samples of known age.

SEA LEVEL DURING THE "LAST INTERGLACIAL" (CORE STAGE 5)

There can be no doubt that "dating" the maximum sea level of the last interglacial at 125,000 ± 10,000 years and confirmation of its wide-spread record are achievements of U-series dating. This episode is commonly viewed as a brief interval, *i.e.*, a few thousand years, during which sea level stood a few meters higher than now. It followed a rapid rise from synglacial low levels, Termination II, estimated by various methods to center at 128 ka, 138 ka, or 145 ka in core V28-238 (Kominz *et al.*, 1979), and was in turn followed by a drop to lower levels. It has been specifically correlated to ^{18}O core substage 5e, the brief peak of which is about 0.5 per mil lighter than any other part of core stage 5.

Statistical errors in $^{230}Th/^{234}U$ values are commonly quoted as ± 0.02, but an interlaboratory calibration (Harmon *et al.*, 1979) suggests that ± 0.04 may be a fairer measure of precision. This is ± 15 ka at 125 ka. $^{230}Th/^{234}U$ values (and ages) of samples used in the study and attributed to the 125 ka event are 0.75 (139 ka) from Windley Key, Florida, 0.69 (124 ka) from Curacao, and 0.66 (118 ka) from Barbados. $^{231}Pa/^{235}U$ was not studied. "It is likely that the observed variation is due in part to both uncertainty derived from counting statistics and different diagenetic modification of the samples." (Harmon *et al.*, 1979). An equivalent range of values and ages has been reported from localities scattered around the world (Table 1).

It should be pointed out that the values 0.66 (118 ka) and 0.63 (105 ka), from two discrete terraces on Barbados (Harmon *et al.*, 1979), are not statistically different; they could not be distinguished by isotopic data alone. We have already noted (Figure 1B) the problem of correlation of the single deposit at Magdalena, Mexico, with one or the

Table 1. Range of $^{230}Th/^{234}U$ ratios (ages) correlated with 125 ka and 105 ka events

125 ka event:	
0.75 (139 ka)	Windley Key[a], Taranto[b], New Guinea[c]
0.72 (134 ka)	Rottnest[d], Jamaica[e], New Guinea[c], Oahu[f]
0.69 (124 ka)	Curacao 6m[a], Barbados III[g]
0.66 (118 ka)	Barbados III[a], Magdalena[h], Oahu[f]
	**********
uncertain:	
0.65 (111 ka)	isolated reef crest, Barbados[g]
	**********
105 ka event:	
0.63 (105 ka)	Barbados II[a,g], New Guinea[c]

[a]Harmon *et al.*, 1979, [b]Dai Pra and Stearns, 1979, [c]Veeh and Chappell, 1970, [d]Szabo, 1979a, [e]Szabo, 1979b, [f]Ku *et al.*, 1974, [g]Mesolella *et al.*, 1969, [h]Omura *et al.*, 1979.

Note that values for Barbados III (0.66) and Barbados II (0.63) are not statistically distinguishable.

other Barbados terrace.

$^{230}Th/^{234}U$ values at both extremes of the range for the "125 ka event" have also been reported from at least five localities: Oahu (Ku *et al.*, 1974), Jamaica (Moore and Somayajulu, 1974), the inner barrier of Gippsland (Marshall and Thom, 1976), New Guinea and Atauro (Chappell and Veeh, 1978) where they represent two discrete "high stands", separated by disconformities recording an interval of lower sea level. In these localities, the 125 ka event is double and, including both "high stands", longer than we had in mind. Individual samples from other localities, where only one "high stand" is recorded, cannot be assigned to a specific "high stand" by ^{230}Th ages alone. In fact, I suppose, in proposing correlation of an isolated "high stand" with the 125 ka event, we may really be correlating it with a "low stand" separating two high stands. Gaven and Vernier (1979) have specifically invoked this possibility to account for the preservation of 125 ka and 105 ka reefs at the same elevation in the Iles Glorieuses. The point to remember is that it is neither a close correlation within the fine structure of the "Last Interglacial" nor a particularly accurate statement of age.

The most complete reference section for the 125 ka event comes from New Guinea and Atauro (Chappell and Veeh, 1978). On these high uplift coasts, the older "high stand" appears to have been lower than the younger. On the low uplift coasts of Oahu, Jamaica, and Gippsland, the two "high stands" are now no more than a meter apart. Both "high stands" on New Guinea, however, were significantly higher than younger "high stands" at 105 ka and 82 ka (Bloom *et al.*, 1974), not recognized on Oahu, Jamaica, and Gippsland. Thus, the New Guinea reference curve for 140 ka to 75 ka (envelop in Figure 2) includes two "high stands" somewhat above present sea level, followed by two younger "high stands" somewhat below present sea level.

Aharon *et al.* (1980) have partially decoupled the New Guinea curve of sea level from ^{18}O fluctuations within core stage 5. I have superimposed their ^{18}O values (circles in Figure 2) on their sea level curve by arbitrary scale. The older "high stand" (134 ka) is isotopically

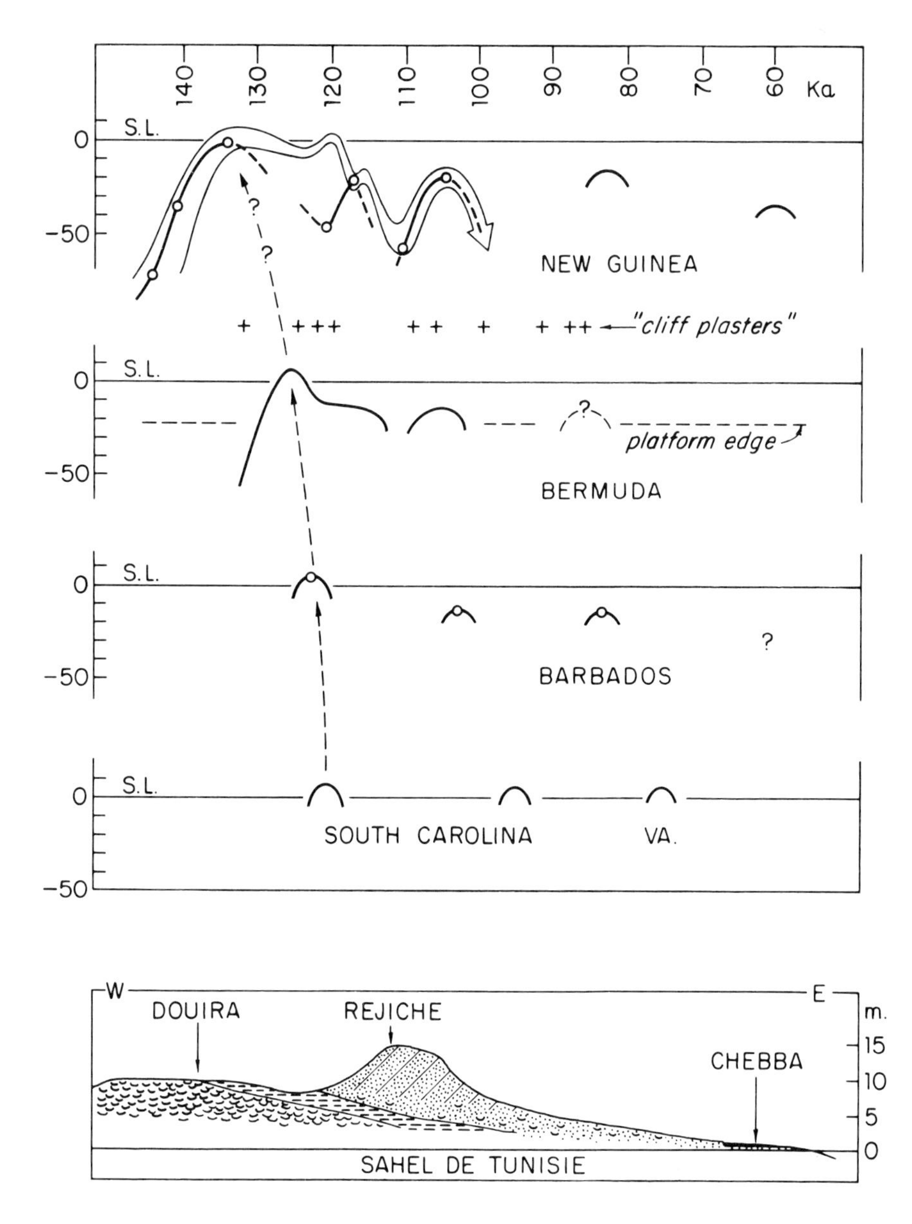

Figure 2 Sea level histories from various localities. Redrawn from sources identified in text.

lightest, the younger (120 ka) 0.5 per mil heavier. Only after sea level has fallen from its second "high stand" did ^{18}O values rise to values only 0.25 per mil heavier than those at 134 ka. Comparable values were regained at 105 ka; values at 82 ka are not reported. Aharon *et al.* (1980) suggest that the decoupling at 120 ka may reflect a massive Antarctic surge.

One may infer that 134 ka on New Guinea, isotopically lightest, includes the peak of core stage 5e, also isotopically lightest. The 125 ka reef on Barbados, also isotopically lightest (Shackleton and Matthews, 1977, Fairbanks and Matthews, 1978), may thus be equivalent to 134 ka on New Guinea. What then of the 105 ka reef on Barbados? Is it equivalent to 120 ka or 105 ka on New Guinea? Probably the latter, but I raise the question to emphasize that the question cannot be answered by ^{230}Th dating alone. $^{230}Th/^{234}U$ activity ratios for 120 ka and 105 ka are not statistically distinguishable. Rather, choice must be made in the context of morphostratigraphic sequences and any other possible argument.

Construction of eolianite dunes on Bermuda implies that the Bermuda platform (-20 m) was submerged for much of the interval 140 to

75 ka (Vacher, 1973, Harmon *et al.*, 1978, 1981). A single peak at about +5 m about 125 ka is recorded by the marine Devonshire Formation. One or more younger maxima are recorded by storm deposits ("cliff plasters") somewhat above sea level. Coral growth implies that the Bahamas platform (-20 m) was submerged for much of the interval 140 to 100 ka, with a single definable maximum at +5.6 m about 125 ka (Neumann and Moore, 1975). ^{230}Th ages from coral in classical Eutyrrhenian deposits near Taranto, Italy, imply that the Mare Piccolo was continuously submerged above +5 m between 156 ka and 90 ka, reaching a single identifiable maximum about 140 ka, now +28 to +35 m.

Harmon *et al.* (1978, 1981) have developed a detailed record of sea level fluctuation on Bermuda (Figure 2) with the additional control of ^{230}Th ages from subaerial speleothems now or formerly submerged. Coral *in situ* from the emergent Devonshire Formation yields ^{230}Th ages 134 ka to 118 ka. A stalactite was submerged to +4 at the same time; bracketing dates are 130 ka and 110 ka. Other speleothems record continuous subaerial deposition at -7 m, 195 ka to 150 ka and 120 ka to 100 ka. One stalagmite may have grown continuously at -15 m between 110 ka and 10 ka.

Two principal episodes of dune-building are inferred from amino acid racemization ages of 105 ka and 85 ka from two different outcrops. There are also redeposited coral in the discontinuous storm deposits ("cliff plasters"), thrown up to present elevations above sea level unknown distances above contemporaneous sea level (as much as 20 m if the stalagmite at -15 m was never submerged. These coral yield one cluster of ^{230}Th ages comparable to those from the Devonshire Formation and two clusters compatible with 105 ka and 85 ka dune-building events.

No discontinuity is recorded in the 125 ka event. The shoulder at -8 m and 120 ka in Figure 2, however, is based on one speleothem age of 119 ± 13, not statistically distinguishable from other "controlling" ages of 110 ± 14, 111 ± 0, and 113 ± 12 at the same depth. The Harrington Formation, 121 ± 9 by ^{230}Th/^{234}U, was interpreted by Vacher (1973) as an emergent "back-beach" deposit. ^{230}Th ages allow a much broader 125 ka peak than that drawn by Harmon *et al.* (1981; Figure 2).

Thus, evidence from the Bahamas, Bermuda, and Taranto implies that sea level was above -20 m for most of the "Last Interglacial" (core stage 5), between 140 ka and 75 ka ago. It rose from synglacial low levels rapidly (Termination II) to levels higher than present about 125 ka; the most complete local records (New Guinea-Atauro) show a double peak, 135 ka and 118 ka, separated by a disconformity recording an intervening low. Thereafter it fell below its present level to at least -20 m by 110 ka. The younger peaks at 105 ka and 82 ka are closely bracketed by the necessity of submerging the Bermuda platform (-20 m) without submerging a stalagmite (-15 m). Similar levels of sea level at 105 ka and 82 ka were estimated by assumptions of uniform uplift on Barbados (Mesolella *et al.*, 1969) and New Guinea (Bloom *et al.*, 1974). I find it difficult, however, to visualize throwing the "cliff plasters" to elevations 20 m above contemporaneous sea level. Were they associated with truly "brief" higher sea level events not recorded in the stalagmite (or in the Barbados and New Guinea reef tracts)? The Antarctic surges envisioned by Hollin (1977, 1980) are just such brief events (n x 100 yr), which might be superimposed on broader peaks (n x 1000 yr) recorded by the construction of dunes on Bermuda and reefs elsewhere.

I have the beginnings of a reference curve (Figure 2), not dissimilar from others (Bloom *et al.*, 1974). Where completely recorded, its characteristic signature is a series of peaks: high - high - low - low. Because ^{230}Th ages have a precision less than the interval between peaks, isolated records of individual sea level events cannot be placed closely in the general curve. Plausible choices will be made in the context of morphostratigraphic sequence, if at all.

Cronin *et al.* (1981) suggest that three "age clusters" of coral from the southeastern U.S. coastal plain may represent 125 ka, 105 ka, and 82 ka events (Figure 2). Two (120 ka, one sample, 94 ka, three samples) are from littoral deposits in South Carolina which have been assigned to a single formation (Wando Formation, McCartan *et al.*, 1980), probably equivalent to the "Princess Anne unit" of Colquhoun (1974) and the Socastee Formation of Dubar *et al.* (1974). Low $^{230}Th/^{232}Th$ ratios (2.0 to 13.0) cast doubt on the reliability of the specific ages. The third "cluster" of four samples from the Norfolk Formation of Virginia ranges from 62 ka to 79 ka; similar ages were reported by Oaks and Coch (unpublished, but quoted by Flint, 1971). The 62 ka sample, which differs from the others by more than one standard deviation, also yields a concordant ^{231}Pa age, suggesting that it may be a fair measure of the average age of the contained uranium (not necessarily the age of the fossil). Amino acid ratios suggest that the Norfolk Formation is indeed younger than the Wando and Socastee Formations (Wehmiller, this paper, and personal communication). The region of southern Chesapeake Bay probably has a recent history of emergence (uplift?) different from that in South Carolina (Mixon *et al.*, 1974; Demarest and Belknap, 1980). Thus, ^{230}Th ages alone, from two separate areas, provide neither firm basis for correlation with a standard sea level history nor support for the suggestion that sea level was at the same elevation at three different times.

The classical Eutyrrhenian (couche à Strombes) of the Mediterranean (= early Ouljian of Morocco) is assuredly the 125 ka event (Brancaccio *et al.*, 1978, Dai Pra and Stearns, 1979). Our postulate of a younger episode 75 ka to 90 ka (Stearns and Thurber, 1965, 1967) from ^{230}Th ages of mollusca alone was premature. Most of the "young" ^{230}Th ages cited (*idem.*, Kaufman *et al.*, 1971, Bernat *et al.*, 1978, Hoang *et al.*, 1978) probably belong to the 125 ka event (Stearns, in press).

Nevertheless, a younger episode (Neotyrrhenian = late Ouljian) can be separated from the Eutyrrhenian locally by firm morphostratigraphic evidence, commonly a disconformity associated with reddish soils. We have no secure age for the Neotyrrhenian. One of the best examples from the Tunisian Sahel (Paskoff and Sanlaville, 1976, 1979, 1980) is illustrated in Figure 2.

The sea level signature "high" (Douira) - "high" (Rejiche) - "low" (Chebba) resembles the 125 ka event (double) plus one of the younger events (105 ka or 82 ka), yet "high" and "low" stands differ in elevation by only 8 or 10 m. Because of contemporaneous uplift, this is a maximum value for the difference in contemporaenous sea levels. Similar relationships at other Neotyrrhenian localities led me to question the postulated 25 m differencc in "high" and "low" sea level peaks during the last interglacial (Stearns, 1976).

Do the differences reflect differences in uplift history on these different coasts? Does the Neotyrrhenian record one or more surges imposed on broader sea level peaks at 105 ka and/or 82 ka? Neotyrrhenian deposits are in fact characteristically poorly sorted coarse littoral conglomerates, contrasting with the more extensive and better differentiated Eutyrrhenian. These are questions which can be raised by ^{230}Th dating, but not answered. The detailed history of sea level during the "Last Interglacial" can be extended only by ^{230}Th ages placed securely in thoroughly documented morphostratigraphic sequences.

SEA LEVEL FLUCTUATIONS DURING THE BRUNHES

Reasonable extrapolations from younger dated terraces (>250 ka) on Barbados and New Guinea through flights of higher terraces, using the assumption of uniform uplift, show that fluctuations of sea level during the last 700 ka have been as numerous as have fluctuations of ^{18}O values in deep-sea cores. At least the major odd-numbered ^{18}O stages are represented, generally by multiple terraces reflecting ^{18}O substages as well. On Barbados, a first text of dating by the accumulation of helium has been tried.

If He^4 produced by alpha emission be quantitatively retained in fossil shell, its accumulation (increase in He-U ratio) is a basis for estimate of age (Fanale and Schaeffer, 1965). Allowance must be made for both initial ("inherited") and modern ("atmospheric") He, unrelated to that produced in the sample. Both of these should be less important sources of error with increasing age. Allowance must further be made for He^4 produced by U and daughter products added *post mortem* from the external environment.

The assumption that He^4 is quantitatively retained is supported by a comparison of He-U ages with stratigraphic ages of coral from the southeastern United States, ranging in age from Eocene to late Pliocene (Bender, 1970, 1973, Blackwelder, 1981). Less satisfactory agreement has been obtained from Miocene coral in the region of Chesapeake Bay; He-U loss is probable (Bender *et al.*, 1979, p. 589).

Eleven reef tracts older than the 125 ka reef on Barbados have been "dated" on the assumptions that He^4 has been quantitatively retained and that a correction can be made for He^4 produced by ^{234}U and ^{230}Th added *post mortem*. Errors are estimated to be ± 10%, equivalent to one core stage for the oldest terraces. Results generally confirm the inference made by extrapolation with the assumption of uniform uplift. The dated terraces are Brunhes, and they record fluctuations in sea level as frequent as ^{18}O substages in the deep-sea record. They are probably not useful for correlations of specific individual terrace remnants, even on Barbados itself.

CONCLUSION

U-series dating is useful, although not as accurate as we might wish (or as some users imply). One must always remember that it "dates" uranium, not fossils. Molluscan "ages" are notoriously unreliable, but corals are also not ideal closed systems. Concordancy of ^{230}Th and ^{231}Pa ages is the only true check on "reliability". ^{231}Pa is commonly not measured, because the analysis is tedious. It should be measured more often. Those of us who are not chemists must also learn to screen the available chemical data for reliability as far as possible.

The sea level maximum of the "Last Interglacial" is securely dated at 125,000 ± 10,000 years ago. It was a double peak, although many localities do not record this. We have the beginnings of a detailed history of sea level between 140 ka and 75 ka, based upon good evidence from a very few localities. There were four sea level maxima, not all of them separately recorded. Their characteristic signature appears to "high" - "high" - "low" - "low". The "highs" equalled or exceeded present sea level. The lows did not, and are not apt to be recorded on "stable" or low-uplift (>0.1 m/ka) coasts. Individual ^{230}Th ages have a precision no greater than the interval between sea level peaks. Reasonable estimates are 135 ka, 118 ka, 105 ka, and 85 ka, but isolated sea level events cannot be assigned to specific positions in the sequence by ^{230}Th dating alone. Estimates of the level of sea level in individual peaks are not secure enough to say whether or not there are significant local differences.

Sea level fluctuations during the Brunhes have been as numerous as ^{18}O fluctuations in deep-sea cores. This has been confirmed on high-uplift coasts by reasonable extrapolations from the 125 ka terrace and the assumption of uniform uplift. ^{230}Th ages may allow assignment to core stage 7 (*ca.* 225 ka), but not to one of its substages. Beyond 250 ka, He-U ages have a precision not much greater than one core stage.

ACKNOWLEDGEMENTS

I am particularly grateful to Barney Szabo and Hoang Chi Trach for their patience with me while trying to learn how better to move through chemical ambiguities. Both have helped me stumble less, but neither is responsible for continuing imperfections in my gait.

REFERENCES CITED

Aharon, P., Chappell, J. and Compston, W., 1980, Stable isotope and sea-level data from New Guinea supports Antarctic ice-surge theory of ice ages: Nature, v. 283, p. 649-651.

Bender, M.L., 1970, Helium-uranium dating of corals (Ph.D. dissert.): New York, Columbia University, 149 p.

____________, 1973, Helium-uranium dating of corals: Geochimica et Cosmochimica Acta, v. 37, p. 1229-1247.

Bender, M.L., Fairbanks, R.G., Taylor, F.W., Matthews, R.K., Goddard, J.G. and Broecker, W.S., 1979, Uranium-series dating of the Pleistocene reef tracts of Barbados, West Indies: Geol. Soc. Amer. Bull., v. 90, p. 577-594.

Bernat, M., Bousquet, J.C. and Dars, R., 1978, Io-U dating of the Ouljian stage from Torre Garcia (southern Spain): Nature, v. 275, p. 302-303.

Blackwelder, B.W., 1981, Stratigraphy of upper Pliocene and lower Pleistocene marine and estuarine deposits of northeastern North Carolina and southeastern Virginia: U.S. Geol. Surv. Bull. 1502-B, 16 p.

Bloom, A.L., Broecker, W.S., Chappell, J.M.A., Matthews, R.K. and Mesolella, K.J., 1974, Quaternary sea level fluctuations on a tectonic coast: new ^{230}Th/^{234}U dates from the Huon peninsula, New Guinea: Quaternary Research, v. 4, p. 185-205.

Brancaccio, L., Capaldi, G., Cinque, A., Pece, R. and Sgrosso, I., 1978, ^{230}Th/^{238}U dating of corals from a Tyrrhenian beach in Sorrentine Peninsula (southern Italy): Quaternaria, v. 20, p. 175-183.

Chappell, J. and Veeh, H.H., 1978, Late Quaternary tectonic movements and sea-level changes at Timor and Atauro Island: Geol. Soc. Amer. Bull., v. 89, p. 356-368.

Colquhoun, D.J., 1974, Cyclic surficial units of the Middle and Lower Coastal Plains, central South Carolina, in Oaks, R.Q. and Dubar, J.R., eds.: Post-Miocene Stratigraphy, central and southern Atlantic Coastal Plain: Logan, Utah State University Press, p. 179-190.

Cronin, T.M., Szabo, B.J., Ager, T.A., Hazel, J.E. and Owens, J.P., 1981, Quaternary climates and sea levels of the U.S. Atlantic Coastal Plain: Science, v. 211, p. 233-240.

Dai Pra, G. and Stearns, C.E., 1977, Sul Tirreniano di Taranto, Datazioni su coralli con il metodo del ^{230}Th/^{234}U: Geologica Romana, v. 16, p. 231-242.

Demarest, J.M. and Belknap, D.F., 1980, Quaternary Atlantic shorelines on Delmarva peninsula - chronology and tectonic implications (abstr.): Geol. Soc. Amer. Abstr. with Progr., v. 12, p. 30.

Dubar, J.R., Johnson, H.S., Thom, B. and Hatchell, W.O., 1974, Neogene stratigraphy and morphology, south flank of the Cape Fear Arch, North and South Carolina, in Oaks, R.Q. and Dubar, J.R., eds., Post-Miocene Stratigraphy, central and southern Atlantic Coastal Plain: Logan, Utah State University Press, p. 179-190.

Fairbanks, R.G. and Matthews, R.K., 1978, The marine oxygen isotope record in Pleistocene coral, Barbados, West Indies: Quaternary Research, v. 10, p. 181-196.

Fanale, F.P. and Schaeffer, O.A., 1965, Helium-Uranium ratios for Pleistocene and Tertiary fossil aragonites. Science, v. 149, p. 312-317.

Flint, R.F., 1971, Glacial and Quaternary Geology, N.Y., Wiley, 892 p.

Gaven, C. and Vernier, E., 1979, Datation Io-U de coraux et Paleogeodynamique du Pleistocene moyen des Iles Glorieuses (Canal de Mozambique): Quaternaria, v. 21, p. 45-52.

Harmon, R.S., Schwarcz, H.P. and Ford, D.C., 1978, Late Pleistocene history of Bermuda: Quaternary Research, v. 9, p. 205-218.

Harmon, R.S., Ku, T.-L., Matthews, R.K. and Smart, P.L., 1979, Limits of U-series analysis: phase 1 results of the uranium-series intercomparison project: Geology, v. 7, p. 405-409.

Harmon, R.S., Land, L.S., Mitterer, R.M., Garrett, P., Schwarcz, H.P. and Larson, G.J., 1981, Bermuda sea level during the last interglacial: Nature, v. 289, p. 481-483.

Hoang, C.T., Lalou, C. and Faure, H., 1974, Les récifs soulevés a l'ouest du Golfe d'Aden (T.F.A.I.) et les hauts niveaux de coraux de la dépression de L'Afar (Ethiopie), géochronologie et paléoclimats interglaciaires: Coll. Intern. du C.N.R.S. No. 219, p. 103-114.

Hoang, C.T., Ortlieb, L. and Weisrock, A., 1978, Nouvelles datations $^{230}Th/^{234}U$ de terrasses marines "ouljiennes" du sud-ouest du Maroc et leurs significations stratigrafique et tectonique: C.R. Acad. Sci. Paris, v. 286 (D), p. 1959-1962.

Hollin, J.T., 1977, Thames interglacial sites, Ipswichian sea levels and Antarctic ice surges: Boreas, v. 6, p. 33-52.

____________, 1980, Climate and sea level in isotope stage 5: an East Antarctic ice surge at 95,000 BP?: Nature, v. 283, p. 629-633.

James, N.P., Mountjoy, E.W. and Omura, A., 1971, An early Wisconsin reef terrace at Barbados, West Indies, and its climatic implications: Geol. Soc. Amer. Bull., v. 82, p. 2011-2018.

Kaufman, A., Broecker, W.S., Ku, T.-L. and Thurber, D.L., 1971, The status of U-series methods of mollusk dating: Geochimica et Cosmochimica Acta, v. 35, p. 1155-1183.

Kominz, M.A., Heath, G.R., Ku, T.-L. and Pisias, N.G., 1979, Brunhes time scales and the interpretation of climatic change: Earth and Planetary Science Letters, v. 45, p. 394-410.

Ku, T.-L., 1968, Protactinium-231 method of dating coral from Barbados Island: Jour. Geophysical Research, v. 73, p. 2271-2276.

_________, 1976, The uranium-series methods of age determination: Ann. Rev. Earth and Planetary Sciences, v. 4, p. 347-380.

Ku, T.-L., Kimmel, M.A., Easton, W.H. and O'Neil, T.J., 1974, Eustatic sea level 120,000 years ago on Oahu, Hawaii: Science, v. 183, p. 959-962.

McCartan, L., Weems, R.E. and Lemon E.M., Jr., 1980, The Wando Formation (upper Pleistocene) in the Charleston, South Carolina, area: U.S. Geol. Surv. Bull. 1502-A, p. 110-117.

Marshall, J.F. and Thom, B.G., 1976, The sea level in the last interglacial: Nature, v. 263, p. 120-121.

Mesolella, K.J., Matthews, R.K., Broecker, W.S. and Thurber, D.L., 1969, The astronomical theory of climatic change: Barbados data:

Jour. Geol., v. 77, p. 250-274.

Mixon, R.B., Hazel, J.E., Rubin, M., Sirkin, L.A. and Szabo, B.J., 1974, Geologic framework of the southernmost Delmarva peninsula, Accomack and Northampton counties, Virginia: Geol. Soc. Amer. Abstr. with Programs, v. 5, p. 56-57.

Moore, W.S. and Somayajulu, B.L.K., 1974, Age determinations of fossil corals using Th^{230}/Th^{234} and Th^{230}/Th^{227}: Jour. Geophysical Research, v. 79, p. 5065-5068.

Neumann, A.C. and Moore, W.S., 1975, Sea level events and Pleistocene coral ages in the northern Bahamas: Quaternary Research, v. 5, p. 215-224.

Omura, A., Emerson, W.K. and Ku, T.-L., 1979, Uranium-series ages of echinoids and corals from the upper Pleistocene Magdalena Terrace, Baja California Sur, Mexico: The Nautilus, v. 94, p. 184-189.

Paskoff, R. and Sanlaville, P., 1976, Sur le Quaternaire marin de la region de Mahdia, Sahel de Sousse, Tunisie: C.R. Acad. Sci. Paris, v. 283 (D), p. 1715-1718.

__________, 1979, Introduction a l'etude du Tyrrhenien en Tunisie: Excursion-table ronde, I.N.Q.U.A., 51 p.

__________, 1980, Le Tyrrhenien de la Tunisie: essai de stratigraphie, C.R. Acad. Sci. Paris, v. 290 (D), p. 393-386.

Shackleton, N.J. and Matthews, R.K., 1977, Oxygen isotope stratigraphy of dated interglacial coral terraces in Barbados: Nature, v. 268, p. 618-620.

Stearns, C.E., 1976, Estimates of the position of sea level between 140,000 and 75,000 years ago: Quaternary Research, v. 6, p. 445-450.

______________, in press, A molluscan revival?: CNRS-Univ. de Paris I Colloque: Niveaux marins et tectonique quaternaires dans l'aire mediterraneenne, Actes.

Stearns, C.E. and Thurber, D.L., 1965, $^{230}Th/^{234}U$ dates of late Pleistocene marine fossils from the Mediterranean and Moroccan littorals: Quaternaria, v. 7, p. 29-42.

______________, 1967, $^{230}Th/^{234}U$ dates of late Pleistocene marine fossils from the Mediterranean and Moroccan littorals: Progr. Oceanography, v. 4, p. 293-305.

Szabo, B.J., 1979a, Uranium-series age of coral reef growth on Rottnest Island, Western Australia: Marine Geology, v. 29, p. M11-M15.

___________, 1979b, ^{230}Th, ^{231}Pa, and open system dating of fossil corals and shells: Jour. Geophysical Research, v. 84, p. 4927-4930.

Szabo, B.J. and Rosholt, J.N., 1969, Uranium-series dating of Pleistocene molluscan shells from southern California - an open system model: Jour. Geophysical Research, v. 74, p. 3253-3260.

Vacher, L., 1973, Coastal dunes of younger Bermuda, in Coates, D., ed., Coastal Geomorphology, Binghamton, State Univ. of New York, Publ. in Geomorphology, p. 355-391.

Veeh, H.H. and Chappell, 1970, Astronomical theory of climatic change: support from New Guinea: Science, v. 167, p. 862-865.

Wehmiller, J.F., Goddard, J.G., Belknap, D.F. and Keenan, E., 1980, Comparison of U-series and amino acid age estimates for Quaternary mollusks, T's Corner, Virginia (Delmarva peninsula): *Geol. Soc. Amer. Abstr. with Programs*, v. 12, p. 88-89.

THE APPLICABILITY OF $^{40}Ar/^{39}Ar$ DATING TO YOUNG VOLCANICS

CHRIS M. HALL and DEREK YORK

ABSTRACT

The $^{40}Ar/^{39}Ar$ step-heating method can be applied successfully to any sample which is suitable for K-Ar dating. The step-heating approach is particularly valuable for Quaternary rocks since it provides data concerning the fundamental assumptions used in the K-Ar technique (*i.e.* atmospheric initial Ar, no excess ^{40}Ar, *etc.*) which are virtually unattainable by any other means. A brief outline of the method is given, along with its advantages and disadvantages for dating Quaternary rocks. The results of a single step-heating analysis for a sample from Mt. Pupuke, N.Z. are presented. This case history demonstrates a few of the method's potential for alleviating problems encountered by the conventional K-Ar technique.

INTRODUCTION

$^{40}Ar/^{39}Ar$ dating is an extremely powerful variant of the standard K-Ar dating technique. Since its development in the mid-1960's (Merrihue, 1965 and Merrihue and Turner, 1966), this method has proven to be a most versatile geochronological tool. Not only can a wide variety of materials be dated using this technique, but also, a vast range of ages can be measured, spanning from less than 50,000 years (Hall and York, 1978 and Hall, 1982) to more than four billion years (York *et al.*, 1972). This paper will concentrate on the special problems and opportunities associated with applying the $^{40}Ar/^{39}Ar$ method to samples of Quaternary age.

Since this application of $^{40}Ar/^{39}Ar$ dating is so very new, it is impossible to give many concrete examples and firm predictions. However, some success has already been achieved and on-going studies can provide us with some basis for outlining the method's potential for Quaternary dating.

Basic Principles

The K-Ar dating method relies on the fact that ^{40}K, who's abundance is 0.01167 atom% of natural K, spontaneously decays by electron capture into ^{40}Ar. Actually, only 10.5% of ^{40}K decays into ^{40}Ar; the rest decays into ^{40}Ca, a common isotope of calcium. The K-Ar method requires the measurement of the ^{40}K concentration from one aliquot of a sample, usually by flame photometry or some other wet chemistry technique. From another aliquot, the radiogenic ^{40}Ar (or $^{40}Ar^*$) concentration is measured by fusing the sample in an ultra-high vacuum system and measuring the evolved Ar isotopes in a mass spectrometer. York and Farquhar (1972) and Dalrymple and Lanphere (1969) discussed details of the experimental techniques that are commonly used.

Assuming that all of the $^{40}Ar^*$ was removed from the rock during its formation, the age of the sample is given by:

$$t = \frac{1}{\lambda} \ln \{1 + \frac{\lambda}{\lambda_e} \frac{^{40}Ar^*}{^{40}K}\} \quad (1)$$

where λ, the total decay constant for ^{40}K, is 5.543×10^{-10} year^{-1} and λ_e, the constant for ^{40}K decay via electron capture into ^{40}Ar, is 0.581×10^{-10} year^{-1} (Steiger and Jäger, 1977). For ages less than about 20 million years, Eq. 1 can be approximated by:

$$t = \frac{1}{\lambda_e} \frac{^{40}Ar^*}{^{40}K} \quad (2)$$

The amounts of ^{40}Ar produced in typical Quaternary volcanic rocks, though small, are easily measurable, and good K-Ar dates as low as 10,000 years have been found (Dalrymple, 1967). Unfortunately, radiogenic ^{40}Ar is not the only ^{40}Ar in a volcanic rock. The argon in the atmosphere, which comprises about 1% of the total atmospheric pressure, is 99.6% ^{40}Ar. Whole-rock basalts and K-feldspar mineral separates typically have about 1×10^{-7} ccSTP/g of trapped or dissolved atmospheric ^{40}Ar (Dalrymple and Lanphere, 1971). The amount of atmospheric ^{40}Ar in a sample can be estimated by measuring the concentration of ^{36}Ar and noting that the atmospheric ^{40}Ar to ^{36}Ar ratio is 295.5. Thus the radiogenic ^{40}Ar in the sample is given by:

$$^{40}Ar^* = {^{40}Ar}_{measured} - 295.5\ {^{36}Ar} \quad (3)$$

Now Eq. 2 can be rewritten as:

$$t = \frac{1}{\lambda_e} \frac{^{36}Ar}{^{40}K} \{(^{40}Ar/^{36}Ar)_{measured} - 295.5\} \quad (4)$$

For Quaternary samples, the difference between $^{40}Ar/^{36}Ar$ in the sample and 295.5 can be extremely small and this is the fundamental factor limiting the precision of the method. Because the ^{36}Ar abundance is always very small, the error in its measurement is the dominant source of error in the calculated age. The error in the age due to an error in the ^{36}Ar peak is approximately:

$$\frac{\Delta t}{t} = \frac{295.5}{\{(^{40}Ar/^{36}Ar_{measured} - 295.5\}} \cdot \frac{\Delta ^{36}Ar}{^{36}Ar} \quad (5)$$

Thus a sample that has 95% atmospheric argon (typical for a basalt less than 100,000 years old) with a ^{36}Ar peak measured with an accuracy of 0.5%, will have an approximate age error of 10%.

The above discussion deals with the conventional K-Ar system and not the $^{40}Ar/^{39}Ar$ method, but for Quaternary rocks, many of the problems are the same. Specifically, the sensitivity of both methods to errors from the ^{36}Ar measurement are very similar, but it is much easier to show this with the conventional K-Ar method than it is with the $^{40}Ar/^{39}Ar$ technique.

THE $^{40}AR/^{39}AR$ METHOD

In the $^{40}Ar/^{39}Ar$ method, both the Ar and the K measurements are performed on the same aliquot of the sample. This is done by irradiating the sample with fast neutrons in a nuclear reactor to produce ^{39}Ar from the reaction ^{39}K (n,p) ^{39}Ar. ^{39}Ar is a radioactive isotope of Ar which does not occur naturally in significant quantities. Since the ^{39}Ar concentration is proportional to the ^{39}K in the sample, and since ^{39}K is in turn proportional to the ^{40}K, the ^{39}Ar peak from the mass spectrometer run is a measure of the ^{40}K in the sample. Thus both the $^{40}Ar^*$ and the ^{40}K concentrations can be determined simultaneously from a single mass spectrometer analysis. In practice, the proportionality constant relating ^{39}Ar with ^{40}K is determined by irradiating a standard sample of known K-Ar age along with the unknown.

Unfortunately, there are unwanted interference reactions which complicate this simple picture. Specifically, ^{39}Ar and ^{36}Ar are both produced from nuclear reactions with Ca, and ^{40}Ar is generated from a slow-neutron reaction with ^{40}K. The production of extraneous ^{36}Ar from Ca is especially worrying for Quaternary rocks, as this isotope is critical for the atmospheric argon correction. Fortunately, neutron reactions with Ca produce much more ^{37}Ar than either ^{36}Ar or ^{39}Ar. ^{37}Ar, like ^{39}Ar, is a short-lived isotope of Ar which does not occur naturally, and hence it can be used to correct for the presence of Ca-derived ^{36}Ar and ^{39}Ar. As pointed out by Turner (1971), by optimizing the neutron flux that the sample receives, it is possible to reduce the Ca interference effects to negligible levels. Even when samples receive a large neutron flux, the ^{37}Ar corrections are very successful in accounting for Ca-derived ^{36}Ar and ^{39}Ar in young basalts, as shown by Dalrymple and Lanphere (1971). The $^{40}Ar/^{39}Ar$ version of Equation 1 is:

$$t_u = \frac{1}{\lambda} \ \ell n \ \{1 + J \ (^{40}Ar^*/^{39}Ar_K)_u\} \tag{6a}$$

where

$$J = \frac{(e^{\lambda ts} - \ )}{(^{40}Ar^*/^{39}Ar_K)s} \tag{6b}$$

$$^{40}Ar^*/^{39}Ar_K = (1-f_1)(^{40}Ar/^{39}Ar)_m - 295.5\ (1-f_2)(^{36}Ar/^{39}Ar)_m - (^{40}Ar/^{39}Ar)_K \tag{6c}$$

$$f_1 = 1/\ |1-(^{37}Ar/^{39}Ar)_{Ca}(^{39}Ar/^{37}Ar)_m| \tag{6d}$$

$$f_2 = f_1\ |1-(^{36}Ar/^{39}Ar)_{Ca}(^{39}Ar/^{36}Ar)_m| \tag{6e}$$

Here the subscripts have the following meanings:

m is a quantity as measured;

u means the unknown, or the sample under study;

s is the standard, or flux monitor;

Ca,K are the calcium and potassium derived isotopes respectively.

Note that ts is the K-Ar age of the standard which is irradiated with the unknown. The values of the interference correction factors are determined by irradiating and then fusing pure K and Ca salts. The values that we currently use (Bottomley, 1982), for samples irradiated in the McMaster University nuclear reactor are:

$$(^{37}Ar/^{39}Ar)_{Ca} = 1536.1 \tag{7a}$$

$$(^{36}Ar/^{39}Ar)_{Ca} = 0.390 \tag{7b}$$

$$(^{40}Ar/^{39}Ar)_K = 0.0156 \tag{7c}$$

Although Equations 6a to 6e are considerably more involved than Equation

4, the crucial correction for young rocks is still that due to atmospheric ^{40}Ar in Equation 6c. From extensive analysis of the error propagation effects from measurements of ^{40}Ar, ^{39}Ar, ^{37}Ar and ^{36}Ar in Hall (1982), it can be shown that errors in the age of Quaternary samples are dominated by errors in the ^{36}Ar peak. For very young samples (a few hundred thousand years or less), Equation 5 is a good approximation of the precision obtainable from $^{40}Ar/^{39}Ar$ dating.

ADVANTAGES OF THE $^{40}AR/^{39}AR$ METHOD

The $^{40}Ar/^{39}Ar$ method totally eliminates the wet chemistry side of the K-Ar Method. In the conventional K-Ar technique, the ^{40}K and $^{40}Ar^*$ concentrations are measured from different aliquots of the same sample. If the sample is not precisely homogeneous, the measured $^{40}Ar^*$ to ^{40}K ratio will be in error, leading to an incorrect age. This is particularly troublesome when working with whole-rock basalts which can have significant K heterogeneity. This problem does not exist with $^{40}Ar/^{39}Ar$ dating, however, since both the K and $^{40}Ar^*$ measurements are done on the same aliquot of the sample.

Although single fusion $^{40}Ar/^{39}Ar$ ages, which are analogous with K-Ar ages, can be measured, the real power of the $^{40}Ar/^{39}Ar$ method is realized only with the step-heating variant of the technique. Instead of fusing the sample in one step, gas fractions are collected, purified, and analysed from a sequence of increasing temperature steps. Since $^{40}Ar^*$ and ^{39}Ar are simultaneously released at each temperature step, it is possible to calculate an apparent age for each gas fraction. An ideal, undisturbed mineral should give essentially the same apparent age over the entire gas release, thus yielding a "plateau" age. A major disturbance in the age spectrum is a good diagnostic indicator of non-ideal behaviour of the sample (*i.e.* argon loss, argon gain, K mobility, multiple phases with different apparent ages, *etc.*). Any single conventional K-Ar date does not have such an internal check, and if the sample's behaviour is reproducible, multiple conventional K-Ar analyses might not detect any problems with a disturbed mineral. For fine-grained whole-rock basalts, the step-heating method can be particularly attractive. As shown in Hall and York (1978), the step-heating technique can degas different minerals at different temperatures, thereby effectively performing a mineral separation on a rock for which mechanical mineral separation is not practical. Also, it was found that the bulk of the atmospheric ^{40}Ar in a basaltic sample is often released at the very highest temperatures. From the ratio of ^{37}Ar to ^{39}Ar in these gas fractions, it can be shown that the minerals that degas at the highest temperatures are deficient in K and have a high Ca to K ratio. The bulk of the ^{39}Ar and $^{40}Ar^*$ is actually released at low to intermediate temperatures (*i.e.* 650-900 degrees Celsius). Therefore, step heating can separate the atmospheric argon and radiogenic argon components, and as can be seen from the discussion on error propagation above, this can improve the precision of the age estimate beyond that obtainable by conventional K-Ar dating.

When dating volcanic tuffs, feldspar mineral separates are the most widely dated material. K-feldspars are particularly useful, and they usually give reliable K-Ar ages for Quaternary and Pliocene volcanics (Evernden and Curtis, 1965). However, $^{40}Ar/^{39}Ar$ dating can make a very real contribution with these samples by directly demonstrating whether or not the feldspars are as well-behaved as one assumes in conventional K-Ar dating. A recent example is the work of McDougall (1981) on the KBS tuff problem. The nearly perfect plateaux from the KBS anorthoclase added considerable weight to the previously determined age of 1.88 Ma determined from conventional K-Ar measurements (McDougall *et al.*, 1980 and Drake *et al.*, 1980). In a study on different material similar results were obtained by Radicati di Brozolo *et al.* (1981). In this study, $^{40}Ar/^{39}Ar$ plateau ages for Quaternary samples were found to be concordant with Rb-Sr isochron ages from the same rocks. The fundamental point is that $^{40}Ar/^{39}Ar$ step-heating ages can be obtained on almost any sample that can be dated by the conventional K-Ar technique, and a few $^{40}Ar/^{39}Ar$ analyses can provide more information than many conventional K-Ar runs on the same material.

Recently, there has been a unique application of the $^{40}Ar/^{39}Ar$ step-heating method to the dating of Quaternary rocks which has no direct analogy in conventional K-Ar dating. Gillespie *et al.* (1982) analysed microcline xenocrysts which had been incorporated within young volcanics. The heat of the eruption had caused partial, but not total, radiogenic Ar loss from the trapped crystals. Conventional K-Ar analyses on these xenocrysts would have yielded ages considerably in excess of the eruption age. However, as shown by Gillespie *et al.* (1982), the lowest temperature fractions from a $^{40}Ar/^{39}Ar$ step-heating run should record the age of the volcanic eruption and this was confirmed by experiment.

DISADVANTAGES OF THE $^{40}AR/^{39}AR$ SYSTEM

Implicit in the $^{40}Ar/^{39}Ar$ age equations is the assumption that the standard mineral of known K-Ar age receives the same neutron dose as the unknown sample. For Quaternary rocks, sample masses are normally large - typically 1 to 10 g, and it is often not possible to have a perfectly uniform neutron flux over the space required to hold the sample and the standard. This problem can be reduced by (1) choosing irradiation locations within the reactor which have low neutron flux gradients and (2) averaging the J values from several standards in the same irradiation can. In any case, uncertainties in the relative neutron doses received by the samples and standards can usually be kept below 1%, and this is almost always less than the expected error in Quaternary ages found by the $^{40}Ar/^{39}Ar$ method.

A much more serious problem is caused by Ar recoil. As shown by Turner and Cadogan (1974), a ^{40}K nucleus which decays into ^{40}Ar might move less than a nanometer, but a ^{39}K nucleus which is converted into ^{39}Ar by an n,p reaction can move up to about 100 nm. ^{37}Ar, ^{39}Ar and ^{36}Ar which are produced from Ca can also be expected to recoil, but the distances they would travel have not yet been accurately estimated.

If the irradiated minerals are homogeneous on a scale of about 50 microns, recoil presents no problems, as any transport of Ar nuclei will average out. Only a very thin (about 0.1 μm) layer surrounding the mineral grain will have any net effect from recoil. Unfortunately, the K-rich phases in volcanic rocks sometimes are (1) very small, and (2) surrounded by K-poor regions. In such cases, there can be a significant net transport of ^{39}Ar from the K-rich phase into the K-poor phase. As noted above, the different minerals degas at different temperatures. Thus some gas fractions will have their apparent ages changed by the irradiation. If the small K-rich phases are not in intimate contact with other minerals, it is even possible to have a net loss of ^{39}Ar.

One can at least check to see if there is a net loss of ^{39}Ar by comparing the integrated or total fusion $^{40}Ar/^{39}Ar$ age, with the conventional K-Ar age from the same sample. If there is only redistribution of ^{39}Ar, the two ages will agree within experimental error. It is our experience (Hall and York, 1978 and Hall, 1982) that whole-rock basalts do retain all of their ^{39}Ar. It is likely that as long as rocks are not crushed to a fine powder, recoil is not likely to affect the integrated $^{40}Ar/^{39}Ar$ age of a sample.

In comparison with flux gradients, recoil effects are more likely to complicate interpretation by altering the structure of the age spectrum derived from a $^{40}Ar/^{39}Ar$ step-heating experiment. However, any explanation of odd age spectra features based upon recoil is severely constrained by mass balance considerations and by the likely recoil ranges. This restricts the use of recoil as an explanation for any non-ideal behaviour by a sample.

A Case History

Figure 1 shows the results from a single $^{40}Ar/^{39}Ar$ experiment. The sample is an olivine basalt from Mt. Pupuke, N.Z. It is the sample GA 2853 that was studied by the conventional K-Ar method by McDougall

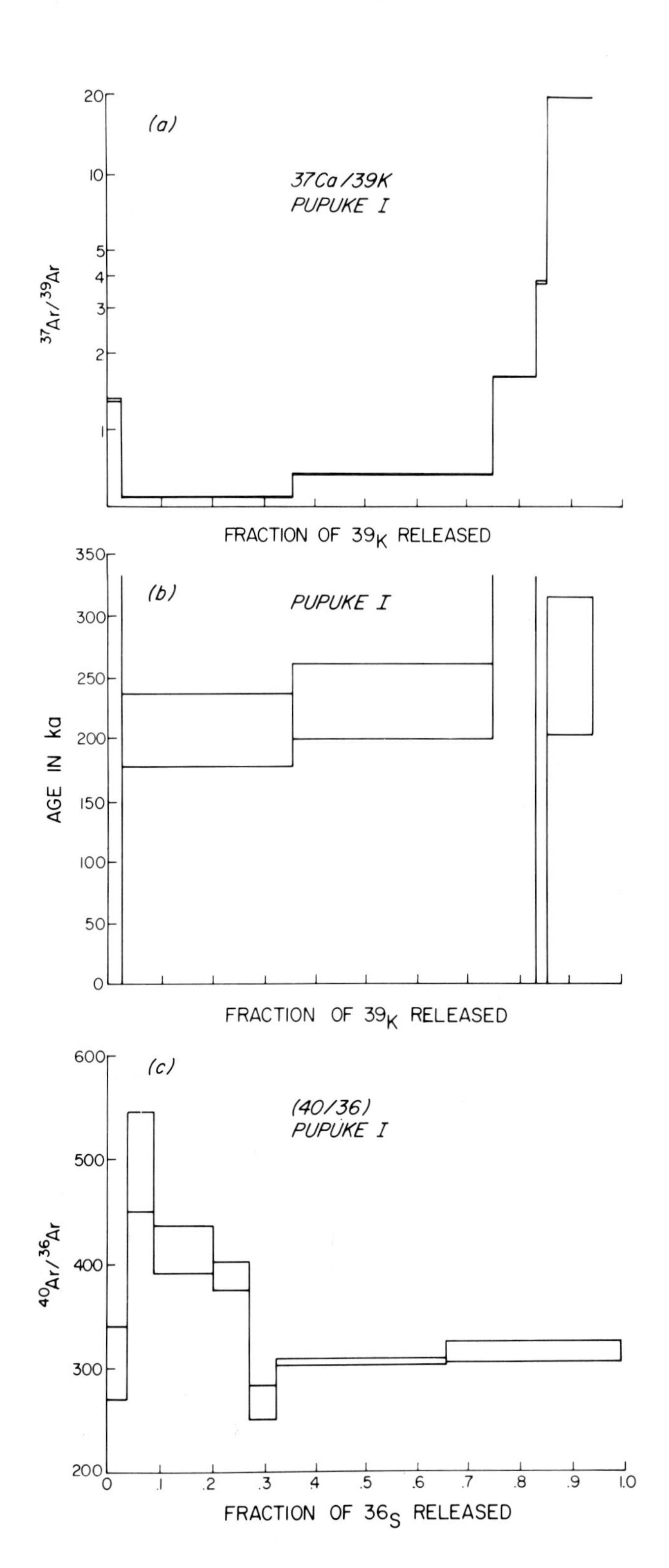

Figure 1 The results of a $^{40}Ar/^{39}Ar$ experiment on sample GA 2853 from Mt. Pupuke, N.Z. The fraction temperatures are in increasing order: 500, 700, 950, 1060, 1170, 1250 and 1600 degrees Celsius.

a) $^{37}Ar_{Ca}/^{39}Ar_K$ *vs* $^{39}Ar_K$ released. The highest temperature fraction is off-scale. It has a value of 39.5.

b) Apparent age *vs* $^{39}Ar_K$ released. The highest temperature fraction is off-scale with an apparent age of 900 ± 420 ka.

c) $^{40}Ar/^{36}Ar$ *vs* ^{36}Ar released.

All isotope concentrations have been corrected for K and Ca interference effects. All error estimates are 1σ.

et al. (1969). Figure 1a is a graph of the ^{37}Ar to ^{39}Ar ratio as a function of ^{39}Ar released. The ^{37}Ar to ^{39}Ar ratio is proportional to the Ca to K ratio. Figure 1a shows that the K-rich phases preferentially degassed in the low and intermediate temperature steps. Figure 1b is a standard $^{40}Ar/^{39}Ar$ age spectrum while figure 1c shows $^{40}Ar/^{36}Ar$ as a function of ^{36}Ar released. The bottom figure graphically illustrates which fractions contain the bulk of the radiogenic ^{40}Ar. The volume of radiogenic ^{40}Ar in a fraction is directly proportional to the area between the fraction's $^{40}Ar/^{36}Ar$ box and a horizontal line running through 295.5 (*i.e.* the atmospheric value of $^{40}Ar/^{36}Ar$). As can be seen from figure 1c, most of the radiogenic ^{40}Ar resides in the same K-rich phases which degas at low and intermediate temperatures. Also note that a large proportion of the ^{36}Ar (and hence atmospheric argon) was released at high temperatures.

It is interesting to note that the conventional K-Ar of GA 2853 was found to be 255 Ka (McDougall *et al.*, 1969), while the integrated ^{40}Ar/^{39}Ar age from the run shown in Figure 1 is 260 ± 29 Ka. However, McDougall *et al.* (1969) interpreted this age, and other ages from the Auckland volcanic field, to be anomalously old due to the presence of excess ^{40}Ar. From Figure 1, it is clear that there is excess ^{40}Ar in the higher temperature fractions. The second fraction (700 degrees C) has an apparent age of 207 ± 29 ka, which may be a better age estimate than that from the conventional K-Ar technique. It appears that the step-heating method has separated the radiogenic ^{40}Ar from the bulk of the initial argon trapped within the sample.

CONCLUSIONS

Any sample which can be analysed by the conventional K-Ar method can also be dated by the ^{40}Ar/^{39}Ar step-heating technique. In a large majority of cases, the ^{40}Ar/^{39}Ar technique yields much more useful information than does the conventional K-Ar system. The step-heating process can effectively separate different mineral phases and it can separate radiogenic ^{40}Ar from much of the sample's contaminating initial Ar.

It may not be possible in all cases to determine a geologically meaningful age from each sample, but in many instances, the ^{40}Ar/^{39}Ar technique can see through problems which would defeat the conventional K-Ar system. Just as importantly, the ^{40}Ar/^{39}Ar method has internal consistency checks which can often determine whether a sample is suitable for dating at all.

REFERENCES CITED

Bottomley, R.J., 1982, ^{40}Ar-^{39}Ar dating of melt rock from impact craters: Ph.D. Thesis, University of Toronto.

Dalrymple, G.B., 1967, Potassium-argon ages of recent rhyolites of the Mono and Inyo craters, California: Earth and Planetary Science Letters, v. 3, p. 289-298.

Dalrymple, G.B. and Lanphere. M.A., 1969, Potassium-Argon Dating, San Francisco, W.H. Freeman, 258 p.

______________, 1971, ^{40}Ar/^{39}Ar technique of K/Ar dating: a comparison with the conventional techniques: Earth and Planetary Science Letters, v. 12, p. 300-308.

Drake, R.E., Curtis, G.H., Cerling, T.E., Cerling, B.W. and Hampel, J., 1980, KBS Tuff dating and geochronology of tuffaceous sediments in the Koobi Fora and Shungura Formations, East Africa: Nature, v. 283, p. 268-372.

Evernden, J.F. and Curtis, G.H., 1965, The Potassium-Argon Dating of Late Cenozoic Rocks in East Africa and Italy: Current Anthropology, v. 6, p. 343-385.

Gillespie, A.R., Huneke, J.C. and Wasserburg, G.J., 1982, Dating Pleistocene Basalts by ^{40}Ar-^{39}Ar Analysis of Granitic Xenoliths: EOS Transactions, American Geophysical Union, v. 63, p. 454.

Hall, C.M., 1982, The Application of K-Ar and ^{40}Ar/^{39}Ar methods to the dating of recent volcanics and the Laschamp event: (Ph.D. Dissert.): University of Toronto.

Hall, C.M. and York, D., 1978, K-Ar and ^{40}Ar/^{39}Ar Age of the Laschamp geomagnetic polarity reversal: Nature, v. 274, p. 462-464.

McDougall, I., 1981, $^{40}Ar/^{39}Ar$ age spectra from the KBS Tuff, Koobi Fora Formation: Nature, v. 294, p. 120-124.

McDougall, I., Maier, R., Sutherland-Hawkes, P. and Gleadow, A.J.W., 1980, K-Ar age estimate for the KBS Tuff, East Turkana, Kenya: Nature, v. 284, p. 230-234.

McDougall, I., Polach, H.A. and Stipp, J.J., 1969, Excess radiogenic argon in young subaerial basalts from the Auckland volcanic field, New Zealand: Geochimica et Cosmochimica Acta, v. 33, p. 1485-1520.

Merrihue, C.M., 1965, Trace-element determinations and potassium-argon dating by mass spectroscopy of neutron-irradiated samples: EOS Transactions, American Geophysical Union, v. 46, p. 125.

Merrihue, C.M. and Turner, G., 1966, Potassium-argon dating by activation with fast neutrons: Journal of Geophysical Research, v. 71, p. 2852-2857.

Radicati di Brozolo, F., Huneke, J.C., Papanastassion, D.A. and Wasserburg, G.J., 1981, $^{40}Ar-^{39}Ar$ and Rb-Sr age determinations on Quaternary volcanic rocks: Earth and Planetary Science Letters, v. 53, p. 445-456.

Steiger, R.H. and Jäger, E., 1977, Subcommission on geochronology: Convention on the use of decay constants in geo- and cosmochronology, Earth and Planetary Science Letters, v. 36, p. 359-362.

Turner, G., 1971, $Argon^{40}-Argon^{39}$ Dating: The optimization of irradiation parameters: Earth and Planetary Science Letters, v. 10, p. 227-234.

Turner, G. and Cadogan, P.H., 1974, Possible effects of ^{39}Ar recoil in $^{40}Ar-^{39}Ar$ dating: Proceedings of the Fifth Lunar Conference, Geochimica et Cosmochimica Acta, Supplement 5, v. 2, p. 1601-1615.

York, D. and Farquhar, R.M., 1972, The Earth's Age and Geochronology: Oxford, Pergamon, 178 p.

York, D., Kenyon, W.J. and Doyle, R.J., 1972, $^{40}Ar-^{39}Ar$ ages of Apollo 14 and 15 samples: Proceedings of the Third Lunar Conference: Geochimica et Cosmochimica Acta, Supplement 3, v. 2, p. 1613-1622.

LEAD-210 DATING OF SEDIMENTS FROM SOME NORTHERN ONTARIO LAKES

R.W. DURHAM and S.R. JOSHI

ABSTRACT

Fourteen sediment cores extracted from twelve small lakes in northern and eastern Ontario were sectioned at 0.5 cm intervals, freeze-dried, and then analyzed for ^{210}Pb, ^{226}Ra, and ^{137}Cs. The excess ^{210}Pb (*i.e.*, total ^{210}Pb less that produced from the decay of *in situ* ^{226}Ra) concentration profiles obtained were used to calculate annual sedimentation rates and thus the age profiles of the cores. The dating model assumes a constant flux of ^{210}Pb at the lake surface from the decay of atmospheric ^{222}Rn, followed by adsorption on suspended particulates which carry the ^{210}Pb to the lake sediments at a constant rate. The ^{210}Pb half-life of 22.26 years allows sediments to be dated back in time about 100 years. ^{137}Cs concentration profiles are used to corroborate the annual sedimentation rates since the horizon for 30 year half-life ^{137}Cs in lake sediments corresponds to about 1958, the onset of large-scale testing of nuclear weapons.

All the cores had very high water content with porosities near the sediment/water interface varying from 0.95 to 0.98 and only one lake showed any compaction of the sediment in the top 10 cm. The annual sedimentation rates were generally very low, ranging from 0.4 to 1.6 mm y^{-1}. One core showed evidence of severe perturbation of the top two cm of sediment.

INTRODUCTION

Precipitation scavenging of the decay products from atmospheric ^{222}Rn results in a natural supply of ^{210}Pb to surface water resources where, following rapid adsorption onto settling particulates, it is incorporated in the accumulating sediments. The radioactive decay of this ^{210}Pb with a 22.26 yr half-life constitutes a geological clock for establishing the chronology of recent sediments. Since the first application of ^{210}Pb measurements in calculating sedimentation rates for recent marine and lake sediments (Krishnaswamy *et al.*, 1971), a number of investigators have successfully utilized the technique for dating purposes (Koide *et al.*, 1973; Robbins and Edgington, 1975; Farmer, 1978; Oldfield *et al.*, 1978; Durham and Joshi, 1980a). The use of the technique is increasing rapidly and various applications include studies of eutrophication in lakes (Allan *et al.*, 1980), the recent history of heavy metal pollution (Christensen and Chien, 1981), the persistence of organotoxics (Brownlee *et al.*, 1977), and soil erosion (Goldberg *et al.*, 1978).

In this paper we describe the dating of sediments from twelve lakes in northern and eastern Ontario using the ^{210}Pb technique. This work is part of a joint study with Fisheries and Oceans Canada of heavy metal and acid rain inputs to these lakes in the recent past and the concomitant impact on diatom and chironomid populations. The sediment cores were sectioned and analyzed for ^{210}Pb and ^{226}Ra to obtain "excess" ^{210}Pb profiles. The "excess" ^{210}Pb is that obtained by subtracting the ^{210}Pb produced from ^{226}Ra decay in detrital material in the sedimenting particulates from the total ^{210}Pb. Application of an exponential radioactive decay model to the excess ^{210}Pb profiles gave sedimentation rates, ^{210}Pb atmospheric fluxes and age profiles of the sediments. Corroboration of the age profiles was obtained by determining ^{137}Cs profiles for the cores. The ^{137}Cs input to the lakes were from nuclear weapons testing in the atmosphere which started on a large scale in 1958. The fallout from these tests peaked in 1963-64 and then decreased sharply following the agreement by most of the nuclear powers to end atmospheric weapons' tests. The ^{137}Cs horizon in sediment cores is thus about 1958.

EXPERIMENTAL TECHNIQUES

Sediment core sampling

Surficial sediment cores were obtained with a Benthos corer at the stations shown in Figure 1. Cores were stored at 4°C until sectioned at 0.5 cm intervals. While sectioning, the outside 0.5 cm of each core was rejected to avoid possible contamination of deeper lying material during coring. Precise sample thicknesses were obtained by dividing the wet mass of the sediment aliquot by its bulk density and area.

Figure 1 Sampling locations in northern Ontario.

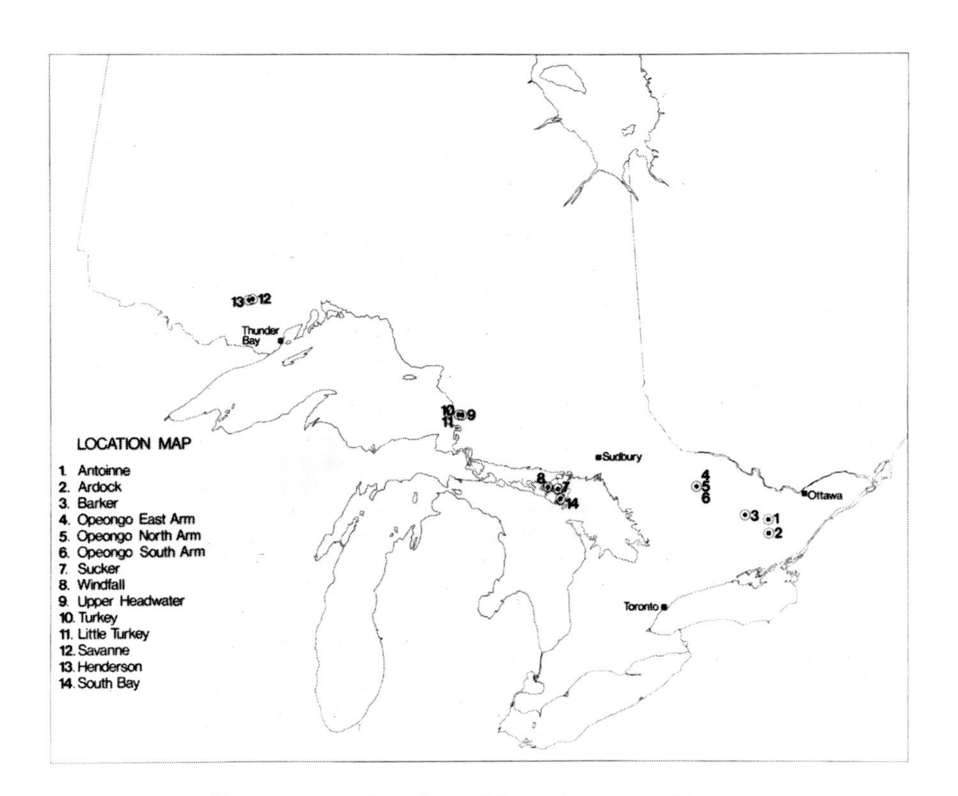

Measurement of sediment porosity

The wet sediment sections were freeze-dried and the water content determined from the weight loss. Porosity (ϕ) values were obtained using the relationship

$$\phi = \frac{M_w/\rho_w}{M_w/\rho_w + M_s/\rho_s}$$

where M_w and M_s are the masses and ρ_w (1 g cm^{-3}) and ρ_s (2.45 g cm^{-3}) are the densities of water and solids in the sediment, respectively. All the cores had very high water content with porosities near the sediment/water interface varying from 0.95 to 0.98 and only one lake

(Lake Opeongo) showed any compaction of the sediment in the top 10 cm.

Measurement of ^{210}Pb, ^{226}Ra, and ^{137}Cs activities

The dried sections were analyzed for ^{210}Pb, ^{226}Ra, and ^{137}Cs using procedures described earlier (Joshi and Durham, 1976). Briefly, the method involved separation of ^{137}Cs from ^{210}Pb and ^{226}Ra with a carbonate fusion following addition of respective carriers and removal of silica by cyclic HNO_3-HF treatments. The Pb and Ba carriers, containing ^{210}Pb and ^{226}Ra, were separated from Ca using 70-72% HNO_3 precipitation and alcohol-ether separation techniques before mutual separation of Pb and Ba using anion exchange. ^{210}Pb and ^{226}Ra were determined by β- and α-counting of Pb and Ba chromate precipitates, respectively, after allowing suitable ingrowth periods for their daughters. ^{137}Cs was removed from the carbonate fusion filtrate by absorption on ammonium molybdophosphate. Following radiochemical purification it was precipitated as $Cs_3Bi_2I_9$ and β-counted.

DETERMINATION OF SEDIMENTATION RATES

Krishnaswamy *et al.*, (1971) have shown that in homogeneously depositing sediments the excess ^{210}Pb concentration A(x) in pCi g^{-1}, at depth x below the sediment/water interface is given by

$$A(x) = \frac{P}{DS} \exp\,[-\lambda x/S], \tag{1}$$

where P is the flux of excess ^{210}Pb at the sediment/water interface (pCi cm^{-2} y^{-1}), λ is the decay constant of ^{210}Pb (0.0311 y^{-1}), D is the *in situ* density of the sediment (g cm^{-3}), and S the sedimentation rate (cm y^{-1}). Robbins and Edgington (1975) refined this model to take into account compression of the sedimentary material using a relationship between compaction and porosity (Athy, 1930). They modified equation (1) to give

$$A(x) = \frac{P}{S_o(1-\phi_o)\rho_s} \exp\,[-(\lambda/S_o)f(x)], \tag{2}$$

where

$$f(x) = \frac{1}{(1-\phi_o)} \int_0^x (1-\phi)dx', \tag{3}$$

with ϕ_o being the porosity at the sediment/water interface and S_o the surface sedimentation rate.

As discussed in another publication (Durham and Joshi, 1980a) we have preferred to use the following expression, also given by Athy (1930), to evaluate equation (3) analytically:

$$\phi_x = \phi_o \exp\,[-\beta x]\,, \tag{4}$$

where β is a constant. Substitution of (4) in (3) gives

$$f(x) = \frac{1}{(1-\phi_o)} [x - \frac{\phi_o}{\beta} (1-e^{-\beta x})]\;. \tag{5}$$

Substitution of (5) in (2) then gives

$$A(x) = \frac{P}{S_o(1-\phi_o)\rho_s} \exp[(-\frac{\lambda}{S_o})(\frac{1}{1-\phi_o})\{x-\frac{\phi_o}{\beta} (1-e^{-\beta x}\}] \tag{6}$$

Linear regression analysis of the log transform of equation (6) gives values of S_o, the linear surface sedimentation rate, P, the ^{210}Pb flux at the sediment/water interface, and ω, the mass sedimentation rate ($=S_o(1-\phi_o)\rho_s$).

Assuming a constant supply of atmospheric ^{210}Pb and a constant sedimentation rate, the age profile of the sediment core is represented by

$$A(x) = A(o) \exp [-\lambda t] \tag{7}$$

Since the sediment surface ^{210}Pb concentration, A(o), corresponds to $P/S_o(1-\phi_o)\rho_s$, it follows from (6) and (7) that the correspondence between the age of a core section and its depth is given by

$$t = \frac{x}{S_o(1-\phi_o)} [1 - \phi_o\{\frac{(1-e^{-\beta x})}{\beta x}\}] \tag{8}$$

As these cores showed little change in porosity with depth in the surface layers there was no compaction and t approximates x/S_o.

RESULTS

^{210}Pb Profiles

The excess ^{210}Pb profiles are given in Figure 2. The errors in the sedimentation rates were evaluated by linear regression analysis of the core data. The regression lines for each core

$$\hat{Y}_i = a + bX_i \tag{9}$$

were obtained by log transformation of the excess ^{210}Pb concentrations to give values of Y_i and calculation of values of f(x) from equation (5) to give X_i values. It can be seen from equation (7) that the slope of the regression line, b, corresponds to λ/S_o, while the intercept, a, when X_i=0, is P/ω. The regression lines are also plotted in Figure 2, while the computed values of P, S_o, and ω are given in Table 1. As discussed earlier (Durham and Joshi, 1981), the standard deviations given in Table 1 were obtained using the following equations for the variances of S_o and ω:

$$\hat{\sigma}^2_{S_o} = \frac{\hat{S}^4_o \sum^n (Y_i - \hat{Y})^2}{\lambda^2(n-2) \sum^n (X_i - \bar{X})^2} \tag{10}$$

$$\hat{\sigma}^2_{\omega} = \hat{\sigma}^2_{S_o} (1-\phi_o)^2 \rho_s^2 \tag{11}$$

Age profiles of the sediment cores have been calculated using equation (8) and are given in Figure 3.

^{137}Cs measurements

The ^{137}Cs profiles obtained for each core are depicted in Figure 4. The profiles of all fourteen cores show the usual shape for lake sediments consisting of a relatively high concentration for the top portion followed by a drop in concentration below a few cm. Since large scale atmospheric testing of nuclear weapons started in 1958, it would be expected that this drop in ^{137}Cs concentration corresponds to a "^{137}Cs horizon" and would coincide with this date. The depths corresponding to 1958, obtained from the age-versus-depth curves in Figure 3, are in reasonably good agreement with the observed ^{137}Cs horizon in each case except for Barker Lake where diffusion of ^{137}Cs is observed in the top few cm.

DISCUSSION

The results presented in Table 1 show that the surface sedimentation rates in these small lakes vary from 0.37 to 1.65 mm y^{-1} (3.9 to 39.3 mg $cm^{-2}y^{-1}$). These values are similar to those obtained earlier

Table 1. Parameters from excess ^{210}Pb and ^{137}Cs profiles of sediment cores

Lake	Location		Porosity of surface sediment (ϕ_o)	Sedimentation rate, S_o (mm y^{-1})	Flux of excess ^{210}Pb, P (pCicm^{-2}y^{-1})	Mass sedimentation rate, $\omega = S_o(1-\phi_o)\rho_s$ (mg cm^{-2}y^{-1})	^{137}Cs Horizon (cm)
	Lat.N.	Long.W.					
Antoinne	45°00'	76°44'	0.982	1.28 ± 0.11	0.143	5.6 ± 0.5	3.2
Ardoch	44°56'	76°52'	0.979	1.04 ± 0.17	0.048	5.3 ± 0.9	1.8
Barker	45°07'	77°23'	0.941	1.17 ± 0.14	0.526	16.9 ± 2.0	Diffuse(2-4)
Opeongo E. Arm	45°42'	78°23'	0.957	0.37 ± 0.04	0.072	3.9 ± 0.4	2.3
Opeongo N. Arm	45°42'	78°23'	0.968	1.55 ± 0.25	0.132	12.3 ± 2.0	3.3
Opeongo S. Arm	45°42'	78°23'	0.969	1.55 ± 0.28	0.092	11.8 ± 2.1	2.5
South Bay	45°39'	81°50'	0.885	1.39 ± 0.24	0.655	39.3 ± 6.8	3.0
Sucker	45°43'	81°52'	0.984	1.65 ± 0.54	0.021	6.5 ± 2.1	2.4
Windfall	45°42'	82°04'	0.985	1.26 ± 0.35	0.051	4.6 ± 1.3	2.6
Upper Headwater	47°04'	84°24'	0.971	0.96 ± 0.06	0.933	6.9 ± 0.4	3.0
Turkey	47°03'	84°25'	0.956	0.67 ± 0.06	0.620	7.0 ± 0.6	3.5
Little Turkey	47°03'	84°25'	0.972	0.79 ± 0.04	0.400	5.5 ± 0.3	3.0
Savanne	48°50'	90°06'	0.948	0.49 ± 0.05	0.068	6.2 ± 0.6	1.5
Henderson	48°49'	90°18'	0.953	1.48 ± 0.35	0.077	17.1 ± 4.0	4.8

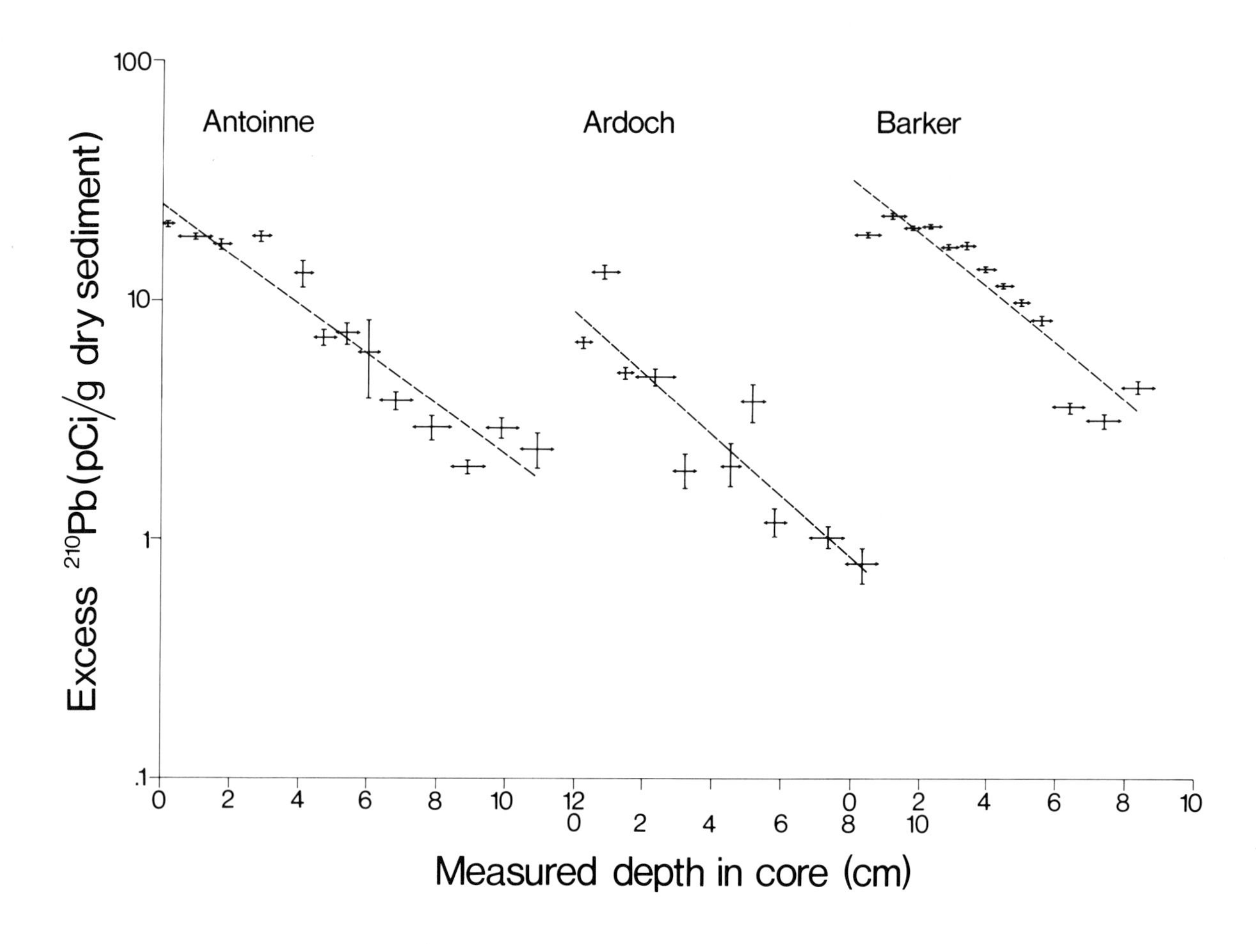

Figure 2a-2e Excess ^{210}Pb profiles in sediment cores.

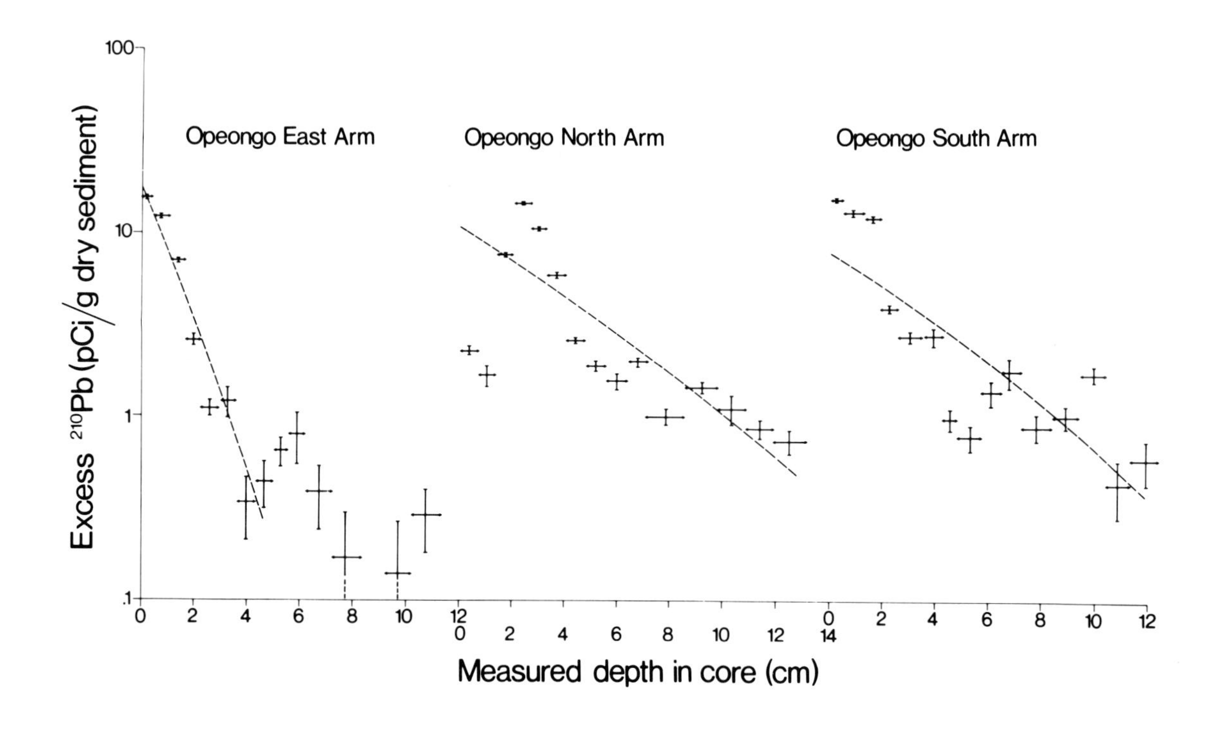

Figure 2b

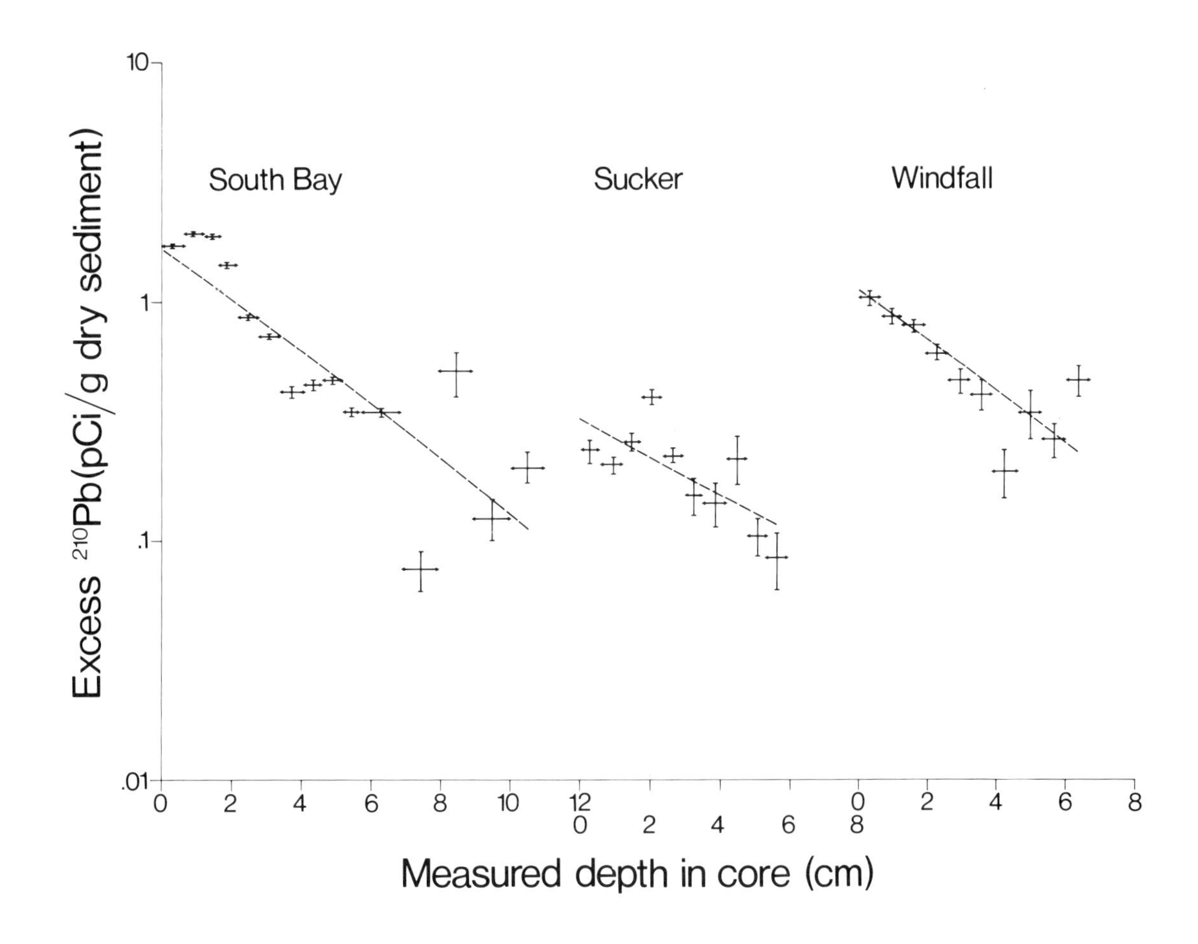

Figure 2c

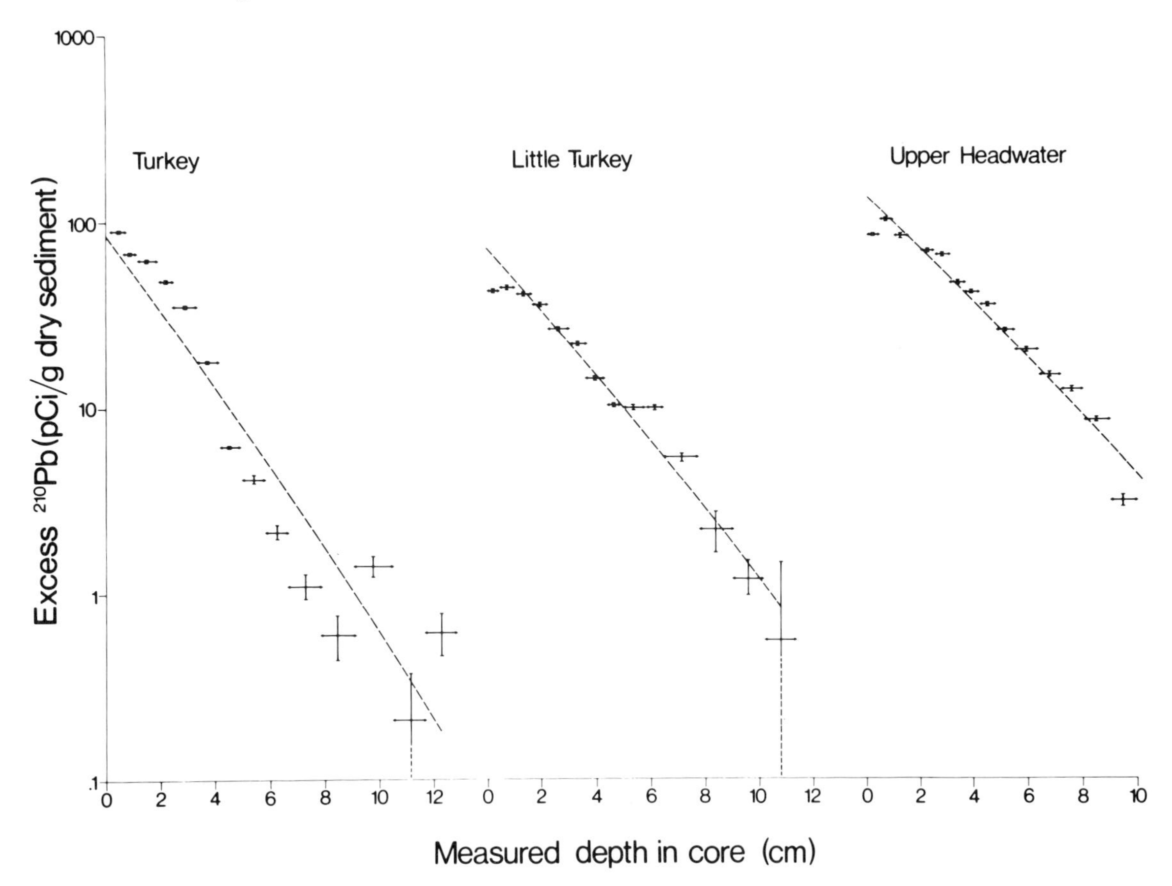

Figure 2d

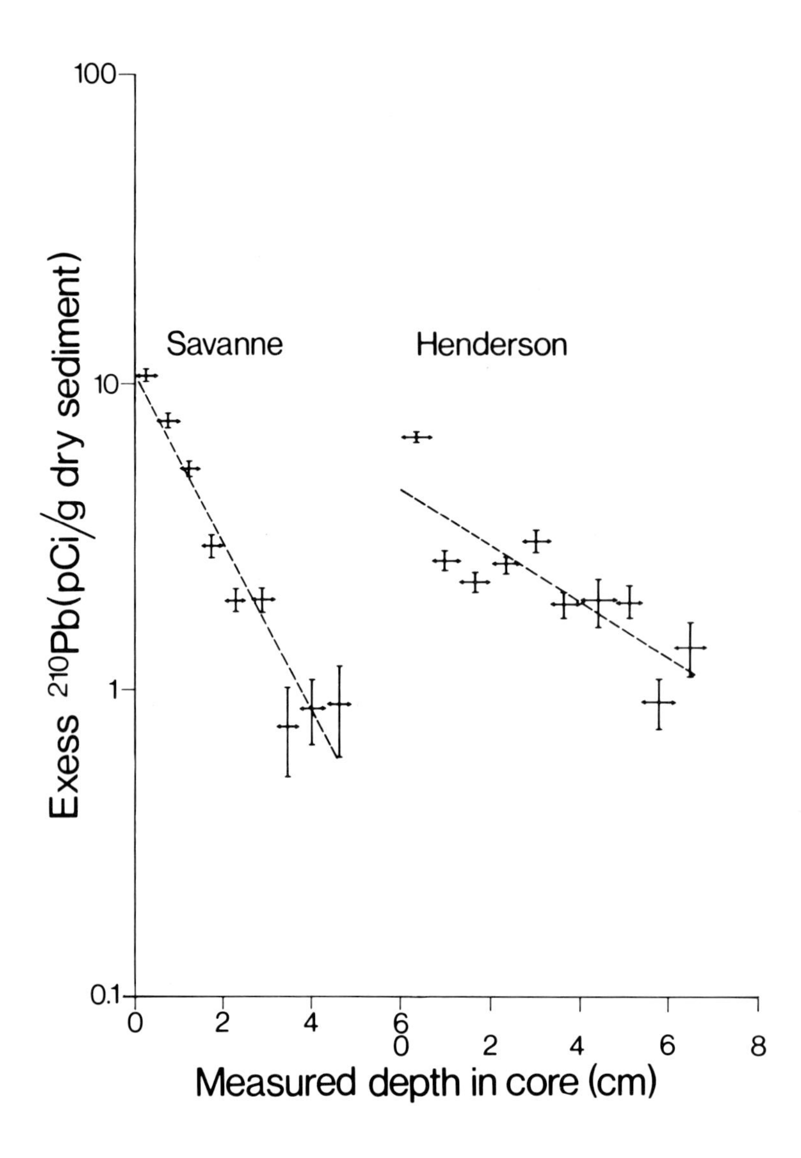

Figure 2e

(Durham and Joshi, 1980b) for small lakes Matagami (0.81 mm y^{-1}; 59 mg $cm^{-2}y^{-1}$) and Quevillon (0.52 mm y^{-1}; 35 mg cm^{-2} y^{-1}) in northwest Quebec, but are considerably lower than those derived for four small prairie lakes (4.7 to 6.6 mm y^{-1}) in Saskatchewan (Durham *et al.*, 1980) where spring snow-melt provides high sediment loading. These sedimentation rates are also similar to those reported for much larger and deeper lakes in the Great Lakes system which have rates ranging from 0.1-2.1 mm y^{-1} (4.8 - 93 mg cm^{-1} y^{-1}) for Lakes Huron (Durham and Joshi, 1980a), Ontario (Farmer, 1978), Michigan (Robbins and Edgington, 1975) and Superior (Kemp *et al.*, 1978). A much wider range of 0.2-8.5 mm y^{-1} (13-204 mg cm^{-2} y^{-1}) was measured for Lake Erie (Nriagu *et al.*, 1979) which is much shallower than the other Great Lakes.

One core (Opeongo north arm) contained very little excess ^{210}Pb in the top 2 cm compared to the next few cm. This may be due to sediment reworking in the recent past whereby older sedimentary material has overlain that deposited earlier. This apparent loss of ^{210}Pb can not be attributed to possible biological or chemical reactions occurring, such as methylation to volatile tetramethyl-lead, in the top few cm since ^{137}Cs - an isotope having different chemical behaviour and biological interactions than ^{210}Pb - also shows a similar drop in concentration in this region (Figure 4).

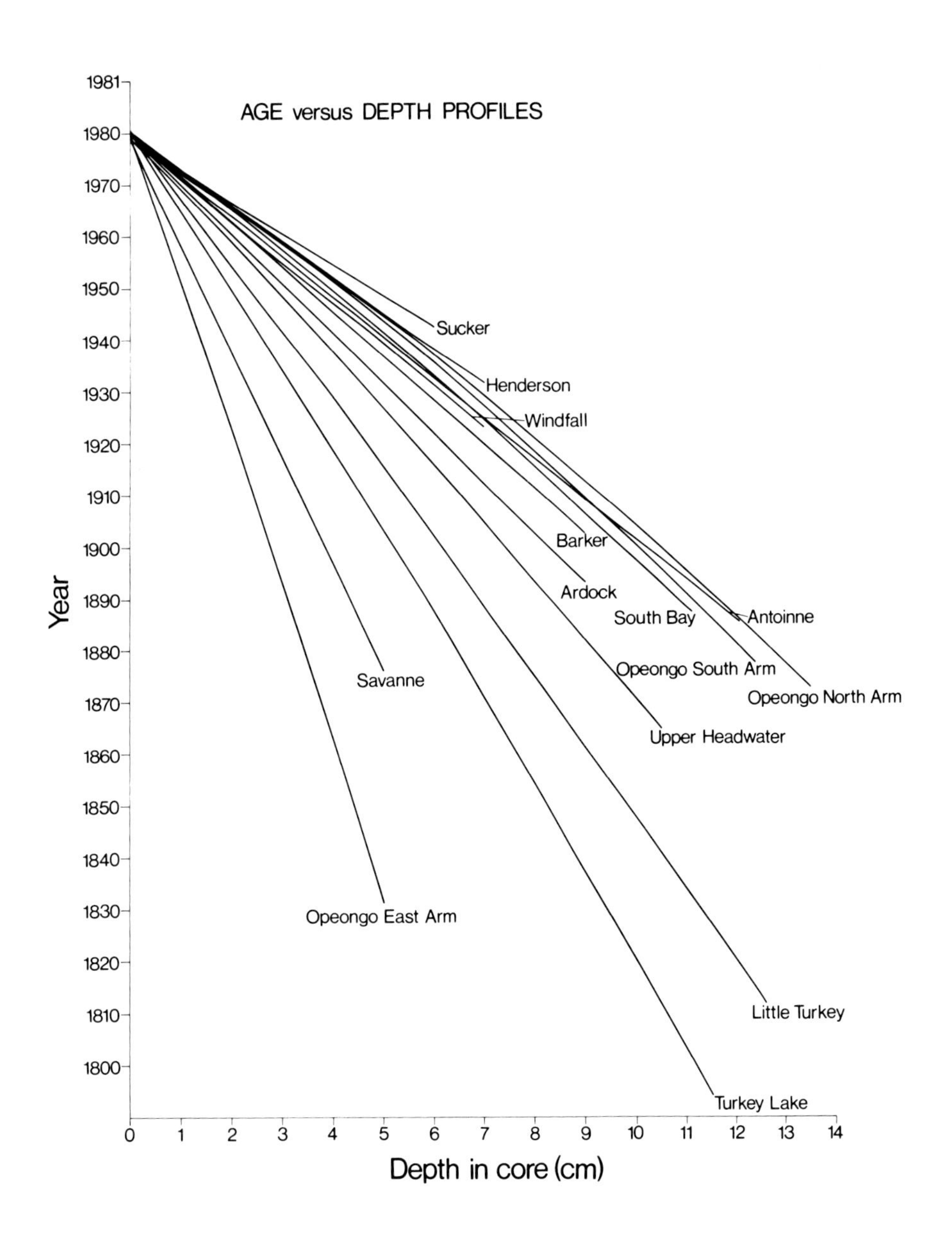

Figure 3 ^{210}Pb age versus depth in core.

The ^{210}Pb fluxes at the sediment/water interface range between 0.021 and 0.933 pCi cm^{-2} y^{-1} for these lakes. Moore and Poet (1976) have calculated direct atmospheric ^{210}Pb fluxes from measurements of ^{210}Pb concentrations in rainfall. They have concluded that the atmospheric ^{210}Pb flux ranges from 0.24 to 0.92 pCi cm^{-2} y^{-1} for the north temperature zone. Previous measurements of ^{210}Pb fluxes in Lake Huron (Durham and Joshi, 1980a), Lake Superior (Durham and Joshi, 1981), Lake Matagami in Quebec (Durham and Joshi 1980b), and those reported by Robbins and Edgington (1975) for Lake Michigan cover a range similar to that reported by Moore and Poet (1976). However, several of the

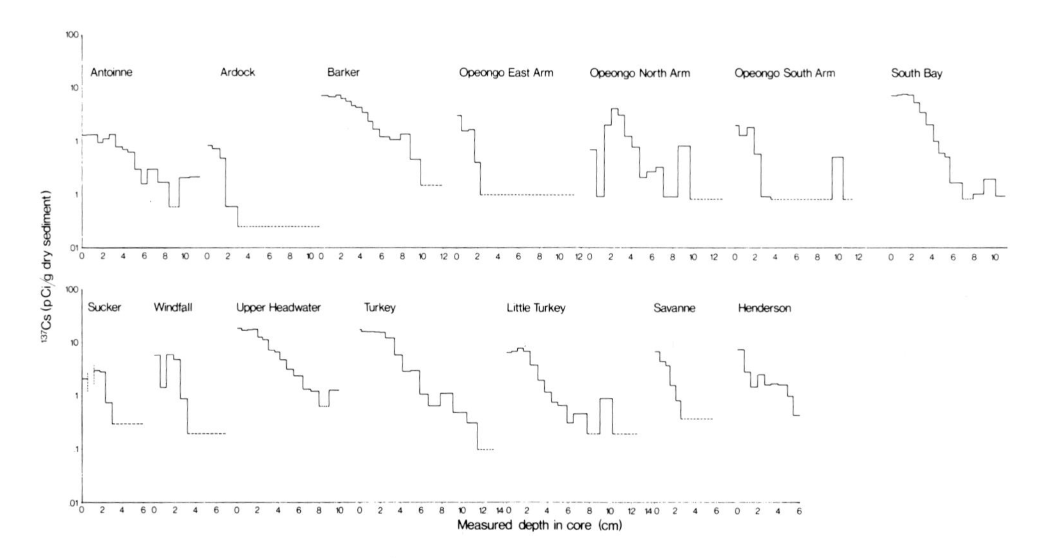

Figure 4 ^{137}Cs profiles in sediment cores.

northern Ontario lakes (Ardoch, Opeongo, Sucker, Windfall, Savanne, and Henderson) display ^{210}Pb fluxes lower than expected from the data of Moore and Poet. It is unlikely that the atmospheric flux of ^{210}Pb would be depressed so much in the vicinity of these lakes. A more probable reason for the lower flux values at the sediment/water interface is complexation of ^{210}Pb by amino acids in the lake water to give soluble lead compounds (Rickard and Nriagu, 1978). The lakes involved in this study are all located in forested areas with the drainage basins providing sources of organic acids from the decay of vegetation.

ACKNOWLEDGEMENTS

We would like to thank Dr. M.G. Johnson and Dr. R.L. Thomas for providing the sediment cores and raison d'être for this study and R. Dawson, L. Rolston, J.A. FitzGerald and S. Livermore for carrying out the analytical work.

REFERENCES CITED

Allan, R.J., Williams, J.D.H., Joshi, S.R. and Warwick, W.F., 1980, Historical changes and relationship to internal loading of sediment phosphorus forms in hypertrophic Prairie lakes: Jour. Environ. Qual., v. 9, p. 199-206.

Athy, L.F., 1930, Density, porosity, and compaction of sedimentary rocks: Bull. Amer. Assoc. Pet. Geol., v. 14, p. 1-24.

Brownlee, B., Fox, M.E., Strachan, W.M.J. and Joshi, S.R., 1977, Distribution of dehydroabietic acid in sediments adjacent to a kraft pulp and paper mill: Jour. Fish. Res. Board Can., v. 34, p. 838-843.

Christensen, E.R. and Chien, N.K., 1981, Fluxes of arsenic, lead, zinc, and cadmium to Green Bay and Lake Michigan sediments: Environ. Sci. Technol., v. 15, p. 553-558.

Durham, R.W. and Joshi, S.R., 1980a, Recent sedimentation rates, ^{210}Pb fluxes, and particle settling velocities in Lake Huron, Laurentian Great Lakes: Chem. Geol., v. 31, p. 53-66.

__________, 1980b, The ^{210}Pb and ^{137}Cs profiles in sediment cores from Lakes Matagami and Quevillon, northwest Quebec, Canada: Can. Jour. Earth Sci., v. 17, p. 1746-1750.

__________, 1981, Sedimentation rates in western Nipigon Bay, Lake Superior using the ^{210}Pb method: Sci. Total Environ. (Submitted).

Durham, R.W., Joshi, S.R. and Allan, R.J., 1980, Radioactive dating of sediment cores from four contiguous lakes in Saskatchewan, Canada: Sci. Total Environ., v. 15, p. 65-71.

Farmer, J.G., 1978, The determination of sedimentation rates in Lake Ontario using the ^{210}Pb dating method: Can. Jour. Earth Sci., v. 15, p. 431-437.

Goldberg, E.D., Hodge, V., Koide, M., Griffin, J., Gamble, E., Bricker, O.P., Matisoff, G., Holden, G.R. and Braun, R., 1978, A pollution history of Chesapeake Bay: Geochim. Cosmochim. Acta, v. 42, p. 1413-1425.

Joshi, S.R. and Durham, R.W., 1976, Determination of ^{210}Pb, ^{226}Ra, and ^{137}Cs in sediments: Chem. Geol., v. 18, p. 155-160.

Kemp, A.L.W., Williams, J.D.H., Thomas, R.L. and Gregory, M.L., 1978, Impact of man's activities on the chemical composition of the sediments of Lakes Superior and Huron: Water, Air and Soil Pollution, v. 10, p. 381-402.

Koide, M., Bruland, K.W. and Goldberg, E.D., 1973, Th-228/Th-232 and Pb-210 geochronologies in marine and lake sediments: Geochim. Cosmochim. Acta, v. 37, p. 1171-1187.

Krishnaswamy, S., Lal, D., Martin, J.M. and Maybeck, M., 1971, Geochronology of lake sediments: Earth Planet Sci. Lett., v. 11, p. 407-414.

Moore, H.E. and Poet, S.E., 1976, ^{210}Pb fluxes determined from ^{210}Pb and ^{226}Ra soil profiles: Jour. Geophys. Res., v. 81, p. 1056-1058.

Nriagu, J.O., Kemp, A.L.W., Wong, H.K.T. and Harper, N., 1979, Sedimentary record of heavy metal pollution in Lake Erie: Geochim. Cosmochim. Acta, v. 43, p. 247-258.

Oldfield, F., Appleby, P.G. and Battarbee, R.W., 1978, Alternative ^{210}Pb dating: results from the New Guinea Highlands and Lough Erne: Nature (London), v. 271, p. 339-342.

Rickard, D.T. and Nriagu, J.O., 1978, Aqueous environmental chemistry of lead, in Nriagu, J.O., ed., The Biogeochemistry of Lead in the Environment, Part A: Elsevier/Holland Biomedical Press.

Robbins, J.A. and Edgington, D.N., 1975, Determination of recent sedimentation rates in Lake Michigan using Pb-210 and Cs-137; Geochim. Cosmochim. Acta, v. 39, p. 285-304.

FISSION-TRACK DATING

NANCY D. NAESER and CHARLES W. NAESER

ABSTRACT

Fission tracks are zones of intense damage that result when fission fragments pass through a solid. Several naturally occurring isotopes undergo spontaneous fission, but only ^{238}U produces a significant number of tracks over geologic time. Spontaneous fission occurs at a known rate, and so by determining the number of tracks and the amount of uranium in a mineral or glass, it is possible to calculate its age.

Many materials contain uranium, but because of factors such as uranium abundance, track retention, and relative abundance, only glass and zircon are routinely dated in Quaternary rocks.

Advantages of fission track dating for Quaternary materials include low contamination and typically small samples needed. Disadvantages include few tracks in samples less than 100,000 years old, and annealing of tracks in glass at ambient surface temperatures.

Quaternary materials dated by fission tracks include volcanic ash, archaeological materials, tektites and impact glass, and natural clinker.

THEORY AND METHODS

A fission track is the zone of intense damage formed when a fission fragment passes through a solid. Several naturally occurring isotopes undergo spontaneous fission, but only ^{238}U has a fission half-life (9.9×10^{15} years) that is sufficiently short to produce a significant number of tracks over geologic time. Yet even the spontaneous fission of ^{238}U is a rare event. More than a million ^{238}U atoms decay by alpha emission for each fission decay. When an atom such as ^{238}U fissions, the nucleus breaks up into two lighter nuclei, one about 90 atomic mass units and the other about 135 a.m.u., with the liberation of about 200 MeV of energy. The two highly charged nuclei recoil in opposite directions and disrupt the electron balance of the atoms in the mineral lattice or glass along their path. This disruption causes the positively charged ions in the lattice to repulse each other and force themselves into the crystal structure, forming the track or damage zone (Fleischer *et al.*, 1975). The new track is only a few angstroms wide and is about 10-20 μm in length. The track is longer in low density minerals and glasses than in dense minerals such as zircon.

A track in its natural state can only be observed with an electron microscope, but a chemical etchant can enlarge the damage zone so that it can be observed in an optical microscope at intermediate magnifications (x200-500) (Figure 1). Common etchants used include: nitric acid (for apatite), hydrofluoric acid (for glass and micas), concentrated basic solutions (for sphene, and basic fluxes (for zircon) (Fleischer *et al.*, 1975, Table 2-2; Gleadow *et al.*, 1976).

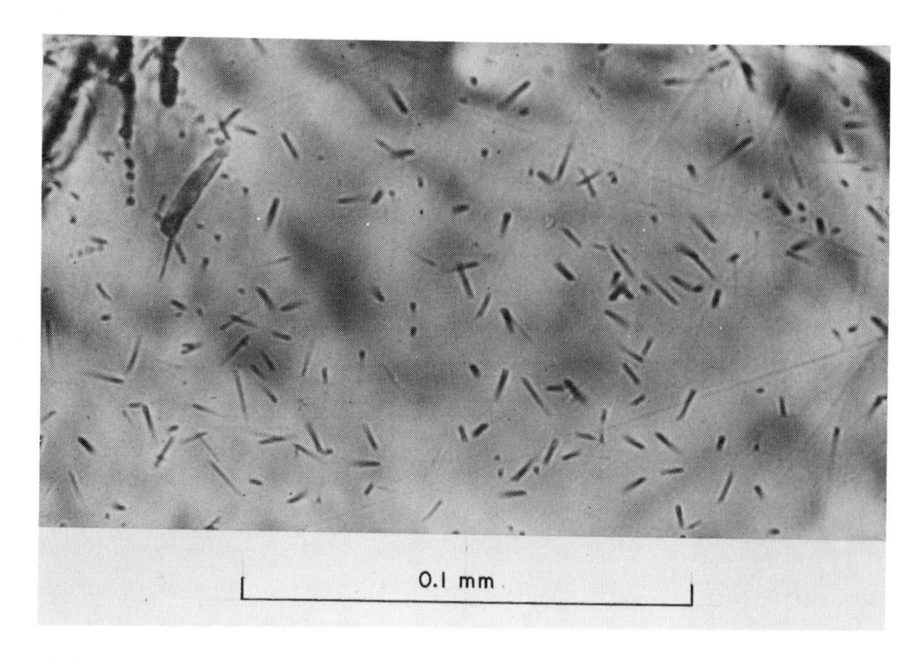

Figure 1 Fission tracks in apatite from the radial dike at Shiprock, New Mexico.

Trace amounts of ^{238}U occur in a number of common minerals and glasses. Because ^{238}U fissions spontaneously at a constant rate, fission tracks can be used to date these materials. The techniques used for dating have been developed by physicists and geologists over the last 20 years. Early development of the method has been reviewed by Fleischer *et al.* (1975) and Naeser (1979). The age of a mineral or glass can be calculated from the amount of uranium and number of spontaneous tracks that it contains. Spontaneous track density is usually determined by: (1) polishing the surface of the specimen, (2) enlarging the fission tracks intersecting this surface by etching, and (3) counting the number of tracks per unit area under an optical microscope, generally at magnifications of x500 to x2500. The relative abundance of ^{238}U and ^{235}U is constant in rocks, and thus the easiest and most accurate way to determine the amount of uranium is to create a new set of fission tracks by irradiating the sample in a nuclear reactor with thermal neutrons, which induce fission in ^{235}U. The resulting induced track density is a function of the amount of uranium in the sample and the neutron dose it received in the reactor. The dose may be determined by including with the samples a standard of known uranium content (Carpenter and Reimer, 1974; Fleischer *et al.*, 1975, Table 4-1).

A fission-track age is calculated using the spontaneous track density from ^{238}U (ρ_s), the neutron-induced track density from ^{235}U (ρ_i), and the thermal neutron fluence (ϕ, in neutrons cm^{-2}), (Price and Walker, 1963; Naeser, 1967):

$$A = \frac{1}{\lambda_{\hat{d}}} \ln \left[1 + \frac{\rho_{\hat{s}} \lambda_{\hat{d}} \sigma I \phi}{\rho_{\hat{i}} \lambda_{\hat{f}}}\right]$$

$\lambda_{\hat{d}}$ = total decay constant for ^{238}U ($1.551 \times 10^{-10} yr^{-1}$)

$\lambda_{\hat{f}}$ = decay constant for spontaneous fission of ^{238}U

($6.85 \times 10^{-17} yr^{-1}$; Fleischer and Price, 1964a)

($7.03 \times 10^{-17} yr^{-1}$; Roberts *et al.*, 1968)[1]

($8.42 \times 10^{-17} yr^{-1}$; Spadavecchia and Hahn, 1967)

[1] Value preferred by authors.

σ = cross-section for thermal neutron-induced fission of ^{235}U ($580 \times 10^{-24} cm^2$)

I = isotopic ratio $^{235}U/^{238}U$ (7.252×10^{-3})-- and

A = age in years

Several factors determine if a sample can be dated by the fission-track method. Firstly, the sample must contain a mineral or glass of appropriate uranium content. In Quaternary samples there must be enough uranium that a statistically significant number of tracks can be counted in a reasonable time. Secondly, tracks must be retained once they are formed, or the apparent age will be anomalously young. Several environmental factors can cause the loss or "annealing" of spontaneous tracks once they are formed (Fleischer *et al.*, 1975; Harrison *et al.*, 1979), but by far the most common cause is heating, which can cause partial to complete fading of spontaneous fission tracks.

Data on the temperatures required for annealing have been determined by (1) extrapolating laboratory heating experiments to geologic time (Naeser and Faul, 1969) and (2) measuring age-decrease with increasing depth and temperature in deep drillholes from areas where the rocks have undergone heating of known duration (Naeser, 1981). Such studies have shown that the annealing temperature depends on the mineral involved--different minerals anneal at different temperatures--and the duration of heating; the longer a mineral is heated, the lower the temperature that is required to anneal its tracks. The track is stable in most non-opaque minerals at temperatures of 80 C or less but fission tracks in natural glasses are affected at much lower temperatures (Seward, 1979; Naeser *et al.*, 1980b).

By dating a mineral it is often possible to determine if it has ever been heated above its critical temperature and when it last cooled below this temperature. Although annealing can cause problems in determining the primary age of samples, it is a powerful method for studies of their thermal history. These studies have been directed mainly at older rocks, for determining uplift rates and thermal history of sedimentary basins and mineralization (*e.g.*, Wagner *et al.*, 1977; Naeser, 1979; Bryant and Naeser, 1980; Naeser *et al.*, 1980a; Briggs *et al.*, 1981), but several Quaternary studies have also made use of annealing.

Etching studies have shown that tracks can be revealed in more than 150 minerals and glasses (Fleischer *et al.*, 1975), but the combination of such factors as typical uranium content, annealing characteristics, and relative abundance results in very few minerals being used for dating. Zircon and glass are the only materials that are dated routinely in Quaternary rocks.

Zircon and glass require different methods because glass from a single source tends to have uniform uranium content whereas zircon crystals from the same source tend to have very inhomogeneous uranium distributions.

Zircon requires the use of the external detector method (Naeser, 1979) (Figure 2). Because uranium can be distributed inhomogeneously both within and between zircon crystals, it is necessary to count the induced tracks produced from the same areas of a crystal in which fossil tracks are counted. In the external detector method the fossil tracks are counted of the crystal and the induced tracks are counted in a detector that covered the crystal mount during neutron irradiation. Either a low-uranium-content (<10 ppb) muscovite or a plastic detector can be used. Six to twelve zircons are counted for most samples.

Glass can be dated by the population method (Naeser, 1979) (Figure 3). Because all of the glass from a single source has a similar uranium concentration, it is possible to determine the fossil and induced track densities from different splits of the sample. One split is mounted in epoxy, polished, and etched for the fossil track density

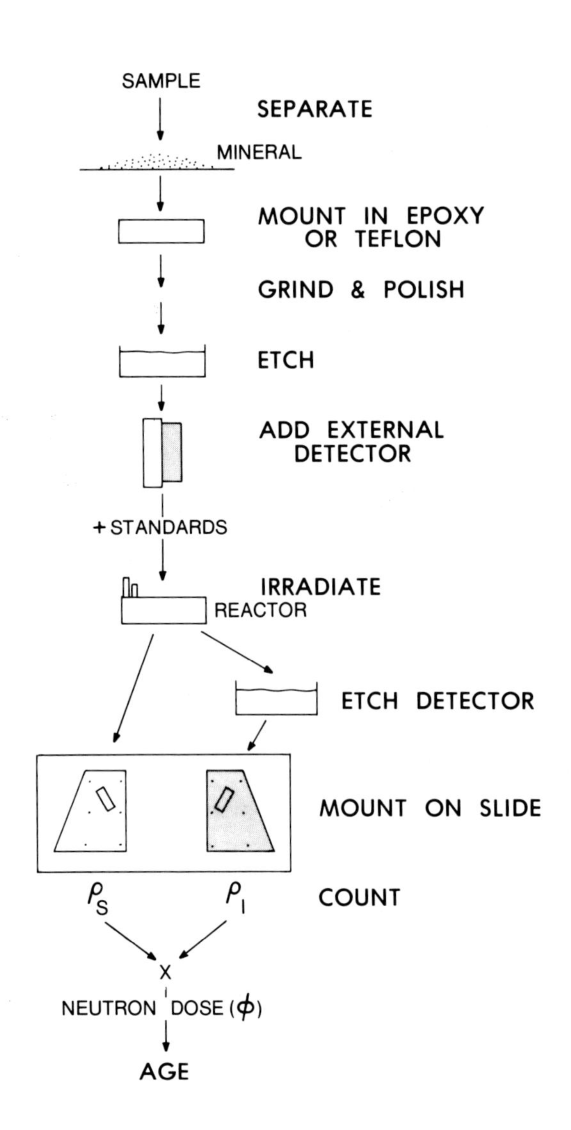

Figure 2 Steps involved in obtaining a fission-track age using the external detector method.

determination. The second split is irradiated, and then mounted, polished, and etched (it is standard practice to etch both groups at the same time). The irradiated split contains both fossil and induced tracks; preirradiation annealing to remove fossil tracks is not recommended because it can alter the etching characteristics and chemistry of glass. The fossil track density (ρ_s) is subtracted from the total track density in the irradiated sample ($\rho_s + \rho_i$) to arrive at the induced track density (ρ_i). The amount of glass that must be counted depends on several factors, including its uranium content, age, and vesicularity.

Naeser (1976) details the laboratory methods.

ADVANTAGES AND LIMITATIONS

One advantage of fission-track dating is that contamination is minimized. In conventional radiocarbon and K-Ar dating, bulk samples must be analyzed. Contamination of a ^{14}C sample with recent carbon results in a younger age, and a few older detrital grains in a K-Ar sample can have a significant effect on a K-Ar age (Naeser *et al.*, 1981). Fission-track dating is a grain-discrete method in which individual grains are scanned and counted. In the course of dating a sample using zircons, an age is obtained on each grain that is counted. Therefore older grains show up as contamination; a grain with a Miocene age in a Pleistocene sample is obvious because of its older age. In addition, primary zircon grains in Pleistocene tephras can generally be identified

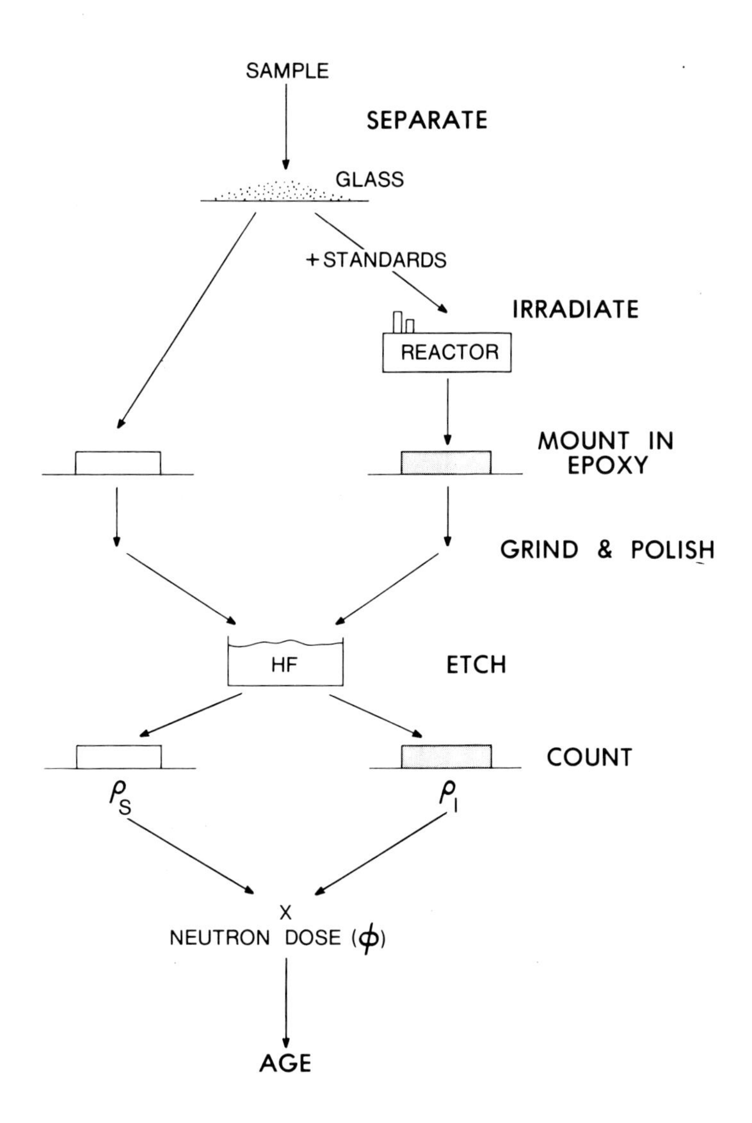

Figure 3 Steps involved in obtaining a fission-track age using the population method.

by the glass adhering to them (Figure 4). Contamination is usually not a problem with glass shards, although N.D. Naeser has recognized older glass in a tephra sample from Saskatchewan. It comprised less than 0.1% of the glass but was obvious because of its higher track density; this glass is shown in Westgate and Gorton (1981, Figure 5).

A major problem of fission-track dating in the Quaternary is that very young samples (<100,000 years) contain very few tracks. This leads to long counting times and to ages with large analytical uncertainties. Herd and Naeser (1974) determined a zircon to be about 100,000 years old, with a 40% standard deviation; in 45 zircons a total of 16 tracks were observed. Briggs and Westgate (1978) reported one glass sample in which they did not see any spontaneous tracks in thousands of shards. Thus for young samples the analytical uncertainty is large, but even then the result might answer a geological question.

Zircons, although preferable to glass, are not present in all samples. In tephras, their presence depends upon the chemistry of the parent magma and the distance downwind from the eruption vent. Experience has shown that acidic tephras have more usable zircons than basic tephras. Zircons that are extremely fine grained (<75μm) are too small to be dated by fission track counting. This is often the case in tephra sampled a long distance from its vent.

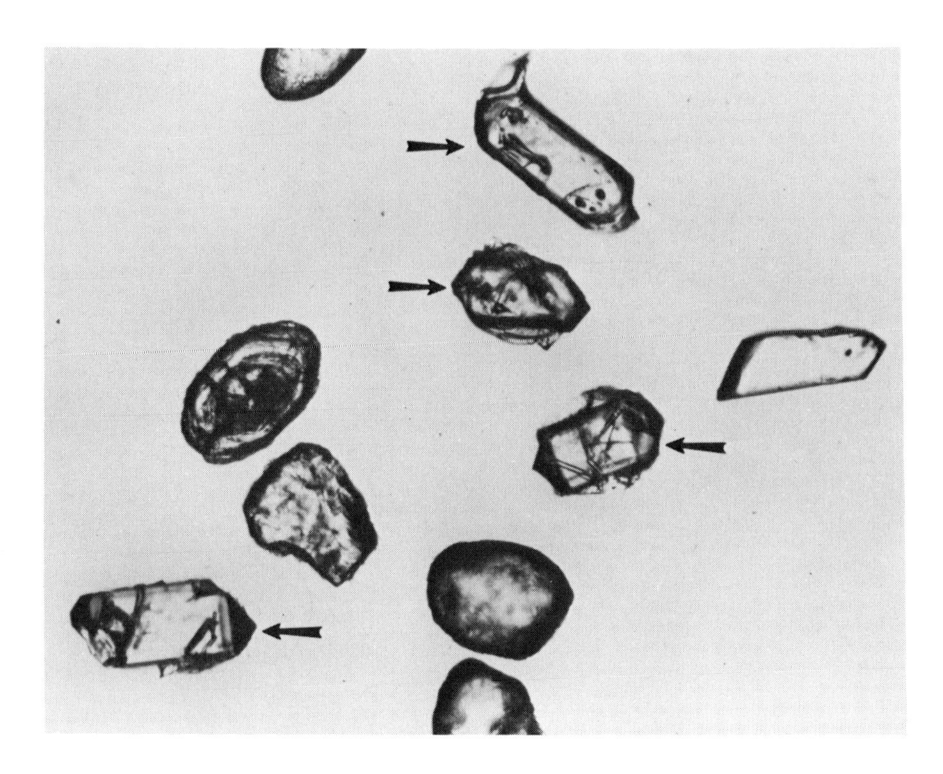

Figure 4 Zircons separated from a contaminated tephra deposit. The primary zircons (arrowed) are glass mantled; the grains without glass are detrital.

Natural glasses have been dated extensively because of their abundance in Quaternary tephra and their archaeological and other applications. However, glasses present special problems, the greatest of which is the ease with which they lose spontaneous tracks by annealing (Fleischer *et al.*, 1965a; Storzer and Wagner, 1969; MacDougall, 1976; Seward, 1979). Hydrated glass, which is found in the typical tephra deposit, is particularly susceptible to annealing (Lakatos and Miller, 1972), but recent work by Naeser *et al.* (1980b) has shown that both hydrated and nonhydrated glass can lose spontaneous tracks at ambient surface temperatures over geologic time. In a study of 14 tephras from upper Cenozoic (<30 m.y.) deposits of the western United States, only one glass had a fission-track age concordant with the fission-track age of coexisting zircon (Figure 5). All other samples had ages that were significantly younger than the zircon ages. Seward (1979) showed that about 60% of the glass fission-track ages of Quaternary tephras in her study in New Zealand were significantly younger than the fission-track ages of the coexisting zircons.

Two procedures are available for treating glass to check for partial annealing and to correct the resulting lowered fission-track ages. These are the track diameter measurement method (Storzer and Wagner, 1969) and the plateau annealing method (Storzer and Poupeau, 1973). The plateau annealing method serves better for Quaternary samples because it is potentially more precise for young glasses of low track density (Naeser *et al.*, 1980b). In this method, different splits of the irradiated and nonirradiated glass are heated together in a furnace for one hour intervals at progressively higher temperatures, and an age is determined after each heating step. If the glass has previously been partially annealed, the age will increase through the lower heating steps until it reaches a plateau, which is usually the primary age. If no annealing has occurred, progressive heating does not affect the age; it remains the same as the age of the untreated glass, although the induced and spontaneous track densities decrease. Figure 6 shows the results of plateau annealing for twelve upper Tertiary glasses, including both obsidians and tephra shards.

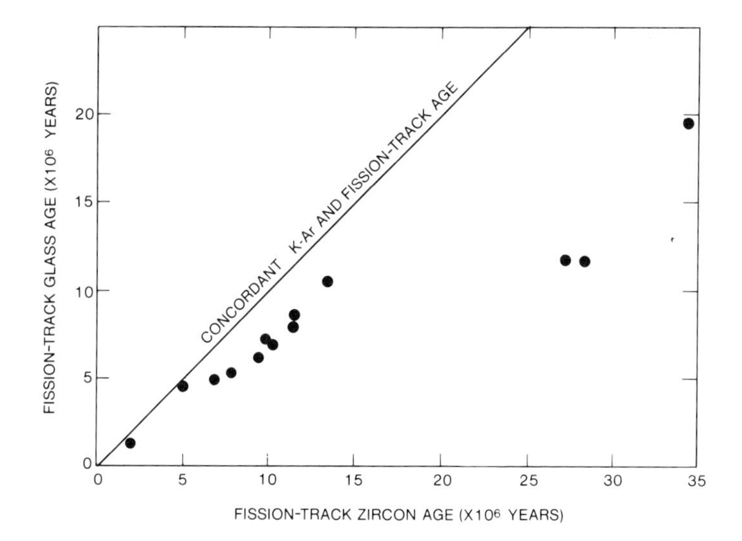

Figure 5 Plot showing the fission-track ages of coexisting glass shards and zircons from upper Cenozoic tephra deposits of the western United States (from Naeser *et al.*, 1980b).

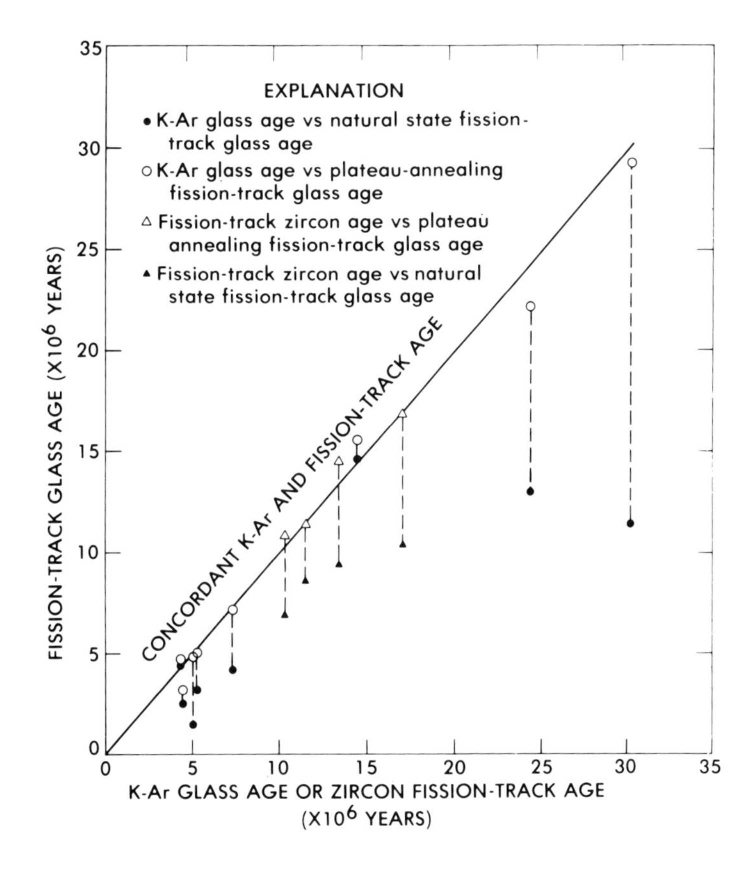

Figure 6 Diagram showing apparent and plateau corrected ages of glasses (obsidians and tephra shards) compared to zircon fission-track or K-Ar ages (from Naeser *et al.*, 1980b).

Both the diameter measurement and plateau annealing methods are difficult to apply to Quaternary glass because the small number of tracks present generally prevents statistically significant measurements. Also, even a corrected age will still be too young if any of the tracks in the glass were totally annealed (MacDougall, 1976; Naeser *et al.*, 1980b).

Because of the ease with which glasses anneal and the uncertainty of correcting their ages, great care must be used in the interpretation of all glass fission-track ages. They should always be considered minimum ages.

Another limitation for dating glass shards relates to their grain size and vesicularity. Large bubble-junction, platy shards are by far the easiest to date. In very fine-grained or pumiceous shards (Westgate and Briggs, 1980, Figure 4), it is difficult to count tracks, and to determine glass area by the conventional method of a grid in the microscope eyepiece. The problem of determining area can be overcome for many samples by use of a point-counting technique analagous to that used in petrographic modal analysis (Seward, 1974; Briggs and Westgate, 1978; Naeser *et al.*, 1982). Even so, dating such glass is very time consuming.

Glass containing abundant microlites may be difficult or impossible to date because of the close resemblance of the etched microlites to fission tracks. This problem is rarely encountered in glass shards, but is common in dating obsidians.

Despite these problems, glass has been used to date Quaternary samples because in some cases, it is the only datable phase present and, in some situations, even minimum dates provide useful information.

APPLICATIONS

Tephrochronology

The major contribution of fission-track dating to Quaternary chronology has been in the field of tephrochronology. Fission tracks have provided the most suitable method for dating tephras, particularly those older than the 40,000 to 50,000-year BP limit of radiocarbon dating. An advantage of the fission-track method, over K-Ar and radiocarbon dating, is that the problem of contamination is greatly minimized (see discussion above). This is particularly important for dating tephras, which commonly contain detrital contaminants. Also, glass shards cannot be dated reliably by the K-Ar method because excess radiogenic argon is often present, producing ages that are too old, and potassium gain and/or argon loss during or after hydration can produce ages that are too young. Therefore, it is not possible to know if K-Ar ages of glass shards are too old, too young, or, possibly fortuitously, correct (Naeser *et al.*, 1981).

During the last decade there have been a number of examples in the literature of fission-track dating of zircon and/or glass in Quaternary tephras (Naeser *et al.*, 1973; Seward, 1974, 1975, 1979; Izett and Naeser, 1976; Aronson *et al.*, 1977; Briggs and Westgate, 1978, Westgate *et al.*, 1978; Gleadow, 1980). Studies on the Pearlette family ash beds and the Bishop ash bed illustrates the use of fission-track dating in tephrochronology.

Volcanic ash beds of the Pearlette family occur in Pleistocene deposits of western North America (Izett *et al.*, 1970, 1972). Before 1970, the Pearlette ash was considered to be a single ash bed, and it was used as a time marker for many midcontinent Quaternary deposits. Then Izett *et al.* (1970, 1972) recognized minor chemical differences among some of the Pearlette ash localities, and found that the ashes could be correlated geochemically with deposits of the three major ash-flow eruptions that had occurred in the region of Yellowstone National Park, Wyoming. These eruptions have been dated at 2.02 ± 0.08 m.y. (Huckleberry Ridge Tuff), 1.27 ± 0.1 m.y. (Mesa Falls Tuff), and 0.616 ± 0.008 m.y. (Lava Creek Tuff) using the K-Ar method (J.D. Obradovich, written communication, 1973). Naeser *et al.* (1973) dated zircons from two of the three types of tephra and obtained ages of 1.9 ± 0.1 m.y.

for ash[1] correlated with the Huckleberry Ridge Tuff and 0.6 ± 0.1 m.y. for ash[2] correlated with the Lava Creek Tuff. These ages matched the ages in the source region and confirmed the geochemical evidence of Izett *et al.* (1970, 1972) that there were three Pearlette ashes rather than just one. Dates obtained for the Huckleberry Ridge ash bed give an excellent example of the problems caused by track annealing in glass (Table 1).

Table 1. Fission-track ages of zircon and glass from the Huckleberry Ridge ash bed

Investigator	Glass[a]	Zircon[a]
Naeser *et al.* (1973, 1980b)	1.3 ± 0.17 m.y.	1.9 ± 0.1 m.y.
Seward (1979)	1.39 ± 0.08 m.y.	1.93 ± 0.16 m.y.
N.D. Naeser (unpub. data)	1.21 ± 0.07 m.y.	1.9 ± 0.25 m.y.

[a]Error shown is ± 1 sigma.

The Bishop ash bed, like the Pearlette family of ashes, is a widespread airfall tephra in the western United States (Izett *et al.*, 1970; Izett and Naeser, 1976). Izett *et al.* (1970) have identified it as the airfall equivalent of the Bishop Tuff, which originated in the Long Valley Caldera in California about 700,000 years ago. Dalrymple *et al.* (1965) reported an average K-Ar age of 0.708 ± 0.015 m.y. (± at the 95% confidence interval) on minerals from the Bishop Tuff in its source area. Both glass and zircon from the Bishop ash have been dated by the fission-track method (Table 2). These ages also show the problem caused by track annealing in glass.

Table 2. Fission-track ages of zircon and glass from the Bishop ash bed

Investigator	Glass[a]	Zircon[a]
Izett and Naeser (1976)	--	0.74 ± 0.03 m.y.
N.D. Naeser (unpub. data)	0.56 ± 0.05 m.y.	--

[a]Error shown is ± 1 sigma.

In several cases fission-track dating has demonstrated that tephras are considerably older than inferred by radiocarbon dating of associated material. For example, the Salmon Springs Drift at its type locality, at Salmon Springs, Washington, consists of two drift sheets separated by about 1.5 m of peat, silt, and volcanic ash (Lake Tapps tephra) (Crandell *et al.*, 1958; Easterbrook *et al.*, 1981). The peat grades downward with decreasing organic content into about one meter of silt, which in turn grades into the volcanic ash (D.J. Easterbrook, written commun., 1980). The peat has been radiocarbon dated at $71,500 \pm^{1700}_{1400}$ years BP by the enrichment method (Stuiver *et al.*, 1978), and the drift sheets were thus considered early Wisconsin in age. However, fission-track data on the ash, and on correlative ash at Auburn, Washington, along with paleomagnetic and tephrochronological evidence, show that the Salmon Springs Drift is much older (Easterbrook *et al.*, 1981) (Table 3).

[1]Huckleberry Ridge ash bed of Izett and Wilcox (1982); = Pearlette type B ash of Izett *et al.* (1973).

[2]Lava Creek B ash bed of Izett and Wilcox (1982); = Pearlette type O ash of Izett *et al.* (1970) and Naeser *et al.* (1973).

Table 3. Comparison of fission-track age of the Lake Tapps tephra, Washington, and radiocarbon age of associated peat

Locality	Dating Method	Material Dated	Age[a] (years BP $\pm 1\sigma$)
Salmon Springs	C-14	peat	$71,500 \pm^{1700}_{1400}$
	F-T	zircon	840,000 ± 210,000
Auburn	F-T	zircon	870,000 ± 270,000
	F-T	glass	660,000 ± 40,000

[a]Radiocarbon age from Stuiver *et al.* (1978); fission-track ages from Easterbrook *et al.* (1981).

Archaeology

Fission tracks are potentially useful for dating a variety of archaeological materials (see review by Wagner, 1978). They have been used (1) to date stratigraphic layers, such as tephra beds associated with archaeological remains (Fleischer *et al.*, 1965c; Aronson *et al.*, 1977; Gleadow, 1980; Steen-McIntyre *et al.*, 1981); (2) to date the material used to manufacture artifacts; and (3) to date heating events in which glasses or minerals were heated to a high enough temperature to totally anneal their spontaneous tracks.

Bigazzi and Bonadonna (1973), Suzuki (1973, 1974), Durrani *et al.* (1971), and others have used fission tracks to determine the age, and in some cases the uranium content, of the obsidian used to make artifacts in order to trace the obsidian back to its geologic source and thus map ancient trade routes. Fleischer *et al.* (1965b) dated a mesolithic knife whose wilted shape indicated that it had been heated to a high temperature at some time after its manufacture; fission tracks dated the heating event at 3700 ± 900 years BP. Watanabe and Suzuki (1969) determined the time of heating of various artifacts from Japan, including a glass glaze on a bowl fragment which they dated at 520 ± 110 years BP. Miller and Wagner (1981) determined both the geological age of obsidian and the time of its manufacture into artifacts for several samples from South America. The youngest archaeological material to be dated thus far is manmade "uranium glass" manufactured within the last 140 years (Wagner, 1976, 1978).

Most archaeological samples are very young and composed of material that is relatively low in uranium, so that track densities are generally extremely low. This means that counting a statistically significant number of spontaneous tracks is a tedious, time-consuming process that often involves repeated steps of polishing, etching, and counting. Some examples of this are discussed by Fleischer *et al.* (1975).

Other

Coates and Naeser (1981) have demonstrated the use of fission-track dating of zircons from sandstones (clinker) heated by natural burning of coal to study the rate of Quaternary landform development in the eastern Powder River Basin of Wyoming.

Fission-track dating and K-Ar dating have helped establish the age and extent of major tektite strewnfields, including the 1-m.y.-old Ivory Coast and 0.7-m.y.-old Australasian fields (see for example Fleischer and Price, 1964b; Storzer and Wagner, 1969; Gentner *et al.*, 1967, 1969). Fission tracks have been particularly useful in dating microtektites from deep sea cores where there is too little material for K-Ar dating (Gentner *et al.*, 1970). One of the youngest natural materials to be dated by the fission-track method is an impact glass from the Kofels structure in Austria, which yielded an age of 8.0 ± 6.0 x 10^3 years (Storzer *et al.*, 1971).

REFERENCES CITED

Aronson, J.L., Schmitt, T.J., Walter, R.C., Taieb, M., Tiercelin, J.J., Johanson, D.C., Naeser, C.W. and Nairn, A.E.M., 1977, New geochronologic and paleomagnetic data for the hominid-bearing Hadar Formation of Ethiopia: Nature, v. 267, p. 323-327.

Bigazzi, G. and Bonadonna, F., 1973, Fission track dating of the obsidian of Lipari Island (Italy): Nature, v. 242, p. 322-323.

Briggs, N.D. and Westgate, J.A., 1978, A contribution to the Pleistocene geochronology of Alaska and the Yukon Territory: fission-track age of distal tephra units, in Zartman, R.E., ed., Short papers of the 4th International Conference on Geochronology, Cosmochronology, and Isotope Geology: U.S. Geological Survey Open File Report 78-701, p. 49-52.

Briggs, N.D., Naeser, C.W. and McCulloh, T.H., 1981, Thermal history of sedimentary basins by fission-track dating: Nuclear Tracks, v. 5, p. 235-237.

Bryant, B. and Naeser, C.W., 1980, The significance of fission-track ages of apatite in relation to the tectonic history of the Front and Sawatch Ranges, Colorado: Geol. Soc. Amer. Bull., v. 91, p. 156-164.

Carpenter, B.S. and Reimer, G.M., 1974, Standard reference materials: calibrated glass standards for fission track use: National Bureau of Standards Special Publication 260-49, 16 p.

Coates, D.A. and Naeser, C.W., 1981, Fission-track ages and landscape development, eastern Powder River Basin, Wyoming: Geol. Soc. Amer., Abstracts with Programs, v. 13, p. 428.

Crandell, D.R., Mullineaux, D.R. and Waldron, H.H., 1958, Pleistocene sequence in the southeastern part of the Puget Sound Lowland, Washington: Amer. Jour. Science, v. 256, p. 384-398.

Dalrymple, G.B., Cox, A. and Doell, R.R., 1965, Potassium-argon age and paleomagnetism of the Bishop Tuff, California: Geol. Soc. Amer. Bull., v. 76, p. 665-674.

Durrani, S.A., Khan, H.A., Taj, M. and Renfrew, C., 1971, Obsidian source identification by fission track analysis: Nature, v. 233, p. 242-245.

Easterbrook, D.J., Briggs, N.D., Westgate, J.A. and Gorton, M.P., 1981, Age of the Salmon Springs Glaciation in Washington: Geology, v. 9, p. 87-93.

Fleischer, R.L. and Price, P.B., 1964a, Decay constant for spontaneous fission of ^{238}U: Physical Review, v. 133, no. 1B, p. 63-64.

__________, 1964b, Fission track evidence for the simultaneous origin of tektites and other natural glasses: Geochimica et Cosmochimica Acta, v. 28, p. 755-760.

Fleischer, R.L., Price, P.B. and Walker, R.M., 1965a, Effects of temperature, pressure, and ionization on the formation and stability of fission tracks in minerals and glasses: Jour. Geophysical Research, v. 70, p. 1497-1502.

Fleischer, R.L., Price, P.B., Walker, R.M. and Leakey, L.S.B., 1965b, Fission track dating of a mesolithic knife: Nature, v. 205, p. 1138.

__________, 1965c, Fission-track dating of Bed I, Olduvai Gorge: Science, v. 148, p. 72-74.

Fleischer, R.L., Price, P.B. and Walker, R.M., 1975, Nuclear Tracks in Solids: Principles and applications: Berkeley, California, Univ. California Press, 605 p.

Gentner, W., Kleinmann, B. and Wagner, G.A., 1967, New K-Ar and fission track ages of impact glasses and tektites: Earth Planetary Science Letters, v. 2, p. 83-86.

Gentner, W., Storzer, D. and Wagner, G.A., 1969, New fission track ages of tektites and related glasses: Geochimica et Cosmochimica Acta, v. 33, p. 1075-1081.

Gentner, W., Glass, B.P., Storzer, D. and Wagner, G.A., 1970, Fission track ages and ages of deposition of deep-sea microtektites: Science, v. 168, p. 359-361.

Gleadow, A.J.W., Hurford, A.J. and Quaife, R.D., 1976, Fission track dating of zircon: Improved etching techniques: Earth Planetary Science Letters, v. 33, p. 273-276.

Gleadow, A.J.W., 1980, Fission track age of the KBS Tuff and associated hominid remains in northern Kenya: Nature, v. 284, p. 225-230.

Harrison, T.M., Armstrong, R.L., Naeser, C.W. and Harakal, J.E., 1979, Geochronology and thermal history of the Coast Plutonic Complex, near Prince Rupert, British Columbia: Canadian Jour. Earth Sciences, v. 16, p. 400-410.

Herd, D.G. and Naeser, C.W., 1974, Radiometric evidence for pre-Wisconsin glaciation in the northern Andes: Geology, v. 2, p. 603-604.

Izett, G.A. and Naeser, C.W., 1976, Age of the Bishop Tuff of eastern California as determined by the fission-track method: Geology, v. 4, p. 587-590.

Izett, G.A. and Wilcox, R.E., 1982, Map showing localities and inferred distributions of the Huckleberry Ridge, Mesa Falls, and Lava Creek ash beds (Pearlette family ash beds) of Pliocene and Pleistocene age in the western United States and southern Canada: U.S. Geol. Survey Misc. Investigations Map MI-1325, scale 1:4,000,000 (In Press).

Izett, G.A., Wilcox, R.E., Powers, H.A. and Desborough, G.A., 1970, The Bishop ash bed, a Pleistocene marker bed in the western United States: Quaternary Research, v. 1, p. 121-132.

Izett, G.A., Wilcox, R.E. and Borchardt, G.A., 1972, Correlation of a volcanic ash bed in Pleistocene deposits near Mount Blanco, Texas, with the Guaje pumice bed of the Jemez Mountains, New Mexico: Quaternary Research, v. 2, p. 554-578.

Lakatos, S. and Miller, D.S., 1972, Evidence for the effect of water content on fission-track annealing in volcanic glass: Earth Planetary Science Letters, v. 14, p. 128-130.

MacDougall, J.D., 1976, Fission track annealing and correction procedures for oceanic basalt glasses: Earth Planetary Science Letters, v. 30, p. 19-26.

Miller, D.S. and Wagner, G.A., 1981, Fission-track ages applied to obsidian artifacts from South America using the plateau-annealing and the track-size age-correction techniques: Nuclear Tracks, v. 5, p. 147-155.

Naeser, C.W., 1967, The use of apatite and sphene for fission track age determinations: Geol. Soc. Amer. Bull., v. 78, p. 1523-1526.

__________, 1976, Fission track dating: U.S. Geol. Survey Open-File

Report 76-190.

__________, 1979, Fission-track dating and geologic annealing of fission tracks, in Jager, E. and Hunziker, J.C., eds., Lectures in Isotope Geology: Berlin, Springer-Verlag, p. 154-169.

__________, 1981, The fading of fission tracks in the geologic environment-data from deep drill holes: Nuclear Tracks, v. 5, p. 248-250.

Naeser, C.W. and Faul, H., 1969, Fission track annealing in apatite and sphene: Jour. Geophysical Research, v. 74, p. 705-710.

Naeser, C.W., Izett, G.A. and Wilcox, R.E., 1973, Zircon fission-track ages of Pearlette family ash beds in Meade County, Kansas: Geology, v. 1, p. 187-189.

Naeser, C.W., Cunningham, C.G., Marvin, R.F. and Obradovich, J.D., 1980a, Pliocene intrusive rocks and mineralization near Rico, Colorado: Economic Geology, v. 75, p. 122-127.

Naeser, C.W., Izett, G.A. and Obradovich, J.D., 1980b, Fission-track and K-Ar ages of natural glasses: U.S. Geol. Survey Bulletin 1489, 31 p.

Naeser, C.W., Briggs, N.D., Obradovich, J.D. and Izett, G.A., 1981, Geochronology of Quaternary tephra deposits, in Self, S. and Sparks, R.S.J., eds., Tephra Studies: NATO Advanced Studies Institute Series C, Dordrecht, Netherlands, Reidel Publishing Company, p. 13-47.

Naeser, N.D., Westgate, J.A., Hughes, O.L. and Pewe, T.L., 1982, Fission-track ages of late Cenozoic distal tephra beds in the Yukon Territory and Alaska: Canadian Jour. Earth Sciences (In Press).

Price, P.B. and Walker, R.M., 1963, Fossil tracks of charged particles in mica and the age of minerals: Jour. Geophysical Research, v. 68, p. 4847-4862.

Roberts, J.A., Gold, R. and Armani, R.J., 1968, Spontaneous-fission decay constant of ^{238}U: Physical Review, v. 174, p. 1482-1484.

Seward, D., 1974, Age of New Zealand Pleistocene substages by fission-track dating of glass shards from tephra horizons: Earth Planetary Science Letters, v. 24, p. 242-248.

__________, 1975, Fission-track ages of some tephras from Cape Kidnappers, Hawke's Bay, New Zealand: New Zealand Jour. Geology Geophysics, v. 18, p. 507-510.

__________, 1979, Comparison of zircon and glass fission-track ages from tephra horizons: Geology, v. 7, p. 479-482.

Spadavecchia, A. and Hahn, B., 1967, Die Rotationskammer und einige Anwendungen: Helv. Phys. Acta, v. 40, p. 1063-1079.

Steen-McIntyre, V., Fryxell, R. and Malde, H.E., 1981, Geologic evidence for age of deposits at Hueyatlaco archaeological site, Valsequillo, Mexico: Quaternary Research, v. 16, p. 1-17.

Storzer, D. and Poupeau, G., 1973, Ages-plateaux de mineraux et verres par la methode des traces de fission: C.R. Acad. Sci., Paris, v. 276, Series D, p. 137-139.

Storzer, D. and Wagner, G.A., 1969, Correction of thermally lowered fission-track ages of tektites: Earth Planetary Science Letters, v. 5, p. 463-468.

Storzer, D., Horn, P. and Kleinmann, B., 1971, The age and origin of Köfels structure, Austria: Earth Planetary Science Letters, v. 12,

p. 238-244.

Stuiver, M., Heusser, C.J. and Yang, I.C., 1978, North American glacial history extended to 75,000 years ago: Science, v. 200, p. 16-21.

Suzuki, M., 1973, Chronology of prehistoric human activity in Kanto, Japan. Part I: Jour. Faculty Science, Univ. Tokyo, sec. V, v. IV, pt. 3, p. 241-318.

__________, 1974, Chronology of prehistoric human activity in Kanto, Japan, Part II: Jour. Faculty Science, Univ. Tokyo, sec. V., v. IV, pt. 4, p. 395-469.

Wagner, G.A., 1976, Radiation damage dating of rocks and artifacts: Endeavour, v. 35, p. 3-8.

__________, 1978, Archaeological applications of fission-track dating: Nuclear Track Detection, v. 2, p. 51-64.

Wagner, G.A., Reimer, G.M. and Jager, E., 1977, Cooling ages derived by apatite fission track, mica Rb-Sr and K-Ar dating: the uplift and cooling history of the Central Alps: Memorie degli Institui di Geologia e Mineralogia dell'Universita di Padova, v. 30, p. 1-27.

Watanabe, N. and Suzuki, M., 1969, Fission track dating of archaeological glass materials from Japan: Nature, v. 222, p. 1057-1058.

Westgate, J.A. and Briggs, N.D., 1980, Dating methods of Pleistocene deposits and their problems: V. Tephrochronology and fission-track dating: Geoscience Canada, v. 7, p. 3-10.

Westgate, J.A. and Gorton, M.P., 1981, Correlation techniques in tephra studies, in Self, S. and Sparks, R.S.J., eds., Tephra Studies: NATO Advanced Studies Institute Series C, Dordrecht, Netherlands, Reidel Publishing Company, p. 73-94.

Westgate J.A., Briggs, N.D., Stalker, A. MacS. and Churcher, C.S., 1978, Fission-track age of glass from tephra beds associated with Quaternary vertebrate assemblages in the southern Canadian Plains: Geol. Soc. Amer. Abstracts with Programs, v. 10, p. 514-515.

USING PALEOMAGNETIC REMANENCE AND MAGNETIC SUSCEPTIBILITY DATA FOR THE DIFFERENTIATION, RELATIVE CORRELATION AND ABSOLUTE DATING OF QUATERNARY SEDIMENTS

RENE W. BARENDREGT

ABSTRACT

Paleomagnetism is used in Quaternary stratigraphic studies as a tool for correlation and relative age dating of equivalent strata or for the absolute dating of deposits. The method is based on the detection of changes in the earth's magnetic field and especially changes of polarity that are recorded by ferromagnetic sediments at the time of deposition.

Dating by paleomagnetic characterization and geomagnetic polarity history is a relatively new technique. The large-scale features of the earth's magnetic field character have been well worked out for the past 5 million years or so. The detailed small-scale features for this period are still being discovered and defined through analysis of terrestrial sediments. Because of the much greater sedimentation rate on land, these are more likely to show short-lived events and record the excursions which ultimately will become useful correlative tools.

Fine-grained sediments, lava flows, and baked pottery are the media most frequently used. Because reversals have occurred repeatedly in the past their identification within incomplete sedimentary records is only possible through comparison with other stratigraphic or radiometric data collected for similar or related sedimentary sequences. Continuously deposited marine or terrestrial sediments which show a high sedimentation rate provide isochrons which can be used for worldwide correlation. The recent flourishing of research activity into the secular variation of the earth's non-dipole field promises to greatly refine and embellish the geomagnetic timetable for the Quaternary.

The only practical way of demonstrating the validity of interpreted magneto-stratigraphy is to show that results are reproducible in widely separated sections with different lithology and sedimentation rates. In Canada where Pleistocene deposits are largely glacial in origin, and were thus episodic, one must be aware that these deposits may only have recorded the earth's magnetic field in short time intervals. Possible subsequent alteration of these sediments by the processes outlined in this paper, must be borne in mind. Great Lake sediments and other glacial lake sediments provide excellent opportunity for magneto-stratigraphic correlation and dating.

INTRODUCTION

Dating of Pleistocene sediments beyond the range of radiocarbon dating as well as the dating of sediments which contain no dateable carbon material has always been a problem. The new and the perfected absolute methods (potassium-argon, fission track, amino acid racemization and accelerated carbon-14) require media which are often lacking in Pleistocene sediments.

A new method of dating based upon the paleomagnetic characterization of sediments and rocks offers great promise for the partial alleviation of this problem. In the past ten years considerable use has been made of this technique in Canada, U.S.A., Japan, The Netherlands, Britain, The U.S.S.R., and elsewhere. Its major value in Pleistocene stratigraphic work will come in the near future, when a more detailed record of the secular variation of the non-dipole field becomes available.

A wide range of materials has been employed with the method. Silt to fine sand-sized sediments work best, but in general any fine-grained sediment can potentially be analyzed. Marine as well as terrestrial sediments have been used. In addition, lava flows serve as excellent recorders of the earth's past magnetic field and can be sampled with relative ease. Baked pottery and other objects can also be used if the clay of which they are made took on a thermal remanent magnetization during the time of last baking. Archaeomagnetic studies provide information about secular changes often more detailed than that which has been obtained from lake sediments, but are limited by the intermittent rise and fall of past civilizations.

GEOMAGNETIC DYNAMO THEORY

It is widely accepted that the magnetic field of the earth and sun are produced by dynamos. In the case of the earth the dynamo action is thought to be produced by motions in the electrically conducting fluid core (Figure 1). Edward Bullard (1972) states that there is no direct demonstration of the existence of these dynamos; the reasons for believing in them are the absence of any other satisfactory theory and rather vague qualitative arguments about their possible properties. The magnetic field for which the theory attempts to account is a somewhat idealized version of the actual field and is described by Bullard as a field that is approximately (±20%) a dipole field, that changes by a large fraction of itself in a few hundred years, that has repeatedly and rather accurately reversed its dipole component, and that the non-dipole part (Figures 2 and 3) and its rate of change have length scales of a few thousand kilometers, are quite complex, have a tendency to drift westward, and are not correlated with geology or geography (except for a systematic tendency to small rates of change over the Pacific). In the dynamo theory the topography of the non-dipole field is probably associated with eddies in the convection patterns in the fluid core. Bullard concludes that there are not many ways in which a complicated changing, reversing field can be produced in a sphere of molten iron shut in a rigid container. The only plausible one is to assume that there are motions in the material and that these act as a self-exciting dynamo. He states that in a bounded, stationary electrically conducting body, any system of electric currents will decay exponentially with its own time constant. The time constant is proportional to the electrical conductivity and the characteristic length representing the distance in which the field changes by an appreciable part of itself. Bullard has calculated this time constant to be 15,000 years for the earth (assuming a radius of 3500 km and a conductivity of about $3x10^5 ohm^{-1}m^{-1}$). As this period is very short from a geological point of view, it is essential to maintain the field, which in a self-exciting dynamo is done by the currents produced by electromagnetic induction resulting from the movement of the conducting material through the field. Bullard goes on to state that it is far from obvious whether a given motion can maintain a field, or whether any motion can. He lists three non-dynamo theorems that prohibit particular types of motion or field in a sphere from acting as dynamos and discusses the work of Herzenberg (1958), Backus (1958), Childress (1969) and Roberts (1970). He states that Roberts has shown that almost all motions spatially periodic in three dimensions act as dynamos. Of three-dimensional motions spatially periodic in two dimensions, he has shown that about half act as dynamos.

It is almost certain that there is motion in the core. Without it there can be no dynamo, and it seems impossible to account for the short time scale of the magnetic variations. If there is motion, there must be forces to maintain it. The core is well protected from external influences, so no great variety of forces can plausibly be supposed to produce the motion. Earlier workers suggested thermal convection

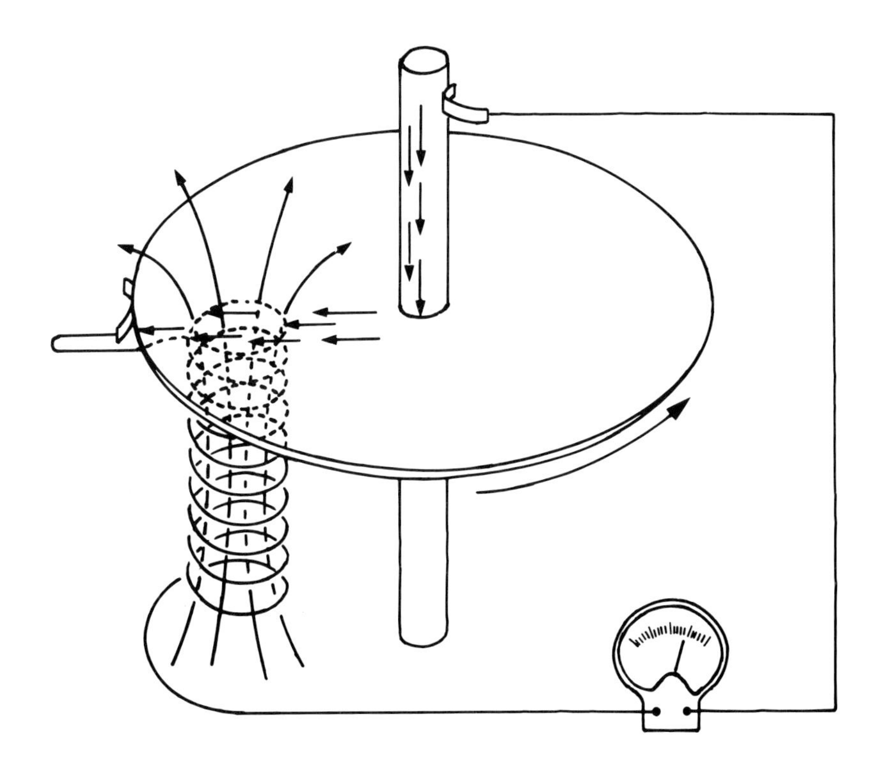

Figure 1 Faraday Disk Dynamo generates electric currents (short arrows) when a copper disk is turned through the magnetic lines of force of a bar magnet or through a coil of wire (as shown). In a self-sustaining dynamo the generated electric currents serve to reinforce the magnetic field of the coil so that no external supply of magnetism is needed beyond that which originally served to trigger the dynamo. The metallic liquid in the core of the earth is believed to flow in such a way as to act as a mechanical dynamo in generating magnetic fields of the earth. (From Carrigan and Gubbins, 1979).

resulting from radioactive heating. Whether, in fact, the radioactivity of the core is high enough and whether the processes for removing heat from the outside of the core are sufficiently effective to produce the temperature gradient needed to initiate convection is unknown. The main requirement is that the temperature gradient exceed the adiabatic; the heat flow required for this greatly exceeds the energy absorbed by the dynamo (Bullard, 1972).

The question of reversals of the earth's magnetic field is a challenge to any theory of the origin of the earth's field. A two-disk dynamo connected together so that each disk feeds current to the coil of the other has been devised to demonstrate that reversals can be produced. Although the disk dynamo is far from anything that can be imagined to exist in the earth's core, there is a certain similarity in the structure of the equations that control it and the equations constructed for magneto-hydrodynamic dynamo theories. If the equations give unstable solutions that flip from one direction to the opposite one, then a reversal has no specific "cause", it is simply a consequence of the unfolding in time of the solution of the equation. It is not certain that this is the real state of affairs; it might be that the reversals were a result of disturbance of the motion by some catastrophic event, such as the fall of a very large meteorite. It is also possible that the cause is statistical and is associated with random fluctuations in the electric currents and the motions or the forces. It is known that random emf's can cause reversals in a disk-dynamo (Bullard, 1955). Cox (1968) has suggested that random fluctuations in the non-dipole part of the field may precipitate a reversal. He has developed a particular model in some detail and has shown that it leads to a Poisson distribution of time intervals between reversals.

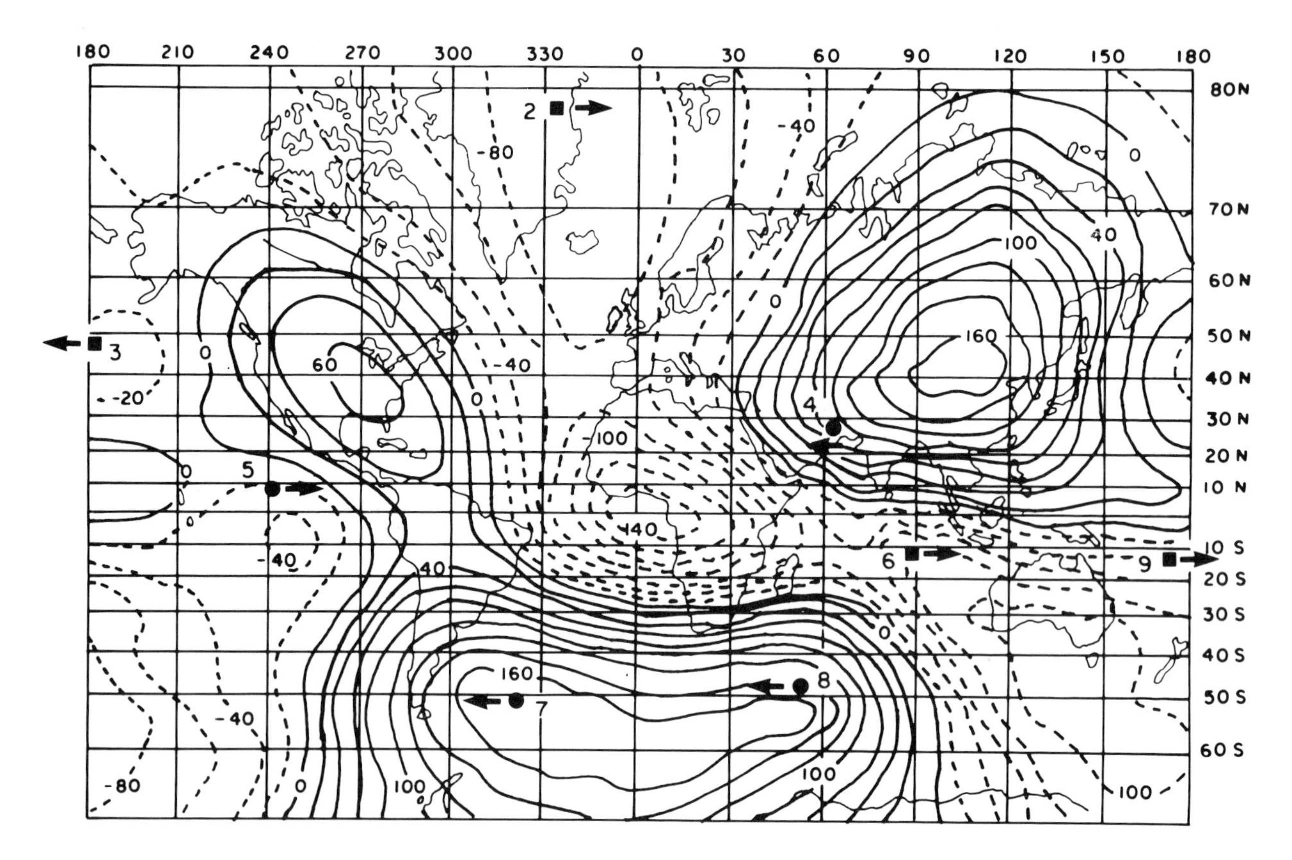

Figure 2 Map of non-dipole field, vertical component, for epoch 1945, Contours labelled in milligauss. Positions are shown of the radial dipoles with which, together with a central dipole, Alldredge & Hurwitz (1964) represented the 1945 field: symbols ● represent positions of dipoles pointing down while symbols ■ represent positions of dipoles pointing up. All the radial dipoles are located at .025 Earth radii. The arrows indicate whether the dipoles drifted eastwards or westwards between epochs 1945 and 1955. (From Creer, 1977).

Comparison with the observed distribution then leads to an estimation of the parameters of the model. Inevitably the model is based on intuitions about the dynamo process rather than on any fundamental theory; it is nonetheless suggestive, and it seems possible, that the reversals of the field are associated with changes in the complex non-dipole field (Bullard, 1972). An excellent discussion of the source of the earth's magnetic field is found in Carrigan and Gubbins (1979).

PALEOMAGNETIC REMANENCE AND SUSCEPTIBILITY ANALYSIS

The earth's magnetic field has two stable states: in its normal state the field is believed to wobble several tens of degrees about a dipole direction which, over the earth's surface, is directed toward the north. In its reversed state, it wobbles about a south polar direction. Transition from one state to another takes an average of 10,000 years. A record of these changes is left behind in the form of what is termed a remanent magnetization of sediments and rocks. In the case of volcanic rocks, this remanence, called thermoremanent magnetism (TRM) is very strong. Because of the sharpness of the transition zones, they provide very useful stratigraphic markers which are easily traceable and mappable in the field.

Two different types of geomagnetic polarity time units can be distinguished on the basis of their duration. The longer of the two is termed a geomagnetic polarity epoch and is defined as a time interval during which the earth's field was entirely or predominantly of one

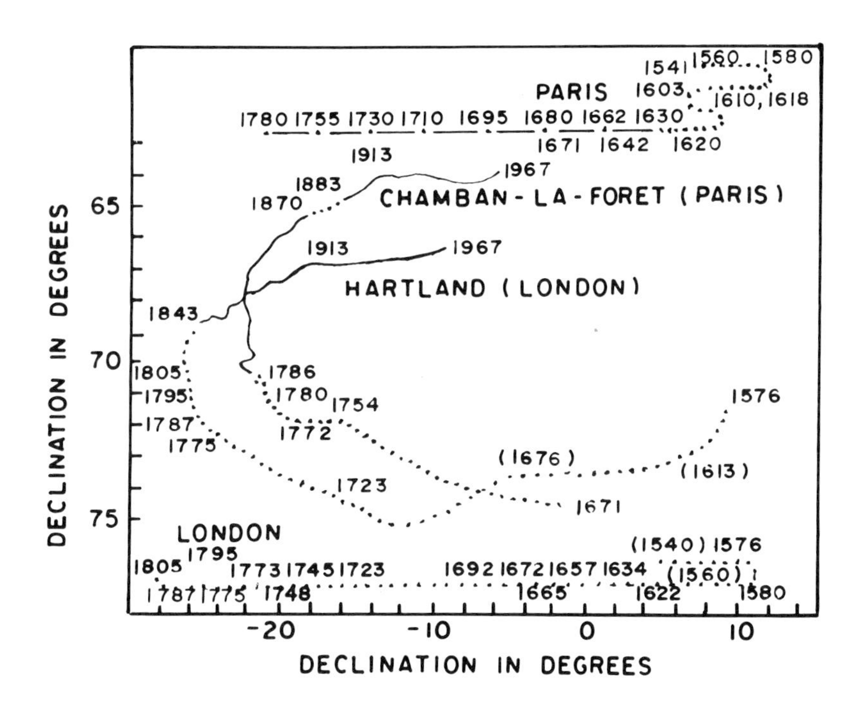

Figure 3 Secular variation of the geomagnetic field direction near London and Paris. All values have been reduced to the present observatories at Hartland (50°59.7'N, 350°31.0'E) and Chambon-la-Forêt (48°01.4'N, 2°15.6' E). The solid curves are taken primarily from continuous observatory records, whereas the dashed curves are based upon historical values given by Gaibar-Puertas (1953). For the dates in parenthesis, the values of the declination were obtained by extrapolation. (From Skiles, 1970).

polarity to that of the epoch. The time scale for geomagnetic polarity epochs and events, constructed by Cox *et al.* (1964) and modified by other workers (see Figure 4) is based on the potassium-argon dates and paleomagnetic measurements from about 60 lava flows from California, Hawaii, Alaska, Europe and Africa.

The lines of force in the earth's magnetic field are directed toward the magnetic poles and the angle of any point between true North and the direction of the field is called the declination. The lines of force are also directed (except at the equator) toward or away from the center of the earth, and the angle above or below the horizontal is called the inclination (see Figure 5). It is along these lines of force that 'memory elements' have been oriented. The memory elements are magnetic domains (local small zones within the ferromagnetic material that have a large uniform spontaneous magnetization within the various iron and titanium grains).

A 'good' sample from the sedimentary record should be fine-grained, strongly magnetized, free from secondary mineralization or weathering and unlikely to have a history of lightning strikes. Fine-grained samples are essential because minute grains of ferromagnetic material in the sediment must be oriented by the earth's ambient magnetic field rather than by the geologic agent that transported and deposited the minerals at the sampling site. Magnetization resulting from these oriented mineral grains is called detrital remanent magnetization (DRM) and conveys a record of the earth's magnetic field at the time of deposition. This DRM may be destroyed or weakened through weathering, secondary mineralization, or lightning strikes. These secondary magnetizations may be an unstable part of the original remanent magnetization and are referred to as a viscous remanent magnetization (VRM). The unwanted secondary magnetizations must be removed to isolate any existent primary magnetization which recorded the ancient earth's field.

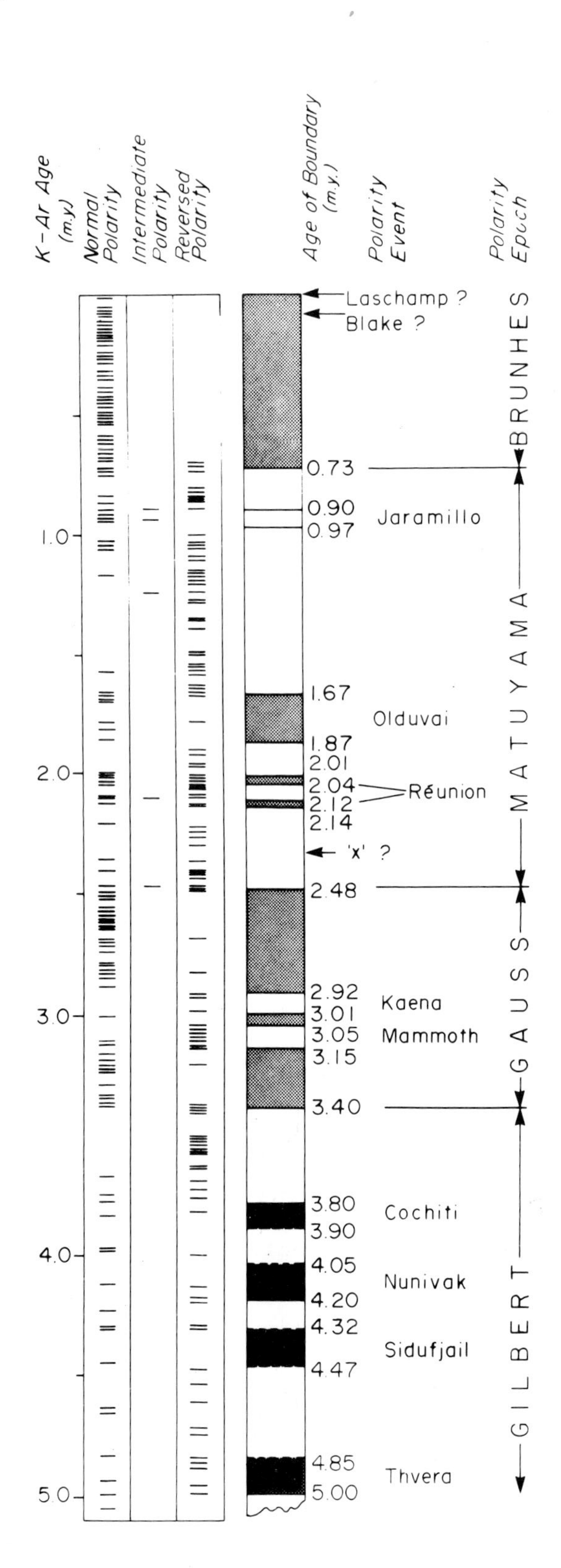

Figure 4 Revised Late Cenozoic polarity time scale reconciled to the new K-Ar constants. Each short horizontal line indicates a potassium-argon age and magnetic polarity determined for one volcanic cooling unit. Stippled pattern indicates periods of normal polarity with coarse stippled pattern indicating events whose limits are poorly defined. Arrows indicate possible brief polarity events. The time scale is based on volcanic cooling units and deep-sea cores. (From Mankinnen and Dalrymple, 1979).

If paleomagnetism is to provide a valuable dating tool, the magnetic record must be stable and of a single component.

It has been shown by As and Zijderveld (1958) and one cannot simply divide samples into magnetically stable and unstable ones, but that each rock usually contains several natural magnetizations of different stability and often different directions. The total natural remanent magnetization (NRM) obtained by simply measuring a sample is the

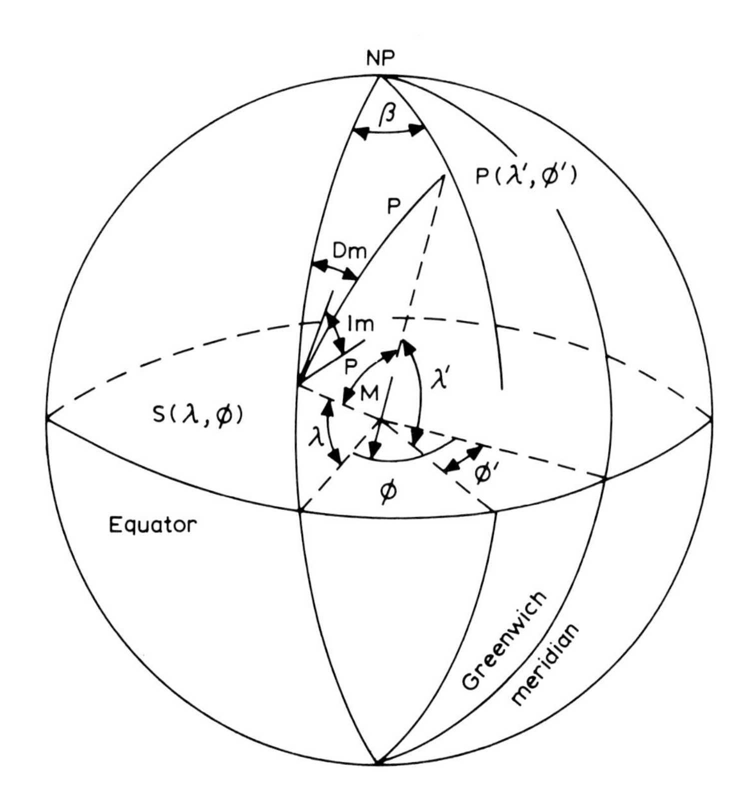

Figure 5 Calculation of paleomagnetic pole from average directions of magnetization (D_m, I_m). The sampling site S has latitude and longitude (λ,ϕ) and the pole P has coordinates (λ',ϕ'). (From McElhinny, 1973).

resultant of these different magnetizations. If a stable magnetization remains after 'quality' cleaning it is inferred to be primary. However, a primary magnetization need not necessarily be stable if, for example, it was recorded by soft multi-domain particles.

Various field and laboratory tests have been devised to determine stability of magnetization and are outlined in all textbooks of paleomagnetism. The most commonly used method involves alternating field (a.f.) demagnetization and is best described by Zijderveld (1975). If a sample is placed in a magnetic field of a certain intensity, a part of its remanent magentization will become unblocked from its fixed position and aligned in this ambient field. Upon intensifying this ambient field, more and more of the remanence will be unblocked and redirected. When the applied field is alternating and is slowly decreased from its peak value the unblocked grains are randomized (eliminated) leaving behind the more stable remanence fractions. The sample can be demagnetized stepwise with increasingly stronger alternating magnetic fields by which fractions with higher and higher unblocking fields are randomized. Complete measurement after each treatment gives the magnetization remaining after each demagnetization step. By vector subtraction the direction and intensity components are determined. The strength of the alternating field entirely eliminating the secondary magnetization lies at the point where the resultant vector stops changing direction. This is an important value since it is just with an alternating magnetic field of that strength, that the sample is said to be magnetically 'cleaned' at which time the remaining vector is <u>assumed</u> to be the DRM. It must be pointed out however that vector subtraction will only yield the direction and intensity of the subtracted vector when a single component is removed. Components of remanence may have overlapping coercivity spectra so that two or more components may be removed simultaneously in the same proportion. Furthermore, 'cleaning' may in some cases leave a stable secondary remanence as for example in sediments where the smallest grains (*i.e.* of magnetite) were finally fixed in position sometime after the main DRM. These smallest grains may be the most stable component through being single domained. Chemical overprint such as deuteric oxidation may similarly leave a stable secondary remanence.

To better understand the means by which sediments become magnetized it is necessary to identify the magnetic minerals responsible for the NRM (see Løvlie *et al.*, 1971). If for example, the principle magnetic mineral is magnetite and the deposits are not substantially weathered, their remanence is most probably a detrital remanent magnetization (DRM) and dates from the time of deposition of the sediment.

In addition to magnetic remanence characteristics, sediments exhibit magnetic susceptibility characteristics which can be used in their differentiation. Magnetic susceptibility is a measure of the degree to which a substance is attracted to a magnet, that is, the ratio of the intensity of magnetization to the magnetic field strength in a magnetic circuit. In a mineral grain this susceptibility may be isotropic or anisotropic depending on the shape and crystallography of the grain. In general the maximum susceptibility of an irregular-shaped grain lies along its greatest dimension and minimum susceptibility across it. Certain exceptions may exist however. The orientation of any particle which has a magnetic anisotropy may be affected by the magnetic field of the earth during deposition. Also, the magnetic particles of a sediment may have a different size or shape distribution from that of the non-magnetic fraction and so may be affected differently by hydrodynamic forces. Magnetite, which is isotopic, has no strong crystallographic anisotropy of susceptibility. Any anisotropy in magnetite is due to the deviation in shape of the grain from a sphere. Therefore, unless most magnetite grains in a sediment are equant, the susceptibility anisotropy will reflect their alignment. In a rock or sediment however, the effects are observed only if either the crystallographic or shape axes of many grains are aligned. A primary fabric is associated with conditions at the time of the deposition and is generally characterized by minimum susceptibility aligned about the normal to the bedding plane. Thus a strong anisotropy in magnetic susceptibility tends to rotate the direction of magnetization into the plane of maximum susceptibility (Uyeda *et al.*, 1963).

Gravenor *et al.* (1973) and Barendregt *et al.* (1976) have shown that bulk magnetic susceptibility values of glacial sediments can be used as a rapid and accurate diagnostic technique for the geologist to: (1) identify the source area of sediments and (2) differentiate deposits. Gravenor and Stupavsky (1974) also showed that there is a significant correlation between the amount of magnetite and the amount of heavy minerals in a till, thus allowing magnetic susceptibility measurements to replace the tedious task of heavy mineral separation, identification and counting which is often required for till differentiation.

ABSOLUTE AGE DATING

In the past, geochronological determinations using paleomagnetism have been carried out almost entirely on deep-sea sediments and lava flows. Although some evidence for reversals of the earth's field was recognized long ago, it was not until 1963 that attempts were made to define the geomagnetic polarity history. Reversals are global phenomenon and thus provide useful marker horizons for the stratigrapher.

Since several reversals have occurred in the last 2 m.y. some other dating tool or stratigraphic control must identify the reversal before an absolute correlation can be made. The polarity time scale and associated nomenclature was first outlined in detail by Cox (1969) and has been refined by a sub-commission of the International Commission on Stratigraphy (International Union of Geological Sciences). A new version of the time scale (Figure 4) has been based on more complete K-Ar data (Mankinen and Dalrymple, 1979). The best radiometric age in the scale obtained by multiple K-Ar dates of ashes and lavas, fixes the base of the Olduvai event at 1.8 ± .1 m.y. BP (Curtis and Hay, 1972). Other dates marking boundaries between normal and reverse epochs are much less accurate.

Opdyke *et al.* (1977) have shown from work in Anza Borrego State Park in California that faunal changes from Blancan to Irvingtonian land mammal ages can be accurately dated using the magnetic record of sediments in which the bones occur. Johnson *et al.* (1975) have shown from work in California, Texas and Kansas, that the oldest Irvingtonian based on the occurrence of Lepus(hare) and other small mammals occurs within the Matuyama reversed polarity epoch in the region of the Olduvai event, while Blancan faunas occur in the Gauss normal polarity epoch and range into the lower Matuyama epoch. This correlation has also been found in Saskatchewan, Canada by Foster and Stalker (1976). Many more fossil ages will undoubtedly become fixed with continued correlation between the paleomagnetic and fossil records in terrestrial sediments (Barendregt and Stalker, 1978).

CORRELATION AND RELATIVE AGE DATING

In addition to absolute dating, paleomagnetic studies can be used to correlate equivalent horizons and to relative age-date horizons provided they show one or more of the following: (1) a secular variation (oscillations of the local magnetic vector originating principally from the non-dipole field[1] which can be recognized over a relatively large area); (2) comparable total intensity oscillations of the NRM; and (3) similar magnetic susceptibility characteristics (Gravenor and Stupavsky, 1974; Barendregt *et al.*, 1976). In these cases magnetostratigraphy affords a convenient and simple means of correlating Pleistocene deposits of either terrestrial or marine origin (see Figure 6). One of the more exciting prospects emerging from such chronologic correlations is calibration of evolution rates and directions of dispersal in restricted mammalian species (Opdyke *et al.*, 1977; Lindsay *et al.*, 1976; and Johnson *et al.*, 1975).

The secular variation record whose basic period generally varies from 4 to 10^4 years (McElhinny and Merrill, 1975) has been shown to be of some use for regional correlations by Stober and Thompson, 1977 (see Figure 6) as well as by Turner and Thompson (1979) and Vitorello and van der Voo (1977). Turner and Thompson (1979) have shown that secular variation is not caused by wobbling of the main geomagnetic dipole as suggested by Kawai and Hirooka (1967) but rather results dominantly from more localized non-dipole changes involving both westward and eastward drift (see also Denham, 1974). They arrive at this conclusion because their 0-7000 year BP secular variation record from Loch Lomond sediments in Scotland does not compare with Japanese archaeomagnetic records or with North American sediment data from this same period. The cause of long or short period oscillations of the geomagnetic secular variation record remains unknown. The oscillations may be due to the main dipole field wobble, westward or eastward drift of non-dipole sources, fluctuations in intensity of stationary non-dipole sources, or combinations of each of these. However, any one of these mechanisms would produce different secular changes in widely separated localities. It is likely that secular variation generally must be due to both non-dipole and dipole changes. The matter is still under intensive discussion in the literature.

Verosub (1979) has shown that the geomagnetic pole can make large shifts of 10° or more in less that 200 years. This implies that a sudden change in the paleomagnetic directions as recorded by sediments

---

[1]The non-dipole field is the result of local complexities in the magnetic field, having an average value of about 5% of the main field at the earth's surface, and showing some eight regions, of continental dimensions displaying positive or negative values with an amplitude of around .15 Oe. The non-dipole field has a westerly drift of some .2 to .3 degrees of longitude per year and is thought to originate from regions near the core-mantle boundary, where local centers of fluid motion may distort the main toroidal field locally. (Figures 2 and 3)

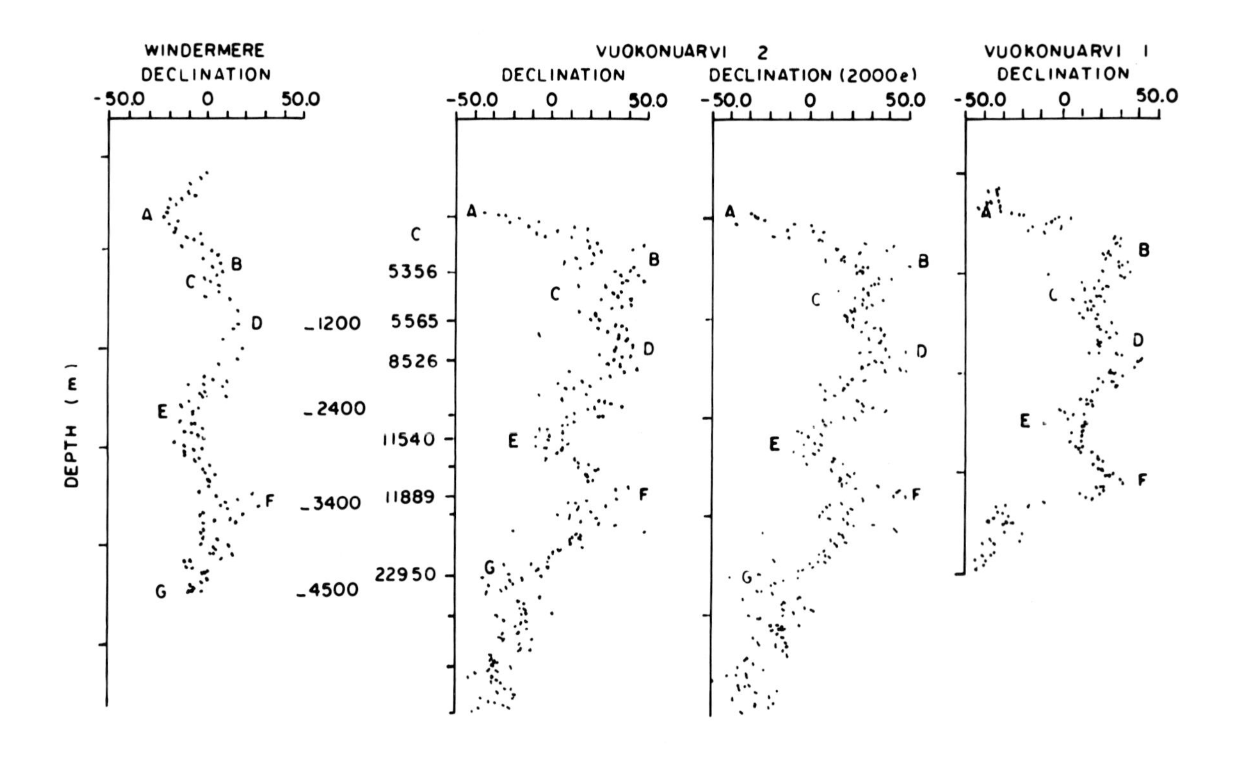

Figure 6 Paleomagnetic relative declination logs for a core from Lake Windermere (England), Vuokonjarvi (Finland) core 2(NRM and partially demagnetized at 200 Oe) and Vuokonjarvi core 1 (200 Oe demag.) Windermere data is from Mackereth (1971). (Figure is from Stober and Thompson, 1977).

may not be interpreted as evidence for the occurrence of a diastem (minor depositional break). Many studies have shown no magnetic anomalies in locations and sediments where they would be expected on the basis of other correlations. Though gaps in the record may be present and dating errors are possible, it is not unlikely that some apparent polarity excursions[1] are only very localized geomagnetic occurrences or are the result of mechanical disturbance of strata. Much more data are needed before conclusions about using secular field oscillations for stratigraphic correlation can be made.

---

[1]The Internal commission on stratigraphy of the International Union of Geological Sciences defines polarity excursion as "a sequence of virtual geomagnetic poles which may reach intermediate latitudes and which may extend beyond 135° of latitude from the poles, for a short interval of time, before returning to the original polarity" (Watkins, 1976)

Correlation on the basis of intensity fluctuations presents yet another relative dating tool. Cox (1968) constructed a curve showing the variation in the dipole moment (intensity) of the geomagnetic field. The composite curve obtained by Opdyke *et al.* (1972) from deep-sea cores is similar. One must of course be confident that intensity changes are not related to lithologic or mineralogical changes. Correlation on the basis of intensity requires correction to a common datum such as the equator, since intensity varies according to latitude. It was found that a gradual drop in sample intensity often corresponds with shallowing of inclination and declination swings. Kean *et al.* (1979) have shown that bog and lake cores from Cedarburg bog and Lake Michigan have intensity records which afford good correlation having similar long and short wavelength features, and both records are correlatable with records from Lake St. Croix described by Lund and Banerjee (1979).

The ratio between natural remanent magnetization (NRM) intensity and initial susceptibility (χ) sometimes referred to as the modified Koenigsberger ratio or Q ratio which in deep-sea sediments often reflects changes in intensity of the geomagnetic field (Harrison, 1966; Opdyke, 1972; Creer, 1974) has been used by others (Thompson, 1975; Levi and Banerjee, 1976) to correlate cores from lake sediments. Thompson concludes that the Q ratio of sediments does not correlate between lakes or in certain cases even within lakes and states that in apparently uniform lake sediments and possibly also in deep sea cores it is important to have shown that NRM intensity and (χ) (initial susceptibility) are due to the same magnetic minerals before using normalized intensities as indicators of ancient field intensities. Levi and Banerjee state that before using a particular sediment core for relative paleointensity determinations, the sediment's remanence properties must be established, because only those sections with similar remanence properties can be compared for relative paleointensities. They found that homogeneity can be established from similarities in the NRM properties of the sediment and from similarities of properties of laboratory induced remanences such as ARM (anhysteretic remanent magnetization) and IRM (isothermal remanent magnetization). On the other hand they found that magnetic susceptibility and saturation magnetization are not favored as normalizing parameters, because they are likely to activate a disproportionately large fraction of the superparamagnetic and multidomain particles which are relatively less important as stable NRM carriers.

The non-dipole portion of the geomagnetic field is presently observed to be drifting westward about .2° per year (Bullard *et al.*, 1950). Drift in a westward direction is in accord with magnetohydrodynamic models in which convective and Coriolis forces control the dominant processes in the fluid core, with electromagnetic coupling occurring between the core and mantle (Denham 1974). However, some features of the field show other than westward motion, as for example at Sitka, Alaska, where the three principle orthogonal field components are all drifting eastward at the present time (Skiles, 1970). Eastward drift has also occurred in the past as can be seen from the paleomagnetic record. These non-dipole magnetic fluctuations may cause the local field vector to move about in a looping fashion, with periods on the order of hundreds to thousands of years. These vector loops may also provide a means of sediment correlation and thus can be considered as a relative dating tool. Working with a suggestion of Runcorn (1959), Skiles (1970) discussed the method for inferring the drift direction using paleomagnetic data from a single site and showed that the study of vector loops could be valuable for gaining insight into the drift of the ancient geomagnetic field. The procedure is to view the unit-vectors of magnetic direction along their axis from negative toward positive, and note whether they tend to trace a clockwise or a counterclockwise loop as time advances (Figure 7). In a normally polarized main field, such loops were shown to correspond to westward or eastward drift respectively, independent of the source polarity or the latitude of the source and observation site (Denham, 1974). Skiles also notes that if westward drift persists through a field reversal as the theories of Bullard *et al.* (1950), Hide (1966) and Skiles (1969) indicate, then during a reversal we should look for a counterclockwise

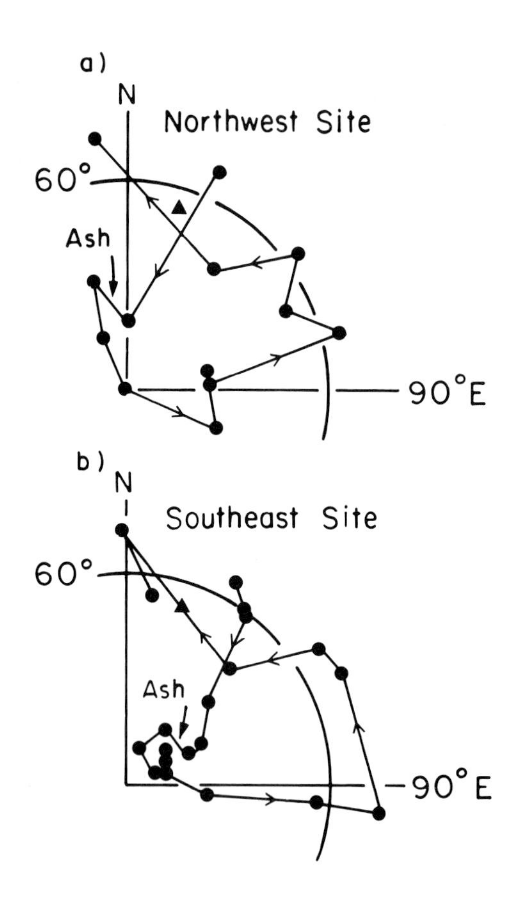

Figure 7 Equal-area stereographic projection of paleomagnetic directions at two sites around Mono Lake from Denham (1974). Each record exhibits counterclockwise rotation of the magnetic vector although details of the two records are different. The triangle represents the present field direction. the arrow denotes the position of a distinct ash layer used in correlating between the two sites. Arrows indicate counterclockwise looping motion as time advances. (From Verosub, 1977).

rotation of the field vector at all observatories (Skiles, 1970). According to Hide (1966) slow eastward motions on the core are theoretically possible, so long as the overall average drift is maintained towards the west. Intuition suggests that the life-span of an eastward drifting source is inversely related to its rate of drift. Motions contrary to the overall westward drift may have a low probability of traversing great distances before dying out. Thus, it is quite possible that the life-spans of the loop and the perturbing source were similar (Denham, 1974). As more detailed sedimentary data become available, distinct events of paleo-secular variation, such as the Mono Lake counterclockwise looping described by Denham (1974) may prove useful for Quaternary stratigraphic correlation over relatively large areas.

Creer *et al.* (1972) found a long-period counterclockwise looping of V.G.P.'s about the geographic pole for their Lake Windermere data. This data furthermore suggests a half-period of about 11,000 years and they state that it is tempting to associate this with the period of precession of the equinoxes, or some 25,800 years. (Due to the gravitational attraction of the moon and the sun on the earth's equatorial bulge, the earth's spin axis describes a cone around the celestial pole with a period of about 25,800 years. As one faces north the precessional motion is counterclockwise, *i.e.* in the opposite sense to the rotation Creer *et al.*, 1972).

Abrahamsen and Readman (1980) also describe a V.G.P. which moves first in a clockwise and then in a counterclockwise direction as time advances which they interpret as indicating either a rather sudden change in the sense of drift of the non-dipole field from westward to eastward, or a fairly rapid growth or decay of local dipole configurations in the outer core. They use this vectorial rotation pattern as a possible dating tool and suggest that the Older Yoldia clays near Norre Lyngby, Denmark, may have an age of 24,000 yrs. BP, similar to the Mono Lake sediments containing similar V.G.P. looping patterns.

PROBLEMS AND LIMITATIONS IN PALEOMAGNETIC ANALYSIS

There are a number of possible reasons why the natural remanent magnetization of sediments should not always faithfully reproduce the directions of the ambient magnetic field in which they were deposited. These are outlined by van Montfrans (1971), Verosub (1975), Verosub and Banerjee (1977) and are summarized here. The reasons include the following: (1) Inclinations may be systematically low due to the preferential alignment of elongated, tabular or flat shaped magnetic particles. Inclination error may also be due to compaction after deposition. (2) The presence of currents during the deposition of the sediment, may result in systematically deviating directions of magnetization. (3) The presence of magnetic material that is too coarse to be aligned by the ambient field during or shortly after deposition and thus fails to record the directions of the magnetic field. (4) Bioturbation after deposition of the sediment results in randomly distributed directions of magnetization or a re-alignment of particles in a field direction different from the one present during original deposition. (5) Collapsing of sediment on a small scale during decalcification. (6) The presence of cryoturbatic structures resulting in randomly distributed directions in these structures. (7) Deformation of sediments as a result of glacial pushing (over-riding), turbidity currents, slumping, liquifaction or some other process which moves sediments away from their original depositional orientation spoils the record. (8) Possible self-reversal, as a result of post-secondary alterations in the ionic ordering of the crystal framework of the magnetic constituents of the sediment (Irving, 1964). (9) Formation of new magnetic minerals during a time when the polarity of the geomagnetic field was different or opposite to the period of original deposition. (10) Instability of the natural remanent magnetization, allowing for viscous components of magnetization to set in. (11) Small inaccuracies in field sampling and preparation of specimens in the laboratory (*i.e.* drying). (12) Errors and misorientations, made during field sampling or labelling of samples. Coring operations may result in breaking and twisting of the sediments which may then be interpreted as an excursion. These errors can usually be traced.

For the above reasons, some sample data must be discarded because the analytical results show:

(1) a remanent magnetization which makes too large an angle with the direction which would result from the field of an axial geocentric dipole, taking into account secular variation;

(2) the sediments were not stably magnetized; and

(3) the present of a stable secondary magnetization.

The glacial sediments of the Quaternary represent special problems in magnetostratigraphic correlation. Just as glaciers picked up, transported and deposited bedrock blocks (Stalker, 1976) they may well have picked up, transported and deposited blocks of frozen glacial sediments and placed them in anomalous stratigraphic positions. Also, over-riding of glacial sediments by renewed glaciation may have induced stress deformation (Stupavsky *et al.*, 1979) and given rise to structures which will yield 'apparent' magnetic excursions, such as those described by Verosub (1975). Other mechanisms of deformation and disturbance of glacial (lake) sediments include seismic activity, turbidity flows, density currents, and periglacial activity such as cryoturbation and congelifluction. Glacial sediments often lack continuous fine-grained sediments or contain large hiatuses or were subject to erosion, all of which represent interruptions in their paleomagnetic record.

Glacial tills also represent special problems (Barendregt *et al.*, 1977). Time stratigraphic lines within tills are often not horizontal due to different rates of erosion and/or deposition. Stupavsky *et al.* (1979) in a discussion of the Meadowcliffe till show a lack of correlation of the remanence characteristics over short interprofile separations leading them to conclude that the till deposition was non-uniform with rapid and variable rates of deposition from point to point within the till. In a study of the Seminary till Gravenor *et al.* (1979) and

Symons *et al.* (1980) have proposed that the thixotropic effect reported by Garnes (1977) for adobe bricks, is a probable cause for some reversed directions. According to them, the lack of correlation in remanence directions between two profiles .7 m apart and the large within-core variation of 13° between specimen pairs, suggest the possibility of remanence resetting caused by the hammer blows on the plastic tube sampler which they used.

It has been suggested that tills exhibit a magnetic remanence that is largely the result of the direction of ice movement, and not due to the presence of a layer of water at the ice/sediment interface which would have allowed the magnetic grains to become aligned in the earth's ambient field. The former has been ruled out by anisotropy of magnetic susceptibility measurements carried out by Gravenor and Stupavsky (1974) who have shown the remanence and the anisotropy of magnetic susceptibility of tills to have distinctly different directions.

Inclination errors are also commonly reported for tills. Laboratory experiments conducted by Verosub *et al.* (1979) showed the partial realignment of artificial sediment slurries on stirring by 'shear-induced liquifaction' to give a post-depositional DRM. In the same study, inclination errors of up to 30° were produced in thin-till slurries while in thick-till slurries no significant errors were noted. The above mentioned variability as well as the shock-induced thixotropic resetting as a result of sampling procedures may explain the far-sided V.G.P. (Virtual Geomagnetic Pole) positions for tills sampled in Ontario (Symons *et al.*, 1980) and in Alberta (Barendregt *et al.*, 1977). Inclination error due to the preferential alignment of irregular-shaped magnetic particles may also explain some of the far-sided V.G.P. positions.

A recurring problem facing the student of Quaternary paleomagnetism today, is the large number of reported excursions which cannot be properly correlated or are based on measurements of doubtful accuracy. Conclusions are sometimes drawn from incomplete data (Opdyke, 1976) or are based on insufficient sampling sites (Kukla and Nakagawa, 1977). For the novice the following guideline might well be useful: "a negative inclination doth not a reversal make".

In several parts of the world, terrestrial paleomagnetic records show excursions during the Brunhes normal polarity epoch. These excursions are reported in great numbers for sediments which are between 8,000 and 20,000 years in age. However, in deep-sea records and the Aegean Sea, there is no evidence for a reversal or excursion of the field during this time (Opdyke *et al.*, 1972). True reversed magnetizations in sediments are caused by a reversal of the main dipole field and should be seen world-wide. Excursions of the field (deviations of the direction of the field greater than the normal secular variation of a few tens of degrees) could also be caused by the main dipole field tilting at a large angle to the rotation axis and then returning to the same orientation as before. Such excursions should also be seen world-wide. However, sources of the non-dipole field could also produce an excursion that could change the direction of the field at one point by 180°, resulting in an apparent reversal at that point. Such conditions would be indistinguishable from a reversal of the dipole field, unless observations at exactly the same time interval from another part of the earth do not show a reversal.

In determining the zone of disturbance of a dipole field caused by a pseudo-reversal (*i.e.* one not involving the main dipole field) at one point in the earth's surface, Harrison and Ramirez (1975) developed a model which placed a vertical dipole at the surface of the earth's core whose magnetic field is opposite to that of the main dipolar field. They show that the areal coverage of the disturbed magnetic field around such a pseudo-reversal can be quite small, such that observations made only a few thousand km away would show no anomalous direction. If this model is applicable there are no contradictory observations of the existence of a reversal at Laschamp and other locations having similar age, provided they are pseudo-reversals caused by non-dipole field sources. Harrison and Ramirez further show that if the vertical dipole model for the non-dipole field is correct, then pseudo-reversals should be much less common at low latitudes than at high latitudes.

Geomagnetic excursions during the Brunhes epoch are more and more being considered as magnetostratigraphic markers. These excursions are apparently sharp movements of the Virtual Geomagnetic Pole (VGP) towards or beyond the equator followed by a return to the stable position, all within the timespan of a few thousand years. Table 1 provides a list of some well documented excursions. A useful discussion of some of these geomagnetic excursions is provided by Verosub and Banerjee (1977) who view with caution the paleomagnetic evidence provided by the various authors.

Table 1. Paleomagnetic Polarity Excursions and Proposed Ages During Brunhes Epoch

Name[a]	Researchers	Proposed age (yrs. BP)
Laschamp event	Bonhommet and Babkins (1967) Bonhommet and Zahringer (1969) Hall, York and Bonhommet (1979) Heller (1980)	8,000-20,000
Mono Lake Excursion	Denham and Cox (1971) Denham (1974)	24,000
Gothenburg Event	Morner *et al.* (1971) Morner and Lanser (1974) Morner (1977)	12,350-12,400
Blake Event	Smith and Foster (1969) Denham *et al.* (1976) Denham *et al.* (1977)	105,000
Lake Mungo Event	Barbetti and McElhinny (1976)	29,500
Lake Biwa Excursions	Nakajima *et al.* (1973) Yaskawa *et al.* (1973)	18,000 and 104,000-117,000
Erieau Excursion	Creer *et al.* (1976)	7,600-14,000
Lake Michigan Excursions	Vitorello and van der Voo (1977)	7,500 and 13,000
Gulf of Mexico Excursions	Freed and Healy (1974) Clark and Kennett (1973)	15,000-18,000 and 30,000-33,000
Maple Hurst Lake and Basswood Road Lake Excursion	Mott and Foster (1973)	12,500
Meadowcliffe Excursion	Stupavsky *et al.* (1979)	30,500
Port Dover Excursion	Morner (1976)	13,300
Maelifell Event	Peirce and Clark (1978)	40,000(?)
Rubjerg Excursion	Abrahamsen and Knudsen (1979)	23,000-40,000
Norre Lyngby Excursion	Abrahamsen and Readman (1980)	23,000-40,000
Kipp Excursion	Barendregt and Stalker (in press)	24,000(?)

[a]Location names are used only to identify the reported excursions and are not necessarily established names.

The Gothenburg event (labelled as a magnetic 'flip' in the literature) is especially problematic since the authors suggest that the event lasted only some tens of years while the inclination switch only took a few years (Morner, 1977). This would make it the most rapid magnetic polarity change known at present. The mechanism for such an abnormally rapid 'flip' is difficult to imagine. Recently Thompson and Berglund (1976) reported finding no evidence for the Gothenburg event in a detailed study of cores from Sweden. They attribute previous reports of anomalous directions as an example of the 'reinforcement syndrome' (Watkins, 1972 and Watkins, 1976). From continuous records of the geomagnetic field for the period 0-16,000 years BP obtained from the sediments of two postglacial lakes in Minnesota and a re-examination of the primary data related to the Gothenburg event and Erieau excursion, Banerjee and Lund (1979) conclude that both excursions might in fact be artifacts of the lithology (see also Banerjee *et al.*, 1979).

Thus it becomes obvious that paleomagnetic records as reported in the literature must be critically weighed. Even from deep-sea cores taken over the past 20 years, there is evidence for 'stable' and 'unstable' cores. According to Verosub and Banerjee (1977): "A stable core is one which gives a clean paleomagnetic record with simple features that are generally consistent with the accepted reversal sequence. An unstable core on the other hand gives a complex paleomagnetic record with many apparent changes in paleomagnetic direction. These cannot be correlated with those in other cores." Harrison (1974) concludes that many of the deep-sea sediment cores show short-period events which cannot be correlated with the presently known time scale of reversals and therefore short-period events are not reliable stratigraphic markers. Other cores in which there is a correlation between direction and/or intensity of magnetization, and climatic indicators, do not seem to have accurately recorded parameters of the Earth's magnetic field. It is hoped that through detailed work on terrestrial sediments, the mechanism whereby sediments become magnetized will be better understood and some of the discrepancies solved.

On land, North American studies of lake sediments have often been made on single samples per horizon, ^{14}C dates are few and frequently have large error limits. Verosub and Banerjee (1977) observe that: "Most disturbing of all is the fact that the observed excursions are frequently observed near lithologic boundaries reflecting a change from glacial to postglacial times with concomitant rapid fluctuations expected in the depositional conditions in the lakes."

Harrison and Ramirez (1975) have shown that the smallest lateral distance over which a geomagnetic fluctuation would be observed is around 1000 km. For this reason, an excursion should be recorded in nearby areas in sediments of similar age and as Verosub and Banerjee (1977) clearly point out: "Failure to detect a paleomagnetic excursion in an adjacent lake or ocean basin is an important result which should not go unreported because it is uninteresting."

Using a similar model to that of Harrison and Ramirez, Denham *et al.* (1976) showed that the Blake event may not have been felt over more than 9% of the earth's surface. On the other hand, Smith (1967) and Verosub and Cox (1971) have shown that paleomagnetic excursions may reflect global geomagnetic phenomena resulting from the change in the relative sizes of the dipole and non-dipole fields.

All this is not to say that localized excursions are not useful. The Quaternary stratigrapher must first come to realize that they can be local in nature or may have a worldwide occurrence, and then be able to determine their age so that they can be used as stratigraphic markers. If an excursion can be shown to be of global extent and dipolar in origin, a very useful marker horizon has indeed been found.

Finally, if each of the reported excursions listed in Table 1 represents a distinct excursion, then the geomagnetic field is much less stable than had previously been thought. Such a high degree of instability when extended back over geological time would produce a magnetic field far more complex than has been observed in the record (Verosub, 1975).

CONCLUDING REMARKS

Dating by paleomagnetic characterization and geomagnetic polarity history is a relatively new technique. The large-scale features of the earth's magnetic field character have been well worked out for the past 5 million years or so. The detailed small-scale features for this period are still being discovered and defined through analysis of terrestrial sediments. Because of the much greater sedimentation rate on land, these are more likely to show short-lived events and record the excursions which ultimately will become useful correlative tools.

The only practical way of demonstrating the validity of interpreted magneto-stratigraphy is to show that results are reproducible in widely separated sections with different lithology and sedimentation rates. In Canada where Pleistocene deposits are largely glacial in origin, and were thus episodic, one must be aware that these deposits may only have recorded the earth's magnetic field in short time intervals. Possible subsequent alteration of these sediments by the processes outlined earlier, must be borne in mind. Great Lake sediments, glacial lake sediments and deposits such as those described by Stalker in western Canada (Foster and Stalker, 1976) provide excellent opportunity for magneto-stratigraphic correlation and dating.

ACKNOWLEDGEMENTS

The author's work in Quaternary paleomagnetism has been supported by grants from the National Research Council of Canada and the Department of Energy, Mines and Resources. I am grateful to Drs. J.H. Foster, K.L. Verosub, D. Packer, and A.M. Stalker for their considerable help in broadening my understanding of paleomagnetism as a useful tool in Quaternary research. I am indebted to Drs. A. Latham and H.C. Palmer for critically reviewing the manuscript and making many helpful comments.

REFERENCES CITED

Abrahamsen, N. and Knudsen, K.L., 1979, Indication of a geomagnetic low-inclination excursion in supposed middle Weichselian Interstadial marine clay at Rubjerg, Denmark: Physics of the Earth and Planetary Interiors, v. 18, p. 238-246.

Abrahamsen, N. and Readman, P.W., 1980, Geomagnetic variations recorded in Older (≧ 23,000 BP) and Younger Yoldia clay (~14,000 BP) at Norre Lyngby, Denmark: Geophys. J. R. Astr. Soc., v. 62, p. 345-366.

Alldredge, L.R. and Hurwitz, L.H., 1964, Radial dipoles as the sources of the Earth's main magnetic field: J. Geophys. Res., v. 69, no. 12, p. 2631-2640.

As, J.A. and Zijderveld, J.D.A., 1958, Instruments and measuring methods in paleomagnetic research: Meded. en Verh. van het K.N.M.I., v. 78, p. 1-56.

Backus, G.E., 1958, A class of self-sustaining dissipative spherical dynamos: Ann. Phys., v. 4, p. 372-447.

Banerjee, S.K. and Lund, S.P., 1979, Gothenburg and Ericau excursions-questions regarding their value as magnetostratigraphic markers: EOS Trans. Am. Geophys. U., v. 60, no. 18, p. 238.

Banerjee, S.K. *et al.*, 1979, Geomagnetic record in Minnesota lake sediments-absence of the Gothenburg and Erieau excursions: Geology v. 7, p. 588-591.

Barbetti, N.F. and McElhinny, M.C., 1976, The Lake Mungo geomagnetic excursion: Phil. Trans. R. Soc., London, v. 281, p. 515-542.

Barendregt, R.W. *et al.*, 1976, Differentiation of tills in the Pakowki-Pinhorn area of southern Alberta on the basis of their magnetic susceptibility: Geol. Surv. Can. Paper 76-1C, p. 189-190.

__________, 1977, Paleomagnetic remanence characteristics of surface tills found in Pakowki-Pinhorn area of southern Alberta: Geol. Surv. Can. Paper 77-1B, p. 271-272.

Barendregt, R.W. and Stalker, A.M., 1978, Characteristic magnetization of some middle Pleistocene sediments from the Medicine Hat area of southern Alberta: Geol. Surv. Can. Paper 78-1A, p. 487-488.

Bonhommet, N. and Babkine, J., 1967, Sur la presence d'aimantations inversées dans la chaines des puys: C.R. Acad. Sci., v. 264, p. 92-94.

Bonhommet, N. and Zahringer, J., 1969, Paleomagnetism and potassium-argon date determinations of the Laschamp geomagnetic polarity event: Earth Planet. Sci. Lett., v. 6, p. 43-46.

Bullard, E.C., 1955, The stability of a homopolar dynamo: Proc. Cambridge Phil. Soc., v. 51, no. 744.

__________, 1972, Geomagnetic dynamos, in: Robertson, E.C., ed., Nature of the Solid Earth, N.Y., McGraw-Hill, 677 p.

Bullard, E.C. *et al.*, 1950, The westward drift of the Earth's magnetic field: Phil. Trans. R. Soc. London, v. 243A-67.

Carrigan, C.R. and Gubbins, D., 1979, The source of the Earth's magnetic field: Scientific American, v. 240, no. 2, p. 118-130.

Childress, S., 1969, A class of solutions of the magnetohydrodynamic dynamo problem, in Runcorn, S.K., ed., The Application of Modern Physics to the Earth and Planetary Interiors, N.Y., Wiley, 629 p.

Clark, H.C. and Kennett, J.P., 1973, Paleomagnetic excursion recorded in latest Pleistocene deep-sea sediments, Gulf of Mexico: Earth and Planet, Sci. Lett., v. 19, p. 267-274.

Cox, A., 1968, Lengths of geomagnetic polarity intervals: J. Geophys. Res., v. 73, p. 3247-3260.

__________, 1969, Geomagnetic reversals: Science, v. 163, p. 237-245.

Cox, A. *et al.*, 1964, Reversals of the earth's magnetic field: Science, v. 144, p. 1537-1543.

Creer, K.M., 1974, Geomagnetic variations for the interval 7,000-25,000 yr. BP as recorded in a core of sediment from Station 1474 of the Black Sea cruise of Atlantis II: Earth Planet. Sci. Lett., v. 23, p. 34-42.

__________, 1977, Geomagnetic secular variations during the last 25,000 years: an interpretation of data obtained from rapidly deposited sediments: Geophys. J.R. Astron. Soc., v. 48, p. 91-109.

Creer, K.M. *et al.*, 1972, Geomagnetic secular variation recorded in the stable magnetic remanence of Recent sediments: Earth Planet. Sci. Lett., v. 14, p. 105-127.

__________, 1976, Late Quaternary geomagnetic stratigraphy recorded in Lake Erie sediments: Earth Planet. Sci. Lett., v. 31, p. 37-47.

Curtis, G.H. and Hay, R.L., 1972, Further geologic studies and K-Ar dating of Olduvai Gorge and Ngorogoro Crater, in Bishop, W.W. and Miller, J.A., eds., Calibration of Hominoid Evolution: Edinburgh, Scottish Academic Press, p. 289-302.

Denham, C.R., 1974, Counterclockwise motion of paleomagnetic directions 24,000 years ago at Mono Lake, California: J. Geomag. Geoelectr., v. 26, p. 487-498.

Denham, C.R. and Cox, A., 1971, Evidence that the Laschamp polarity event did not occur 13,300-30,400 years ago: Earth Planet. Sci. Lett., v. 13, p. 181-190.

Denham, C.R. *et al.*, 1976, Blake polarity episode in two cores from the Greater Antilles outer ridge: Earth Planet. Sci. Lett., v. 29, p. 422-434.

__________, 1977, Paleomagnetism and radiochemical age estimates for Late Brunhes polarity episodes: Earth and Planet. Sci. Lett., v. 35, p. 384-397.

Foster, J.H. and Stalker, A.M., 1976, Paleomagnetic stratigraphy of the Wellsch valley site, Saskatchewan: Geol. Surv. Can. Paper 76-1C, p. 191-193.

Freed, W.K. and Healy, N., 1974, Excursions of the Pleistocene geomagnetic field recorded in Gulf of Mexico sediments: Earth and Planet. Sci. Lett., v. 24, p. 99-104.

Gaibar-Puertas, C., 1953, Varacion secular del campo geomagnetico: Tortosa, Spain: Observ. del Ebro, Memo No. 11.

Garnes, K.P., 1977, The magnitude of the paleomagnetic field: a new non-thermal, non-detrital method using sun-dried bricks: Geophys. J. of the Roy. Astron. Soc., v. 48, p. 315-329.

Gravenor, C.P. and Stupavsky, M., 1974, Magnetic susceptibility of the surface tills of southern Ontario: Can. J. Earth Sci., v. 11, No. 5, p. 658-663.

Gravenor, C.P. *et al.*, 1973, Paleomagnetism and its relationship to till deposition: Can. J. Earth Sci., v. 10, p. 1068-1078.

__________, 1979, DRM errors in Pleistocene tills including the Seminary Till, Scarborough, Ontario: EOS-Trans. Am. Geophys. U., v. 60, no. 18, p. 246.

Hall, C.M. *et al.*, 1979, $^{40}Ar/^{39}Ar$ dating of the Laschamp event and associated volcanism in the chains des puys: EOS Trans. Am. Geophys. U., v. 60, no. 18, p. 244.

Harrison, C.G.A., 1966, The paleomagnetism of deep-sea sediments: J. of Geophys. Res., v. 71, p. 3033-3043.

__________, 1974, The paleomagnetic record from deep-sea sediment cores: Earth Sci. Rev., v. 10, p. 1-36.

Harrison, C.G.A. and Ramirez, E., 1975, Areal coverage of spurious reversals of the earth's magnetic field: J. of Geomag. and Geoelec., v. 27, p. 139-151.

Heller, F., 1980, Self-reversal of natural remanent magnetization in the Olby-Laschamp lavas: Nature, v. 284, p. 334-335.

Herzenberg, A., 1958, Geomagnetic dynamos: Phil. Trans. R. Soc., A-250, p. 543-585.

Hide, R., 1966, Free hydromagnetic oscillations of the Earth's core and the theory of the geomagnetic secular variation: Phil. Trans. R. Soc. London, 259A, p. 615.

Irving, E., 1964, Paleomagnetism and its Application to Geological and Geophysical Problems: Wiley, 399 p.

Johnson, N.M., *et al.*, 1975, Magnetic polarity stratigraphy of Pliocene-Pleistocene terrestrial deposits and vertebrate faunas, San Pedro Valley, Arizona: Geol. Soc. Am. Bull., v. 86, p. 5-12.

Kawai, N. and Hirooka, K., 1967, Wobbling motion of the geomagnetic dipole field in historic time during these 2000 years: J. Geomag. and Geoelec., v. 19, p. 217-227.

Kean, W.F. *et al.*, 1979, Paleomagnetic records from the Cedarburg bog and from Lake Michigan: EOS Trans. Am. Geophys. U., v. 60, no. 18, p. 238.

Kukla, G. and Nakagawa, H., 1977, Late Cenozoic magnetostratigraphy-comparisons with Bio-, Climato- and Lithozones: Quat. Res., v. 7, p. 283-293.

Levi, S. and Banerjee, S.K., 1976, On the possibility of obtaining relative paleointensities from lake sediments: Earth Planet. Sci. Lett., v. 29, p. 219-226.

Lindsay, E.H. *et al.*, 1976, Preliminary correlation of North American land mammal ages and geomagnetic chronology: Univ. Michigan Papers on Paleontology, no. 12, p. 111-119.

Løvlie, R. *et al.*, 1971, Magnetic properties and mineralogy of four deep-sea cores: Earth Planet. Sci. Lett., v. 15, p. 157-168.

Lund, S.P. and Banerjee, S.K., 1979, Secular variation from sediments of Kylen Lake, Minnesota between 4,000 and 16,000 b.p.: EOS Trans. Am. Geophys. U., v. 60, no. 18, p. 243.

Mackereth, F.J.H., 1971, On the variation in direction of the horizontal component of remanent magnetization in sediments: Earth and Planet. Sci. Lett., v. 12, p. 332-338.

Mankinnen, E.A. and Dalrymple, G.B., 1979, Revised geomagnetic polarity time scale for the interval 0-5 m.y. b.p.: J. Geophys. Res., v. 84, no. B2, p. 615-626.

McElhinny, M.W., 1973, Paleomagnetism and Plate Tectonics: Cambridge University Press, 358 p.

McElhinny, M.W. and Merrill, R.T., 1975, Geomagnetic secular variation over the past 5 m.y.: Rev. Geophys. Space Phys., v. 13, p. 687-708.

Morner, N.A., 1976, Paleomagnetism in deep-sea core A 179-15-a reply: Earth Planet. Sci. Lett., v. 29, p. 240-241.

____________, 1977, The Gothenburg magnetic excursion: Quat. Res., v. 7, p. 413-427.

Morner, N.A. and Lanser, J.P., 1974, Gothenburg magnetic 'flip': Nature, v. 251, p. 408-409.

Morner, N.A. *et al.*, 1971, Late Weichselian paleomagnetic reversal: Nature, v. 234, p. 173-174.

Mott, R.J. and Foster, J.H., 1973, Preliminary paleomagnetic studies of freshwater lake sediment cores of Late Pleistocene time: Geol. Surv. Can., Report of Activities, Paper 73-1, Part B, p. 149-153.

Nakajima, T.K. *et al.*, 1973, Very short geomagnetic excursion 18,000 yrs. B.P.: Nature, v. 244, p. 8-10.

Opdyke, N.D., 1972, Paleomagnetism of deep-sea cores: Rev. Geophys. and Space Phys., v. 10, no. 1, p. 213-249.

____________, 1976, Discussion of paper by Morner and Lanser concerning the paleomagnetism of deep-sea core A 179-15: Earth and Planet. Sci. Lett., v. 29, p. 238-239.

Opdyke, N.D. *et al.*, 1972, The Paleomagnetism of two Aegean deep-sea cores: Earth and Planet. Sci. Lett., v. 14, p. 145-159.

____________, 1977, The Paleomagnetism and magnetic polarity stratigraphy of the mammal-bearing section of Anza-Borrego State Park, California: Quat. Res., v. 7, p. 316-329.

Peirce, J.W. and Clark, M.J., 1978, Evidence from Iceland on geomagnetic reversal during the Wisconsinan Ice Age: Nature, v. 273, p. 456-458.

Roberts, G.O., 1970, Spatially periodic dynamos: Phil. Trans. R. Soc. London, Ser. A, v. 266, no. 1179, p. 555-558.

Runcorn, S.K., 1959, On the theory of the geomagnetic secular variation: Ann de Geophys., v. 15, no. 87.

Skiles, D.D., 1969, The reflection and refraction of magnetohydrodynamic waves at a liquid-solid interface, with applications to the Earth's core: Ph.D. Thesis, University of California, Berkeley.

____________, 1970, A method of inferring the direction of drift of the geomagnetic field from paleomagnetic data: J. of Geomag. Geoelect., v. 22, No. 4, p. 441-462.

Smith, P.J., 1967, The intensity of the ancient geomagnetic field: a review and anlysis: Geophys. J., v. 12, p. 321-362.

Smith, J.D. and Foster, J.H., 1969, Geomagnetic reversal in Brunhes normal polarity epoch: Science, v. 163, p. 565-567.

Stalker, A.M., 1976, Megablocks, or the enormous erratics of the Albertan prairies: Geol. Surv. Can. Paper 76-1C, p. 185-188.

Stober, J.C. and Thompson, R., 1977, Paleomagnetic secular variation studies of Finnish lake sediment and the carriers of remanence: Earth Planet. Sci. Lett., v. 37, p. 139-149.

Stupavsky, M. *et al.*, 1979, Paleomagnetic stratigraphy of the Meadowcliffe till, Scarborough Bluffs, Ontario: a late Pleistocene excursion?: Geophys. Res. Lett., v. 6, no. 4, p. 269-272.

Symons, D.T.A. *et al.*, 1980, Remanence-resetting by shock-induced thixoptropy in the Seminary till, Scarborough, Ontario: Geol. Soc. Amer. Bull., v. 91, p. 593-598.

Thompson, R., 1975, Long period European geomagnetic secular variation confirmed: Geophys. J.R. Astron. Soc., v. 43, p. 847-859.

Thompson, R. and Berglund, B., 1976, Late Weichselian geomagnetic "reversal" as possible example of the reinforcement syndrome: Nature, v. 259, p. 490-491.

Turner, G.M. and Thompson, R., 1979, Behavior of the earth's magnetic field as recorded in the sediment of Loch Lomond: Earth Planet. Sci. Lett., v. 42, p. 412-426.

Uyeda, S. *et al.*, 1963, Anisotropy of magnetic susceptibility of rocks and minerals: J. Geophys. Res., v. 68, p. 279-291.

Van Montfrans, H.M., 1971, Paleomagnetic dating in the North Sea basin: Earth Planet. Sci. Lett., v. 11, p. 226-235.

Verosub, K.L., 1975, Paleomagnetic excursions as magnetostratigraphic horizons: a cautionary note: Science, v. 190, p. 48-50.

______________, 1977, The absence of the Mono Lake geomagnetic excursion from the paleomagnetic record of Clear Lake, California: Earth Planet. Sci. Lett., v. 36, p. 219-230.

______________, 1979, Paleomagnetic evidence for the occurrence of rapid shifts in the position of the geomagnetic pole: EOS Trans. Am. Geophys. U., v. 60, no. 18, p. 244.

Verosub, K.L. and Cox, A., 1971, Changes in the total magnetic energy external to the earth's core: J. Geomag. and Geoelec., v. 23, p. 235-242.

Verosub, K.L. and Banerjee, S.K., 1977, Geomagnetic excursions and their paleomagnetic record: Rev. Geophys. and Space Phys., v. 15, no. 2, p. 145-155.

Verosub, K.L. *et al.*, 1979, The role of water content in the magnetization of sediments: Geophys. Res. Lett., v. 6, p. 226-228.

Vitorello, I. and van der Voo, R., 1977, Magnetic stratigraphy of Lake Michigan sediments obtained from cores of lacustrine clay: Quat. Res., v. 7, p. 398-412.

Watkins, N.D., 1972, Review of the development of the geomagnetic polarity time scale and discussion of prospects for its finer definition: Geol. Soc. Am. Bull., v. 83, p. 551-574.

______________, 1976, Polarity group sets up guidelines: Geotimes, v. 21, p. 18-20.

Yaskawa, K. *et al.*, 1973, Paleomagnetism of a core from Lake Biwa(I) J. Geomagn. Geoelec., v. 25, p. 447-474.

Zijderveld, J.D.A., 1975, Paleomagnetism of the Esterel rocks: Utrecht State University, Ph.D. dissertation.

PALEOMAGNETIC DATING OF QUATERNARY SEDIMENTS: A REVIEW

M. STUPAVSKY and C.P. GRAVENOR

ABSTRACT

Paleomagnetic dating of Quaternary sediments is a secondary dating method that is based on matching polarity transitions, excursions and secular variations of the paleomagnetic field recorded in the sediments with their radiometrically dated equivalents in the magnetic polarity time scale.

Approximately 10 polarity reversals (epochs and events) are known to have occurred during the Quaternary and, in many cases, these have been used successfully to date deep-sea and terrestrial sediments. In contrast, only a few sediments have been dated successfully by their paleosecular variation and polarity excursion records. The limited success rate of the use of these small amplitude time variable characteristics of the paleomagnetic field recorded in sediments results from the introduction of "magnetic noise" of comparable amplitude into the sediment magnetic record during sedimentation or sampling.

At present, the probability of successfully dating a sedimentary sequence by paleomagnetic methods is low (approximately 20% for deep-sea cores). The factors involved in this low success rate include, sedimentation processes which create distortions in the record, unrecognized sediment deformation after remanence acquisition, remanence resetting caused by shock induced thixotropy (*e.g.*, during sampling or by earthquakes), diagenetic production of secondary minerals and magnetic remanence instability. Nevertheless, the success rate is being improved significantly by the use of better sampling methods, sampling strategy, magnetic cleaning, and screening and smoothing of sediment remanence data to reject "magnetic noise".

These improved procedures are designed to identify and reject spurious observations and refine the geomagnetic calendar. This refinement of the geomagnetic time scale is an ongoing process and relative to other sediment dating techniques that span the Quaternary. Successful paleomagnetic dating has the highest relative and absolute time resolution.

INTRODUCTION

Paleomagnetic dating of sediments is a secondary dating method in that it is dependent upon a radiometrically dated magnetic polarity time scale and master secular variation curves to determine the magnetic remanence age of a sediment (Creer *et al.* 1976a; Cox, 1969; Dodson *et al.*, 1977; Lund *et al.*, 1976; Mankinen and Dalrymple, 1979; Thompson, 1975, 1978).

Thus, the first requirement in the dating process is to measure paleomagnetic field variations in rocks that have been dated radiometrically and, from this information, build up global paleomagnetic polarity time scale and regional master secular variation curves. The paleomagnetic variations in rocks of unknown age are then measured and the results matched against polarity time scale or master secular variation curves in order to determine the paleomagnetic and absolute age of the rocks in question.

It is the purpose of this report to provide a "state-of-the-art" review of the paleomagnetic dating of Quaternary sediments and to show the potential and limitations of the technique as an aid in dating and correlating sediments which have been deposited over the past 1.8 million years.

PALEOMAGNETIC FIELD TIME VARIATIONS

Paleomagnetic studies on rocks ranging in age from the earliest Precambrian to the Quaternary show that the Earth's magnetic field exhibits polarity reversals at frequent intervals throughout geologic time. Time intervals lasting approximately 10^6 years during which the magnetic polarity is primarily of one polarity are called epochs and named after early workers in geomagnetism (Brunhes, Matuyama, Gauss, Gilbert,......).

Table 1. Geomagnetic Field Variations

Process	Origin	Geographic extent	Amplitude	Time Scale (y)
micropulsations	external	continental	$\ll$ 1% of	10^{-4}
solar storms	meteorological	extent -	main dipole	10^{-3}
diurnal		global	field	10^{-2}
annual				10^{0}
11 yr				10^{1}
secular variation	non dipole field	continental extent	$\Delta D \sim 30°$ $\Delta I \sim 15\%$	$10^2 - 10^3$
polarity excursions	main dipole field	global	$\Delta D \gg 30°$ $\Delta I \gg 15\%$	$10^2 - 10^3$
polarity events	"	"	$\Delta D \sim 180°$ $\Delta I \sim 200\%$	$10^3 - 10^5$
polarity epochs	"	"	"	$10^5 - 10^7$

Note: ΔD, (ΔI) is the variation in the declination (inclination) of the geomagnetic field direction.

Within a polarity epoch, the paleofield may have one or more global polarity reversals, lasting the order of 10^4 to 10^5 years, termed events, and/or one or more global or regional large paleofield amplitude deviations from the mean time average value, lasting for less than 10^4 years, which are termed excursions. At all times, the Earth's magnetic field displays variations of small amplitude, the order of 30° in declination and 10° in inclination, which last from about 500 to 3000 years per cycle, which is termed secular variation and is the smallest variation in the Earth's magnetic field which can be used for dating purposes.

These paleomagnetic field variations, that range in time span from 10^6 to a few hundred years, determine the possible plaeomagnetic dating resolution for a particular sedimentary record. Thus, the age of a sediment and its rate of sedimentation, along with other factors which will be discussed, combine to determine which of the features of the paleomagnetic field can be used for dating purposes.

DATING OF RECENT SEDIMENTS

Thick sequences of Recent sediments which contain abundant amounts of datable organic matter are commonly found on most parts of the Earth. As a result, these sediments have been examined extensively in the past decade to determine the regional paleosecular variations in the magnetic field (*e.g.*, Lund and Banerjee, 1979). These studies have been used to derive several dated paleosecular variation master curves which can be used for regional correlation and dating purposes (Thompson, 1975, 1978; Dodson *et al.*, 1977; Creer *et al.*, 1976a; Banerjee *et al.*, 1979).

Thus, the age of a Recent sediment can be determined by measuring its secular variation record and matching the record with the regional master curves. However, successful matching and, hence, dating, is dependent upon a number of factors, such as:

(1) the paleofield is accurately recorded in the sediment

(2) the magnetic record is stable with time

(3) no deformation of the sediment has taken place after the paleofield record was set

(4) the magnetic record shows several secular variation cycles to achieve a reliable match

(5) the magnetic record has not been altered by earthquakes, by deformation of the specimen during sampling or by new magnetic remanence carriers created by diagenetic chemical reactions.

In many cases, one or more of these conditions is not met, with the result that the "unknown" magnetic records cannot be successfully matched with the master secular variation curves and, hence, cannot be dated by paleomagnetic means (Lund and Banerjee, 1979). Further information on sampling techniques, remanence measurement, data smoothing and presentation, analysis to test for remanence stability and the analytical objective method for correlating a magnetic record with master curves is contained in two recent articles by Lund and Banerjee (1979) and Thompson (1979).

DATING OF PLEISTOCENE SEDIMENTS

Paleomagnetic dating of Pleistocene sediments, such as those found in lake bottoms, dry-lake sediments, ocean-bottom sediments and glacial deposits, is possible by the use of polarity changes and excursions known to exist in the geomagnetic time scale. Successful dating by the use of these features requires that:

(1) the sediment accurately records one or more of these features

(2) the record has not been altered by sediment deformation after the paleofield record was set or by remanence resetting by earthquakes or during sampling

(3) the recorded features can be unambiguously correlated with the polarity time scale.

If the Pliocene-Pleistocene boundary is placed at about 1.8 Mya, then as shown in Figure 1, the early Pleistocene up to 0.73 Mya is in the Matuyama Epoch within which there are eight dated points corresponding to magnetic polarity changes. In the post 0.73 Mya Brunhes Epoch, there are two dated points, the Blake Event at 110,000 ya and the Lake Mungo Event at 30,000 ya. In addition, there are several other potential points within this time frame which might be used for dating purposes but these still require confirmation on a global scale.

The status of various excursions and events that have been detected in sediments, volcanic rocks and baked sediments deposited over the past 300,000 years has been critically reviewed by Verosub and Banerjee (1977). They point out that in order to date a sediment,

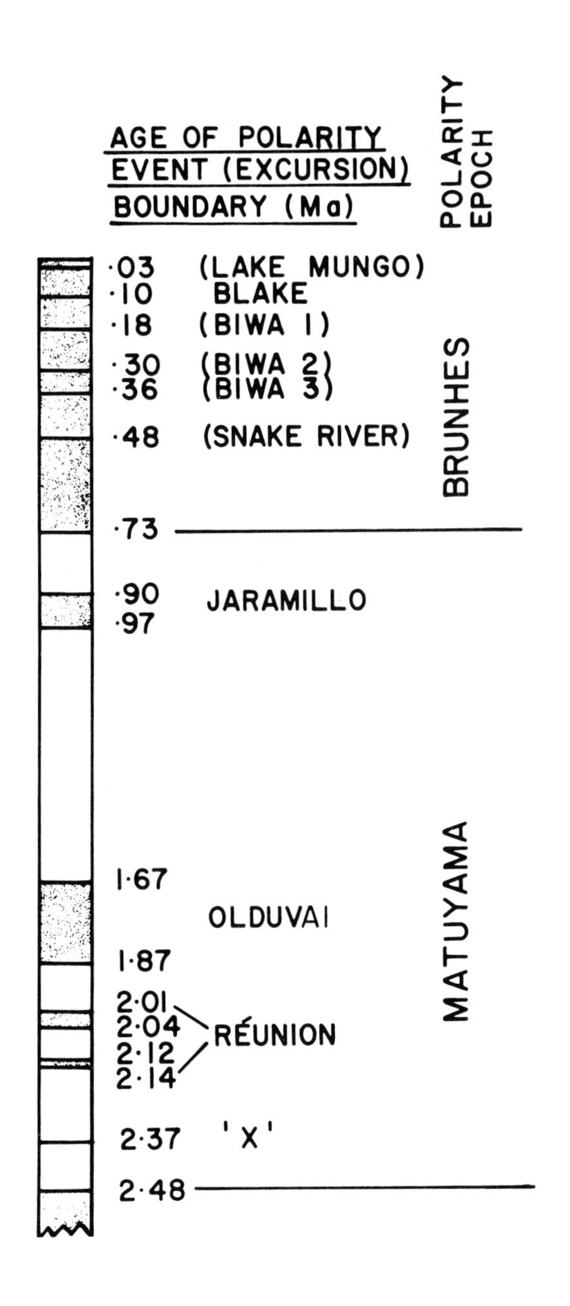

Figure 1 Polarity reversal time scale for the Quaternary. (After, Mankinen and Dalrymple, 1979)

events and excursions in the sedimentary record must have internal consistency. In order to meet this requirement, the excursion or event with the same characteristics must be found repeatedly in the same sediment. In addition, it is desirable to show that the event or excursion has spatial consistency in sediments of the same age and it has been suggested that regional geomagnetic field fluctuations should be observed over lateral distances exceeding 1000 km (Harrison and Ramirez, 1975; Verosub and Banerjee, 1977).

Many of the excursions listed in Table 2 lack the requirement of spatial consistency. For example, the Starno event (Noel and Tarling, 1975), the Gothenburg event (Mörner and Lanser, 1974), the Erieau excursion (Creer *et al.*, 1976b), an excursion in Lake Michigan sediments (Vitorello and Van der Voo, 1977), have not been observed in time equivalent sediments from Recent lake sediments in Minnesota (Banerjee *et al.*, 1979). This does not mean that all of these events or excursions are spurious but rather that spatial consistency has yet to be demonstrated so that they can be accepted for dating purposes.

In a given sedimentary record, only one event or excursion may be present and, obviously, this can present difficulties in matching the event or excursion with accepted polarity changes found in other sedimentary records. In some instances, however, the characteristic signature of a global event or excursion may be useful in identifying which event or excursion has been found. Thus, a polarity excursion recorded in sediments from widely separated geographic locations should give similar virtual geomagnetic poles (VGP). For example, the excursion recorded in the Meadowcliffe Till at Toronto, Ontario, dated at approximately 30,000 ya (Stupavsky *et al.*, 1979), yields a similar VGP to that recorded in some New England tills (Soloyanis and Brown, 1979) and both give VGP which correlate well with those given by the Lake Mungo excursion recorded in baked sediments in Australia (Barbetti and McElhinny, 1976).

Table 2. Reported Polarity Excursions in the Past 500,000 Yr.

Locality	Years BP	
Southern Sweden	∿ 2800	Noel and Tarling, 1975
Southern Sweden	12,077-12,103	" " " "
Gothenburg, Sweden	12,350-12,400	Mörner and Lanser, 1974
Czechoslovakia	12,000	Bucha, 1973
Lake Michigan cores	8000-10,000	Vitorello and Van der Voo, 1977; Creer *et al.*, 1976a
Lake Erie cores	8000-14,000	Creer *et al.*, 1976b
Lake Biwa, Japan	18,000	Yaskawa *et al.*, 1973; Yaskawa, 1974
Imuruk Lake, Alaska	18,000	Noltimier and Colinvaux, 1976
Gulf of Mexico	15,000-18,000	Clark and Kennett, 1973
Rubjerg, Denmark	23,000-40,000	Abrahamsen and Knudsen, 1978
Mono Lake, California	25,000	Denham and Cox, 1971; Denham, 1974; Liddicoat and Coe, 1979
Lake Mungo, Australia	28,000-31,000	Barbetti and McElhinny, 1976
Gulf of Mexico	30,000-33,000	Freed and Healy, 1974
Toronto, Ontario	∿ 30,000	Stupavsky *et al.*, 1979
New England, U.S.A.	∿ 30,000	Soloyanis and Brown, 1979
Laschamp, France	∿ 33,000	Bonhommet and Zahringer, 1969; Valladas *et al.*, 1977
Washington, U.S.A.	> 45,000	Easterbrook, 1975; Easterbrook and Othberg, 1975
Lake Biwa, Japan	49,000	Yaskawa *et al.*, 1973; Yawkawa, 1974
deep-sea cores (Blake event)	104,000-117,000	Smith and Foster, 1969; Denham, 1976; Creer *et al.*, 1980
Lake Biwa, Japan	∿ 110,000)	
Lake Biwa, Japan	176,000-186,000)	Kawai *et al.*, 1972; Creer
Lake Biwa, Japan	292,000-298,000)	*et al.*, 1980
Snake River	480,000	Champion *et al.*, 1979

To sum up, an excursion or event that lacks internal and spatial consistency and has a signature which cannot be correlated with known excursions is probably unrelated to a geomagnetic polarity excursion. In many instances, these spurious excursions or events likely result from problems related to disturbance of the sediment which have been mentioned above and will be discussed in more detail.

SAMPLING AND MEASUREMENT PROCEDURES

Terrestrial sediments may be sampled by a variety of techniques but, in order to provide evidence of internal consistency, it is normal practice to continuously sample two or more time-equivalent profiles. Samples are usually obtained by cutting out oriented small cubic specimens ($\simeq$ 2.2 cm on a side) or oriented blocks (10 cm or more on a side) from which smaller cubes or cylinders are later cut out or drilled out in the laboratory. If the sediment is sufficiently soft and moist, the smaller cubes can be obtained in the field by pressing hollow plastic cubes into the sediment. The vertical spacing between consecutive samples will depend, in part, on the sedimentation rate but, in general, 10 cm spacing intervals may be used if this amount of sediment was deposited over approximately 50 to 100 years.

When sampling lake or ocean bottom sediments, long continuous cores are obtained using one of the many coring devices available. The relative merits of these coring devices has been discussed in detail by Lund and Banerjee (1979) and Thompson (1979). The remanence of the core may be measured continuously on a long core magnetometer (Molyneaux *et al.*, 1972; Dodson *et al.*, 1974), or by removing specimens by pressing hollow plastic cubes into the core at intervals of about 2 to 5 cm.

If two samples are removed from each time-equivalent position within the core and two parallel time-equivalent cores are used then it should be possible, from remanence variance analysis, to segregate that fraction of the observed sediment remanence which is due to variations of the paleomagnetic field from that fraction which results from disturbance of the sediment or other "noise" in the magnetic record. A simple estimate of the paleomagnetic field variation (signal) content that is recovered from the sediment record is the difference of mean between consecutive horizon specimens and mean within horizon sample angular deviation.

Any differential variation between the profiles in the fidelity of the sediment as a recorder of the paleofield and sediment deformation after remanence acquisition is revealed by a non-zero angular deviation between time-equivalent samples from two profiles (Symons *et al.*, 1980). An analytical cross-validation smoothing of the remanence data from a single profile has been devised by Clark and Thompson (1978) to recover the paleomagnetic signal and its confidence limits from the observed sediment record. It should be pointed out, however, that the correlatable between core variation may be spurious if it originates from regional climate-controlled sedimentologic factors which can alter the accuracy with which the sediment recorded the paleomagnetic field (Harrison, 1974; Verosub, 1977).

Over the past few years, the results of paleomagnetic work on sediments has become more reliable as various tests are used to determine the nature of sediment remanence and its stability. These include:

(1) alternating field (AF) demagnetization to remove viscous remanent magnetization (VRM) acquired during storage in the ambient magnetic field

(2) storage tests in the ambient laboratory magnetic field to determine the rate of VRM acquisition

(3) measurement of magnetic susceptibility to determine the amount of magnetic minerals present in a specimen and to determine Koenigsberger ratios

(4) measurement of the anisotropy of magnetic susceptibility (which is a measure of magnetic fabric) to identify deformation in sediments (Verosub, 1977; Løvlie and Holtedahl, 1980)

(5) the measurement of saturated isothermal remanent magnetization

(SIRM) to identify magnetic minerals

(6) shock tests to determine the significance of remanence re-setting from shock induced thixotropy during sediment sampling (Symons *et al.*, 1980).

After all due care has been taken in the sampling and measurement procedures, the remanence data may be screened to select only homogeneously (reliably) magnetized specimens for interpretive purposes. The screened data is smoothed by either using a simple moving mean direction of a number of adjacent sample directions or using an analytical cubic spline smoothing technique developed by Clark and Thompson (1978). Finally, the smoothed declination (if available) and inclination of the remanence is plotted against stratigraphic position.

Alteration of the Paleomagnetic Record by Sedimentation Processes

The processes by which a sediment acquires a remanent magnetization are complex and have been reviewed by Verosub (1977). The processes involved are given in Table 3. The initial detrital remanent magnetization (DRM) is developed when the magnetic moments of the remanence carriers are oriented in the magnetic field as they fall through the water column. The remanence thus acquired may vary significantly from the magnetic field due to a number of sedimentologic factors, such as the action of bottom currents, sediment grain size distribution and slope of bedding planes. After initial sedimentation, diagenetic effects, such as bioturbation and dewatering, may result in distortion or loss of part or all of the original DRM and the development of post-depositional remanent magnetization that is parallel to the ambient field present when the disturbance took place.

Table 3. Origin of NRM of Sediments

Magnetization	Process	Accuracy of record of the geomagnetic field	Time interval of NRM acquisition
detrital remanent magnetization (DRM)	settling through a water column	poor, $\Delta I \sim 20°$ $\Delta D \sim 40°$	virtually instantaneous
post deposition remanent magnetization (PDRM)	shear induced liquefaction during bioturbation, sediment deformation, earthquakes, sediment coring	excellent	10^3 - 10^4 after deposition depending on the rate of dewatering in deep-sea sediment; 10 - 100 y after deposition in lake sediments
chemical remanent magnetization (CRM)	chemical production of magnetic minerals in the sediment	excellent	over the age of the sediment

Note: ΔI = 'inclination error' = deviation of the inclination of the NRM from the ambient magnetic field

ΔD = 'declination error' = deviation of the declination of the NRM from the ambient magnetic field

As a result of these and other factors (see Table 4), the remanent magnetization of a sediment may consist partly of the initial DRM and postdepositional remanent magnetization. Thus, while varved clays may retain up to 100% of the initial DRM, up to 100% of the remanent magnetization in bioturbated sediments may have been acquired after deposition.

Table 4. Sediment NRM = f (DRM, PDRM, CRM)

Sediment type	NRM magnetization type	Record of paleomagnetic field
glacial varved deposits:	NRM $\approx$ 100% DRM	poor record of the paleomagnetic field. Sedimentologic factors produce spurious variations of comparable amplitude to paleosecular variations. (Figure 2)
good deep-sea sediment cores: good lake sediment cores:	NRM $\approx$ 100% PDRM	excellent record of paleomagnetic field retaining the sequence of polarity excursions, events and epochs in accord with the polarity reversal time scale. (Figures 3-4)
poor deep-sea sediment cores: poor lake sediment cores:	NRM $\approx$ 30-100% CRM	poor record of paleomagnetic field. The sequence of polarity excursions, events and epochs has been overprinted by intense and stable CRM acquired in the Recent geomagnetic field.
	NRM $\approx$ 30-100% PDRM	poor record of paleomagnetic field. Paleomagnetic field record has been completely or partially overprinted by recently acquired PDRM as a result of shock-induced thixotropy during recent earthquakes or during sediment sampling. (Figure 5, Table 5)

In addition to those syn-depositional difficulties, post-depositional deformations present a different set of problems. For example, varved clays usually retain their original DRM during post-depositional deformation which can be proven by a fold test (Graham, 1949; Johnson *et al.*, 1948; Verosub, 1975). On the other hand, in other sediments the remanence may be acquired after deformation (Keen, 1963). Between these two extremes, some sediments may acquire part of their RM before deformation and after deformation (Niitsuma, 1977).

Sediment deformation may be detected in some cases by visual examination of the sediment, X-ray analysis or by measuring the anisotropy of magnetic susceptibility (Løvlie and Holtedahl, 1980). Quite often, sediment deformation is local in character and, if two parallel cores have been taken and a large amplitude paleofield variation is

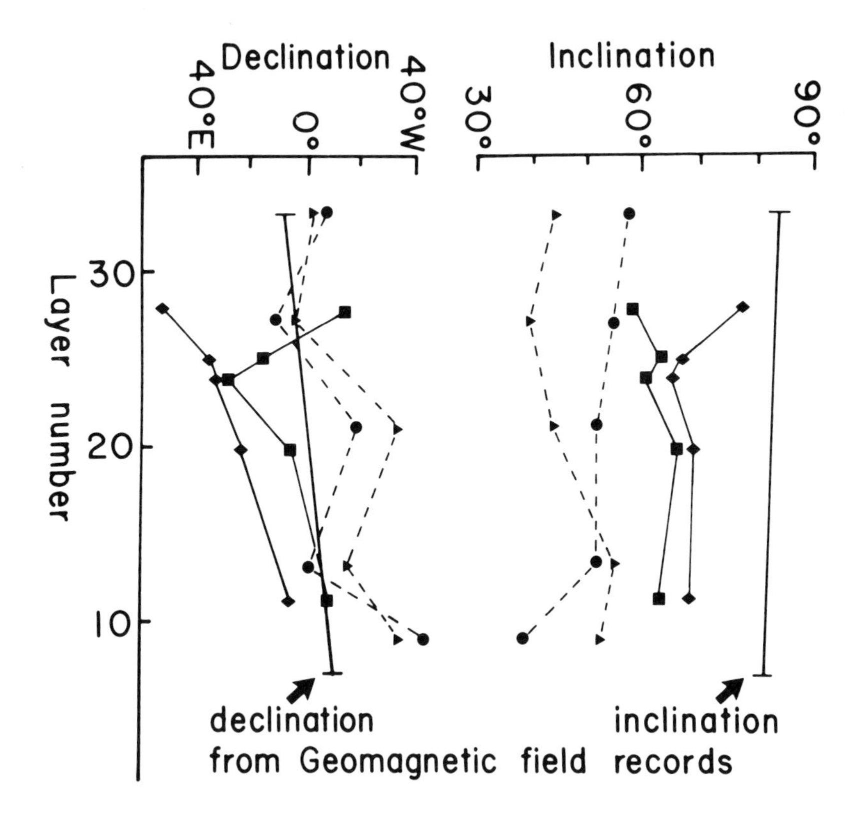

Figure 2 Remanence inclination and declination of varved sediments deposited between 1700 to 1800 AD from Hagavatn, Iceland. The two sites were about 100 m apart.
♦, site 1, ochre, ▲, site 1, grey; ■ , site 2, ochre, ●, site 2, grey. These varves provide a poor record of the paleomagnetic field. The observed variations are clearly spurious (After Griffith *et al.*, 1960).

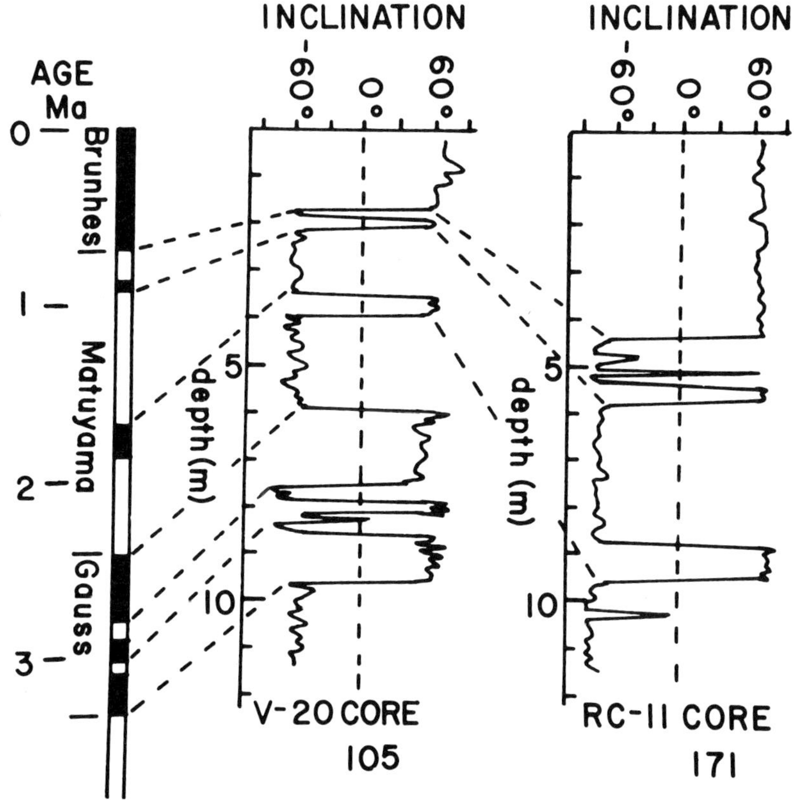

Figure 3 Remanence inclinations from two 'good' deep-sea sediment cores (V-20 core 105; RC-11 core 171). The sequence of polarity reversals are in accord with the polarity reversal time scale. (After Opdyke, 1972)

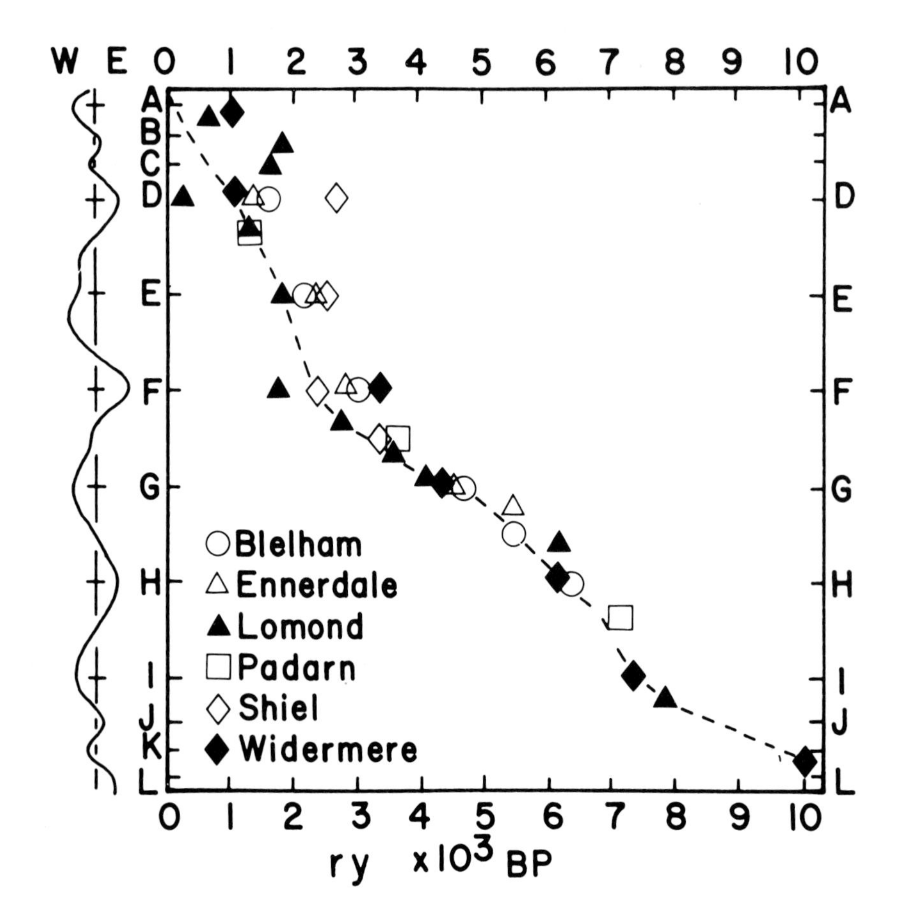

Figure 4 Remanence declination variations (A-L) *vs* radiocarbon ages for 6 British lakes. Dashed line shows the preferred ages of the geomagnetic fluctuations (A-L). The good within lake and between lake paleomagnetic correlation suggests that these sediments provide a good record of the paleomagnetic field. (After Thompson, 1979)

found in one core but not at the same time-equivalent position in the second core, then the spurious results can be rejected.

REMANENCE RESETTING BY EARTHQUAKES AND SAMPLING PROCEDURES

Laboratory studies indicate that sediments acquire a stable RM about 10 days after deposition when the water content is reduced below a certain critical level (Løvlie, 1974). On deposition, sediments have a high water content and, hence, the magnetic torque on most of the remanence carriers exceeds the viscous drag force and, consequently, the magnetic moments are aligned parallel to the magnetic field. As a sediment loses water and the viscosity increases, the magnetic torque on large multi-domain carriers which have the lowest magnetic moment per unit volume, falls below the viscous drag force, and the remanence of these carriers is fixed. As dewatering proceeds, the magnetic torque on single domain carriers which have the highest magnetic moment per unit volume falls below viscous drag forces and their remanence is fixed. Thus, most of the post-depositional RM is acquired when the sediment is in the final stages of dewatering.

When a sediment is subjected to a shock, such as an earthquake or during sampling, its viscosity may momentarily decrease from 10^4 to 10 poises (Francis, 1971). During this shock induced thixotropy, the magnetic moment of some carriers, particularly those which have high magnetic moment per unit volume, are reoriented in the ambient field. Thus, the sediment RM may be partially or wholly reset dependent on the fractions of carriers which are realigned.

Remanence resetting by earthquake induced thixotropy was first proposed to account for the observation that many deep-sea cores which span several million years are magnetized normally (Francis, 1971). He suggested that deep-sea sediments found in earthquake-prone areas may have their remanence reset at about 400 y intervals.

Games (1977) showed that the RM found in mud bricks is acquired when the bricks undergo thixotrophy during injection into a mould. Similarly, laboratory studies show that some sediments are very susceptible to remanence resetting when subjected to mechanical shock (Symons *et al.*, 1980; Stupavsky *et al.*, 1979). Consequently, any sampling process which produces mechanical shock or an effect similar to injecting clay into moulds may cause remanence resetting. Sediments which are susceptible to remanence resetting by shock can be readily identified by a simple laboratory mechanical shock test (Symons *et al.*, 1980). This laboratory test should be used on a routine basis so that, if required, sampling procedures can be modified to prevent remanence resetting during sampling (Stupavsky *et al.*, 1979).

CHEMICAL PRODUCTION OF REMANENCE CARRIERS

Diagenetic chemical reactions in sediments are well known and, in some instances, these reactions may produce remanence carriers which give the sediment a chemical remanent magnetization (CRM). In addition to the production of new carriers, old carriers which carried the original DRM may be destroyed. It is evident that if a significant amount of CRM is developed in a sediment that the sediment will be unsuitable for dating purposes. For example, it has been found that the reason that some deep-sea cores are unsuitable for dating is because of the presence of excess amounts of CRM (Harrison, 1974).

INHERENT LIMITATIONS ON THE ACCURACY OF PALEOMAGNETIC DATING ARISING FROM THE NATURE OF THE NRM OF SEDIMENTS

The inherent limitations on sediment dating are:

(1) The NRM is acquired over a variable period of time after deposition. In deep-sea sediments, PDRM is acquired 10^3-10^4 y after deposition depending on the rate of dewatering. In lake sediment, it is acquired 10-100 y after deposition.

(2) The accuracy of the paleomagnetic field recorded in the sediment NRM is variable. Thus, the accuracy of secular variation sediment dating that is based on small amplitude variations of the paleomagnetic field is poor.

(3) The paleomagnetic field recorded by the sediment NRM is subject to magnetic overprinting by the Recent geomagnetic field. Magnetic overprinting is a serious problem. In deep-sea sediments, $\sim$ 80% of cores have been overprinted by CRM or PDRM acquired in the Recent geomagnetic field and are thus unsuitable for dating. In lake sediments, magnetic overprinting problems are likely even more serious. Even partial magnetic overprinting during sampling will introduce spurious variation in the sediment record of comparable magnitude to the paleosecular variation by which the sediment is to be dated.

SUCCESS RECORD OF PALEOMAGNETIC DATING OF QUATERNARY SEDIMENTS

The limitations and problems discussed in this paper have an effect on the success record of absolute age paleomagnetic dating of sediments. For example, the absolute paleomagnetic age of a sediment refers to the time when the sediment acquired its stable RM. Thus, if the sedimentation rate is low and the RM is fixed after the effects of bioturbation and dewatering, then the age obtained is likely much younger than the true age of the sediment. In deep-sea sediments, sediment rates are the order of 1 cm/10^3 y and the depth of bioturbation may be the order of 10 cm (Gartner, 1972). In this case, the paleomagnetic age would be about 10^4 y younger than the age of deposition.

The success of polarity-reversal paleomagnetic dating of slowly deposited oceanic sediments which span periods of about 10^6 years is well documented and is well below unity. For example, from an analysis of 216 deep-sea cores that penetrate the Brunhes-Matuyama boundary, Harrison (1974) showed 43 or 20% are suitable for dating. The remaining 80% of the cores either showed no polarity reversals or the polarity-reversal pattern did not correlate with the polarity-reversal time scale.

The frequency of successful polarity-reversal dating of terrestrial sediments is not known but is probably not significantly higher than the 20% of deep-sea sediments. Nevertheless, the low success rate emphasizes the need to exercise considerable care in observing the limitations imposed by difficulties arising from the nature of sedimentation and sampling procedures.

Aside from the well documented Brunhes-Matuyama boundary and the Blake, Jaramillo and Olduvai events, at present there are only two polarity excursions within the Quaternary which can be used for dating purposes. For example, polarity excursion dates of about 29,000 y have been obtained from the Meadowcliffe Till at Toronto and from some New England tills (Stupavsky *et al.*, 1979; Soloyanis and Brown, 1979). The excursive VGP recorded by these tills correlate very well with the Lake Mungo VGP which has been dated at about 29,000 y (Barbetti and McElhinny, 1976). Finally, although many attempts have been made to date Recent sediments by paleosecular variation, the only explicit dating of Recent sediments has been made by Thompson (1979) in his test of paleosecular variation dating.

TIME RESOLUTION OF PALEOMAGNETIC DATING

Although one of the purposes of this paper is to point out the limitations of paleomagnetic dating, it is appropriate to also consider certain of the positive aspects of this dating technique. Of all the sediment dating techniques that span the Quaternary, paleomagnetic dating probably has the best relative and absolute time resolution. For example, if proper procedures are used and regional master secular variation curves developed, it has been shown that it is possible to provide dates on sediments of unknown age which are in good agreement with accepted radiocarbon dates (Thompson, 1979).

Polarity changes and global polarity excursions provide an unparalleled opportunity to correlate and date Quaternary terrestrial and deep-sea sediments on a global scale. The time required for a polarity transition is about 10^3 y (Fuller *et al.*, 1979) and for a polarity excursion about 10^2 - 10^3 y. Thus, the age of any sediment which records a global excursion (*e.g.*, the Lake Mungo excursion) or a polarity change (*e.g.*, the Blake event or the Brunhes-Matuyama boundary) can be determined with a relative accuracy of between 10^2 and 10^3 y.

Obviously, the absolute paleomagnetic age of a sediment is limited by the accuracy of the polarity time scale. At present, the time scale for the past 5 my is based on 354 K-Ar dates on volcanic rocks (Mankinen and Dalrymple, 1979). The ages of the polarity reversals have either been directly measured from volcanic rocks that record polarity reversals or are known within tight limits. As new excursions are verified and dated, it is evident that the potential of paleomagnetic dating will improve.

RECOMMENDATIONS FOR ROUTINE SCREENING OF SEDIMENT DATA

In examining the various published paleosecular variation studies on sediments for this review article, a serious deficiency is observed in the data treatment procedure. In most publications, all the measured data is accepted for interpretation. No routine screening of the data is carried out to reject unreliably magnetized data before interpretation despite the strong evidence that a significant fraction may be unreliably magnetized so that its interpretation has no paleomagnetic significance. The evidence includes large remanence variations within and between time equivalent samples, erratic remanence direction changes of large amplitude between approximately time equivalent samples, and the remanence variations of pilot specimens during step demagnetization. The rejection of the unreliable data through a suitable screening procedure will increase the validity of any interpretation.

In rock paleomagnetic studies, the remanence data is routinely screened at the core and site level to select only the reliably magnetized data for interpretation. This two-tiered screening procedure that requires remanence homogeneity at the core and site levels may reject more than 50% of the data as unreliable in some studies.

The screening criteria devised for rock paleomagnetic studies are directly applicable to sediment studies.

SPECIMEN LEVEL SCREENING

The remanence measurement procedure on the various magnetometers is such that each Cartesian remanence component is measured several times. From these redundant measurements, the within specimen remanence variation can be computed. Only homogeneously magnetized specimens having remanence angular standard deviations smaller than some chosen value (*e.g.*, 5°) are accepted for interpretation (Harrison, 1980; Briden and Arthur, 1981; Lowrie *et al.*, 1980). This within specimen remanence variation provides a measure, σ_N, of the spurious remanence variations found in rapidly deposited sediments.

CORE LEVEL SCREENING

The remanence angular deviation between two specimens from a single core; *i.e.*, two time equivalent samples if the coring is horizontal may also be used as a screening criterion. Thus, only homogeneously magnetized cores or samples having an angular deviation between the directions of the two specimens or samples which is smaller than some chosen value (*e.g.*, 5°) are accepted for interpretation. This within core remanence variation also provides a measure for σ_N of the spurious remanence variations in rapidly deposited sediments.

FURTHER REMANENCE VARIANCE ANALYSIS OF SEDIMENT DATA

The measured sediment remanence consists in part of the paleomagnetic field variations (signal) and spurious variations (noise) arising from variable sedimentation conditions, variable degree of partial remanence overprinting during sediment thixotropy, variable degree of CRM acquisition, *etc*. An estimate of the signal to noise content of the record may be determined by computing the between consecutive core (specimen) angular remanence deviation. When the consecutive samples are separated by more than about 100 years in time, then this provides a measure of the signal variations, σ_S, together with noise variations, σ_N. The mean signal variation content, σ_S, for the whole profile is the difference of the mean between core (specimen) and the mean within core (specimen) angular remanence variation. The mean signal to noise ratio $\bar{\sigma}_S/\bar{\sigma}_N$ provides a measure of the reliability of the sediment's remanence. High values (> 2) indicate good remanence reliability. Conversely, low values (< 1) indicate the sediment record is unsuitable for paleomagnetic interpretation since the measured variations are almost completely spurious. Applying this type of remanence variance analysis to the data in Figure 5, (Table 5), a low $\bar{\sigma}_S/\bar{\sigma}_N \simeq 0.4$ is obtained that indicates the observed variations are largely spurious and have no paleomagnetic significance.

OTHER EFFECTIVE SCREENING CRITERION

When the sediment is sampled at two time equivalent profiles, the angular remanence variation between time equivalent samples from the two

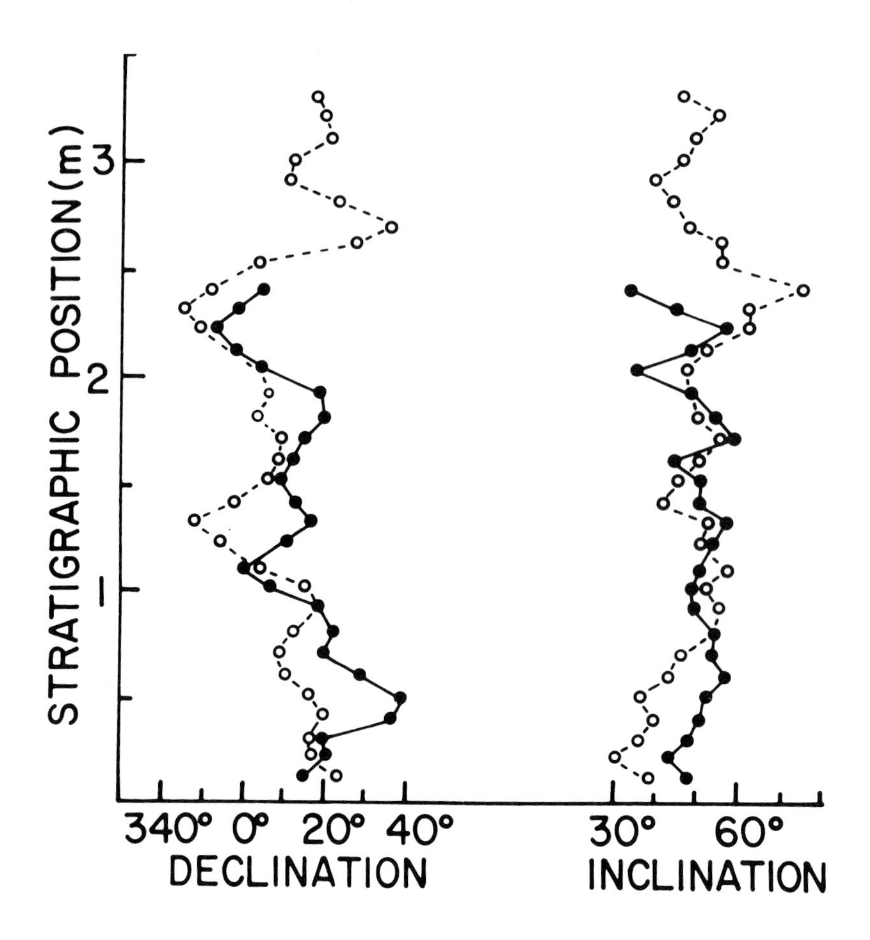

Figure 5 Remanence variations from the Seminary Till, Toronto. This sediment provides a poor record of the paleomagnetic field. Variance analysis of the data (Table 5) shows that the observed variations are spurious. (After Symons *et al.*, 1980).

******

Table 5. Remanence Angular Variance Analysis

Remanence variation	Number of cores	Mean angular deviation (degrees)	Standard deviation (degrees)
within core	43	13	11
between consecutive cores	48	18	10

mean paleosecular variation estimate = mean between - within core deviation
= mean ((signal + noise) - noise)
= 5°, s.d. = 21°

Thus the observed variation has no dating value because it is mostly spurious.

profiles provides a measure of σ_N which may be used also an an effective screening criterion. Such screening ensures good between profile remanence correlation.

The most suitable and effective sediment screening procedure still needs to be determined. However, it is apparent to the authors that routine screening and detailed remanence analysis of sediment data can significantly improve its validity.

SUMMARY

Throughout this paper, we have attempted to demonstrate certain of the limitations inherent in the paleomagnetic dating of Quaternary sediments. These limitations are related, in part, to the complex syn- and post-depositional processes by which the remanent magnetization is recorded in a sediment. Other limitations relate to the relatively small number of well-documented polarity excursions and events which have taken place over the past 1.8 my.

Although the success rate for the paleomagnetic dating is low, this will undoubtedly improve as more attention is paid to those factors which effect the acquisition of an accurate RM in sediments. In addition, better instrumentation is now available which will help to isolate the original remanent magnetization in sediments.

In short, the paleomagnetic dating of sediments is going through "growing pains" which have been encountered in other methods of dating sediments. It has the inherent advantage of being independent of finding datable material, such as organic matter and tephra and, where successful, has the highest relative and absolute time resolution of any other dating method. Hence, despite the limitations, paleomagnetic dating of Quaternary sediments has a bright future and will play a strong role in the correlation and dating of Quaternary sediments.

ACKNOWLEDGEMENTS

We would like to thank those people who were kind enough to send us reprints, preprints and comments on their most recent studies.

REFERENCES CITED

Abrahamsen, N. and Knudsen, K.L., 1978, Indication of a geomagnetic low-inclination excursion in supposed middle Weichselian interstadial marine clay at Rubjerg, Denmark: Physics of the Earth and Planet. Interiors, v. 18, p. 238-246.

Banerjee, S.K., Lund, S.P. and Levi, S., 1979, Geomagnetic record in Minnesota lake sediments - Absence of the Gothenburg and Erieau Excursions: Geology, v. 7, p. 588-591.

Barbetti, N.F. and McElhinney, M.W., 1976, The Lake Mungo geomagnetic excursion: Phil. Trans. R. Soc. London, v. 281, p. 515-542.

Bonhommet, N. and Zähringer, J., 1969, Paleomagnetism and potassium - argon age determinations of the Laschamp geomagnetic polarity event: Earth Planet. Sci. Lett., v. 6, p. 43-46.

Briden, J.C. and Arthur, G.R., 1981, Precision of measurement of remanent magnetization: Can. J. Earth Sci., v. 83, p. 527-538.

Bucha, V., 1973, The continuous patterns of variation of the geomagnetic field in the Quaternary and their causes: Stud. Geophys. Geod., v. 17, p. 218-231.

Champion, D., Dalrymple, B., Kuntz, M. and Doherty, D., 1979, Reversed polarity lava flows within a Late Pleistocene volcanic sequence from Snake River Main: a possible reversed event within the Bruhnes normal polarity epoch: EOS, v. 60, p. 814.

Clark, H.C. and Kennett, J.P., 1973, Paleomagnetic excursion recorded in latest Pleistocene deep-sea sediments, Gulf of Mexico: Earth Planet. Sci. Lett., v. 19, p. 267-274.

Clark, R.M. and Thompson, R., 1978, An objective method for smoothing paleomagnetic data: Geophys. J.R. Astr. Soc., v. 52, p. 205-213.

Cox, A., 1969, A paleomagnetic study of secular variation in New Zealand: Earth Planet Sci. Lett., v. 6, p. 257-267.

Creer, K.M., Gross, D.L. and Lineback, J.A., 1976a, Geomagnetic variations recorded by Late Pleistocene lacustrine sediments from Lake Michigan: Geol. Soc. Am. Bull., v. 87, p. 531-540.

Creer, K.M., Anderson, T.W. and Lewis, C.F.M., 1976b, Late Quaternary geomagnetic stratigraphy recorded in Lake Erie sediments: Earth Planet. Sci.Lett., v. 31, p. 37-47.

Creer, K.M., Readman, P.W. and Jacobs, A.M., 1980, Paleomagnetic and paleontological dating of a section at Gioia Tauro, Italy: Identification of the Blake event: Earth Planet. Sci. Lett., v. 50, p. 289-300.

Denham, C.R., 1974, Counter-clockwise motion of paleomagnetic directions 24,000 years ago at Lake Mono, California: J. Geomag. Geoelectr., v. 26, p. 487-494.

__________, 1976, Blake polarity episode in two cores from the Greater Antilles outer ridge: Earth Planet. Sci. Lett., v. 29, p. 422-434.

Denham, C.R. and Cox, A., 1971, Evidence that the Laschamp polarity event did not occur 13,300-30,400 years ago: Earth Planet. Sci. Lett., v. 13, p. 181-190.

Dodson, R.E., Fuller, M.D. and Pilant, W., 1974, On the measurement of the remanent magnetism of long cores: Geophys. Res. Lett., v. 1, p. 185-188.

Dodson, R.E., Fuller, M.D. and Kean, W.F., 1977, Paleomagnetic records of secular variation from Lake Michigan sediment cores: Earth Planet. Sci. Lett., v. 34, p. 387-395.

Easterbrook, D.J., 1975, Excursions of the geomagnetic field during the Brunhes epoch: IAGA Program and Abstracts, XVI General Assembly SM4-11, p. 174.

Easterbrook, D.J. and Othberg, K., 1975, Paleomagnetism of Pleistocene sediments in the Puget Lowland, Washington: in Easterbrook, D.J. and Sibrava, V., eds., Project 73/1/24 Quaternary Glaciations in the Northern Hemisphere. Report No. 3. On the session in Bellingham, Washington, U.S.A., p. 189-207.

Francis, T.J.G., 1971, Effect of earthquakes on deep-sea sediments. Nature, 233, p. 98-102.

Freed, W.K. and Healy, N., 1974, Excursions of the Pleistocene geomagnetic field recorded in the Gulf of Mexico sediments: Earth Planet. Sci. Lett., v. 24, p. 99-104.

Fuller, M.D., Williams, I.S. and Hoffman, K.A., 1979, Paleomagnetic records of geomagnetic field reversals and the morphology of the transitional fields: Rev. of Geophys. and Space Phys., v. 17, no. 2, p. 179-203.

Games, K.P., 1977, The magnitude of the paleomagentic field: A new non-thermal, non-detrital method of using sun-dried bricks: Royal Astron. Soc. Geophys. J., v. 48, p. 315-329.

Gartner, S., 1972, Late Pleistocene calcareous nannofossils in the Caribbean and their interoceanic correlation: Palaeogeog., Palaeoclimatol., Palaeoecol., v. 12, p. 169-191.

Graham, J.W., 1949, The stability and significance of magnetism in sedimentary rocks: J. Geophys. Res., v. 54, p. 131-167.

Griffith, D.H., King, R.F., Rees, A.I. and Wright, A.E., 1960, The remanent magnetism of some recent varved sediments: Proc. Roy. Soc. 'A', v. 256, p. 359-383.

Harrison, C.G.A., 1974, The paleomagnetic record from deep-sea sediment cores: Earth Science Reviews, v. 10, p. 1-36.

_________________, 1980, Analysis of the magnetic vector in a single rock specimen: Geophys. J.R. Astr. Soc., v. 60, p. 489-492.

Harrison, C.G.A. and Ramirez, E., 1975, Areal coverage of spurious reversals of the earth's magnetic field: J. Geomag. Geoelect., v. 27, p. 139-151.

Johnson, E.A., Murphy, T. and Torreson, O.W., 1948, Pre-history of the Earth's magnetic field: Terr. Mag., v. 53, p. 349-372.

Kawai, N., Yaskawa, K., Nakajima, T., Torii, M. and Horie, S., 1972, Oscillating geomagnetic field with a recurring reversal discovered from Lake Biwa: Proc. Japan Acad., v. 48, p. 186-190.

Keen, M.J., 1963, The magnetization of sediment cores from the eastern basin of the North Atlantic Ocean: Deep-Sea Res., v. 10, p. 607-622.

Liddicoat, J.C. and Coe, R.S., 1979, Mono Lake geomagnetic excursion: J. Geophys. Res., v. 84, p. 261-271.

Løvlie, R., 1974, Post-depositional remanent magnetization in a re-deposited deep-sea sediment: Earth Planet. Sci. Lett., v. 21, p. 315-320.

Løvlie, R. and Holtedahl, H., 1980, Apparent paleomagnetic low-inclination excursion in a pre-consolidated continental shelf sediment: Physics of the Earth and Planet. Interiors, v. 22, p. 137-143.

Lowrie, W., Channell, J.E.T. and Heller, F., 1980, On the credibility of remanent magnetization measurement: Geophys. J.R. Astr. Soc., v. 60, p. 493-496.

Lund, S.P., Bogdan, D., Levi, S., Banerjee, S.K. and Long, A., 1976, Continuous geomagnetic record for the past 10,000 years from Lake St. Croix, Minnesota: EOS Trans. Amer. Geophys. Union, v. 57, p. 10.

Lund, S.P. and Banerjee, S.K., 1979, Paleosecular variations from lake sediments: Rev. Geophys. and Space Phys., v. 17, no. 2, p. 244-249.

Mankinen, E.A. and Dalrymple, G.B., 1979, Revised geomagnetic polarity time scale for the interval 0-5 m.y. BP: J. Geophys. Res., v. 84, no. B2, p. 615-626.

Mörner, N.A. and Lanser, J.P., 1974, Gothenburg magnetic 'flip': Nature, v. 251, p. 408-409.

Molyneaux, L., Thompson, R., Oldfield, F. and McCallen, M.E., 1972, Rapid measurements of the remanent magnetization of long cores of sediment: Nature, v. 237, p. 42-43.

Niitsuma, N., 1977, Remanent magnetization of slumped marine sedimentary rocks: in Rock Magnetism and Paleogeophysics: v. 4, published by the Rock Magnetism and Paleogeophysics Research Group in Japan, p. 44-52.

Noel, M. and Tarling, D.H., 1975, The Laschamp geomagnetic 'event': Nature, v. 253, p. 705-707.

Noltimier, H.C. and Colinvaux, P.A., 1976, Geomagnetic excursion from Imuruk Lake, Alaska: Nature, v. 259, p. 197-200.

Opdyke, N.D., 1972, Paleomagnetism of deep-sea cores: Rev. of Geophys. and Space Phys., v. 10, no. 1, p. 213-249.

Smith, J.D. and Foster, J.H., 1969, Geomagnetic reversals in Brunhes normal polarity epoch: Science, v. 163, p. 565-567.

Soloyanis, S.C. and Brown, L.L., 1979, Late Pleistocene magnetic stratigraphy recorded in some New England tills: Geophys. Res. Lett., v. 6, no. 4, p. 265-268.

Stupavsky, M., Gravenor, C.P. and Symons, D.T.A., 1979, Paleomagnetic stratigraphy of the Meadowcliffe Till, Scarborough Bluffs, Ontario: A Late Pleistocene excursion?: Geophys. Res. Lett., v. 6, no. 4, p. 269-272.

Symons, D.T.A., Stupavsky, M. and Gravenor, C.P., 1980, Remanence resetting by shock-induced thixotropy in the Seminary Till, Scarborough, Ontario, Canada: Geol. Soc. Amer. Bull., v. 91, Part 1, p. 593-598.

Thompson, R., 1975, Long period European geomagnetic secular variation confirmed: Geophys. J.R. Astr. Soc., v. 43, p. 847-859.

____________, 1978, European paleomagnetic secular variation, 13,000-0 BP: Pol. Arch. Hydrobiol., v. 25, p. 413-418.

____________, 1979, Paleomagnetic correlation and dating, in Berglund, B.E., ed., Paleohydrological changes in the temperate zone in the last 15,000 years. IGCP 158 B Lake and Mine Environments.

Valladas, G., Gillot, P.Y., Paupeau, G. and Reyss, J.L., 1977, Thermoluminescence dating of recent volcanic rocks. Laschamp magnetic event dating, Fifth ECOG Conference, Pisa, Italy.

Verosub, K.L., 1975, Paleomagnetic excursions as magnetostratigraphic horizons: A cautionary note: Science, v. 190, p. 48-50.

____________, 1977, Depositional and post-depositional processes in the magnetization of sediments: Rev. Geophys. Space Phys., v. 15, p. 129-145.

Verosub, K.L. and Banerjee, S.K., 1977, Geomagnetic excursions and their paleomagnetic record: Rev. Geophys. Space Phys., v. 15, p. 145-155.

Vitorello, I.R. and Van der Voo, R., 1977, Magnetic stratigraphy of Lake Michigan sediments obtained from cores of lacustrine clays: Quat. Res., v. 7, p. 398-412.

Yaskawa, K., 1974, Reversals, excursions and secular variations of the geomagnetic field in the Brunhes normal polarity epoch, in Horie, S., ed., Paleolimnology of Lake Biwa and the Japanese Pleistocene, 2nd Issue, p. 77.

Yaskawa, K., Nakajima, T., Kawai, N., Torii, M., Natshuhara, N. and Horie, S., 1973, Paleomagnetism of a core from Lake Biwa: J. Geomag. Geoelect., v. 25, p. 447-474.

THE PRESENT STATUS OF OBSIDIAN HYDRATION DATING

FRED TREMBOUR and IRVING FRIEDMAN

ABSTRACT

Obsidian hydration dating is being used by both archaeologists and geologists to date events ranging in age from a few hundred to several million years. The method requires that a measurement of hydration thickness or the depth of penetration of water into obsidian be measured, and a rate of hydration be known. The measurement can be made optically, using a thin section of the sample and an ordinary microscope or it can be made by more sophisticated methods using particle accelerators. The rate of hydration is a function of the chemistry of the obsidian and of the temperature (s) that the sample was exposed to. This paper will be mainly concerned with methods to measure or estimate the rates of hydration as a function of temperature, and devices to measure or record integrated temperature. These latter devices include the Pallmann sucrose inversion cell, and the modified Ambrose cells developed by a group at the USGS in Denver. Data will be given comparing the performance of these various types of temperature integrating devices. We will also discuss the problem of direct measurement of hydration rates on obsidian powder.

Formation and Hydration of Obsidian

The natural glass called obsidian owes its origin either to the rapid cooling of volcanically erupted acidic magmas of rhyolitic composition (approx. 70 to 78% SiO_2) or to the welding together of explosively erupted glass shards. When thus exposed to the conditions at the Earth's surface the solidified glass begins to react with its environment and eventually forms perlite. From its inception this change is characterized by the absorption of ambient moisture which builds up as a distinct and progressively thickening hydration layer or rind. While the initial H_2O content of most obsidians is from 0.1 to 0.3%, the saturation level after hydration reaches about 3 to 4% by weight.

Any fresh surface formed later, as by fracture or cracking, on the pristine obsidian will proceed to acquire a hydration layer of its own in the same way. The hydration layers, Figure 1, are firmly adherent to the parent glass and resistant to chemical dissolution under neutral conditions. The hydrated layer is under strain and exhibits strain birefringence under polarized light (Figure 2). The buildup of strain usually causes the hydration rim to spall off after about 50 micrometers of thickness have formed; however the authors have measured rinds as deep as *ca.* 100 micrometers on the geological cortex of some obsidian samples.

Because its isotropic nature favors controllable shaping of sharp edged tools and weapons, by skilled chipping, obsidian was a desired material among stone age people. It was widely used in the industries of all early cultures that inhabited most of the earth's volcanic regions, as far back as several hundred thousand years ago in the paleolithic of East Africa.

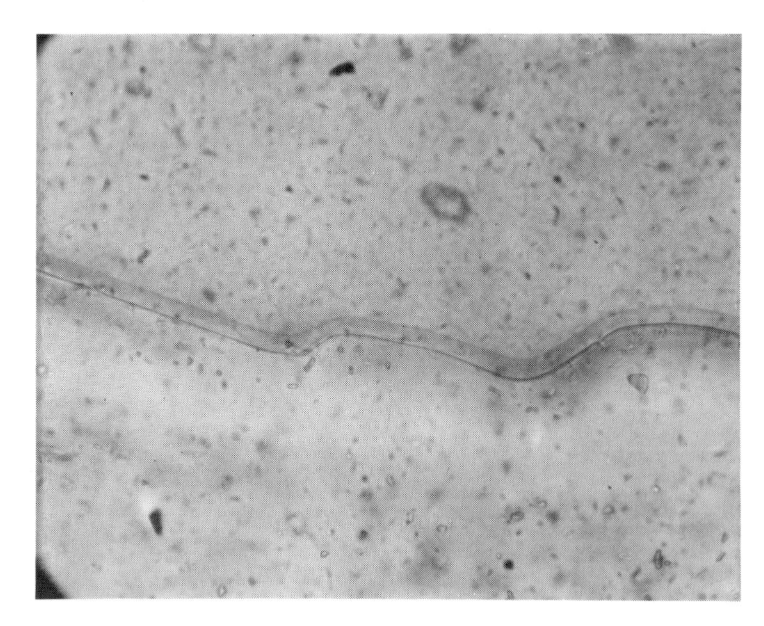

Figure 1 Hydration rind in plain light. The rind is 4 micrometers thick.

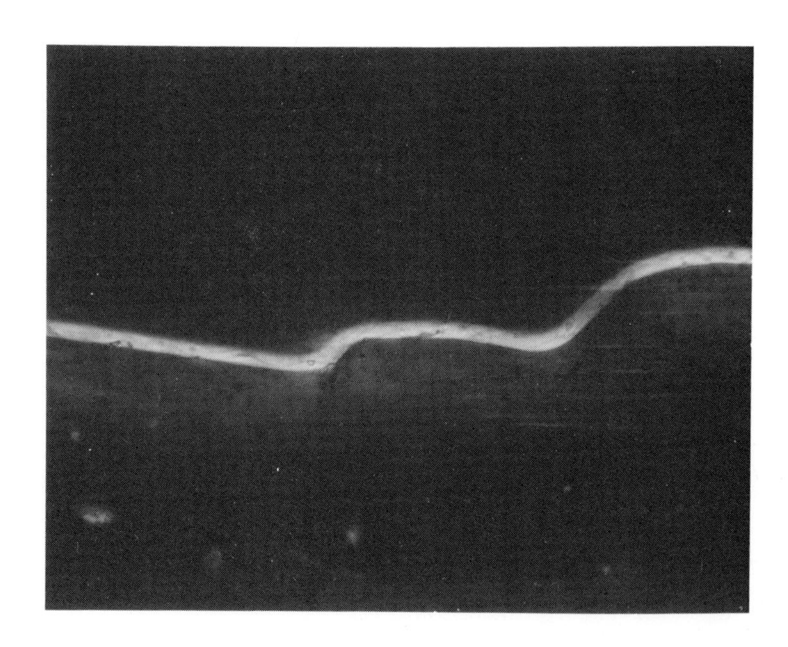

Figure 2 Hydration rind in polarized light.

Hydration Measurement and Dating

In a 1960 publication, Friedman and Smith reported a method of assessing the age of measurable hydration rinds on obsidian, and in a companion article Evans and Meggers (1960) discussed a first application of the method to archaeological lithics from Ecuador. The key relationship between hydration depth (D) and time (t) at any given temperature was found by Friedman and Smith to be:

$$D = kt^{1/2} \quad (1)$$

where k is a constant that incorporates the effect of chemical composition of the material and temperature on the diffusion rate. For a given obsidian and a given temperature the hydration rate is usually expressed in terms of micrometers ^{2}/1000 years.

The basic measurement of rind depth as micrometers (μm) is performed on a sample in the laboratory. In brief, a slice about 1 millimeter thick is cut and broken from the selected obsidian surface with two normal and parallel cuts of a think diamond saw. The removed slice, which can be as small as 2 x 2 millimeters, is mounted on a glass slide and made into a thicker than standard petrographic thin-section (*ca.* 0.1 mm thick) by common lab procedures for examination by transmitted light in a petrographic microscope. The hydration layer can be identifield by its bright birefringence in polarized light (crossed nicols) (Figure 2).

Accurate rind measurement from edge to internal boundary can then be made in plain light illumination with suitable eyepiece accessories such as a Vickers split-image micrometer or a filar micrometer. With care, useful thickness measurements down to a limit of 0.2 μm can be achieved by the optical method. A good laboratory set-up allows the whole procedure including preparing this section and measuring the rim to be carried out in a matter of 10 or 15 minutes. Some very dark-colored natural glasses call for extra thin grinding (to ~0.05 mm) to secure sufficient light transmission for measurement.

Researches with a variety of materials have shown that the intrinsic hydration rates of obsidians vary over a range as wide as 20 to 1, (Friedman and Long, 1976). As a consequence, for example, two pieces of obsidian from the same findspot with identical hydration amounts will necessarily be of the same age only if they are of the same chemical composition.

The method is useful in the time interval of a few hundred years to several million years. Since the inception of hydration dating, many investigators, especially archaeologists, have entered the field and their work has led to scores of publications. Relatively few of these have dealt with geological problems (*e.g.* Friedman *et al.*, 1973), and these papers will be discussed later. In general most studies have involved direct attempts at dating archaeological contexts that contained suitable obsidian artifacts (Clark, 1961; Johnson, 1969; Layton, 1972), but others have been devoted to improvement and extension of the method itself (Lee *et al.*, 1974; Tsong *et al.*, 1978). The present authors published a review article in 1978 (Friedman and Trembour) and another was written by Michels and Tsong in 1980.

In this treatment we shall consider some current research trends. Our chief interest has been the proper evaluation of the two factors that determine the hydration rate of any obsidian piece in question: (1) the effective hydration temperature at its field location and (2) the composition of the glass.

The Temperature Factor

According to the Arrhenius equation

$$k = Ae^{\frac{-E}{RT}}$$

the diffusion rate, k, rises exponentially with increasing temperature, T, and this relationship has been experimentally verified (Figure 3). In the equation A is a constant, E is the activation energy of the hydration process (calories per mole), and R is the gas constant (calories per degree per mole). Under conditions prevailing in nature the obsidian hydration rate increases roughly 10% for each 1°C increase in temperature. Hence, under fluctuating temperature conditions (diurnal and annual) the effective hydration temperature (EHT) of the range is not the arithmetic temperature mean but an integrated value at some higher temperature level (Figure 4).

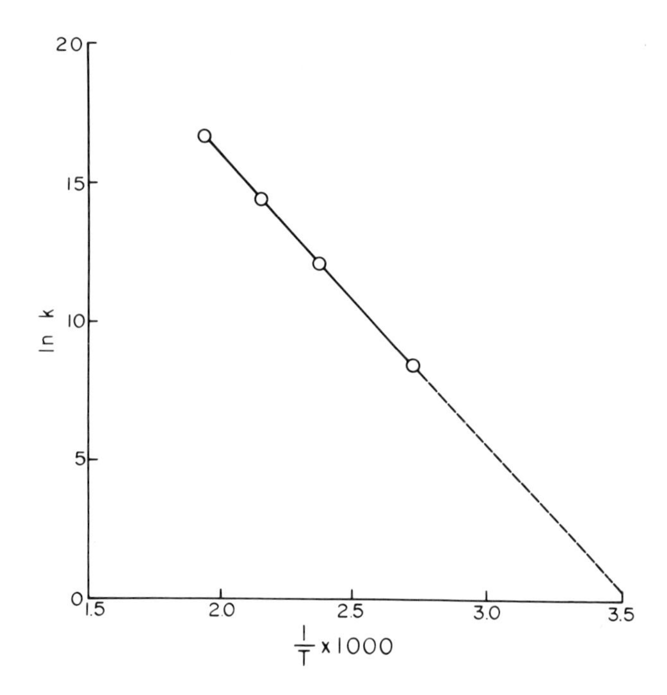

Figure 3 Arrhenius plot for the experimental data for obsidian from Kerlingfjoll, Iceland. (from Friedman and Long, 1976, Figure 3).

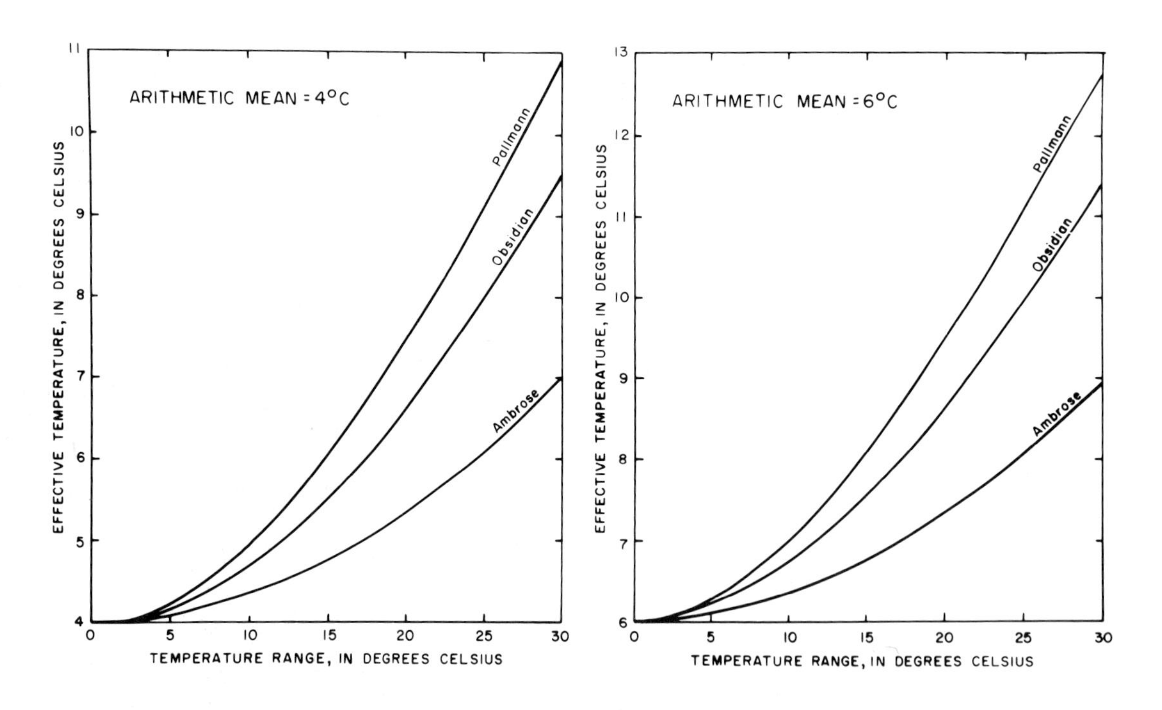

Figure 4 From (Norton and Friedman, 1981, Figure 1). Plots of the temperature range experienced by a sample versus the effective temperatures calculated for the Pallmann reaction, the obsidian hydration reaction and the Ambrose diffusion process. The curves were calculated from the Arrhenius equation and assume a sinusoidal temperature variation within the range, and respective reaction activation energies of 27kcal/mole, 19.7kcal/mole and 11.7kcal/mole.

It is this EHT level that was experienced by the sample where it was found which must be evaluated closely if a reliable age conversion from hydration rind depth is to be obtained.

When the obsidian hydration technique was first developed, temperatures based on the gross thermal subdivisions provided by the earth's climatic zones were used. Researchers quickly saw that ever narrower definitions of temperature were demanded to resolve recurring inconsistencies that arose, *e.g.*, in the comparison of hydration dates with ^{14}C dates or well-established cultural age indices. Published long-term data from nearest weather stations were soon resorted to as a basis for improved estimations. Additional refinements have been applied to these area estimates of mean annual air temperature to compensate for site altitude, findspot depth (at or) below surface, seasonal snow cover, geothermal effects and others (Friedman and Long, 1976). While this approach remains the best feasible way to cope with the temperature factor in many instances, the obvious alternative of making present day on-site measurements at points of interest has been gaining attention. This way of investigating the microenvironment, so to speak, on behalf of hydration dating seems essential for dealing with such still unapprehended sources of error as soil conductivity, albedo, plant and tree cover, slope of terrain, wind chill and more. How are measurements of this kind at precise field spots carried out?

The need here, of course, is for really low-cost, self-contained, no-maintenance field installations that provide an integrated parameter of continuously varying temperatures over a long period of time. We have worked with two different types of compact sensors that fit this description (Figure 5).

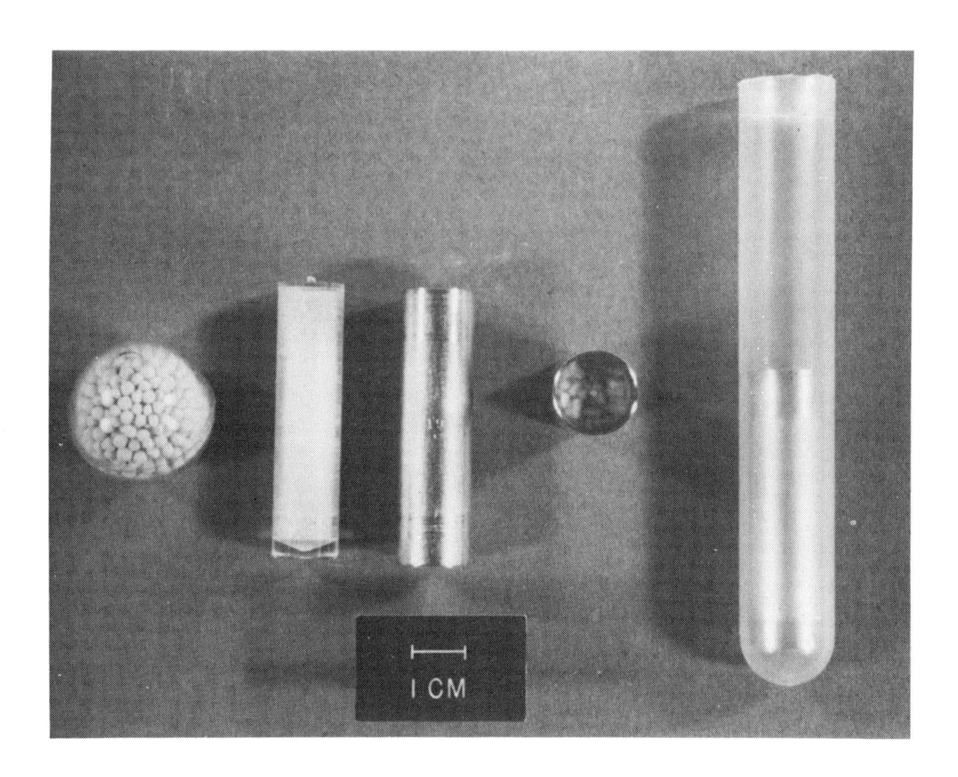

Figure 5 Photo of Ambrose cells. From left to right. (1) original spherical cell described by Ambrose. (2) all plastic modified Ambrose cell. (3) metal cell with plastic end windows on metal cell. (5) metal cell sealed in plastic tube filled with water.

One is the Pallmann glass cell in which a sucrose solution with controlled pH inverts at a temperature dependent rate into other sugar forms, and the degree of inversion is measured as optical rotation by polarimetry at the end of the field run (O'Brien, 1971). With careful laboratory calibration, temperatures can be measured to ± .05°C. The other type is the plastic (acrylic resin) diffusion cell originated by Ambrose (1976) which absorbs and traps water inside its plastic walls at a temperature dependent rate and permits the uptake rate of water to be determined easily as weight gain over time. In both sensing methods, calibration of rate of sugar inversion or of weight change *vs.* temperature is done in the laboratory. The energy source for field operation in both cases is simply the thermal ambience at the site. Both types of cell are plain tubular objects a few centimeters in length and about 15 millimeters in diameter, and both require only a protective cover of sturdy plastic pipe for emplacement at the site. The one field visit necessary after deployment is for retrieval following the allotted time (commonly a year); then a single laboratory measurement for each cell provides the data for integrated temperature conversion. Sub-freezing conditions do not interrupt the operation. Figure 6 shows typical calibration curves for an Ambrose diffusion cell.

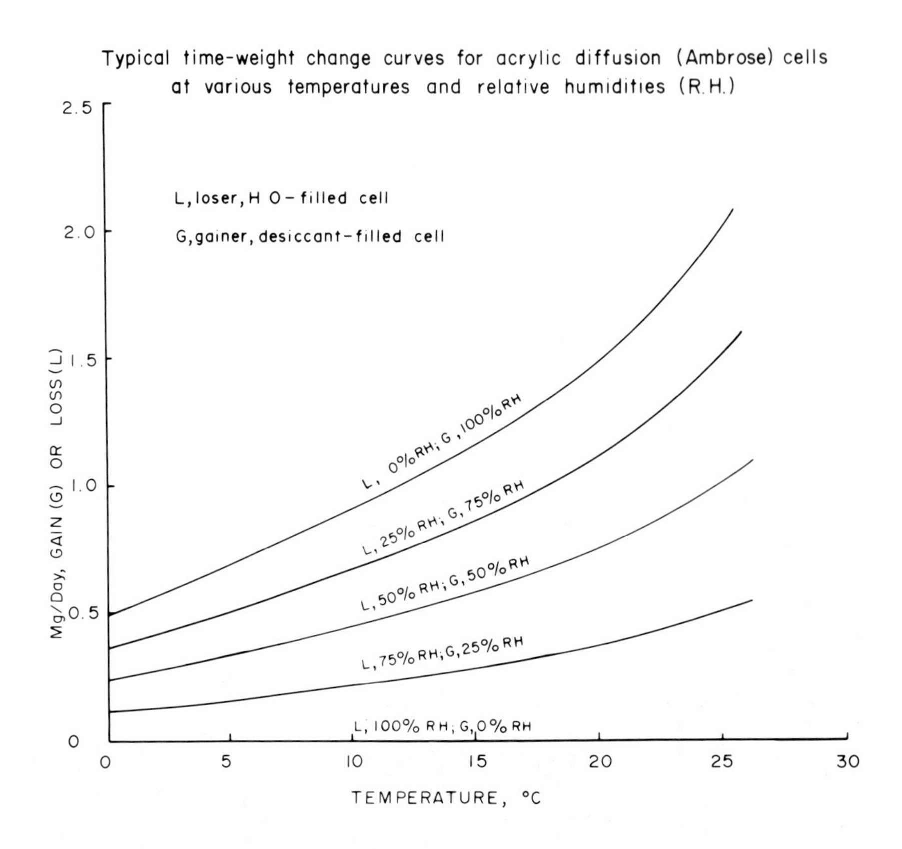

Figure 6 Typical time-weight change curves for acrylic diffusion (Ambrose) cells at various temperatures and relative humidities (R.H.).

The use of the Pallmann technique, including details of a method of emplacement of the two meter long probes using explosives is given in a recent publication by Norton and Friedman (1981).

Beginning in May 1979 Trembour and F. Smith (unpublished) developed a variation of the Ambrose method in which the plastic cell is filled with pure water. The cell is surrounded by a desiccant. The water diffuses <u>out</u> of the cell at a rate which is temperature dependent. Filling the cell with water rather than with desiccant allows the use of a smaller cell and longer service life.

To achieve standardized field performance the diffusion cells must be precisely machined and assembled from raw material stock such as extruded plastic rod. Otherwise, variability in a manufactured lot will cause the need for extensive individual cell calibration after completion. However, experiments with certain commercially available plastic containers have shown some promise for the application of molded products as cell bodies also. An example is the (Wheaton[1]) 43 mm-long x 12 mm-diameter polypropylene vial with screw cap. Water-filled and sealed with a proprietary plastic glue (such as Scotch Grip by 3M) this cell variety has shown a low diffusion loss rate that could make it useful for very long field runs and/or high ambient temperatures.

In attempting to date obsidian hydration that is older than about 10,000 years BP we need to correct present day temperatures to take into account climatic fluctuations in the past. In their paper comparing obsidian hydration dates of rhyolite flows with ^{14}C and K-Ar dates on the same flows, Friedman and Obradovich (1981) estimated the duration and intensity of past temperature changes based upon the $^{18}\delta O$ composition of foraminifera from dated deep sea cores.

Another approach was used by Pierce *et al.*, 1976, in dating cracks in obsidian caused by glacial transport. In this research, K-Ar dates on obsidian flows found near the glacial deposits were used to calibrate the hydration rate. This calibration of hydration rate allowed these authors to date the last two glaciations in the Yellowstone Park area. Figure 7 illustrates the percussion cracks in obsidian caused by glacial transport. Figure 8 shows the curve relating hydration thickness to age from which the dating of the glacial events was obtained.

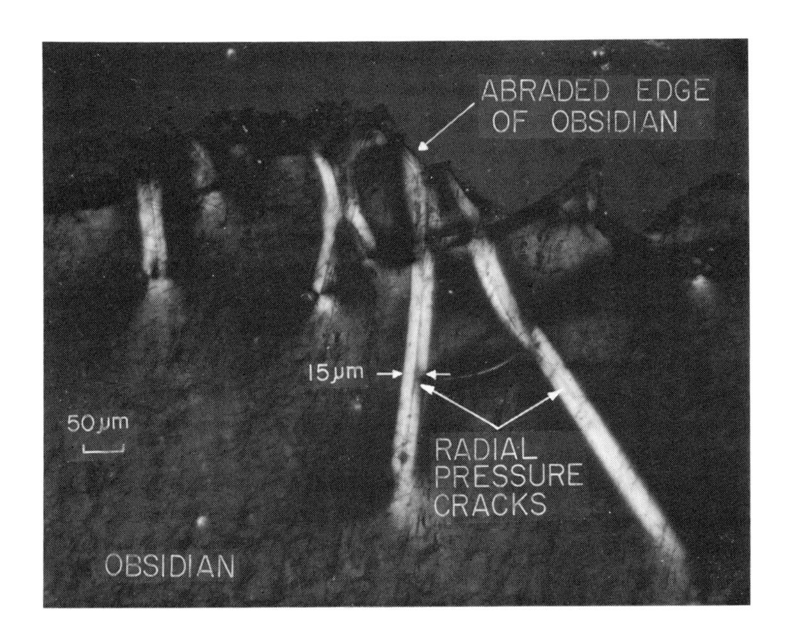

Figure 7 Photomicrograph of thin section of obsidian pebble showing hydration along cracks. Cracks such as these probably formed during glacial abrasion when the obsidian pebble impinged against another rock fragment. (From Pierce *et al.*, 1976).

The Chemical Composition Factor

Obsidian is composed mainly of 10 elements. The sequence in decreasing concentration, listed as oxides, is silica (SiO_2), alumina (Al_2O_3), potassium oxide (K_2O) sodium oxide (Na_2O), iron oxide (Fe_2O_3 and FeO), calcium oxide (CaO), magnesia (MgO), titanium dioxide (TiO_2) and water (H_2O+).

[1] Any use of trade names is for descriptive purposes only and does not imply endorsement by the U.S. Geological Survey.

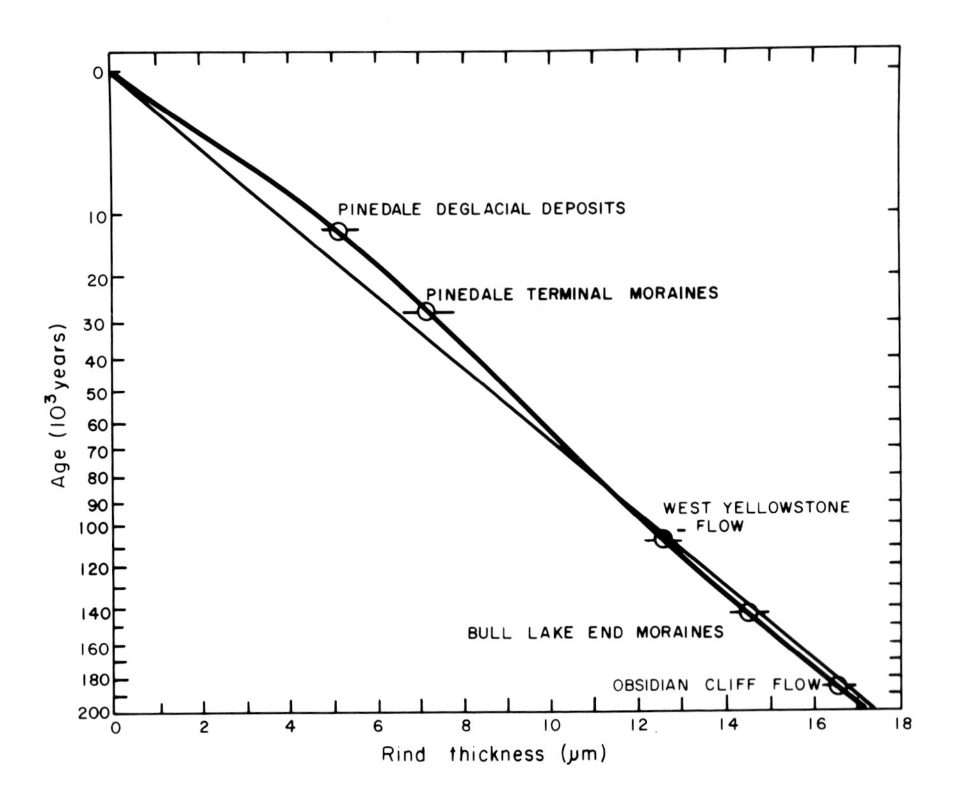

Figure 8 The rate of obsidian hydration for the West Yellowstone Basin is determined by hydration rinds on cooling cracks of the West Yellowstone flow, K-Ar dated as 179,000 ± 3,000 years old. The black line gives the average rate based on the two dated flows. The white line is drawn to account for variation in hydration rate due to climatic change during the late Quaternary. (From data by Pierce *et al.*, 1976).

The minor or trace element analyses of obsidian have also been studied extensively by modern investigators - often at low levels of parts per million - for "fingerprinting" purposes to identify geologic source of artifact material. The information serves to trace ancient transport and cultural exchange patterns. The analytical methods used are commonly of the radiation emission types such as optical spectroscopy, neutron activation and X-ray fluorescence (Cann et al., 1970). Natural glasses cover a wide range in chemical composition and at present a comprehensive understanding of the water diffusion process for all natural glasses, including obsidian, is still lacking (*e.g.* Findlow, 1977). Recent research (*e.g.*, Haller, 1963; Doremus, 1975; Michels and Bebrich, 1971; Hench and Clark, 1978) has brought out new features of hydration diffusion such as ion exchange reactions and concentration-dependent as well as independent diffusion patterns for specific glasses. For the time being it might be prudent to suspect that glasses of quite different chemical composition (*e.g.*, basalt-rhyolite) have different hydration mechanisms.

Observing differing sets of hydration amounts among stratified flake collections from Valley of Mexico archaeological sites, Clark (1961) deduced that the green-colored variety hydrated faster than the gray one under like conditions because of chemical differences between the two obsidian source deposits.

Suzuki (1973) attempted a correlation of hydration rate and chemistry of obsidians from five different sources of the Kanto district of Japan. Although he considered his findings highly provisional he favored the ratio of the K_2O/Al_2O_3 percentages as the best among three chemical parameters that he investigated.

In 1976 Friedman and Long published their "chemical index" based on studies of rhyolitic glass source samples from 9 widely separated North American proveniences and one from Iceland. Their best correlation with experimental hydration data was expressed as a function of the following sum of chemical content percentages: SiO_2 - 45(CaO + MgO) - $20H_2O+$. In reviewing efforts of the kind just described one may suppose that the empirical results achieved correlations useful within a limited subgroup of all obsidians while not illuminating the underlying causes of differences among them. However, a recent study by Michels *et al.* (1981), involving experimental hydration and chemical analyses on a selection of 12 Sardinian artifact flakes and source samples from the Monte Arci flow system on the island, confirmed the chemical index of Friedman and Long (1976). The hydration rate predicted by the chemical index differed from their laboratory result, 4.0 $(\mu m)^2$/1000 years at 19°C, by a mere 1%.

CONCLUSIONS

In closing this section of the review we wish to mention the application of substitute parameters for precise chemical analysis. An example is refractive index of obsidian which has been resorted to for hydration rate approximations when warranted to save time and cost. Using a few grains of powdered obsidian, a series of index-calibrated immersion oils and glass slides and covers, the specimen is examined in a microscope at low magnification with illumination by the focal masking technique. Using index oils of high dispersion (Wilcox, 1964) and observing interface Becke-line phenomena between glass particle and oil, a match of refractive indices accurate to .001 can be obtained in a matter of 5 to 15 minutes per specimen. This measurement can be converted to estimated hydration rate at say 10°C from the experimental correlation shown in Figure 9. Here again we think that this empirical short-cut might best be used to obtain a first approximation of the glass's hydration rate only if its refractive index falls within the range (1.483-1.494) of a group investigated by Friedman and Long.

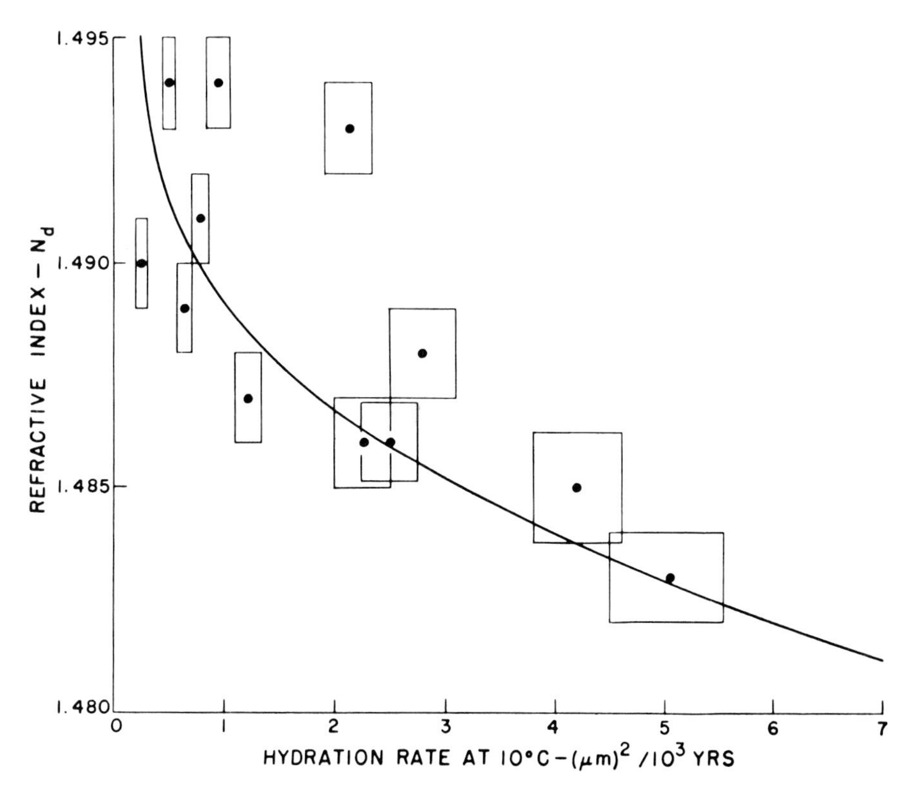

Figure 9 Hydration rate at 10°C as a function of refractive index (Nd) for obsidian samples from 12 sources. (from Friedman and Long, 1976, Figure 6).

Current laboratory work by the authors includes the application of obsidian powder samples to accelerated hydration tests at controlled temperatures in the normal ambient range. The procedures are adapted from those proposed by Ambrose (1976). The powder samples are 1 to 3 gms, 250 to 400 sieve mesh size and about 1 sq. gram in surface area. The powders are hydrated in a closed tube for a period of several days or weeks between sample weighings on an electrobalance that is sensitive to 10^{-6} gm. The experiments require 2 or 3 spaced weight readings to yield the linear calibration plots of the kind shown from Ambrose (1976, Figures 5.4).

The directions of the research discussed in the preceding paragraph have been paralleled to some extent by the developments in hydrogen profile measurements by nuclear reaction techniques (as by Laursen and Lanford, 1978). This method is especially suited for the detection of very thin hydration deposits on artifacts of low age and slow hydration rate, such as abound for example in New Zealand (Leach, 1977). It may open new frontiers for dating as well as for research in the mechanism of hydration.

REFERENCES CITED

Ambrose, W.R., 1976, Intrinsic hydration rate determinations, in Taylor, R.E., ed., Advances in Obsidian Glass Studies: Noyes Press, p. 81-105.

Cann, J.R., Dixon, J.E. and Renfrew, C., 1970, Obsidian analysis and the obsidian trade, in Brothwell, D. and Higgs, E., eds., Science in Archaeology: Praeger Publishers, p. 578.

Clark, D., 1961, The application of the obsidian dating method to the archaeology of central California: (Ph.D. Dissert.), Stanford Univ., 160 p.

Doremus, R.H., 1975, Interdiffusion of hydrogen and alkali ions in a glass surface: Journal of Non-crystalline Solids, v. 19, p. 137.

Evans, C. and Meggers, B.J., 1960, A new dating method using obsidian: Part II: American Antiquity, v. 25 p. 523-537.

Findlow, F.J., 1977, A revision in the Government Mountain-Sitgreaves Peak, Arizona, obsidian hydration rate: The Kiva, v. 43, no. 1, p. 27-28.

Friedman, I. and Long, W., 1976, Hydration rate of obsidian: Science, v. 191, p. 347-352.

Friedman, I. and Obradovich, J., 1981, Obsidian hydration dating of volcanic events: Quaternary Research, v. 16, p. 37-47.

Friedman, I. and Smith, R.L., 1960, A new dating method using obsidian, Part I: American Antiquity, v. 25, p. 476-493.

Friedman, I. and Trembour, F.W., 1978, Obsidian: the dating stone: American Scientist, v. 66, no. 1, p. 44-51.

Friedman, I., Pierce, K.L., Obradovich, J.D. and Long, W., 1973, Obsidian hydration dates glacial loading, Science, v. 180, p. 733-734.

Haller, W., 1963, Concentration-dependent diffusion coefficient of water in glass: Physics and Chemistry of Glasses, v. 4, p. 217-220.

Hench, L.L. and Clark, D.E., 1978, Physical chemistry of glass surfaces: Journal of Non-crystalline Solids, v. 28, p. 83.

Johnson, LeRoy, Jr., 1969, Obsidian hydration rate for the Klamath Basin of California and Oregon: Science, v. 165, p. 1354-56.

Laursen, T. and Lanford, W.A., 1978, Hydration of Obsidian: Nature, v. 276, p. 153-156.

Layton, T.N., 1972, Lithic chronology in the Fort Rock Valley, Oregon: Tebiwa, v. 15(2), p. 1-20.

Leach, B.F., 1977, New perspectives on dating obsidian artefacts in New Zealand: New Zealand Journal of Science, v. 20, p. 123-138.

Lee, R., Leich, D., Tombrello, T., Ericson, J. and Friedman, I., 1974, Obsidian hydration profile measurements using a nuclear reaction technique: Nature, v. 250, p. 44-47.

Michels, J.W. and Bebrich, C.A., 1971, Obsidian hydration dating, in Michael, H.N. and Ralph, E.K., eds., Dating Techniques for the Archaeologist, MIT Press, p. 164-221.

Michels, J.W. and Tsong, I.S.T., 1980, Obsidian hydration dating: a coming of age, in Schiffer, M.B., ed., Advances in Archaeological Method and Theory, Academic Press, vol. 3, p. 405-444.

Michels, J.W., Atzeni, E., Tsong, I.S.T. and Smith, G.A., 1981, Sardinian Archaeology and Obsidian Dating. Report, Materials Research Laboratory, Pennsylvania State University.

Norton, D.R. and Friedman, I., 1981, Ground temperature measurements, III. Ground temperatures in and near Yellowstone National Park. U.S. Geological Survey Prof. Papers 1203, Chap. A., p. 1-11.

O'Brien, P.J., 1971, Pallmann method for mass sampling of soil, water or air temperatures: Geol. Soc. Amer. Bull., v. 82, p. 2927.

Pierce, K., Obradovich, J. and Friedman, I., 1976, Obsidian hydration dating and correlation of Bull Lake and Pinedale glaciations near West Yellowstone, Montana: Geol. Soc. Amer. Bull, v. 87, p. 701-710.

Suzuki, M., 1973, Chronology of prehistoric human activity in Kanto, Japan. Journal of the Faculty of Science, Univ. of Tokyo (5), v. 4, p. 241-318.

Tsong, I.S.T., Houser, C.A., Yusef, N.A. Messier, R.F., White, W.B. and Michels, J.W., 1978, Obsidian hydration profiles measured by sputter-induced optical emission: Science, v. 201, p. 339-341.

Wilcox, R.E., 1964, Immersion liquids of relatively strong dispersion in the low refractive index range (1.46-1.52): American Mineralogist, v. 49, p. 683-688.

THERMOLUMINESCENCE DATING OF QUATERNARY SEDIMENTS

M. LAMOTHE, A. DREIMANIS, M. MORENCY and A. RAUKAS

ABSTRACT

In any geological environment, natural radiation induces free electrons in minerals that can be trapped into lattice defects. They may escape upon heating and recombine with holes at luminescent centers. Energy will then be released in the form of light. By recording the thermoluminescence (TL) of a mineral, the last drainage of the traps can be dated, assuming a constant radiation level, by the following equation:

$$\text{AGE (years)} = \frac{\text{EQUIVALENT DOSE (rads)}}{\text{DOSE-RATE (rads/year)}}$$

The equivalent dose is the dose that can produce the natural TL level and is found by irradiation from known beta or gamma sources. The dose-rate is computed from the weight of the radioactive elements in the sample to which may be added a small cosmic-ray contribution.

A specific geological problem is the determination of the initial TL level at the time of sedimentation. It appears that, during the sedimentary cycle, exposure to sunlight bleaches all but a fraction of the primary TL. This residual signal may give apparent ages of many thousands of years. Simulation of the sunlight process may be achieved by means of a sunlamp, permitting an evaluation of this primary level. The "sedimentary" equivalent dose is thus the difference between this "residual" and the total natural dose. Stability of the equivalent dose is insured by the plateau test.

Two main methods are now used in geology:

a) The "quartz-inclusion" method in which large inclusions of quartz (40-70 μm or 88-125 μm) are extracted and etched in HF to eliminate the short range alpha-induced TL. The age equation is then dependent only on the beta and gamma contribution.

b) The "fine-grain" method in which determinations are made on polymineralic fine silts (4-11 μm) that have received the full alpha dosage. Because of low efficiency in inducing TL compared to the beta and gamma radiation, a correction factor is applied to the alpha contribution, the total dose-rate being expressed in beta equivalent, in the age equation.

INTRODUCTION

The light emitted by a crystal when heated is termed Thermoluminescence (TL). This thermally stimulated process has been used for dating archaeological objects for over two decades. Its successful application in geochronology is more recent.

The TL dating technique is part of a group of methods based on solid state physics, together with Electron Spin Resonance (ESR) and Thermally Stimulated Current (TSC). They measure trapped electrons in minerals, this being mainly a consequence of the lithospheric decay.

As early as 1945, Randall and Wilkins suggested that electrons in traps have a "Maxwellian" distribution of thermal energies thus establishing the physical basis of the phenomenon. A few years later, Daniels *et al.* (1953) proposed many potential uses of the TL process including correlation and dating. Research was then directed towards calcareous materials and met with many difficulties. Indeed, in 1966, at the Thermoluminescence of Geological Materials Conference (McDougall, 1968), few papers reported age determinations of geological samples.

One of the major problems was the occurrence of non-radiation-induced TL ("Spurious"). Aitken *et al.* (1963) suggested that heating in an oxygen-free atmosphere of an inert gas (nitrogen or argon) would eradicate this problem. The technique then evolved quite rapidly. Today the TL dating method is used on a routine basis in archaeology as research is now directed towards refinement of various aspects of the technique (PACT, 1978).

In geology, we must note the pioneer work of Shelkoplyas (1971) which kept alive the interest towards the technique in spite of methodological misconceptions. Recently, new hopes have been generated by a series of investigations (Hütt and Raukas,1977; Wintle and Huntley, 1980; Wintle, 1981) proposing solutions to the specific problems of geological dating.

We shall outline here the general principles of the thermoluminescence process, review the age equations of the two techniques commonly used in TL dating, followed by a discussion of the applications to Quaternary geology. Finally, some concepts related to the upper and lower terms of the general age equation (*e.g.* the equivalent dose and the dose-rate) are illustrated by the senior author's own work on the Upper Thorncliffe sediments (Middle Wisconsin). Detailed results of the TL dating project of the Toronto Pleistocene will be published at a later date.

This paper is restricted to sediments. References to other rock types will be found in review papers of Valladas (1979), Wintle (1980) and Whippey (1980).

THE TL PROCESS

An exhaustive treatment of this subject can be found in Levy (1974), Aitken (1974) and Fleming (1979). Most minerals previously affected by ionizing radiation will emit light when heated. Since this is an irreversible phenomenon the mineral will not emit light when heated again. Moreover, at moderate doses, the intensity of the light is directly related to the irradiation dose (Figure 1). Therefore, TL can reflect the total radiation a material has received.

Radiation

In the natural environment, the radiation comes mainly from the lithospheric decay with a minute contribution from the cosmic rays. Alpha (α) and beta (β) rays are particle-like radiations whereas the gamma (γ) rays are part of the electromagnetic spectrum. They differ largely in terms of electron-hole pairs production efficiency, the main consequence being that the alpha contribution, because of its small range ($\simeq$22 μm), is restricted to the surface of quartz grains, if free of internal radioactivity. The beta and gamma contributions may be here considered equivalent. In a simplified scheme, non luminescent and some luminescent minerals (feldspars, calcite), surrounding the quartz grains, emit the radiation.

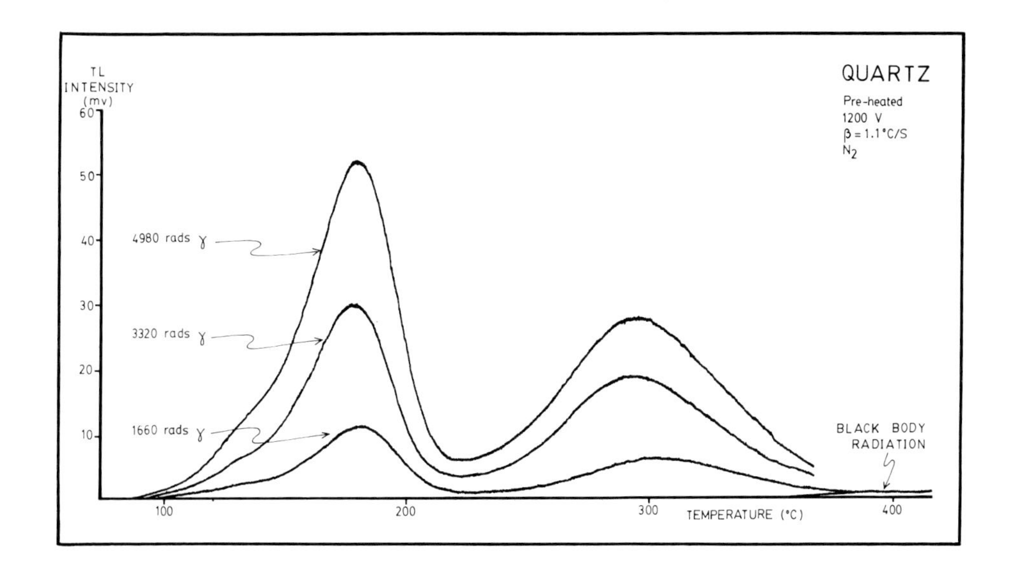

Figure 1 Artificial glow curves for quartz pre-heated at 700°C for 3 hours showing that thermoluminescence and irradiation are directly related. Note that quartz exhibits two peaks and one shoulder. The low-temperature peak has faded as measurements were carried 8 months after irradiation.

Storage of Energy

The luminescent minerals (quartz, feldspars, calcite) in sediments absorb incoming radiation. They behave as natural dosimeters if ionizing radiation can expel an electron from its normal orbit, and if this electron becomes trapped at a lattice charge disequilibrium site, thereby called an electron-trap. On the other hand, by mutual exchanges of peripheral electrons between atoms, an "absence of electron" (a "hole") will travel in the lattice and eventually be trapped. Immediate recombination of the electron with the parent atom or neighbouring hole is nevertheless, by far, a more frequent process. The traps are vacancies, impurities, interstitial atoms, *etc.* (Marfunin, 1979); for instance, it has been suggested that aluminium impurities act as hole traps in quartz, with lithium, sodium and sometimes germanium as charge compensators together with oxygen vacancies acting as electron traps (McMorris, 1971).

Release of the Energy

When the natural dosimeter is heated, the electrons gain enough energy to escape from the traps. The electron release rate is temperature dependent and is expressed by the Randall-Wilkins equation:

$$\frac{dn}{dt} = -nse^{-E/kT}$$

Where n = trap charge concentration at time t

s = "attempt to escape" frequency

k = Boltzman constant

E = Activation energy for thermal charge release

T = temperature

The band model is an energy representation of this process (Figure 2). Having gained freedom, the electron wanders in the conduction band. If the electron recombines at a luminescent center, usually a hole trap, energy will then be released in the form of light. This conduction band charge transfer is not the only possible route for the electron. For instance, radiationless recombination (without light emission) may

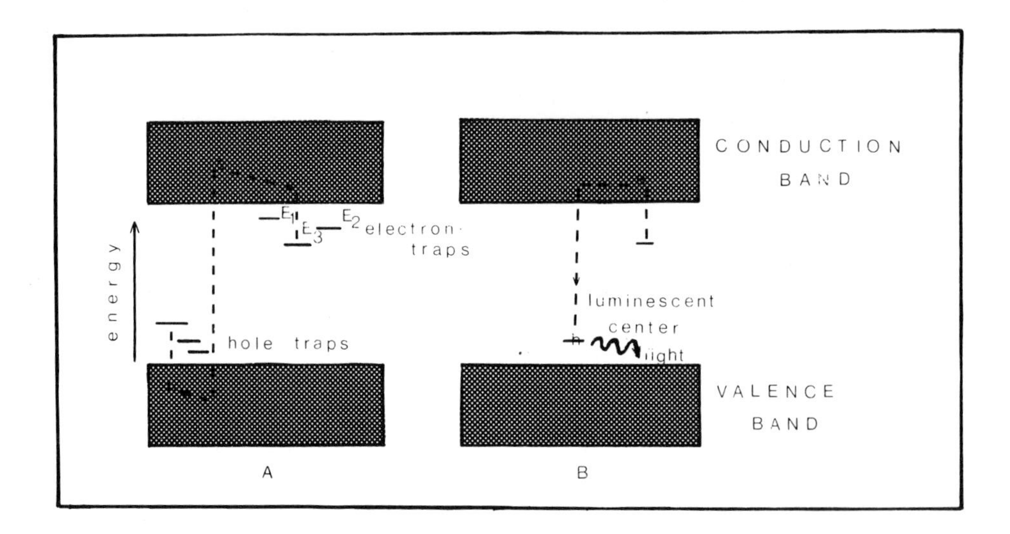

Figure 2 The band model, modified from Aitken (1974). In the natural environment (A), ionization brings electrons in the conduction band; they may be trapped in crystal's defects at different "depths" corresponding to different energy levels. Upon thermal excitation (B), the electron is ejected from the trap and wanders in the conduction band. It may recombine at a luminescent center and energy will then be encountered as light. Many other routes are possible.

occur at "killer" centres.

In a simplified way, minerals receive ionizing radiation. They behave as natural dosimeters if they can store this energy over geological time and release it when heated.

THE AGE EQUATION

From the preceeding section, it follows that if the radiation dose-rate and the number of traps are constant, the light emitted upon heating may be a measure of time.

Therefore, if one can find the dose that can reproduce the natural TL level (ED) and compute the annual dose from the radioactive elements found in the sample, an age relationship can be established:

$$\text{AGE (years)} = \frac{\text{EQUIVALENT DOSE (rads)}}{\text{DOSE-RATE (rads/years)}}$$

Due to different microdosimetries, two techniques have been developed:

The Quartz Inclusion Technique (Fleming, 1970)

This method makes use of quartz grains (40-70 µm) isolated from the matrix by heavy liquids and/or magnetic separation. By etching the sample in HF, the other light minerals (mainly calcite and feldspars) are removed and moreover, the original surface of the quartz grains is dissolved, thereby eliminating the alpha dose contribution.

$$AGE = \frac{ED_i}{D_\beta + D_\gamma + Dc}$$

Where ED_i : equivalent dose for inclusion dating

D_β, D_γ, Dc: dose-rate from the β, γ and cosmic rays.

The Fine-Grain Technique (Zimmerman, 1971)

This method involves very fine grains (4-11 μm) sedimented on small aluminium disks. In this fashion, as compared to loose sand grains, samples are found to be more homogeneous and easier to handle (particularly for irradiation purposes). For this particle size, the alpha dosage attenuation is negligible and the gamma and beta contribution importance is reduced. An efficiency factor is then introduced in the age equation and defined as:

$$a = \frac{\text{TL per rad of } \alpha \text{ radiation}}{\text{TL per rad of } \beta \text{ radiation}}$$

so that the dose-rate is expressed in beta equivalent. For sediments, the alpha radiation is generally 10 times less efficient in inducing TL compared to beta radiation.

The age equation in fine-grain dating is then:

$$\text{AGE} = \frac{ED_{fg}}{aD_\alpha + D_\beta + D_\gamma + D_c}$$

APPLICATION TO QUATERNARY GEOLOGY

Specific problems related to the application of the technique to Quaternary geology are discussed in Dreimanis *et al.* (1978). They concern the inherited signal at time of sedimentation, the type of material used for dating and the age limits.

Inherited thermoluminescence

Archaeological dating assumes complete drainage of the previously acquired TL due to firing of the ceramic. For secondary detrital deposits, this thermal event has not likely happened. However, Morozov (1969), Shelkoplyas (1971) and Vlasov *et al.* (1978b) suggested that sunlight bleaches part of the initial TL signal. This may be attributed to photon interaction with trapped electrons (Huntley, pers. commun.). Indeed, Vlasov *et al.* (1978a) encountered a residual TL for sediments naturally exposed to sunlight from river bars and terraces which yield apparent TL ages of up to 9,400 years (Table 1). Thus an initial low-level of non-bleached thermoluminescence (residual) is inherited at the time of sedimentation. From that time on, in a continuous manner, the radioactive elements supply electrons to trapping centers until the sediment is remobilized in the sedimentary cycle (Figure 3).

Residual contributions must be subtracted from the natural glow curves in order to date the last sedimentation event. By means of a sunlamp, Wintle and Huntley (1980) proposed several methods to simulate the sunlight exposure. TL dates on oceanic sediments and loesses (Wintle 1981) obtained by this technique were found to be in good agreement with stratigraphical evidence. However, whether or not the sunlamp bleaching reproduces exactly the last sunlight exposure to be dated remains open. The zero point problem also impedes assessment of error limits in TL dating of sediments. It is probably in the order of ± 20%.

Material

In Quaternary sedimentary deposits, due to its resistance to the weathering processes, quartz is a stable ubiquitous mineral. Being almost free of internal radioactivity and easy to isolate from the other minerals in inclusion dating by routine laboratory techniques, it is a target dosimeter. Quartz typically exhibits a glow curve with three peaks (Figure 1) and the high temperature part of the glow curve is stable over geological time (Fleming, 1969).

Feldspar shows a very high TL sensitivity. However, it exhibits

Table 1. Residual Thermoluminescence of Sediments Naturally Exposed to Sunlight

(From Vlasov *et al.*, 1978a)

Sample	Area	Type of deposits	ED (rads)	TL AGE (years)
20/1	Issik-Viliskaya Basin	Modern bar of Tong River	4200	4700
20/2	"	Recent beach deposits Tongsky Gulf	2300	1900
21	"	" 10m from strandline	800	900
22	"	" 25m from strandline (Upper Holocene)	1500	1400
24	"	Lower Pleistocene terrace	9400	5400
25	"	Middle Pleistocene lacustrine sediments	4100	3400
K1	Klyazi Valley	Alluvial sediments surface	5400	2800
K2	"	same, 1 cm lower	6400	2800
K3	"	same, 10 cm lower	6900	3900
B1	Moscow Valley	Alluvial sediments surface	15000	8400
B2	"	"lower"	16900	9400

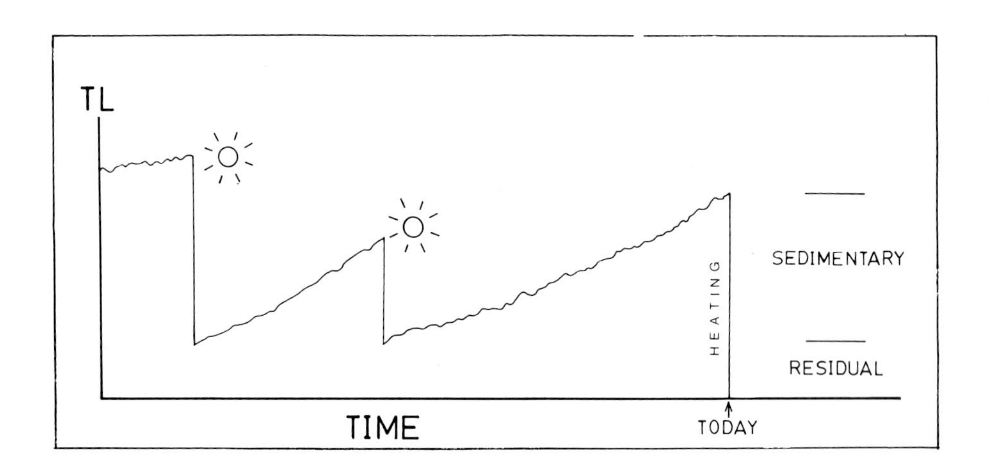

Figure 3 Hypothetical TL level versus time. Any event permitting the sediment to be exposed to sunlight (*e.g.* remobilization in the sedimentary cycle) reduces the thermoluminescence to a residual level. Today's heating drains the sample.

"anomalous fading" (next section) and moreover, extraction may be difficult.

In calcite, which is also used in TL dating, sample inhomogeneity in the distribution of internal radioactive elements in respect to luminescent centers in calcite gave rise to discrepancies from the uranium disequilibrium series dates of up to ± 200% in a speleothem study by Wintle (1978). Walton and Debenham (1980) demonstrated this inhomogeneity during light emission in natural calcites using a high-gain image intensifier.

Isolation of specific minerals is rarely achieved in fine-grain dating, so that measurements are made on polymineralic fractions. The shape of the glow curve is then deprived of acute peaks (Figure 4). Isolation of quartz is a difficult and time consuming task. However, Berger *et al.* (1980) achieved a degree of purity of 95% quartz with a hydrofluorosilicic acid treatment.

TL dating has been performed on many types of sediments: (a) oceanic (Huntley and Johnson, 1976; Wintle and Huntley, 1980); (b) glacio-lacustrine (Troitsky *et al.* 1979, this paper); (c) loesses (Li *et al.*, 1977, Shelkoplyas 1971; Wintle 1981); (d) soils (Shelkoplyas, 1971; and (e) tills (Troitsky *et al.*, 1979), with varying degrees of success. Because of obvious exposure to sunlight during its transport, loess should prove to be the best type of sediment to be dated.

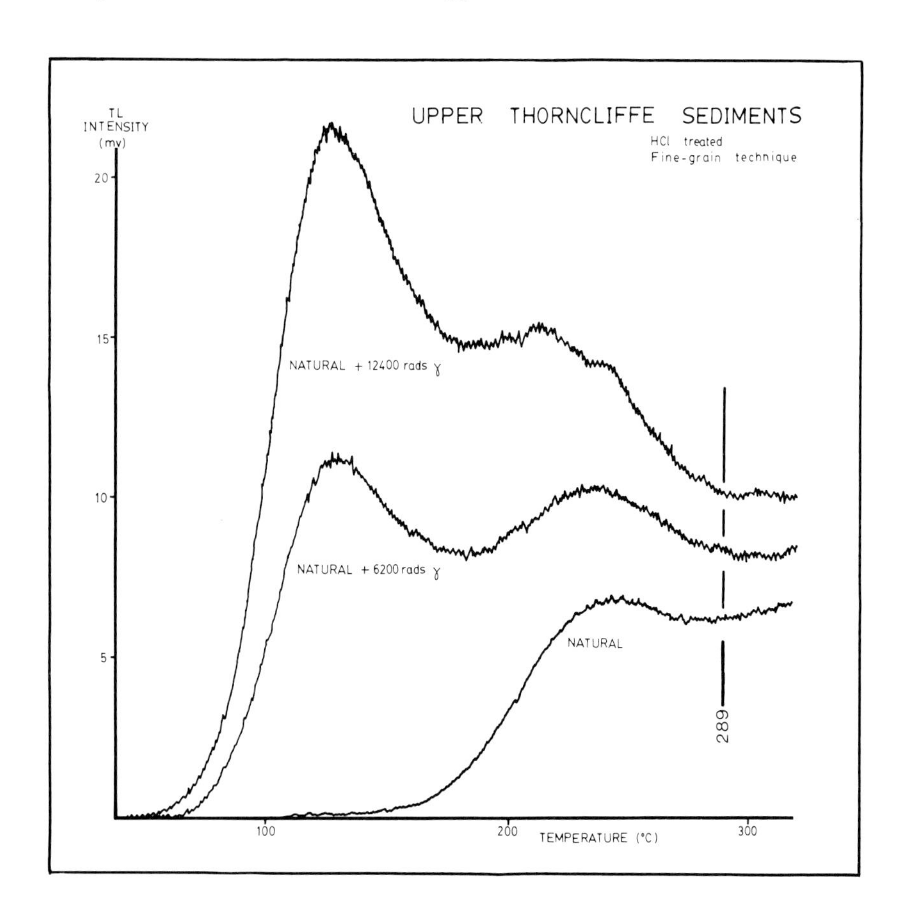

Figure 4 Natural glow curve of Upper Thorncliffe sediments and glow curves of the same sediments to which has been added known irradiation doses.

Age Limits

Theoretically, the upper limit of the TL dating method depends on the stability of the electrons in the traps.

The lifetime of these charges is given by:

$$t = s^{-1}e^{E/kT}$$

Fleming (1969) found lifetime values for quartz, at room temperature, of 3.5 hours for the 110°C peak, of 3,000 years for the 325°C peak and of 4×10^7 years for the 375°C peak (heating rate: 20°C/sec). On the other hand, Hütt *et al.* (1977) suggested, on the basis of saturation doses achieved in quartz from various sedimentary deposits, that the TL range should cover 10^6 years.

Experience shows that a practical upper limit may be set at 250,000 years, mainly because of early saturation of the traps. Hütt and Smirnov (manuscript) hope to extend the range to 500,000 years by using feldspars, on the basis of TL growth curve studies on this mineral.

A lower limit may actually be set at 5000 years because of increased importance of the residual signal for young samples.

THE EQUIVALENT DOSE

The upper term in the TL age equation, the equivalent dose (ED) is defined as the artificial dose that can simulate the natural TL level. It is expressed in rads.

Determination of the ED

The calculation of the ED is normally achieved by the additive method in which the artificial dose is gradually added to the natural sample; the equivalent dose is found at the interception of the TL growth curve with the abcissa (Figure 4 and 5).

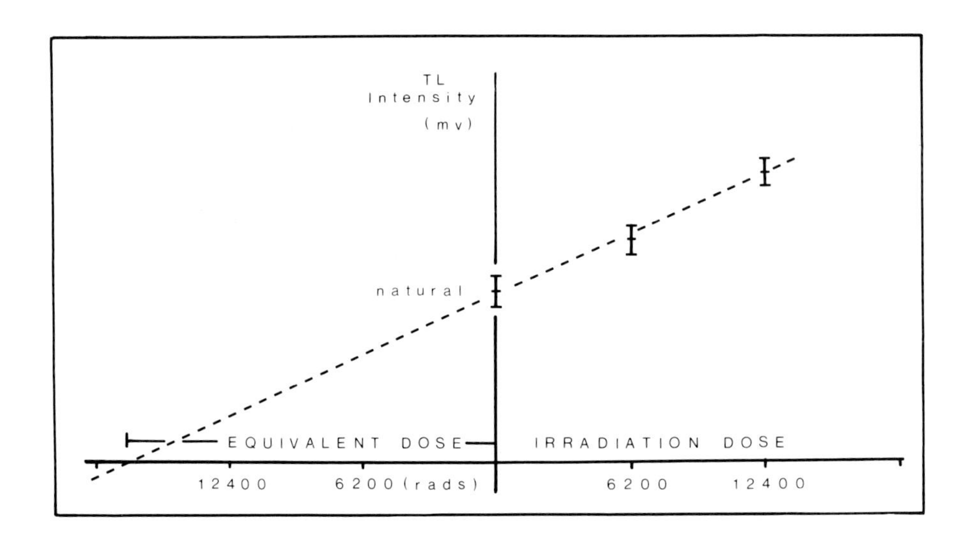

Figure 5 The additive method used to find the equivalent dose (ED).

For sediments, the following data have to be found (Figure 6):

1) the natural TL intensity

2) the hypothetical TL intensity at time of sedimentation by exposing sets of natural samples under a Sylvania sunlamp (*e.g.* one hour; 275 watts; distance: 40 cm)

3) the TL intensity acquired by such light-exposed samples when artificially irradiated (6200 rads γ; 12,400 rads γ).

For the Upper Thorncliffe sediments, the ED_1 found at the interception of the natural level with the TL growth curve was 8,800 rads at 289°C. In the preliminary state of the dating project, we considered as a maximum value for the ED, the projection of the TL growth curve to the TL intensity given by sets of samples exposed to the sunlamp for a very long time (18 hours: ED_{18}; Figure 7).

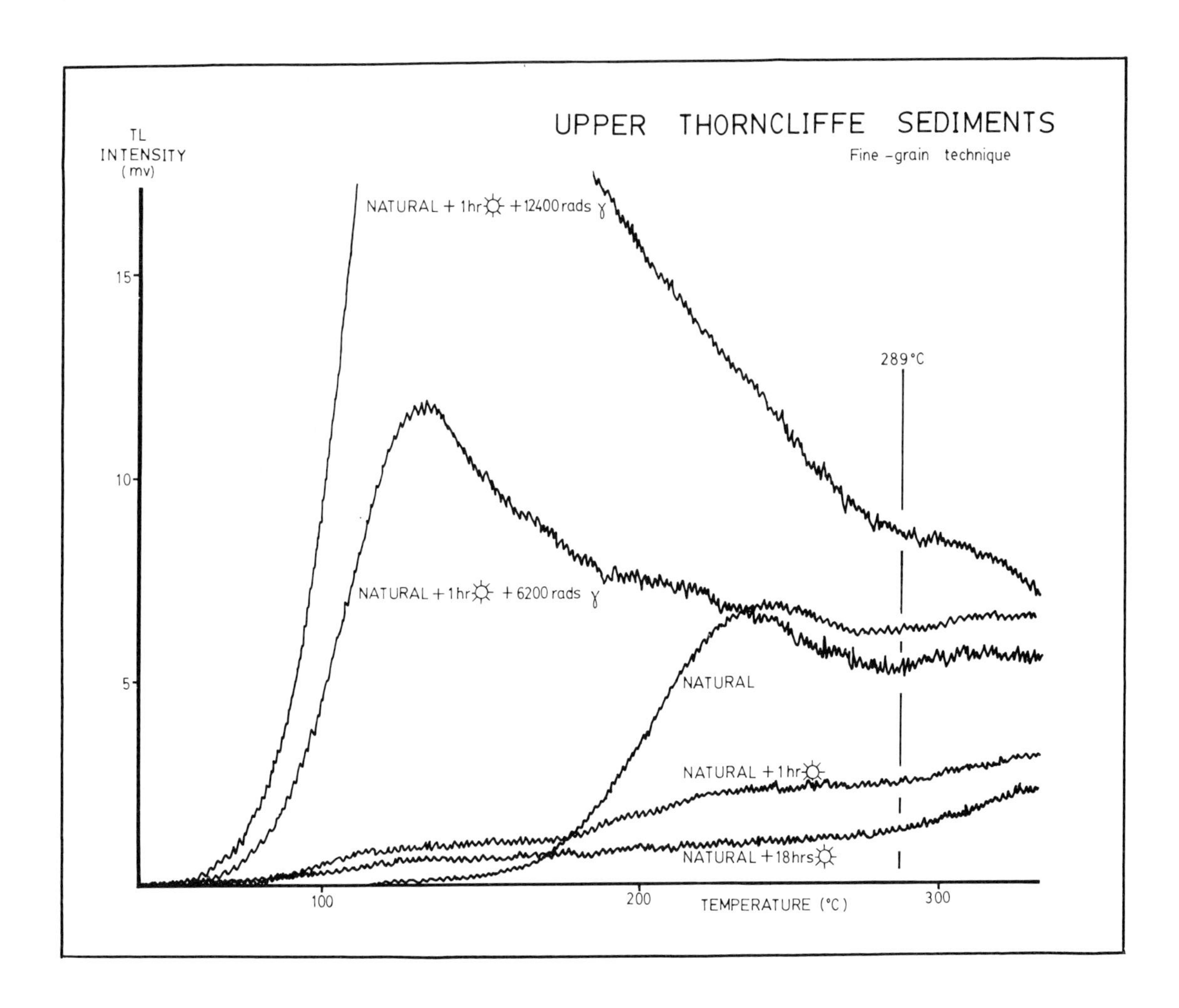

Figure 6 Glow curves of Upper Thorncliffe sediments showing the various steps to find the sedimentary dose.

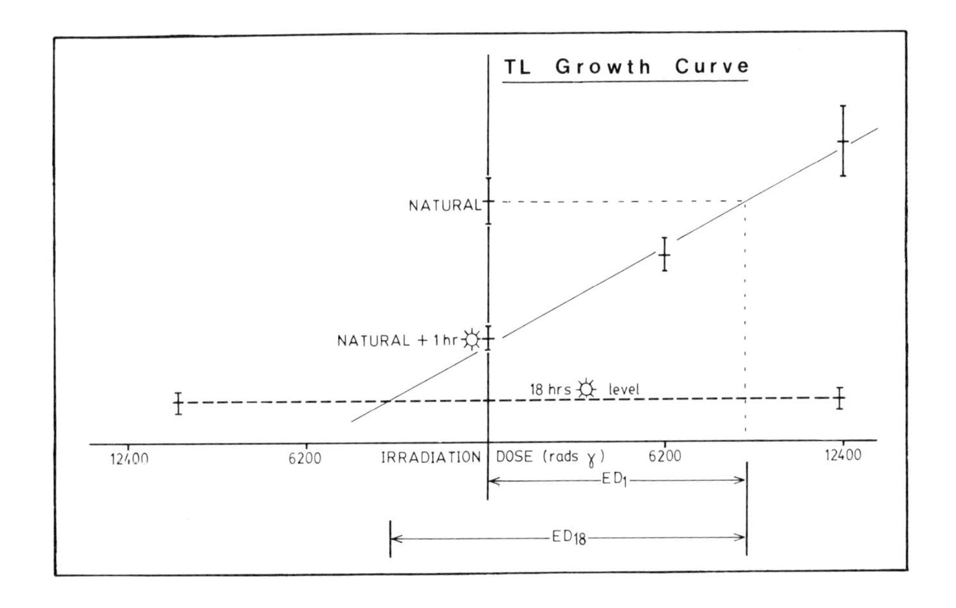

Figure 7 The TL growth curve, at 289°C; values of ED_1 and ED_{18} determined as shown.

At the Institute of Geology, in Tallin (Estonia), the equivalent dose is found by means of an iterative equation (Hütt and Smirnov, manuscript):

$$I: I_o\{1 - e^{-\alpha (ED + AD)} + \beta(ED + AD) \}$$

Where ED = equivalent dose

AD = artificial dose

I_o = natural level

I = artificial level

α and β = constants connected with TL kinetics

Stability of the ED

As a function of temperature, the ED normally increases to a plateau value beginning in this case around 260°C (heating rate: 1.1°C/sec). This Plateau Test (Figure 8) (Aitken, 1974) identifies the stable part of the glow curve, where leakage of electrons should not have happened during geological time. A sample which fails to exhibit a plateau and, yet, have suffered leakage from traps of any depth by wave-mechanical "tunnelling". This phenomenon can be important. Indeed, Wintle (1973) reported 40% fading in 15 hrs. for feldspars. By keeping a sample at room temperature for a period of a few months, one can check if the sample shows this kind of fading. As an example, sample 340-343 of the RC8-39 core from the Wintle and Huntley (1980) study revealed 14% fading over 20 days so that the TL intensity was considered to be minimum. The age of the sample must be greater than what is effectively measured (> 76,000 BP).

Upper Thorncliffe sediments showed a good plateau for both equivalent doses (ED_1 = 8835 rads γ ; ED_{18} = 12,400 rads γ ; Figure 8). As a first approximation, we propose an average sedimentary dose of 10,600 rads γ, to be used in the age equation.

Linearity of the TL growth curve

At moderate dose ($\leq$10,000 rads γ), the TL growth curve is fairly linear. For low doses, the growth curve is, in some cases, slowed down because of competition between TL producing and non-producing traps (Tite, 1966). This leads to underestimation of the equivalent dose, to which must then be added a supralinearity correction (I). Wintle (1981) computed such a correction for loesses. However, as these samples were not heated at time of sedimentation, we should check if such a sensibility change could be generated by solar bleaching.

On the other hand, the TL growth curve flattens at high doses and may even saturate. Troitsky *et al.* (1979) had many saturation problems of this kind so that many of their dates in Spitsbergen were considered as "greater than".

DOSE-RATE

The lower term in the TL age equation may be defined as the annual radiation a sample has received from neighbouring radioactive elements and the atmospheric cosmic rays. It is expressed in rads-year^{-1}.

Potassium -40, and the radioactive chains of Thorium-232, Uranium-238 and 235, and, in some cases, Thorium-230 and Protactinium-231, are the contributing isotopes. The first is easily determined by atomic absorption. The others may be found by gamma-ray spectrometry, neutron activation analysis or, more commonly, by alpha-counting (Huntley, 1977). Bell (1979) computed the most recent values of dose-rate contributions for unit quantity of each element. As dose-rates are expressed in beta-equivalent, the correction factor ($\alpha D_\alpha + D_\beta + D_\gamma + D_c$, as previously defined) must be determined in fine grain dating. Typical

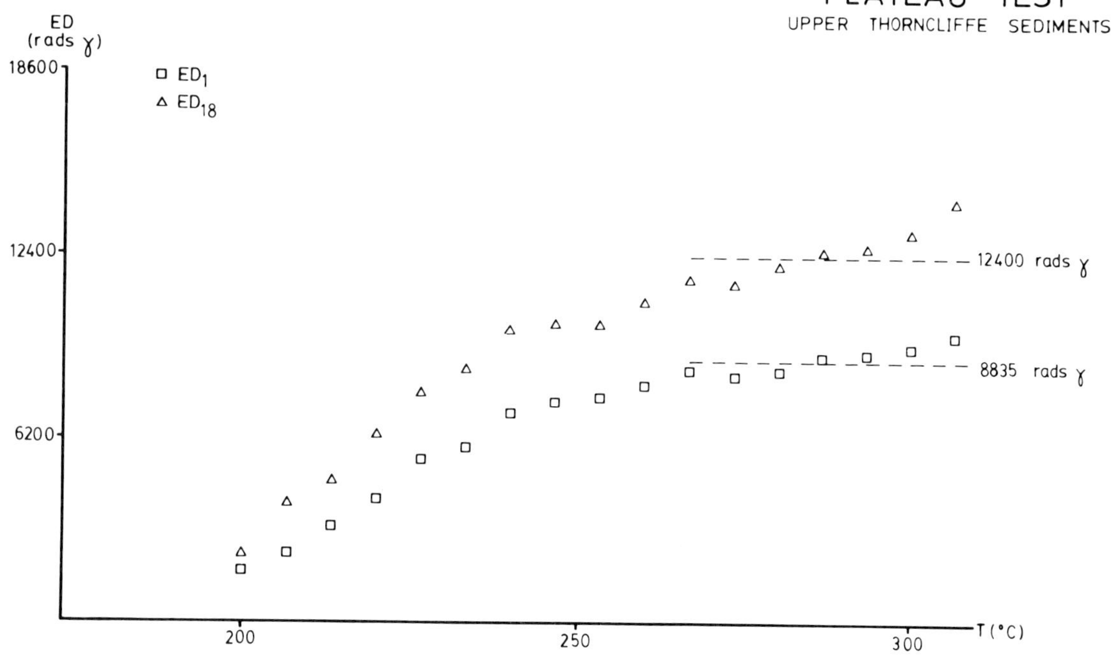

Figure 8 The plateau test for Upper Thorncliffe sediments; an average value of 10,600 rads γ was used in the age equation.

dose-rate values in sediments are of the order of few hundred millirads per year.

Thus, the dose-rate value is computed easily. The possibility remains, however, that this dose-rate could have fluctuated during geological time because of various factors the most critical being the <u>water content</u> of the sediment. It acts as a barrier for all three types of radiation at various degrees. It decreases the effective dose-rate by a factor of:

$$\frac{1}{1 + H\Delta}$$

where H = ratio related to the stopping power of water

Δ = water to solid weight ratio in the sample.

For sediments found in the unsaturated zone, the evaluation of the average water content can be fairly speculative. For the Upper Thorncliffe sediments, the *in situ* Δ value was 9%, the saturation Δ value was 18%. As this sequence was under glacial lake waters and then underneath the Laurentide Ice Sheet, we may suggest that these silty sands were at saturation for most of the time, giving an average water content of 15%. We believe the reader can imagine the dangers of such speculations for samples of unknown age!

On the other hand, Radon-222 is a gaseous product highly mobile in dry soil. Its depletion is important because 98% of the γ contribution in the uranium series is produced beyond radon. However, water greatly prevents its mobility. The radon emanation is roughly counterbalanced by the ground water dilution effect and by the incoming radon from the lower levels.

Disequilibrium of the radioactive chains, removal of radioactive elements by the groundwater and even inhomogeneity of the radiation field (Sutton, 1979) are also possible.

To limit those uncertainties to a lesser extent, phosphor dosimetry (*e.g.* $CaSO_4$-Dy) at the sampling site can be done over periods of many months to account for seasonal variations. This is now done as a routine by the Tallin TL laboratory and Hütt *et al.* (1978) reported significant differences between *in situ* measurements and laboratory determinations from the weight of the radioactive elements, of up to 50% in non-homogeneous profiles.

The dose-rate measured for Upper Thorncliffe sediments is dominated by potassium which contributes two-thirds of the total 324 mrads/year value. The cosmic-ray estimation of 14 mrads/year was taken from Wintle (1981). An α efficiency factor of 0.1 has been postulated.

SOME RECENT TL DATINGS OF GLACIGENIC DEPOSITS

The applicability of the TL dating method may be tested by TL age determinations of samples from sections that have been dated by some other methods. Two such test cases have been discussed recently from areas in USSR where the relative stratigraphy is known from palynologic studies: (1) Pasva section in the district of Archangelsk (Deviatova *et al.*, 1981) and (2) the Kraslava area of S.E. Latvia (Meirons *et al.*, 1981). The TL dates were obtained at the TL laboratory of the Institute of Geology in Tallin, by using the quartz inclusion technique.

Pasva section

This deposit is on River Vaga in northern USSR, in an area that was not covered by the ice sheet of the last or Valdai (= Wisconsin) Glaciation. Two adjoining sections were investigated palynologically, and they contained peat and clay of Mikulino (= Sangamon) Interglacial age overlain by alluvial, lacustrine and glacio-lacustrine sediments of the Valdai Glaciation, and underlain by glacio-lacustrine sediments and till of probable Riss (= Illinoian) age. The TL and ^{14}C dates of both sections are listed in Table 2.

Table 2. TL and ^{14}C dates, Pasva sections. (After Deviatova *et al.*, 1981)

Section No.	Depth (m)	Material	TL or ^{14}C date Lab. No.	date, BP	AGE
1	1.7	Silty sand	Tln-TL-1	26,500	VALDAI
1	2.7	Clay	Tln-TL-5	40,000	
1	8.2	Silty sand	Tln-TL-2	45,000	
1	9.3	Sand	Tln-TL-6	32,000	
1	11.25	Sand wedge injected into peat below	Tln-TL-3	62,000	
2	11.90-11.95	Woody peat	Tln-215	34,500±1100	MIKULINO INTER-GLACIAL
2	12.20-12.25	Wood	Tln-216	36,500± 750	
2	12.45-12.50	Woody peat	Tln-217	34,000± 750	
1	12.45-12.50	Woody peat	Tln-226	>49,700	
2	19.8	Till	Tln-TL-4	151,000	PRE-MIKULINO

Note: ^{14}C dates have lab. Nos. without TL.

The ^{14}C dates Tln-215, Tln-216, and Tln-217 are definintely too young, because of contamination, and the infinite date of >49,700 yrs BP does not specify the absolute age, since the Mikulino Interglacial is beyond the ^{14}C dating range. The TL dates of the post-Mikulino sediments appear reasonable for the Middle to Early Valdai, except for Tln-Tl-6 that is out-of-line in the sequence of the downward increasing ages. The TL date Tln-TL-4 of the underlying till (151,000) also appears reasonable, if the till is of Riss age as suggested by the sedimentological and palynological sequences from till to the overlying sediments.

Kraslava Area

This area is well known for its interglacial deposits since the mid-19th century. The stratigraphic cross-section through the area (Meirons *et al.*, 1981, p. 29) is based upon a considerable number of test drillings and natural exposures investigated by several authors.

The last or Baltic Glaciation (= Valdai = Würm = Wisconsin) is represented by one to two reddish-brown till sheets and some stratified drift deposits. A silty sediment underlying this till at Židini (15 km W. of Kraslava) is probably of Early Baltic age; it has been dated by TL as 97,150 years old (Tln-TL-45).

The grey till of probable Kurzeme Glaciation (= Riss = Illinoian) has been dated from two test drillings:

- at Židini (Tln-TL-42): >68,800 yrs BP

- at Robežnieki (Tln-TL-49): 106,250 yrs BP.

The absence of warm-climate interglacial deposits of the Felicianova (= Mikulino = Eemian = Sangamon) age either above or below this grey till makes its stratigraphic position uncertain. However, if it belongs to the Kurzeme Glaciation, then the finite TL date of 106,250 yrs BP is a few tens of thousands of years too young. The stoneless clay right underneath the till at Židini produced a strange TL date - 5150 yrs BP (Tln-TL-43) - probably due to some mistreatment of the clay sample. This date does not help to finalize the overlying till date (Tln-TL-42: >68,800 yrs BP), but it warns of the possibility of some erratic and completely unreasonable dates, as in any dating procedure.

The next lower layer dated was the famous Kraslava or, more specifically, Adomova Interglacial deposit palynologically investigated and considered to be of Pulvernieki Interglacial (= Likhvin = Mindel-Riss = Yarmouth) age by the majority of its interpreters. Its TL date is >161,550 yrs BP (Tln-TL-48). This infinite date is in the right relative order, in comparison with the dates of the overlying tills. If we dismiss the obviously wrong (Tln-TL-43) date of 5150 years BP, all the others appear in the right relative order: the deeper the layer, the older the date, but half of them are not finite and their absolute values remain uncertain.

Southern England

A third set of recently published TL dates on terrestrial materials are the six age determinations by Wintle (1981) of loesses in S. England that are most probably of Late Devensian age. They range from 14.5 to 18.8 thousand years BP ± 20%. Wintle (1981, p. 480) mentions that they are "easily distinguished from older loesses", and therefore these dates appear to be in the right order of magnitude.

Recently published finite TL dates from three areas either agree with their expected ages, or occasionally, some are erroneous. The absolute values quoted in Soviet literature may be too old as some residual TL was probably still present during deposition of the material dated. Whenever several stratigraphic units of some age differences are dated, *e.g.* at Pasva and Kraslava, most dates are in the right relative age sequence. This is in agreement with the conclusions on the various dates published and discussed by Dreimanis *et al.* (1978).

CONCLUSION

The thermoluminescence age equation of Quaternary sediments is then:

$$AGE = \frac{\text{SEDIMENTARY DOSE}}{\begin{array}{l} W_{K_2O}\{(.0682/1+H_\beta\Delta) + (.0205/1+H_\gamma\Delta)\} \\ +W_U\{(.2783a/1+H_\alpha\Delta) + (.0146/1+H_\beta\Delta) + (.0127/1+H_\gamma\Delta)\} \\ +W_{Th}\{(.0740a/1+H_\alpha\Delta) + (.0029/1+H_\beta\Delta) + (.0050/1+H_\gamma\Delta)\} \\ +Dc/1+H_\gamma\Delta \end{array}}$$

where W: weight in per cent for K_2O and in ppm for U and Th; the values introduced are the α, β and γ contributions for unit quantity of each element in mrads/year (Bell, 1979; Wintle and Huntley, 1980).

Therefore, with a sedimentary dose of 10,600 rads, for a dose-rate of 324 mrads/year, a TL date of 32,700 ± 5500 BP for the Upper Thorncliffe sediments is suggested. This value correlates reasonably well with radiocarbon dating (Table 3 and Figure 9).

Table 3. Thermoluminescence dating of Upper Thorncliffe sediments.

W_{K_2O} = 2.67 ± .05%	SEDIMENTARY DOSE
W_U = 1.6 ± 0.3 ppm	10,600 rads
W_{Th} = 2.9 ± 0.8 ppm	
D_c = 14 mrads/year	DOSE-RATE
H_α = 1.50	324 mrads/year
H_β = 1.25	
H_γ = 1.00	
Δ ≃ 15%	
	TL AGE = 32,700 years

Still uncontrolled variables in TL dating are the determination of the residual signal at time of sedimentation and the fluctuations of the water content. Studies of modern environment or calibration of sun-lamp exposure from chronologically controlled sites may tide over the former. Careful sampling should limit the latter. As TL dating is performed on detrital materials, it provides an overwhelming advantage over many other techniques, and should play a major role in deciphering the time-stratigraphy of the late Pleistocene.

ACKNOWLEDGEMENTS

The authors wish to thank D.J. Huntley who made determinations of the radioactive elements by alpha-counting and who collaborated with M. Lamothe over a one week visit to his laboratory at Simon Fraser University. C. Hillaire-Marcel and P. Pagé kindly reviewed the manuscript.

This research was supported by a "Fonds F.C.A.C. pour l'aide et le soutien à la recherche-Québec" scholarship to M. Lamothe and N.S.E.R.C. grant A4251 to A. Dreimanis. Micheline Lacroix typed the manuscript.

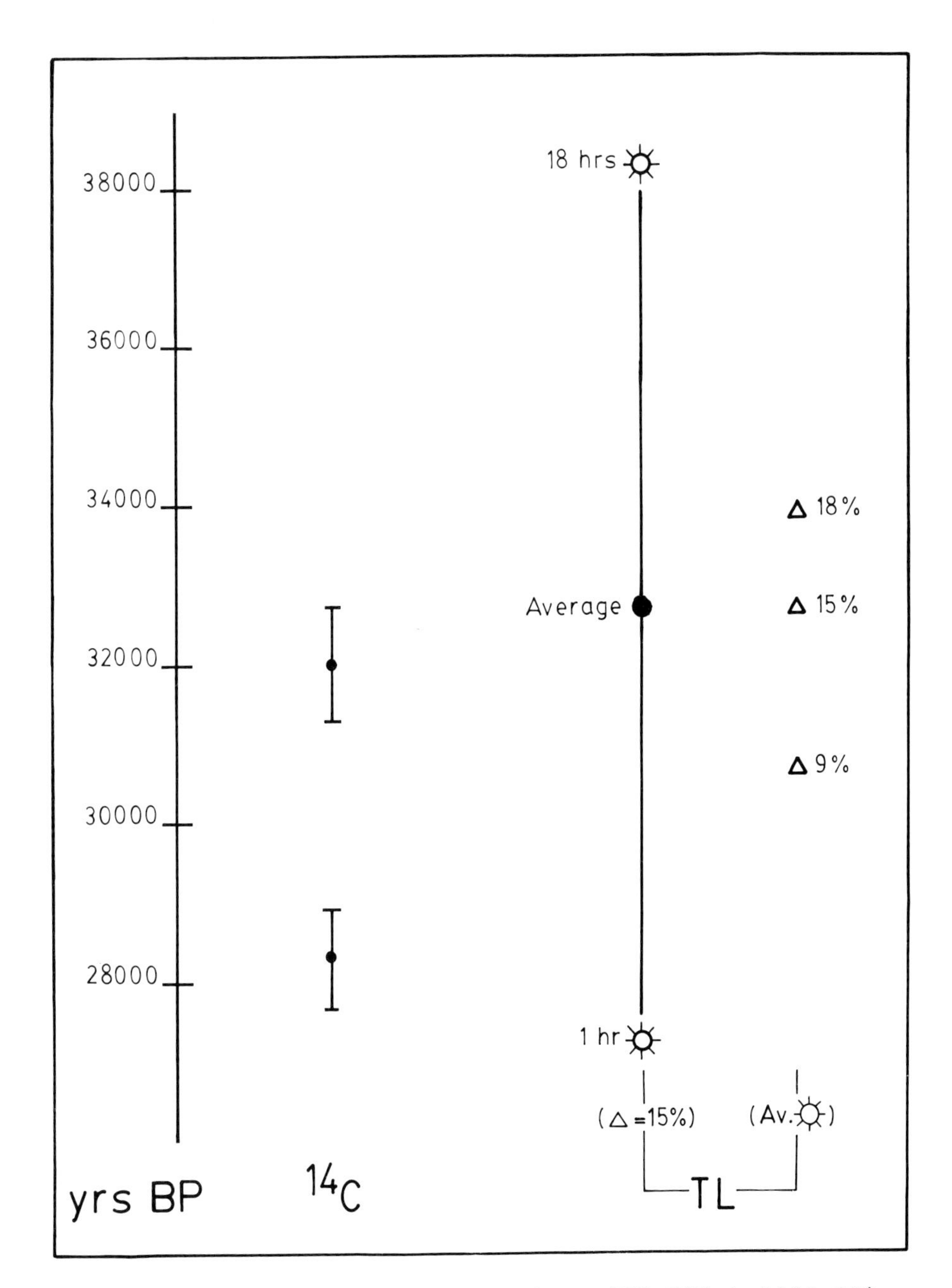

Figure 9 Thermoluminescence date (32,700 ± 5500 BP) versus radiocarbon dates (28,300 ± 600 GSC-1082; 32,000 ± 690 GSC-1221) for Upper Thorncliffe sediments; error bars for the TL date are arbitrarily chosen as the ages given by ED_1 and ED_{18}, respectively; variations in the TL date for different water to solid weight ratios (9, 15 and 18%) are also shown.

Michèle Laithier's drafting greatly enhanced the paper.

APPENDIX

The sample dated was taken at the Hi section of Karrow (1967), in a one metre layer of silts, found between the Meadowcliffe and Leaside tills. This unit represents probably the "number 9 layer" of Mörner (1971). These silts should lie between the 28,300 and 32,000 BP radiocarbon dated peaty sands of Berti (1975). The peaty layers were not observed at time of sampling. The sediments were protected from sunlight in an opaque bag.

REFERENCES CITED

Aitken, M.J., 1974, Physics and Archaeology: Clarendon Press, Oxford,

291 p.

Aitken, M.J., Tite, M.S. and Reid, J., 1963, Thermoluminescent dating progress report: Archaeom., v. 6, p. 65-75.

Bell, W.T., 1979, Thermoluminescence dating: radiation dose-rate data: Archaeometry, v. 21, p. 243-245.

Berti, A.A., 1975, Paleobotany of Wisconsinian Interstadials, Eastern Great Lakes Region, North America: Quaternary Research, v. 5, p. 591-619.

Berger, G.W., Mulhern, P.J. and Huntley, D.J., 1980, Isolation of silt-sized quartz from sediments: Ancient TL, n. 11, p. 8-9.

Daniels, F., Boyd, C.A. and Saunders, D.F., 1953, Thermoluminescence as a research tool: Science, v. 117, p. 343-349.

Deviatova, E.I., Raukas, A.V., Raiamaie, R.A. Hütt, G.I., 1981, Verkhnepleistotsenovii razrez Pasva (r. Vaga, Arkhangelskaia oblast) i ego stratigraficheskoe znachenie: Bulleten Komissii po izucheniiu chetvertichnogo perioda, No. 51, p. 38-50 (in Russian).

Dreimanis, A., Hütt, G., Raukas, A. and Whippey, P.W., 1978, Dating methods of Pleistocene deposits and their problems: I Thermoluminescence dating: Geoscience Canada, v. 5, p. 55-60.

Fleming, S.J., 1969, The acquisition of radioluminescence by ancient ceramics, Unpublish. D. Phil. Thesis, Oxford Univ.

_____________, 1970, Thermoluminescence dating: refinement of the quartz inclusion method: Archaeom., v. 12, p. 135-146.

_____________, 1979, Thermoluminescence techniques in archaeology: Clarendon Press, Oxford, 227 p.

Huntley, D.J., 1977, Experiences with an alpha counter: Ancient TL, no. 1, p. 3-6.

Huntley, D.J. and Johnson, H.P., 1976, Thermoluminescence as a potential means of dating siliceous ocean sediments: Canadian Journal of Earth Sciences, v. 13, p. 593-596.

Hütt, G. and Raukas, A., 1977, Potential use of the thermoluminescence method for dating Quaternary deposits: Bull. Comm. Stud. Quat. Per., v. 47, p. 77-86 (in Russian).

Hütt, G. and Smirnov, A.V., manuscript, Thermoluminescent dating of sediments in the Soviet Union, 12 p.

Hütt, G., Smirnov, A.V. and Punning, Y-M. K., 1978, *In situ* dosimetry by means of $CaSO_4$-Dy for determination of the annual dose-rate: Tallin' 78, p. 124-125 (in Russian).

Hütt, G., Vares, K. and Smirnov, A.V., 1977, Thermoluminescent and dosimetric properties of quartz from Quaternary deposits, Izv. USSR: Chem. Geol. ser., v. 26, p. 275-283 (in Russian).

Karrow, P.F., 1967, Pleistocene Geology of the Scarborough Area: Ontario Department Mines, Geol. Rep. 46.

Levy, P.W., 1974, Physical Principles of thermoluminescence and recent developments in its measurement: N.Y. Upton, Brookhaven Lab., 18 p.

Li, J.L., Pei, J.X., Wang, Z.Z. and Lu, Y.C., 1977, A preliminary study of both thermoluminescence of the quartz, powder in loess and the age determination of the loesses layers: Kexue Tangbau, v. 22, p. 498-502.

McDougall, D.J., 1968, Thermoluminescence of Geological Materials: London, Acad. Press.

McMorris, D.W., 1971, Impurity color centres in quartz and trapped electron dating: ESR, TL studies: Jour. of Geophys. Res., v. 76, p. 7875-7887.

Marfunin, A.S., 1979, Spectroscopy, Luminescence and radiation centers in minerals: N.Y., Springer-Verlag.

Meirons, Z., Punning, J.M., Hütt, G., 1981, Results obtained through the TL dating of South-East Latvian Pleistocene deposits, Eesti NSV Teaduste Akadeemia Toimetised, Geologia, 30/1, p. 28-33 (in Russian, with Estonian and English summaries).

Morozov, G.V., 1969, The dating of Quaternary Ukranian sediments by thermoluminescence: XIIIth International Quaternary Association Congress.

Morner, N.-A., 1971, The Plum Point Interstadial: age, climate and subdivision: Canadian Journal of Earth Sciences, v. 8, p. 1423-1431.

Pact, 1978, A specialist seminar on thermoluminescence dating, Journal of the European Study Group on Physical, Chemical and Mathematical techniques applied to Archaeology, 2 volumes, Oxford.

Randall, J.T. and Wilkins, M.H.F., 1945, Phosphorescence and electron traps: Proc. Roy. Soc. of London, v. A184, p. 366-407.

Shelkoplyas, V.N., 1971, Thermoluminescence method in Quaternary deposits dating, in Zubakov, V.A. and Kotchegura, V.V., eds., Chronology of the glacial age: Leningrad, p. 115-159 (in Russian).

Sutton, S., 1979, Thermoluminescence dating of ancient heated rocks: A progress report and sample request: S.A.S. Newsletter, v. 3, no. 2, p. 1-2.

Tite, M.S., 1966, Thermoluminescent dating of ancient ceramics: a reassessment: Archaeom., v. 9, p. 155-169.

Troitsky, L., Punning, J.M., Hütt, G. and Rajamae, R., 1979, Pleistocene glaciation chronology of Spitsbergen: Boreas, v. 8, p. 401-407.

Valladas, G., 1979, La datation des roches par la thermoluminescence, Applications; Bull. de l'Association Française pour l'étude du Quaternaire, p. 43-52.

Vlasov, V.K., Kulikov, O.A. and Karlov, N.A., 1978a, Determination of the residual TL in quartz from surficial deposits: Tallin '78, p. 23-25 (in Russian).

__________, 1978b, The zero-point problem in thermoluminescence dating: Tallin 78, p. 26-28 (in Russian).

Walton, A.J. and Debenham, N.C., 1980, Spatial distribution studies of thermoluminescence using a high-gain image intensifier: Nature, v. 284, p. 42-44.

Whippey, P.W., 1980, Applications of thermoluminescence to problems in Geology, manuscript, 27 p.

Wintle, A.G., 1973, Anomalous fading of thermoluminescence in mineral samples: Nature, v. 244, p. 143-144.

__________, 1978, A thermoluminescence dating study of some Quaternary calcite: potential and problems: Canadian Journal of Earth Sciences, v. 15, p. 1977-1986.

__________, 1980, Thermoluminescence dating: a review of recent applications to non-pottery materials: Archaeometry, v. 22-2, p. 113-122.

__________, 1981, Thermoluminescence dating of late Devensian loesses in southern England: Nature, v. 289, p. 479-480.

Wintle, A.G. and Huntley, D.J., 1980, Thermoluminescence dating of ocean sediments: Canadian Journal of Earth Sciences, v. 17, p. 348-360.

Zimmerman, D.W., 1971, Thermoluminescent dating using fine grains from pottery: Archeom., v. 13, p. 29-52.

RELATIVE AND ABSOLUTE DATING OF QUATERNARY MOLLUSKS WITH AMINO ACID RACEMIZATION: EVALUATION, APPLICATIONS AND QUESTIONS

JOHN F. WEHMILLER

ABSTRACT

Amino acids are entrapped in living skeletal carbonates as components of the structural protein upon which the carbonate phase forms. After death of the organism, the protein undergoes a complex array of reactions, involving hydrolysis into short-chain polypeptides and free amino acids, decomposition and/or leaching, and racemization of the amino acids in both free and bound forms. The racemization reaction involves conversion of each amino acid from its original 100% L form ("left-handed") into an equilibrium mixture (usually 50-50) of D ("right-handed") and L amino acids. The chronological utility of amino acid racemization has been studied in several genera of mollusks from more than 200 Quaternary marine localities along both the Atlantic and Pacific coasts of the United States. In these studies, both relative and absolute age estimates have been proposed for a large number of previously undated localities. In addition, a number of criteria for the evaluation of the method have been developed and tested, using samples with some form of absolute or relative age control. These criteria include:

1) acceptable precision (5 to 10% depending upon the amino acid) for multiple analyses of shells of the same genus from the same outcrop;

2) mineralogical and structural preservation;

3) ability of each genus of interest to consistently demonstrate increasing D/L values in samples of increasing relative age (Such as vertical terrace sequences or in superposed strata);

4) ability of each genus of interest to achieve racemic equilibrium in "old" samples (early Pleistocene to Miocene, depending on temperature);

5) consistent relative rates of racemization of different amino acids in the same sample;

6) increasing D/L values with decreasing latitude (increasing temperature) in samples of known age equivalents.

Genera that do not meet one or more of these criteria must only be used with caution in chronological applications. Multiple genera should be employed in chronological studies because systematic generic effects on apparent racemization kinetics do exist and because only rarely would one genus be found in abundance at all localities of interest. The use of multiple genera provides cross-checks on age estimates, and occasionally reveals ambiguities that would be unrecognized if only single genera were used.

The simplest application of amino acid racemization is as a relative stratigraphic tool for closely spaced (but discontinuous) localities that can be assumed to have had similar or identical temperature histories. Results from southern California and the mid-Atlantic coastal plain provide good examples of this approach.

If amino acid data are available for at least one absolutely dated locality from within local "aminostratigraphic" sequences, then absolute age estimates can be derived from the D/L values using one of several kinetic models of diagenetic racemization that are under consideration. Correlation of separate aminostratigraphic sequences between two widely-spaced regions (hence with different temperature histories) requires estimates of the differences in temperature histories as well as a quantitative kinetic model of the temperature dependence of racemization in samples of equal age. These kinetic models, and their inherent assumptions, have been calibrated and successfully tested in a few cases, mostly with samples of known age from a broad latitude range (35° N to 25° N) along the Pacific coast of North America. On the Atlantic coast, some conflicts between U-series coral ages and both relative and kinetic model age estimates have been encountered.

INTRODUCTION

Amino acids are found in fossil Quaternary mollusks as the remains of the original protein upon which the calcareous matrix was formed. Diagenesis of this protein-aceous material involves hydrolysis of high molecular weight components, loss of amino acid material by chemical destruction and/or leaching and diffusion, and racemization of the amino acids. This latter process is the conversion of the original L-amino acids (*levo*, or "left-handed") into a mixture of D- (*dextro*, or "right-handed") and L-amino acids. In principle, the enantiomeric ratio (D/L value) for each amino acid retained by the fossil will increase with time from an initial value of 0.0 to an equilibrium value that is usually 1.0. The general pattern of these diagenetic reactions is shown in Figure 1.

The extent of the racemization reaction is usually determined on the total amino acid mixture, though several laboratories also routinely analyze the free amino acid component for enantiomeric ratios. As the measured D/L value is the combined result of a number of contributing reactions (with variable rates of hydrolysis, racemization, leaching or decomposition for different amino acids in different molecular weight components), it is difficult to predict theoretically the rates of racemization for different amino acids in different genera or sample types. From a variety of field and laboratory experiments, it is known that D/L values in fossils are a measure of the age of the fossil, but that these D/L values are also dependent on temperature, amino acid, genus (or sample type), and contamination or other physical-chemical diagenetic effects. In this respect there are many similarities between racemization and obsidian hydration dating methods.

If well-preserved samples of a single genus are available, then the racemization dating method can be used to determine relative ages for closely spaced but discontinuous deposits that can be reasonably assumed to have had similar temperature histories. The racemization method can be used as an absolute dating tool if suitable calibration is available along with an appropriate model for the overall kinetics of racemization in the sample type being used. Temperature assumptions are an inherent component of either relative or absolute applications of the method.

Enantiomeric ratios in Quaternary mollusks have been used as dating tools for coastal deposits of the Atlantic (Hare and Mitterer, 1967; Mitterer, 1974; 1975; Belknap, 1979; Belknap and Wehmiller, 1980), Pacific (Wehmiller *et al.*, 1977; 1978a; Kennedy, 1978; Lajoie *et al.*, 1979; 1980; Wehmiller and Emerson, 1980), and Arctic (Miller *et al.*, 1977; Miller and Hare, 1980a) coasts of North America, as well as the coast of Great Britain (Miller *et al.*, 1979). Many of these studies have developed basic information that aids in the evaluation of the reliability of the method, such as the demonstration of typical precision

Figure 1 Amino Acid Diagenesis in Fossils

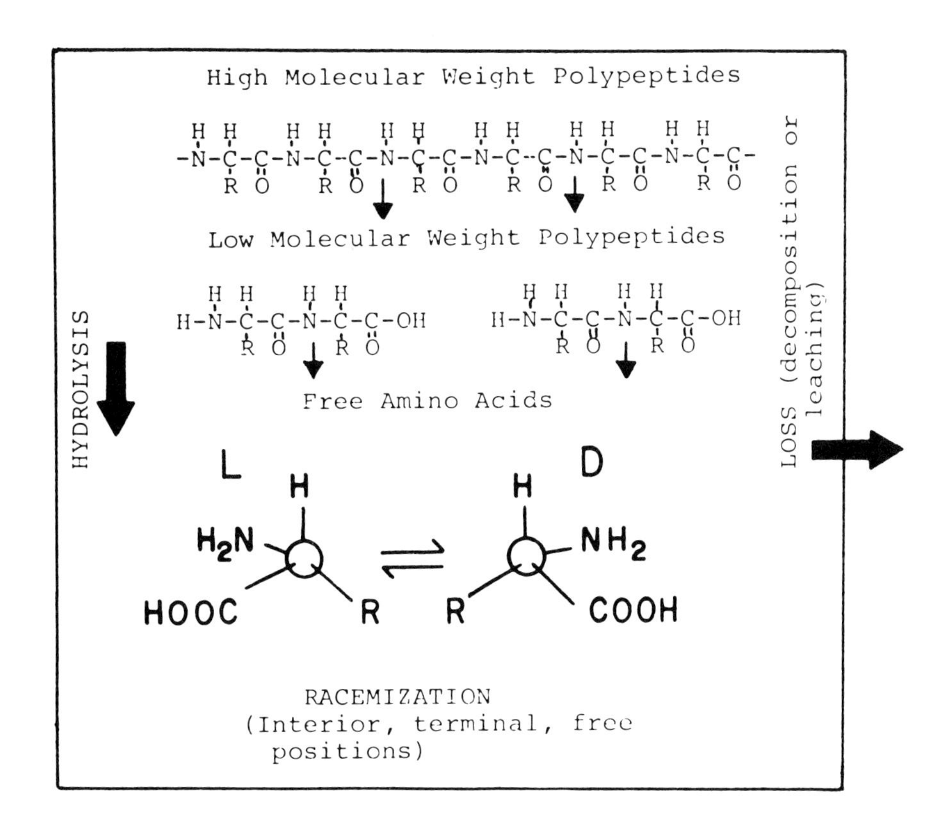

D/L Increases from 0.0 to 1.0
Free/Bound Increases from ∿0 to >10
Common Amino Acids: R =

$-H$	Glycine	$-CH_2CH(CH_3)_2$	Leucine
$-CH_3$	Alanine	$-CH(CH_3)\ (C_2H_5)$	Isoleucine
$-CH(CH_3)$	Valine	$-CH_2CH_2COOH$	Glutamic acid
$-CH_2COOH$	Aspartic Acid	$-CH_2(C_6H_5)$	Phenyl-alanine

```
          H
          |
          N    H
       /    \ /
   H2C       C-COOH
    |        |              Proline
    |        |
   H2C ----- CH2
```

of multiple analyses (Kvenvolden *et al.*, 1979), intrageneric and intergeneric relative apparent rates of racemization (Lajoie *et al.*, 1980), and the "ability" of several genera to show increasing D/L values in samples of known increasing stratigraphic age (Hare and Mitterer, 1967; Wehmiller *et al.*, 1977; 1978a; Lajoie *et al.*, 1980; Belknap, 1979). Studies of samples of equal age (as established by radiometric calibration) along north-south coastlines have been particularly useful in developing a framework for regional correlation, and for quantification of the temperature effect on racemization (Wehmiller *et al.*, 1977; Wehmiller and Belknap, 1978; Wehmiller and Emerson, 1980). These latter studies have also been important in discussions of optional kinetic models for racemization in Quaternary mollusks (Wehmiller, 1981a). The present paper will review some of the conclusions developed in these previous studies and will discuss some unresolved questions about the reliability of the method. A comprehensive discussion of the amino acid dating method is found in Hare, P.E., *et al.* (1980).

LOCALITIES AND SAMPLES

Atlantic and Pacific coast Quaternary localities from which molluscan enantiomeric ratio data are available are shown in Figure 2.

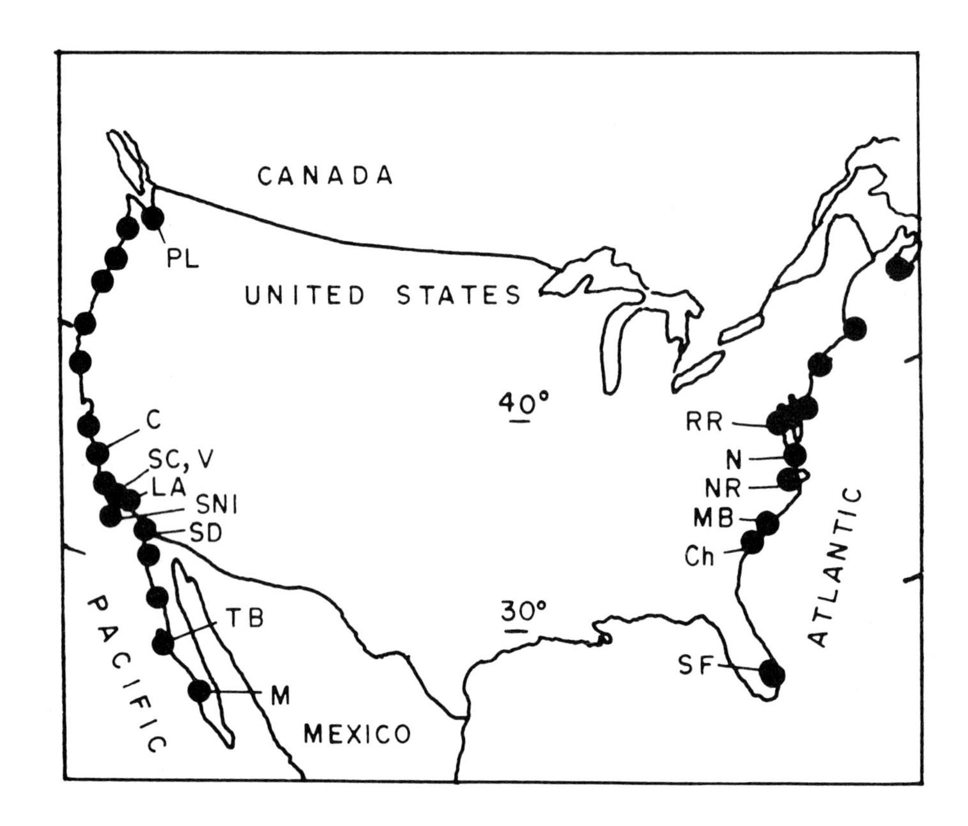

Figure 2 Quaternary fossil localities for which amino acid enantiomeric ratio data are available. Labelled localities have radiometric or stratigraphic control.

Pacific Coast

PL	Puget Lowland: Holocene radiometric loc.
C	Cayucos: 125 ka U-series loc.; Figs. 4, 6.
SC	Sea Cliff: Holocene radiometric loc.; strat. loc.; Figs. 4, 6.
V	Ventura: Strat. loc., Figs. 4, 6.
LA	Los Angeles Basin: strat. loc., Figs. 4, 6.
SNI	San Nicolas Island: 120 ka U-series loc.; strat. loc.; Figs. 4, 6.
SD	San Diego: 120 ka U-series loc.; strat. loc., Figs. 5, 6.
TB	Turtle Bay: strat. loc., Fig. 6.
M	Magdalena Bay: 120 ka U-series loc., Fig. 6.

Atlantic Coast

RR	Rappahanock Riv.: 190 ka U-series loc., Fig. 9.
N	Norfolk: 74 ka U-series loc., Fig. 9.
NR	Neuse River, Flanner Beach Fm., Fig. 9.
MB	Myrtle Beach; strat locs. Fig. 9.
Ch	Charleston: 90 ka U-series locs., Fig. 9.
SF	South Florida: 120 U-series locs., Fig. 9.

Specific locality information is found in previously-cited references for these areas. A few of the localities shown in Figure 2 have absolute age control (U-series dates on corals, or ^{14}C) for rigorous calibration of amino acid kinetics in a given region, and a few of the localities shown in Figure 2 actually represent a series of units with unambiguous stratigraphic relationships. Samples from these types of localities are particularly important in the evaluation of the amino acid dating method.

Molluscan genera that have been employed most extensively in these studies are:

Family Veneridae:	*Protothaca*	(Pacific coast)
	Saxidomus	(Pacific coast)
	Tivela	(Pacific coast)
	Chione	(Pacific and Atlantic coasts)
	Mercenaria	(Atlantic coast)
Family Tellinidae:	*Macoma*	(Pacific and Atlantic coasts)

Other genera that have been analyzed less frequently include: *Epilucina, Cumingia, Tegula, Diodora, Polinices, Hiatella, Mya, Spisula, Busycon, Crassostrea,* and *Ostrea.*

ANALYTICAL METHODS

Methods employed for enantiometric ratio determination rely upon either gas chromatographic resolution of the derivatives of the D- and L-amino acids (Kvenvolden *et al.*, 1972; Wehmiller *et al.*, 1977; Hoopes *et al.*, 1978; Smith and Wonnacott, 1980; Wehmiller and Emerson, 1980), or liquid chromatographic resolution of amino acids, either in mixtures (Hare and Gil-Av, 1979) or individually, either derivatized (Bada and Man, 1980) or underivatized (Hare and Mitterer, 1967; Hare, 1969; Miller and Hare, 1980a). This latter approach has been applied to the resolution of L-isoleucine from D-alloisoleucine (produced by the racemization of L-isoleucine at the α-carbon), and it was the recognition of the presence of D-alloisoleucine in fossils that led to the proposal that racemization reactions be used in geochronology (Hare and Mitterer, 1967). As D-alloisoleucine/L-isoleucine determinations can be performed with standard amino acid analyzers, this analytical method has been most frequently used by most workers. Recent developments in instrumentation (see Hare and Gil-Av, 1979, and references therein) have made this method highly sensitive, rapid, and quite inexpensive. The instrumentation has even been made portable so that it may be used in the field (Miller and Hare, 1980b).

In addition to the advantages listed above for the D-alloisoleucine/L-isoleucine method, this analytical approach also permits quantification of all the amino acids in the analyzed fossil. Its principle disadvantage is that it yields D/L information for only one amino acid in the sample, though it has usually been assumed that the extent of racemization of isoleucine was representative of the extent of racemization of all amino acids in a sample.

D-alloisoleucine/L-isoleucine values have proven to be quite variable (15% uncertainty) in interlaboratory comparisons (Kvenvolden, 1980), and a significant range (between about 1.05 and 1.40) in the "equilibrium" D-alloisoleucine/L-isoleucine value has been reported (Hare and Mitterer, 1967; Mitterer, 1974; Masters and Bada, 1977), though most of the reported equilibrium values are within 10% of 1.30.

Gas chromatrographic methods offer the general advantage of yielding enantiomeric ratio data on several amino acids (six or more) in a sample. As each amino acid racemizes at its own rate, multiple D/L values in a single sample can be used to evaluate internal consistency. These methods are, however, somewhat more time-consuming and expensive than is the isoleucine method, and it is much more difficult to obtain quantitative analyses by gas chromatography.

Two gas chromatographic methods have been employed in this laboratory:

1) NTFA-(+)-2-butyl method, in which the mixture of amino acids to be analyzed is esterified with an optically active alcohol (thereby introducing a second center of asymmetry into the amino acid molecule). Resolution of this mixture of derivatized amino acids is accomplished with chromatographic columns coated with the liquid phases OV225, Carbowax 20M, or UCON 75H 90,000. Examples of the chromatographic results of these procedures are found in Kvenvolden *et al.*, 1972; Hare and Hoering, 1973; Kvenvolden *et al.*, 1979; Belknap and Wehmiller, 1980.

2) NTFA-isopropyl method, in which the mixture of amino acids is esterified with an optically inactive alcohol. Resolution of this mixture of derivatized amino acids is accomplished with a chromatographic column coated with an optically active phase. Examples of this approach are found in Smith and Wonnacott (1980) and Wehmiller and Emerson (1980).

In both analytical approaches, the elution time for a particular D- or L-amino acid is a function of its solubility in the liquid phase coating the interior of the column and the volatility of the derivative.

Detection of the amino acid derivatives eluted from the chromatographic columns in the above methods has usually been by flame ionization detectors (FID), which ionize and detect all eluted carbon-containing molecules. Nitrogen-specific detectors (NPD) permit the specific detection of amino acids, often with several orders of magnitude more sensitivity than with flame ionization detectors.

Using a variety of derivatives, columns, and detectors, it is possible to determine the D/L value for each of seven amino acids (leucine, glutamic acid, alanine, valine, proline, phenylalanine, and aspartic acid) by at least two gas chromatographic schemes, thereby reducing the possibility that measured D/L values have been affected by interfering chromatographic peaks. Reported D/L values are usually the ratios of D- and L- peak heights or areas.

Figure 3 shows a high-resolution chromatogram, obtained in our laboratory, of NTFA-(+)-2-butyl esters of a 100 meter glass capillary column, coated with OV225, using simultaneous flame ionization (FID) and nitrogen-specific (NPD) detectors. The extreme high resolution of this method permits baseline separation of D-alloisoleucine and L-isoleucine. This separation has always been difficult by gas chromatography, thereby preventing confident comparisons of gas chromatographic and liquid chromatographic analyses (Kvenvolden, 1980). On-going work in this laboratory indicates that D-allisoleucine/L-isoleucine values determined by the method shown in Figure 3 are within 5% of the D-alloisoleucine/L-isoleucine values determined by the conventional liquid chromatographic methods. The use of dual detection systems, as shown in Figure 3, occasionally has revealed the presence of non-nitrogenous compounds, some of which actually co-elute with individual amino acid peaks, thereby invalidating the measured FID D/L value.

Interlaboratory comparison of gas chromatographic methods has revealed a range of uncertainty between about 5% and 15% for different amino acids (Kvenvolden, 1980). This variability (partially related to different analytical methods) is somewhat more than would be expected for multiple analyses in a single laboratory: regular repeated analysis of a homogeneous powdered fossil mollusk in this laboratory has revealed a range of 3 - 5% for leucine, valine, alanine, and glutamic acid D/L values. Uncertainties for proline, phenylalanine, and aspartic acid are between 8 and 15%. Comparable precision is reported by most workers for multiple analyses of single molluscan genera at a single outcrop (Kvenvolden *et al.*, 1979; Miller and Hare, 1980a).

The sensitivities of the methods outlined above are such that as little as 5 mg of carbonate material can be analyzed, given typical abundances of amino acids in most Pleistocene fossils. It is important, however, that the analysis be representative of the sample being studied,

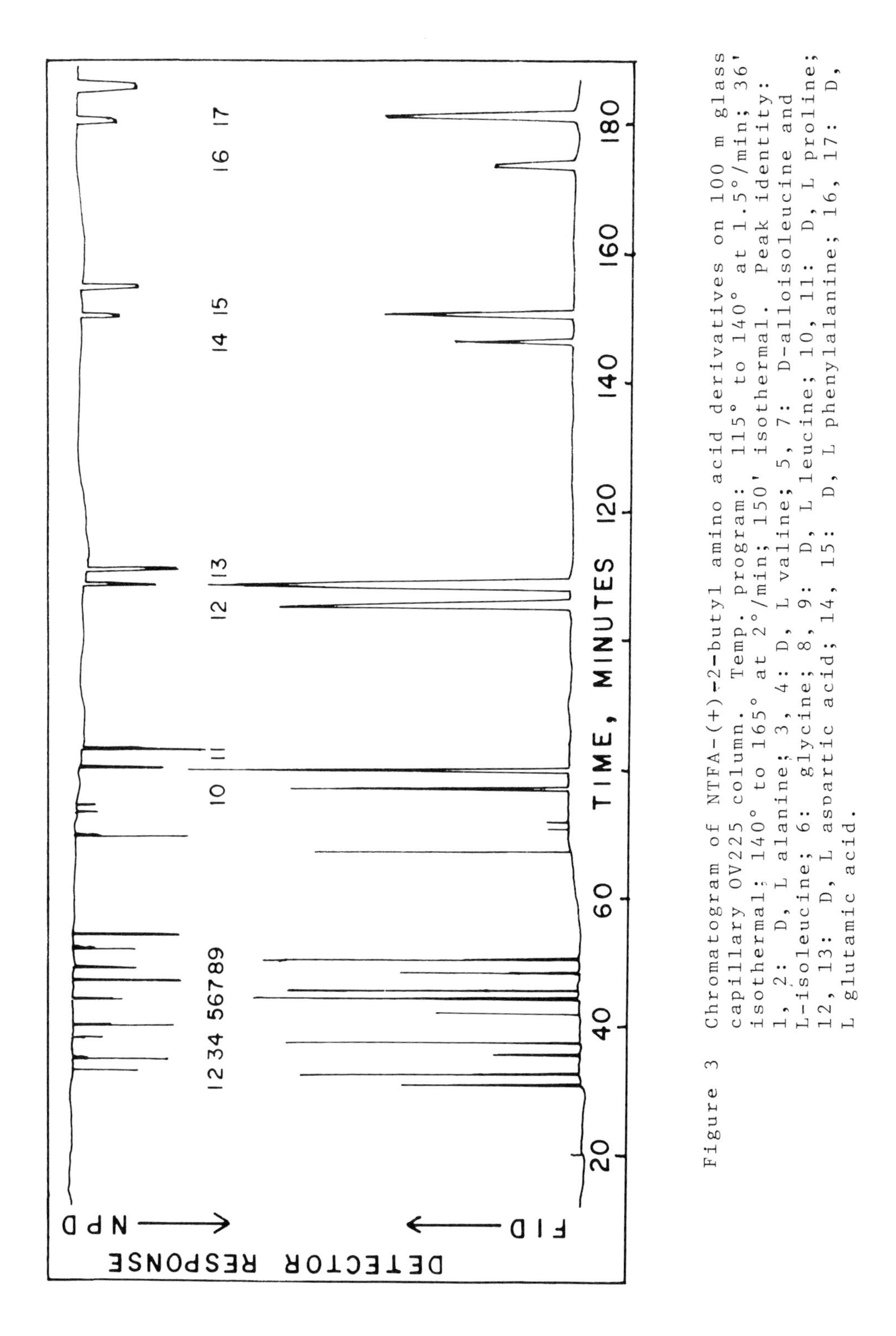

Figure 3 Chromatogram of NTFA-(+)-2-butyl amino acid derivatives on 100 m glass capillary OV225 column. Temp. program: 115° to 140° at 1.5°/min; 36' isothermal; 140° to 165° at 2°/min; 150' isothermal. Peak identity: 1, 2: D, L alanine; 3, 4: D, L valine; 5, 7: D-alloisoleucine and L-isoleucine; 6: glycine; 8, 9: D, L leucine; 10, 11: D, L proline; 12, 13: D, L aspartic acid; 14, 15: D, L phenylalanine; 16, 17: D, L glutamic acid.

---

as inhomogeneities in D/L values have been observed when small (<0.2 gm) fragments have been analyzed from large (>30 gm) bivalves. At least part of this problem can be eliminated if individual layers of mollusks are analyzed (Hare and Mitterer, 1969). As a working rule, we have found that a fragment representing at least 5% of the weight of the shell should be analyzed, even though analytical sensitivity might permit samples several orders of magnitude smaller. In all cases, samples should be extensively cleaned (mechanically and with dilute acid and base washes) to remove contamination and altered surficial layers.

Sample preservation should be evaluated to the fullest extent possible. Macroscopic examination of most bivalves can identify gross leaching or physical alteration, but microscopic and mineralogic examination are also important. It has generally been assumed (often with supportive evidence) that the primary effect of poor sample preservation would be to lower the enantiomeric ratios from their true values, because contamination would introduce amino acids with low D/L values and leaching would remove primarily free amino acids which are the most extensively racemized amino acids in the sample. Though this assumption is probably valid in most cases, a few examples have been encountered where "poorly preserved" samples (leached, chalky, or recrystallized) have yielded higher D/L values than expected (Wehmiller *et al.*, 1976).

In addition to the above chemical and physical criteria for sample selection, it is important that samples be collected from great enough depths so that significant exposure to high temperatures in the shallow soil zone has not occurred. The nature of this exposure is highly dependent on local climates, moisture, and vegetative cover. It has been shown (Wehmiller, 1977) that exposure to high temperatures at depths of less than 1 meter (or on the soil surface) can significantly affect the measured extent of racemization, if the duration of exposure is long in proportion to the absolute age of the sample. This problem is especially important for Holocene samples, archaeological samples, and all samples found in arid climates (Wehmiller *et al.*, 1979).

INTER- AND INTRAGENERIC EFFECTS ON RACEMIZATION

Individual amino acids have their own characteristic rates of racemization in aqueous solutions (Schroeder and Bada, 1976; Williams and Smith, 1977). Similarly, a regular pattern of characteristic relative rates of racemization of the different amino acids is observed in different genera of mollusks, corals, and foraminifera (Lajoie *et al.*, 1980; Wehmiller, 1980; Wehmiller *et al.*, 1976; Kvenvolden *et al.*, 1973). These intrageneric relative apparent rates of racemization are important in the evaluation of the internal consistency of any given sample analysis. The uniformity of these relative intrageneric rates among different fossil types suggests a common diagenetic pathway for the calcified proteins in these samples (Wehmiller, 1980).

Though intrageneric relative racemization kinetics might be similar or identical in different molluscan genera, intergeneric effects on racemization among these genera are potentially significant (Wehmiller *et al.*, 1977; Belknap, 1979; Lajoie *et al.*, 1980). Comparison of individual amino acid apparent rates of racemization between coexisting samples of different genera has revealed that at least two kinetic groups of mollusks exist: the "slow-racemizing" group, comprised of the venerids *Protothaca, Chione, Saxidomus, Tivela,* and *Mercenaria,* and the "fast-racemizing" group comprised of other bivalves or gastropods (*Epilucina, Cumingia, Macoma, Busycon, Tegula*) (Wehmiller *et al.*, 1977; Belknap, 1979; Lajoie *et al.*, 1980). Additional studies of oyster samples coexisting with samples from the above groups (Kvenvolden *et al.*, 1979; Belknap, 1979) suggest that oysters (either *Crassostrea* or *Ostrea*) form a still slower racemizing group. These relative intergeneric rates of racemization have been duplicated in laboratory pyrolysis experiments (Keenan and Wehmiller, in prep.) and have also been observed in foraminifera (King and Hare, 1972; King and Neville, 1977). Wehmiller (1980) has suggested that these intergeneric kinetic effects can be explained by the relative abundance of stable peptide bonds in the various genera, with the slow-racemizing groups containing a greater proportion of stable bonds which hydrolyze more slowly, thereby producing free amino acids (which are more extensively racemized) more slowly. The three categories of mollusks that can be recognized by their relative racemization kinetics can also be grouped identically according to their relative abundances of aspartic acid (Wehmiller, 1980), which apparently form stable peptide bonds in calcareous matrices (Hare *et al.*, 1975). Thus there appears to be a "chemical taxonomy" that roughly parallels the classical taxonomic organization of the mollusks that have been studied.

Generic effects on racemization kinetics complicate the chronologic application of the method because results from different genera may not be directly comparable, unless the genera involved are known to belong to the same kinetic group. Nevertheless, such generic effects provide useful cross-checks on the relative age estimates that might be developed from one genus. As it is rare for a single genus to be abundant at all the localities of potential interest, multiple genera would need to be employed for any regional chronological study. In order to establish the framework for such a study, relative intergeneric kinetics for all the genera of interest should be established using coexisting samples from selected outcrops (see Lajoie *et al.*, 1980, for example). Occasional examples of inversions of intergeneric relationships have been encountered (Wehmiller *et al.*, 1977; 1978a; Lajoie *et al.*, 1980; Yerkes *et al.*, 1980; Wehmiller, unpubl.). Usually these are minor inversions among genera of the same kinetic group, but at least one example of a major inversion between groups has been encountered (Wehmiller, unpubl.; discussion in Yerkes *et al.*, 1980). In several cases the most likely explanation for these inversions is reworking (which can also be recognized by multiple analyses of samples of a single genus), but in a few cases the reason for the inversion is more enigmatic, and the derived age estimate is less certain (Yerkes *et al.*, 1980).

STRATIGRAPHIC EVALUATION AND APPLICATION

The simplest approach to evaluation or application of amino acid enantiomeric ratios is as a stratigraphic tool ("aminostratigraphy" - see Miller and Hare, 1980a) in a local region where it can be reasonably assumed that all samples (which might be at present-day temperatures within 1° C of each other) have had similar or identical temperature histories. The "resolving power" (ability to distinguish aminostratigraphic units) of this approach depends upon the accuracy of this temperature assumption. Examples of this approach, from the Pacific coast of the United States, are summarized in Figure 4. Other similar examples can be found in Hare and Mitterer (1967); Mitterer (1974; 1975); Miller *et al.* (1979); Karrow and Bada (1980); and Miller and Hare (1980a).

In those cases where a clear, unambiguous stratigraphic relationship exists among the analyzed fossil localities, the reliability of amino acid enantiomeric ratios as relative dating tools can be evaluated. The best examples of this type of evaluation are the results for vertical terrace sequences at San Nicolas Island, Palos Verdes Hills, San Joaquin Hills, Point Loma (San Diego), and Sea Cliff (all southern California) (Wehmiller *et al.*, 1977; Lajoie *et al.*, 1979; 1980), and superposed stratigraphic sequences in the Los Angeles Basin (Wehmiller *et al.*, 1977; Lajoie *et al.*, 1980), near Ventura (Wehmiller *et al.*, 1978a, b; Lajoie *et al.*, 1979), and in San Francisco Bay (Atwater *et al.*, 1981). In these cases it has been possible to show that several genera (*Protothaca, Saxidomus, Chione, Tivela, Macoma, Tegula,* and *Ostrea*) exhibit steadily increasing D/L values in samples of known increasing age. Some of the deposits are old enough so that certain samples have reached racemic equilibrium (D/L values at "effective" unity, ≥ 0.95). In all but a very few cases, all of these genera have exhibited proper enantiomeric ratio trends when tested within a rigorous stratigraphic framework.

Some genera (particularly the gastropods *Diodora* and *Polinices*) have repeatedly shown decreasing D/L values with increasing stratigraphic age, thereby failing the most basic test of the method. Occasionally other genera (*Protothaca, Epilucina,* and *Tivela*) have shown decreasing D/L values with increasing age, but these anomalous results can usually be ascribed to poor sample preservation. Even well-preserved *Diodora,* though, have failed to meet these simple tests of reliability.

In many cases where stratigraphic relationships among discontinuous outcrops are unclear, amino acid enantiomeric ratios can be used to propose an aminostratigraphic framework (Mitterer, 1974; 1975;

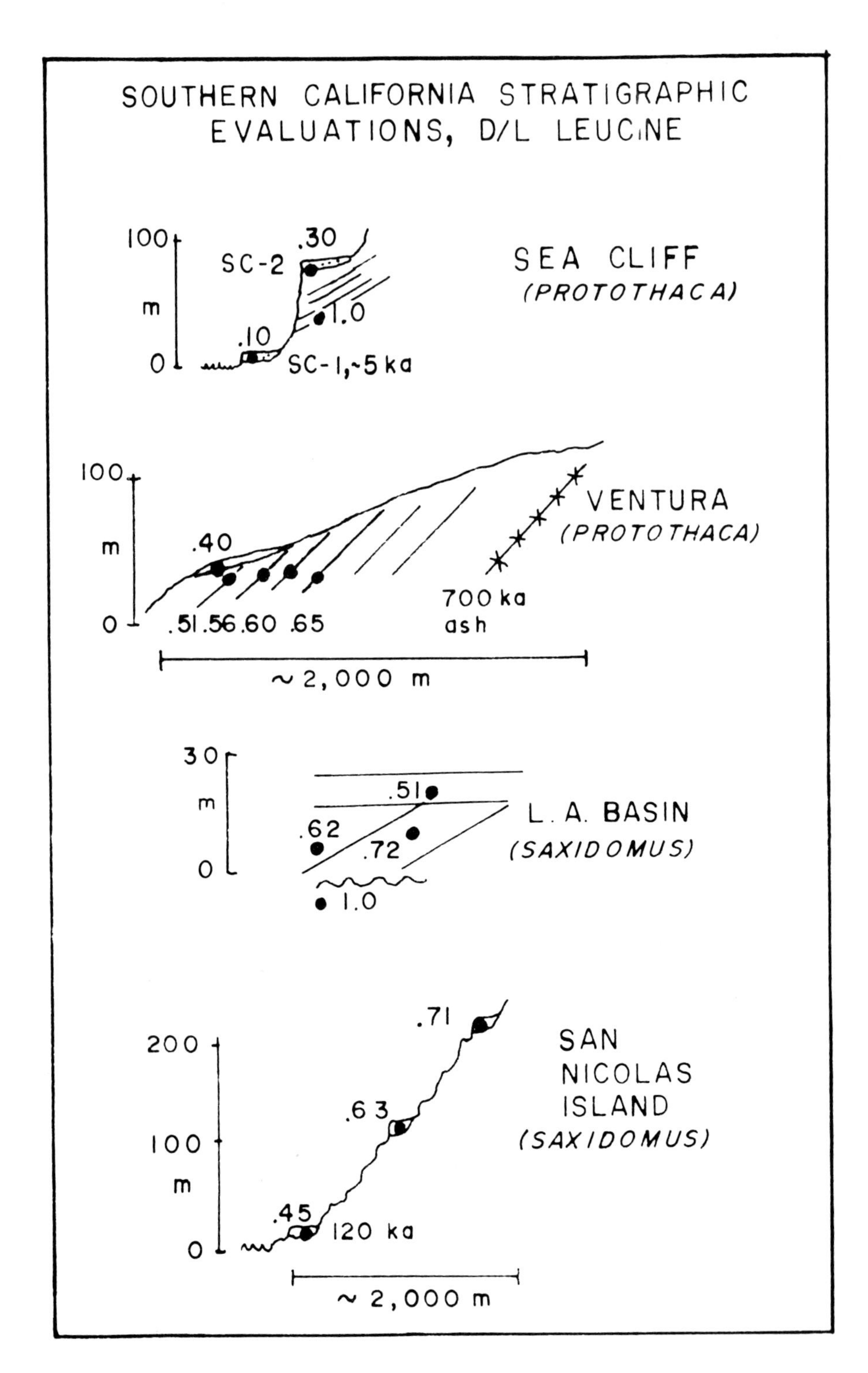

Figure 4 Stratigraphic tests, southern California. D/L leucine trends represent those for other amino acids. Ventura section includes *Macoma* data converted to equivalent *Protothaca*.

Wehmiller *et al.*, 1977; Miller *et al.*, 1979; Lajoie *et al.*, 1979; Belknap and Wehmiller, 1980; Karrow and Bada, 1980). Figure 5 shows how a number of discrete (all statistically significant) aminostratigraphic "events" are recorded in the area of San Diego, California (Wehmiller *et al.*, 1977; Lajoie *et al.*, 1979; Wehmiller, 1981b), and in northeastern South Carolina (Belknap, 1979; Belknap, 1980). In both of these cases, multiple depositional events would be expected on the basis of conventional geomorphic and stratigraphic information, but the number of events would be difficult to specify. Amino acid data appear to contribute to answering this question. However, in the case of the South

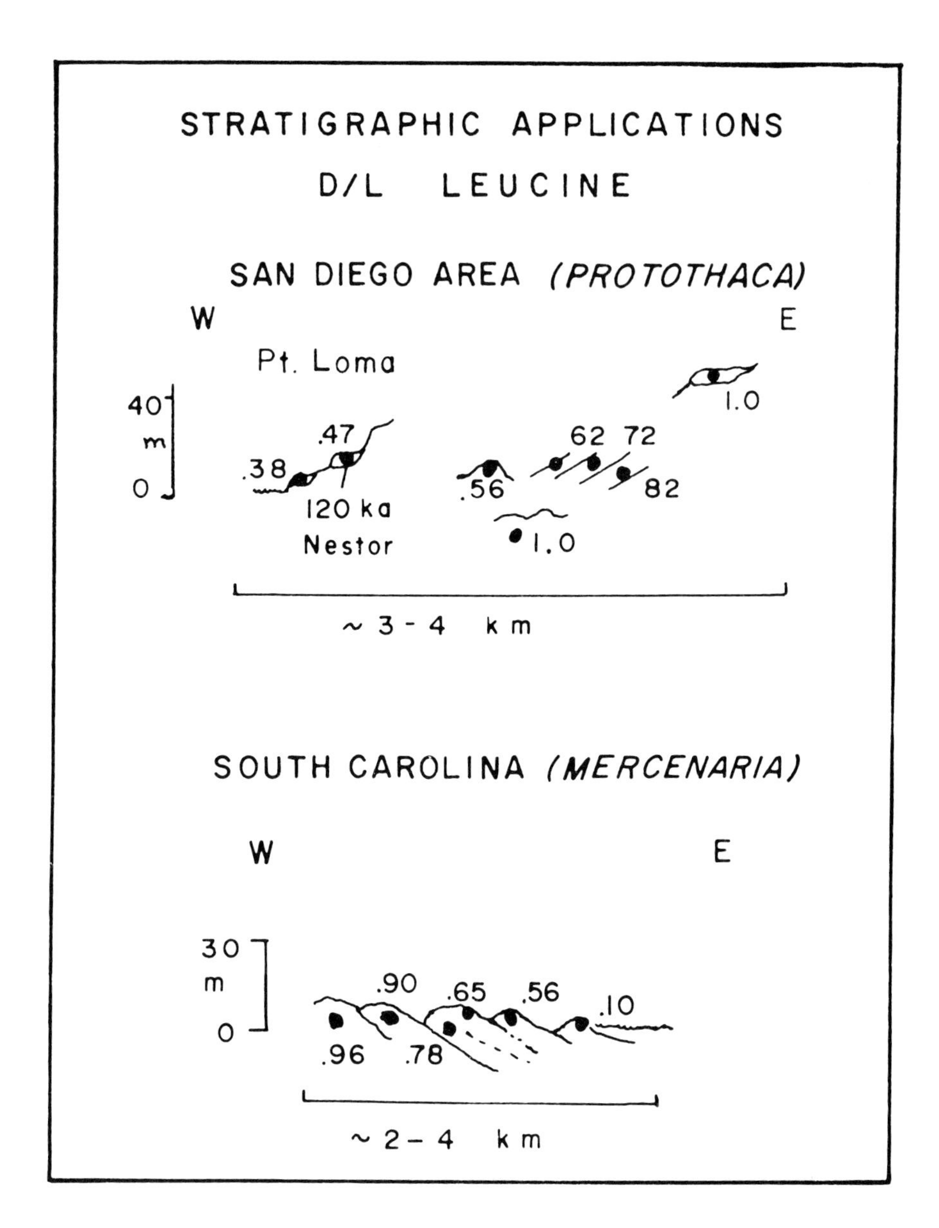

Figure 5 Apparent aminostratigraphic groups, D/L leucine, San Diego, California and South Carolina (Myrtle Beach area). Other amino acids follow similar trends. San Diego area includes the 120 ka Nestor Terrace, on Pt. Loma (see Figs. 6 and 7).

Carolina work, the number of apparent aminostratigraphic events is not in agreement with at least one lithostratigraphic interpretation.

REGIONAL CORRELATIONS - EVALUATION AND APPLICATION

A logical extension of the local aminostratigraphic approach to evaluation and application is the regional approach, involving comparisons of aminostratigraphic sequences at different latitudes (with different present-day mean annual temperatures). Quaternary marine terrace localities along the Pacific coast have proven to be an ideal setting in which to study this aspect of racemization, as several widely spaced localities with good chronologic control (U-series dates on corals) are available for documentation of the latitudinal (temperature) gradient of enantiomeric ratios in samples of equal age.

Figure 6 (modified from Wehmiller and Emerson, 1980, and Emerson *et al.*, 1981) shows this relationship for one amino acid, leucine, between latitudes 25° N and 35° N along the Pacific coast. Other amino

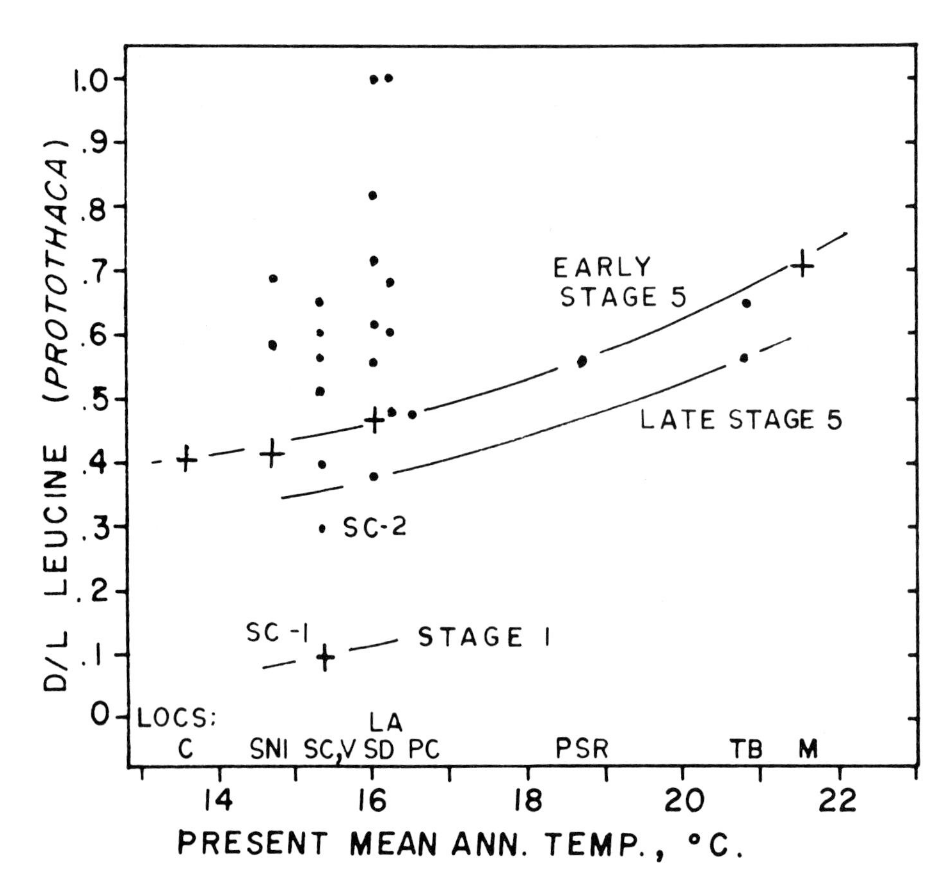

Figure 6 Plot of D/L leucine in *Protothaca* (or other genera converted to equivalent *Protothaca*) *vs.* present mean annual temperature, Pacific coast localities. Present temperature is a smooth function of latitude, localities range from approx. 35° N to 24° N. Other amino acids show similar trends. Early Stage 5 isochron connected to 120 ka localities with U-series control; late Stage 5 isochron connected to two data points for terraces with stratigraphic relationships to 120 ka locs. (Emerson *et al.*, 1981).

+ Locality with 120 ka U-series control or Holocene ^{14}C control
• Locality with no radiometric control

Locality abbreviations as in Figure 2;
PC = Punta Camalú (Valentine, 1980)
PSR = Punta Santa Rosalillita (Woods, 1980).

acids would show a similar trend. D/L leucine values are plotted *vs.* present mean annual temperature (because temperature is the control of kinetics), though it is known that present temperatures are not representative of the long-term temperature history of the samples. Because present-day temperatures are a nearly linear function of latitude in this region, it is possible to assume that differences in long-term temperature histories would be similar in magnitude to present-day temperature differences.

The solid line in Figure 6 connects D/L leucine data for four localities which represent early Stage 5, roughly 120,000 to 130,000 years BP (120 - 130 ka), according to U-series coral dates at these localities. The trend of this line is consistent with the hypothesis that D/L values should increase in samples of equal age at lower latitudes (higher temperatures): consequently these observations become an additional test of the validity of enantiomeric ratios as age or temperature indices. The slope of the line can only be predicted by kinetic modelling and temperature assumptions (Wehmiller and Emerson, 1980), but in those cases where several calibration localities are available, such kinetic models are not necessary for estimation of ages if the D/L data for intermediate localities fall on or near the interpolated 120 ka isochron. It is only necessary to assume that localities with present-day temperatures intermediate to those of calibration localities have always had intermediate temperatures, *i.e.*, that latitudinal gradients of temperature have not changed significantly (though temperatures certainly would have dropped) during Pleistocene climatic cycles.

Shown in Figure 6 are the results for four calibration localities (converted by the intergeneric regressions of Lajoie *et al.*, 1980 to equivalent genera) and the data for Pacific coast stratigraphic sequences shown in Figures 4 and 5. Results from Baja California localities have been discussed by Woods (1980), Valentine (1980), Wehmiller and Emerson (1980), and Emerson *et al.* (1981). The late Stage 5 isochron, roughly parallel to the early Stage 5 isochron and connecting data points for terraces at San Diego and Turtle Bay, is supported by both stratigraphic and faunal relationships with nearby terrace deposits that show slightly higher degrees of racemization, and which are either calibrated with or correlated to early Stage 5 U-series coral dates (Kern, 1977; Lajoie *et al.*, 1979; Emerson *et al.*, 1981).

Data for one locality (Sea Cliff, Second Terrace - SC-2; see Figure 4) plotted in Figure 6 fall well below the late Stage 5 isochron but above the data for Holocene marine terrace samples (SC-1). Sea Cliff 2 is estimated to be a Stage 3 (40 - 50 ka) terrace by both qualitative and kinetic model age estimates (Wehmiller *et al.*, 1978a, b; Lajoie *et al.*, 1979). Collectively the ages and elevations of these two terraces (see Figure 4) imply unusually rapid uplift rates, up to approximately 6 m/1000 years (Lajoie *et al.*, 1979).

KINETIC MODELLING

Estimation of absolute ages for data points in Figure 6 that lie well above or below the calibrated isochrons requires a kinetic model that can quantify age and temperature relationships for calibrated and uncalibrated racemization data. The development and testing of kinetic models requires the availability of both Holocene and Pleistocene calibration samples, and paleoclimatic information with which to estimate the temperature history (usually referred to as the Effective Quaternary Temperature, or EQT) for Pleistocene samples. Age estimates proposed for undated localities also rely upon the assumption, discussed previously, that present temperature differences between localities are a measure of the differences in EQT for these localities - *i.e.*, that gradients of EQT have been similar to present temperature gradients. The validity of this assumption depends upon the intensity and uniformity of climatic change in a region of study.

Recent discussion (Wehmiller, 1981a; Kvenvolden *et al.*, 1981) of kinetic modelling has centered on the kinetics of leucine racemization in the venerid genera *Protothaca* and *Saxidomus*, two closely related mollusks that have very similar relative kinetics (Lajoie *et al.*, 1980). Further evidence (Lajoie *et al.*, 1980; Wehmiller, 1980; unpublished) indicates that whatever kinetic model is appropriate for *Protothaca* and *Saxidomus* should also be appropriate for other venerids such as *Chione* and *Mercenaria*.

Figure 7 summarizes the basic issue of leucine kinetic modelling. Figure 7 shows the variable $(X_E - X)/X_E$ for leucine, in a logarithmic format, where $X = D/(D+L)$ at any time T and $X_E = D/(D+L)$ at equilibrium; for leucine, $X_E = 0.50$. Traditional discussions of racemization kinetics have used this (or similar) graphical format because simple first order reversible racemization (L D) would be represented by a straight line in this format. Deviations from linearity would then indicate more complex kinetics. Two models are shown in Figure 7:

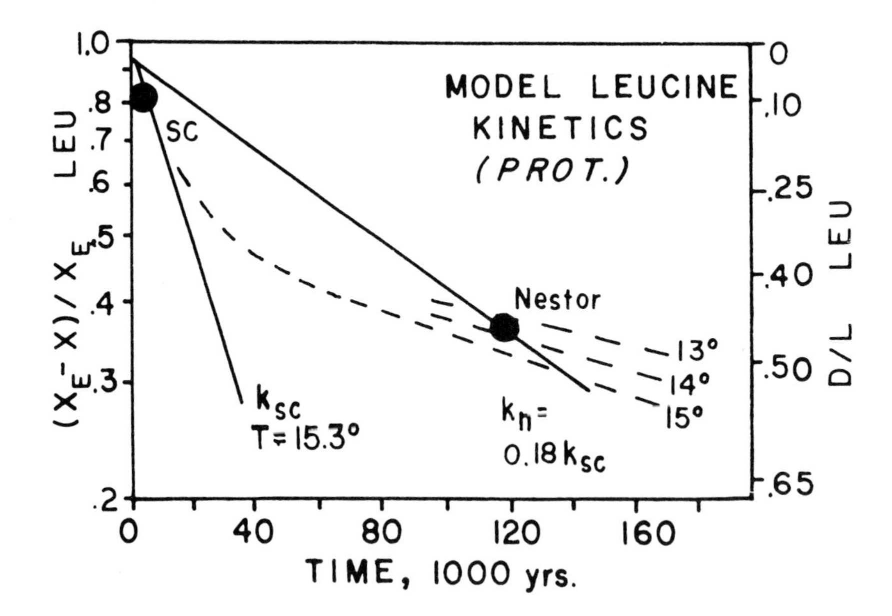

Figure 7 Comparison of kinetic model options for leucine racemization in *Protothaca*. X = D/(D+L) at time T, X_E = X at equilibrium. For leucine, X_E = 0.50. $(X_E - X)/X_E$ would be a linear function of time, in this format, if racemization kinetics were simple first-order reversible. Two data points, SC-1 (Figs. 4 and 6) and Nestor (Figs. 5 and 6) are used to compare models because these points represent Holocene and late Pleistocene calibrated localities with similar present temperatures. Solid lines represent Model B, the extended linear model of Kvenvolden *et al.* (1979). Dashed lines represent model A, the non-linear model of Wehmiller *et al.* (1977). Model B rate constants for SC-1 and N (k_{sc} and k_n), in combination with present temperature of Sea Cliff (15.3°C) would require that the Effective Quaternary Temperature (EQT) of Nestor be about 5° C. Model B would require that the EQT of Nestor be about 13.5° C. See text, Wehmiller (1981b) and Kvenvolden *et al.* (1981), for further discussion.

Model A, the "non-linear" model of Wehmiller *et al.* (1977); Wehmiller and Belknap (1978), which has two linear components with a transition zone between. The early portion of the racemization curve is linear to a D/L leucine value of about 0.2, with a steady decrease in slope so that beyond D/L leucine of about 0.35, the slope is about 10% of the initial slope. This model was derived from actual kinetics observed in foraminifera (Wehmiller and Hare, 1971; Kvenvolden *et al.*, 1973) and requires extrapolation over a rather small temperature range (10 - 15°) to be applicable to the ambient temperatures of molluscan racemization.

Model B, the "extended linear" model of Kvenvolden *et al.* (1979; 1981) which is linear up to a D/L leucine value of about 0.55, with lesser slopes (poorly defined) beyond this D/L value. This model has been derived from kinetics observed in high temperature (140 - 160° C) pyrolysis experiments and requires extrapolation over large temperature ranges (*ca.* 130°) in order to be applicable to the ambient temperatures of molluscan racemization.

The temperature dependence (*ca.* 18%/° C) is assumed to be the same for both models, as there does not appear to be significant disagreement over the activation energy for leucine racemization (see discussion in Wehmiller, 1981a).

D/L leucine results for *Protothaca* in two southern California calibration localities (with similar present-day temperatures) are shown in Figure 7. SC-1 is the Holocene (^{14}C dates) terrace locality shown in Figures 4 and 6. Nestor is the 120 ka upper terrace on Point Loma, San Diego (Figures 5 and 6). The Nestor data, when compared with the SC-1 data in a linear model format (Model B, solid lines, Fig. 7), indicate that the Nestor rate constant (calculated with equation 1 of Kvenvolden *et al.*, 1979) is approximately 20% of the rate constant observed for the Holocene SC-1 samples. Using the accepted temperature dependence of leucine racemization, this apparent linear rate constant for the Nestor samples would imply an EQT about 10° C less than the present temperature (15.3° C) at SC-1. As the EQT is the integrated kinetic effect of all temperatures to which the Nestor samples have been exposed, it can be shown (Wehmiller *et al.*, 1977; Wehmiller, 1981a) that a 10° C cooler EQT would require that full-glacial temperatures at the Nestor locality were as low as 0° C, lower by at least 8° C than most glacial age temperature estimates for coastal southern California (Johnson, 1977; Peterson *et al.*, 1979). Similar temperature ambiguities arise when other Holocene-Pleistocene calibration samples are compared (Wehmiller *et al.*, 1977; Wehmiller, 1981a).

The non-linear model (Model A) is shown as dashed lines in Figure 7. Kinetic pathways for temperatures of 13, 14 and 15° C are shown. The Nestor data indicate that a temperature of about 13.5° C would be the EQT value inferred from the non-linear model for this locality. This EQT value, and the associated extrapolation of the non-linear model, would form the basis for age estimation of more extensively racemized samples from the same temperature region. A reduction of about 2.5° C in EQT for the Nestor locality (relative to its present temperature of 16°) implies a full-glacial temperature reduction of 4 - 5° C, much smaller than the 15° C full-glacial reduction required by Model B and more consistent with paleoclimatic information for southern California (Johnson, 1977; Peterson *et al.*, 1979).

The conflict between the two models discussed above concerns the position of the break in slope of the kinetics. There is general agreement that the break in slope, wherever it is located, is related to the changing abundances of different molecular weight components in the fossils, with each component having its own rate of racemization (see, for example, Wehmiller and Hare, 1971; Bada and Schroeder, 1972; Kriausakul and Mitterer, 1978; 1980; Wehmiller, 1980). In addition, it is known that the initially rapid rate of racemization (roughly 10X that of free amino acids in solution) is significantly affected by hydrolysis and the production of extensively racemized free amino acids (Wehmiller and Hare, 1971).

Model A has observed kinetics in deep-sea sediments as its basis, and though it requires extrapolation over a small temperature range, it relies upon the assumption that foraminifera and mollusks have similar kinetic pathways. Model B has high temperature molluscan kinetics as its basis, hence does not depend upon assumptions regarding the similarity of foraminifera and mollusk racemization kinetics. Nevertheless Model B relies upon the assumption that the kinetics observed at temperatures of roughly 150° C are an accurate model of natural diagenetic kinetics at ambient temperatures. Until this issue is resolved, significant differences in age estimates or paleoclimatic conclusions will be derived from the two kinetic models outlined above.

Several additional approaches can be taken to evaluate these two kinetic models. One is to compare age estimates derived from each model for pre-120 ka samples (those that plot well above the 120 ka isochron in Figure 6). These age estimates rely upon the basic assumption that EQT values have been similar through time in a region of similar present temperatures. Though rigorous independent age control for samples in this age range is not frequently available, non-linear model age estimates appear more consistent with geological constraints for the Ventura, San Nicolas Island, and Los Angeles Basin stratigraphic sequences shown in Figure 4 (Wehmiller *et al.*, 1978a; Lajoie *et al.*, 1980).

The format of Figure 6 can also be used to test the two models shown in Figure 7. This approach is shown in Figure 8, in which 120 ka isochrons are shown for the two models for effective temperature equal to present temperatures and for 5° C lower effective temperatures. The lower position of the lower-temperature isochron reflects the slower kinetics that would be expected at lower temperatures. The positions of the linear model isochrons are approximate because the kinetics beyond D/L leucine of 0.55 are poorly defined by this model. Superimposed upon these two model isochron pairs is the 120 ka calibrated isochron from Figure 6. It is clear that this calibrated isochron plots in the kinetic-temperature range that is more consistent with the non-linear model (model A) of Figure 7. Figure 8 reinforces the point made previously that the extended linear model (model B) can only be reconciled with Holocene and Pleistocene calibration data if extremely cold effective temperatures are invoked for the late Pleistocene paleoclimate of coastal western North America.

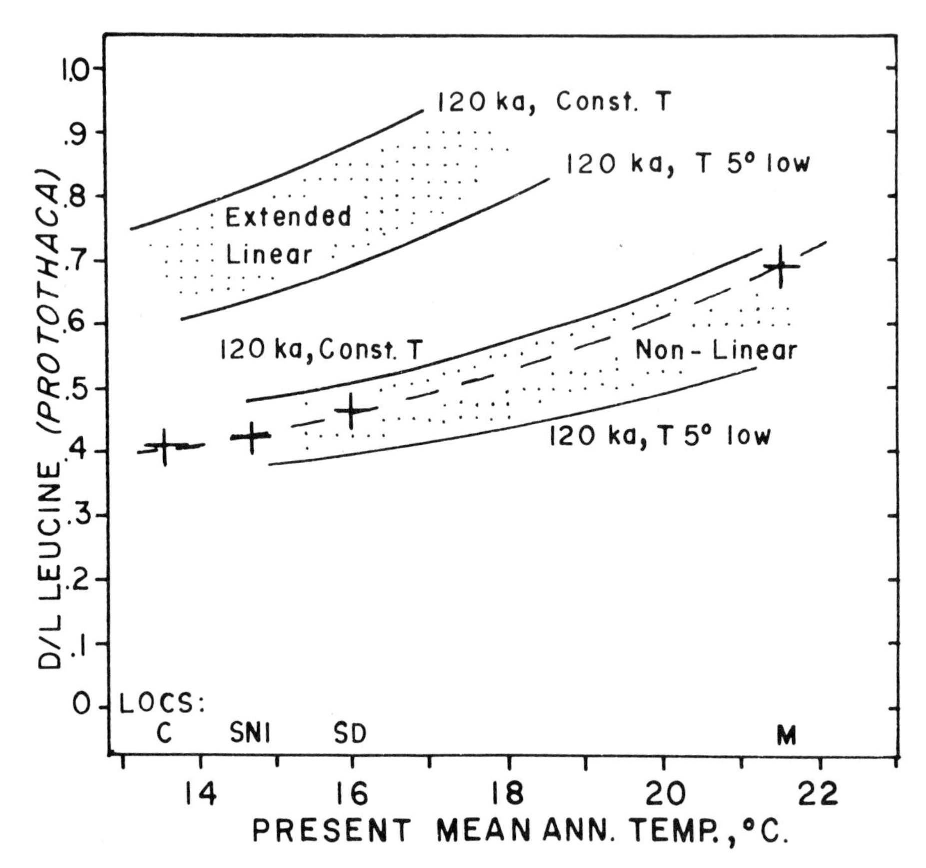

Figure 8 Comparison of kinetic models shown in Figure 7 in the format of Figure 6, Pacific coast *Protothaca*. Only 120 ka calibrated data points from Figure 6 are shown. The two models (Model A, extended linear; Model B, non-linear) have been used to predict the positions of 120 ka isochrons for two selected temperatures: equal to present temperatures (const. T), or 5° cooler (5° low). The Model A isochrons both plot higher than the Model B isochrons because of the greater long-term rate of racemization predicted by Model A. Isochrons for cooler temperature fall below those for the warmer temperature because of reduced rates of racemization at lower temperatures. The 120 ka calibrated isochron from Figure 6 falls within the envelope of non-linear model isochrons, for temperatures between approximately 1 - 3° lower than present temperatures, with greater reductions in temperatures at more northerly latitudes.

The calibrated 120 ka isochron and the non-linear model isochrons shown in Figure 8 can be used to infer gradients of Effective Quaternary Temperature. The relation of the U-series calibrated isochron to the 120 ka model isochrons for 0 and 5° temperature reductions indicates that greater temperature reductions have occurred at more northerly latitudes - *i.e.*, that latitudinal temperature gradients have been steeper than present gradients during the late Pleistocene.

SIGNIFICANT QUESTIONS

Because of the consistency of enantiomeric ratio trends with stratigraphic and latitudinal controls for several venerid genera on the Pacific coast, it has been assumed that the venerid *Mercenaria*, common to Atlantic coastal plain localities, would serve as an equally reliable fossil for aminostratigraphic applications. In fact, this massive, well-preserved bivalve has been shown to meet all of the criteria discussed previously for establishing the credibility of a given genus (Hare and Mitterer, 1967; Mitterer, 1974; 1975; Belknap, 1979; Belknap and Wehmiller, 1980). Nevertheless, recent work on the Atlantic coastal plain has revealed some significant conflicts between apparent aminostratigraphy and independent stratigraphic or radiometric data (Belknap, 1979; Belknap and Wehmiller, 1980). These conflicts have far-reaching implications for the general reliability of the amino acid racemization dating method.

These conflicts relate to the comparison of enantiomeric ratios (in *Mercenaria*) and U-series dates on solitary corals (as reported by Cronin *et al.*, 1981) from the same localities. Figure 9 demonstrates this comparison in the same format as Figure 6, for localities between Virginia and Florida (see Figure 2 for locations). It can be seen in Figure 9 that a smooth latitudinal trend of increasing enantiomeric ratios with decreasing latitude for samples of equal (Stage 5) age, as seen on the Pacific coast, is not seen on the Atlantic coastal plain (points 6, 2 and 1). At least part of this difference in enantiomeric ratio trends along the two coasts might be explained by greater steepening of the Atlantic coastal plain latitudinal gradients during the late Pleistocene because of the proximity of continental ice sheets. Nevertheless, a dramatic reversal of the expected latitudinal trend of enantiomeric ratios is seen for points 7 and 5, in Figure 9. The trend for these two points represents a serious conflict between U-series and amino acid data, and even the younger age (*ca.* 200 ka) proposed by Cronin (1980) for the Flanner Beach locality (point 5) could not reconcile the inversion in enantiomeric ratios for points 5 and 7. In addition, the different D/L leucine values (0.28 *vs.* 0.65) for the 74 ka and 190 ka localities (points 6 and 7, respectively, Figure 9) is far greater than would be expected from Pacific coast samples (with similar temperatures) with this age difference.

Though local temperature variations (up to 10%) might be expected to introduce slight variations around smooth latitudinal trends in enantiomeric ratios (Wehmiller *et al.*, 1979), major inversions and deviations like those shown in Figure 9 raise serious questions about the temperature assumptions inherent to any relative or absolute chronologic application of amino acid enantiomeric ratios. Present temperature trends along both coasts are nearly linear functions of latitude (between 25° N and 50° N) and the Pacific coast data certainly indicates that this trend has not been grossly altered during the late Pleistocene. If the U-series dates for all the localities shown in Figure 9 are correct, and if there are no unknown chemical effects on racemization in *Mercenaria*, (all of the points shown in Figure 9 represent multiple analyses with good precision), then extreme temperature variations (both temporally and latitudinally) must be invoked to reconcile the amino acid data with the U-series data. Both latitudinal inversions of temperature gradients and significant <u>local</u> variations in Effective Quaternary Temperatures (more than *ca.* 1° C) contradict the basic temperature assumptions of aminostratigraphy.

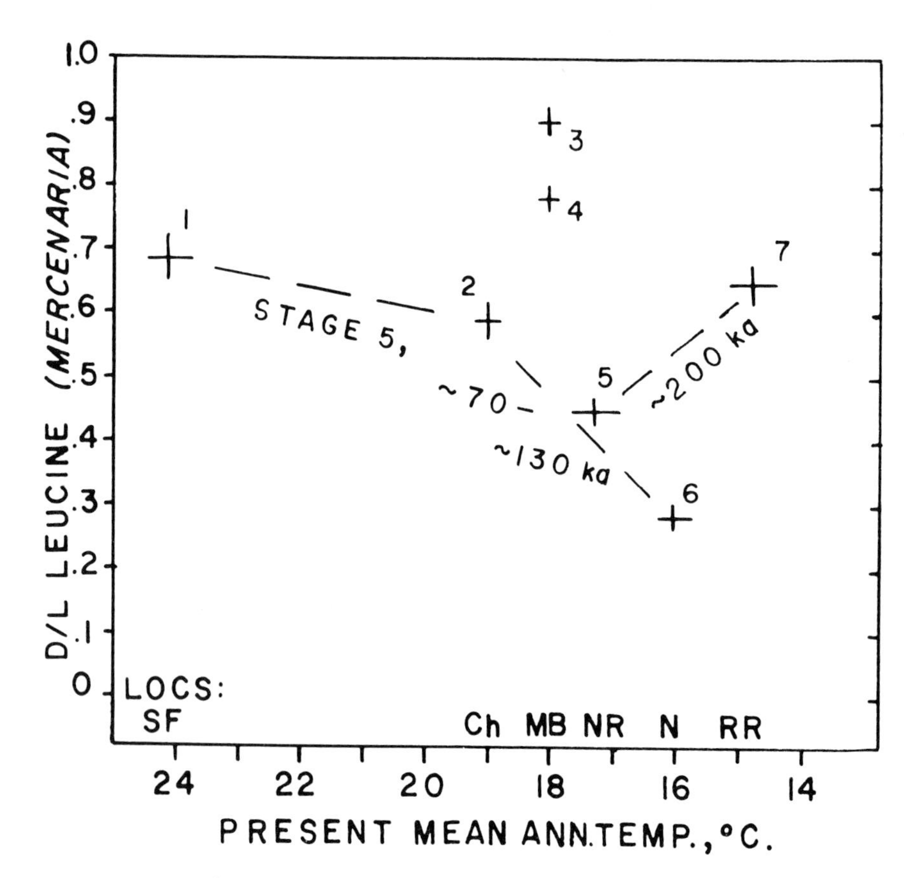

Figure 9 D/L leucine values in Mercenaria, Atlantic coastal plain localities. Temperature is a smooth function of latitude between 25° N and 39° N, the region shown here. Localities as given in Fig. 2. Explanation of data points:

1 South Florida 120 ka calibrated data, converted from results of Mitterer (1975).
2 Charleston, S. C. ∿95 ka calibrated data, locs. 9-12 of Cronin *et al.* (1981).
3 Myrtle Beach, S. C. Waccamaw Fm., reversed magnetic polarity (Cronin, 1980).
4 Myrtle Beach, S. C. Canepatch Fm., ≥ 440 ka, loc. 8 of Cronin *et al.* (1981).
5 Neuse River, N. C. Flanner Beach Fm., >500 ka, loc. 7 of Cronin *et al.* (1981) or ∿200 ka (Cronin, 1980).
6 Norfolk, Va. 74 ± 4 ka calibrated data, loc. 2 of Cronin *et al.* (1981).
7 Rappahanock River, Va. ∿190 ka U-series coral data, loc. 1 of Cronin *et al.* (1981).

Data points 2, 3 and 4 are consistent with stratigraphic relationships (Fig. 5). Dashed lines are correlations dictated by the radiometric data given above.

CONCLUSIONS

Amino acid racemization in Quaternary mollusks has been used in a number of situations for both relative and absolute age estimation. A number of criteria developed in these studies are universally applicable to the evaluation of any new genus or sample type that might be employed in chronostratigraphic applications of amino acid racemization:

1) mechanical, mineralogical, and chemical preservation; absence of contamination;

2) good analytical precision, single and/or multiple genera;

3) stratigraphic and thermal consistency; ability of samples to demonstrate increasing racemization with increasing age, at constant temperature; or, for isochronous samples, with increasing temperature.

4) ability to reach equilibrium D/L values, and to maintain them.

Occasional inversions of intergeneric relations have been encountered, and a few genera of gastropods have been rejected because they consistently violate several of the criteria listed above. Nevertheless, the vast majority of samples studied from the Pacific coast (mostly of the bivalve family Veneridae) appear totally consistent with the few unambiguous stratigraphic or latitudinal tests that are available.

Stratigraphic and latitudinal frameworks have permitted the testing and evaluation of kinetic models of racemization. These models are necessary to quantify age-temperature relationships between calibrated and uncalibrated localities. Models of racemization kinetics in molluscan genera, and the resolution of the question of the similarity of molluscan and foraminifera kinetics, should aid in understanding the overall mechanism of diagenetic racemization in calcified matrices. Presently available kinetic models for molluscan racemization result in significantly different age and temperature conclusions, hence they cannot be applied without thorough consideration of their implications.

In the application of any kinetic model, one must make two related assumptions that are inherent to amino acid racemization dating in all forms:

1) present temperature differences are a measure of past temperature differences, and localities with similar temperatures today, though with different apparent ages, would be assumed to have had similar Effective Quaternary Temperatures;

2) Effective Quaternary Temperatures can be estimated using paleoclimatic information in combination with racemization data for Holocene and late Pleistocene calibration samples.

In the absence of any independent calibration data for a region, amino acid enantiomeric ratios can be used for both relative and absolute age estimation if Effective Quaternary Temperatures can be independently estimated, and if one of the kinetic models discussed herein can be assumed to be valid for the genera being considered.

Mercenaria, a robust venerid mollusk that is common in Atlantic coastal plain Quaternary localities, has been utilized for its reliability in amino acid racemization applications. Though fewer stratigraphic tests have been available from the Atlantic coast, most of the available data has suggested that *Mercenaria* conformed to stratigraphic tests as well as Pacific coast venerids. U-series (solitary coral) dating of Atlantic coastal plain localities, compared with amino acid enantiomeric ratios, has raised some major conflicts (independent of kinetic model issues) that challenge some of the basic temperature assumptions of simple stratigraphic applications of amino acid racemization. The resolution of these conflicts, and a better understanding of their implications, will be important steps in the further development of the racemization dating method.

ACKNOWLEDGEMENTS

This research has been supported by grants from the U.S. Geological Survey, numbers 14-08-0001-G-248 and 4-08-0001-G-592. Many of the observations and publications discussed herein have resulted from collaboration and discussion with the following individuals, all of whom are gratefully acknowledged for their efforts and contributions: D.F. Belknap, T.M. Cronin, J.M. Demarest, W.K. Emerson, E.M. Keenan, G.L. Kennedy, K.R. Lajoie, L. McCarten, R. Morris, R.B. Mixon, and J.P. Owens.

Special appreciation goes to D.F.B. for the use of some unpublished field and laboratory data in the construction and discussion of Figure 9.

REFERENCES CITED

Atwater, B.F., Ross, B.E. and Wehmiller, J.F., 1981, Stratigraphy of late Quaternary estuarine deposits and amino acid stereochemistry of oyster shells beneath San Francisco Bay, California: Quaternary Research, v. 16, p. 181-200.

Bada, J.L. and Man, E.H., 1980, Amino acid diagenesis in Deep Sea Drilling Project cores: kinetics and mechanisms of some reactions and their applications in geochronology and in paleotemperature and heat flow determination: Earth-Science Reviews, v. 16, p. 21-55.

Bada, J.L. and Schroeder, R.A., 1972, Racemization of isoleucine in calcareous marine sediments: kinetics and mechanisms: Earth and Planetary Science Letters, v. 15, p. 1-11.

Belknap, D.F., 1979, Application of amino acid geochronology to stratigraphy of late Cenozoic marine units of the Atlantic coastal plain (Ph.D. Dissert.): Newark, Del., University of Delaware, 532 p.

___________, 1980, Amino acid racemization: application to stratigraphy of the south-central Atlantic coastal plain: Geol. Soc. Amer. Abstracts with Programs, v. 12(4), p. 170.

Belknap, D.F. and Wehmiller, J.F., 1980, Amino acid racemization in Quaternary mollusks: examples from Delaware, Maryland and Virginia: in Hare, P.E., Hoering, T.C. and King, K., eds., Biogeochemistry of Amino Acids, Wiley, p. 401-414.

Cronin, T.M., 1980, Biostratigraphic correlation of Pleistocene marine deposits and sea levels, Atlantic coastal plain of the southeastern United States: Quaternary Research, v. 13, p. 213-229.

Cronin, T.M., Szabo, B.J., Ager, T.A., Hazel, J.E. and Owens, J.P., 1981, Quaternary climates and sea levels of the Atlantic coastal plain: Science, v. 211, p. 233-240.

Emerson, W.K., Kennedy, G.L., Wehmiller, J.F. and Kennan, E., 1981, Age relations and zoogeographic implications of late Pleistocene marine invertebrate faunas from Turtle Bay, Baja California Sur, Mexico: The Nautilus, v. 95(3), p. 105-116.

Hare, P.E., 1969, Geochemistry of proteins, peptides and amino acids: in Eglinton, G. and Murphy, M.T., eds., Organic Geochemisty, Methods and Results, N.Y., Springer-Verlag, p. 438-463.

Hare, P.E. and Mitterer, R.M., 1967, Non-protein amino acids in fossil shells: Carnegie Inst. Wash. Yearbook, v. 65, p. 362-364.

_________, 1969, Laboratory simulation of amino acid diagenesis in fossils: Carnegie Inst. Wash. Yearbook, v. 67, p. 205-208.

Hare, P.E. and Hoering, T.C., 1973, Separation of amino acid optical isomers by gas chromatography: Carnegie Inst. Wash. Yearbook, v. 72, p. 690-694.

Hare, P.E. and Gil-Av, E., 1979, Separation of D and L amino acids by liquid chromatography: use of chiral eluants: Science, v. 204, p. 1226-1228.

Hare, P.E., Hoering, T.C. and King, K., Jr., eds., 1980, Biogeochemistry of Amino Acids, N.Y., Wiley, p. 1-558.

Hare, P.E., Miller, G.H. and Tuross, N.C., 1975, Simulation of natural hydrolysis of proteins in fossils: Carnegie Inst. Wash. Yearbook, v. 74, p. 609-612.

Hoopes, E.H., Peltzer, E.T. and Bada, J.L., 1978, Determination of amino acid enantiomeric ratios by gas liquid chromatography of the N-trifluoroacetyl L-prolylpeptide methyl esters: Jour. Chromatographic Science, v. 16, p. 556-560.

Johnson, D.L., 1977, The late Quaternary climate of coastal California: evidence for an ice-age refugium: Quaternary Research, v. 8, p. 154-179.

Karrow, P.F. and Bada, J.L., 1980, Amino acid racemization dating of Quaternary raised marine terraces in San Diego County, California: Geology, v. 8, p. 200-204.

Kennedy, G.L., 1978, Pleistocene paleoecology, zoogeography and geochronology of marine invertebrate faunas of Pacific northwest coast (San Francisco Bay to Puget Sound) (Ph.D. Dissert.): Davis, Calif., Univ. California, 824 p.

Kern, J.P., 1977, Origin and history of upper Pleistocene marine terraces, San Diego, Calif.: Bull. Geol. Soc. America, v. 88, p. 1553-1566.

King, K. Jr. and Hare, P.E., 1972, Species effects on the epimerization of isoleucine in fossil planktonic foraminifera: Carnegie Inst. Wash. Yearbook, v. 71, p. 596-598.

King, K., Jr. and Neville, C., 1977, Isoleucine epimerization for dating marine sediments: importance of analyzing monospecific foraminiferal samples: Science, v. 195, p. 1333-1335.

Kriausakul, N. and Mitterer, R.M., 1978, Isoleucine epimerization in peptides and proteins: kinetic factors and application to fossil proteins: Science, v. 201, p. 1011-1014.

__________, 1980, Comparison of isoleucine epimerization in a model dipeptide and a fossil protein: Geochimica et Cosmochimica Acta, v. 44, p. 753-758.

Kvenvolden, K.A., 1980, Interlaboratory comparison of amino acid racemization in a Pleistocene mollusk, *Saxidomus giganteus*: in Hare, P.E., Hoering, T.C. and King, K., Jr., eds., Biogeochemistry of Amino Acids, N.Y., Wiley, p. 223-232.

Kvenvolden, K.A., Peterson, E. and Pollock, G., 1972, Geochemistry of amino acid enantiomers: gas chromatography of their diastereoisomeric derivatives: in von Gaertner, H. and Wehner, H., eds., Advances in Organic Geochemistry 1971: N.Y., Pergamon, p. 387-401.

Kvenvolden. K.A., Peterson, E., Wehmiller, J.F. and Hare, P.E., 1973, Racemization of amino acids in marine sediments determined by gas chromatography: Geochimica et Cosmochimica Acta, v. 37, p. 2215-2225.

Kvenvolden, K.A., Blunt, D.J. and Clifton, H.E., 1979, Amino acid racemization in Quaternary shell deposits at Willapa Bay, Washington: Geochimica et Cosmochimica Acta, v. 43, p. 1505-1520.

__________, 1981, Age estimations based upon amino acid racemization: reply to comments of J.F. Wehmiller: Geochimica et Cosmochimica Acta, v. 45, p. 265-267.

Lajoie, K.R., Kern, J.P., Wehmiller, J.F., Kennedy, G.L., Mathieson, S.A., Sarna-Wojcicki, A.M., Yerkes, R.F. and McCrory, P.A., 1979, Quaternary shorelines and crustal deformation, San Diego to Santa Barbara, Calif.: in Abbot, P.L., ed., Geological Excursions in the Southern California Area, San Diego, Dept. of Geology, San

Diego State Univ., p. 3-15.

Lajoie, K.R., Wehmiller, J.F. and Kennedy, G.L., 1980, Inter- and intrageneric trends in apparent racemization kinetics of amino acids in Quaternary mollusks: in Hare, P.E., Hoering, T.C. and King, K., Jr., eds., Biogeochemistry of Amino Acids: N.Y., Wiley, p. 305-340.

Masters, P.M. and Bada, J.L., 1977, Racemization of isoleucine in fossil mollusks from Indian middens and interglacial terraces in southern California: Earth and Planetary Science Letters, v. 37, p. 173-183.

Miller, G.H., Andrews, J.T. and Short, S.K., 1977, The last interglacial-glacial cycle, Clyde Foreland Baffin Island, N.W.T.: stratigraphy, biostratigraphy, and chronology: Canadian Jour. Earth Science, v. 14, p. 2824-2857.

Miller, G.H., Hollin, J.T. and Andrews, J.T., 1979, Aminostratigraphy of U.K. Pleistocene deposits: Nature, v. 281, p. 539-543.

Miller, G.H. and Hare, P.E., 1980a, Amino acid chronology: integrity of the carbonate matrix and potential of molluscan fossils: in, Hare P.E., Hoering, T.C. and King, K., Jr., eds., Biogeochemistry of Amino Acids: N.Y., Wiley, p. 415-443.

__________, 1980b, Amino acid geochronology: a portable instrument for field use: Geol. Soc. Amer. Abstracts with Programs, v. 12(7), p. 484.

Mitterer, R.M., 1974, Pleistocene stratigraphy in southern Florida based on amino acid diagenesis in fossil *Mercenaria*: Geology, v. 2, p. 425-428.

__________, 1975, Ages and diagenetic temperatures of Pleistocene deposits of Florida based upon isoleucine epimerization in *Mercenaria*: Earth and Planetary Science Letters, v. 28, p. 275-282.

Peterson, G.M., Webb, T. III, Kutzbach, J.E., van der Hammen, T., Wijmstra, T.A. and Street, F.A., 1979, The continental record of environmental conditions at 18,000 yr BP: an initial evaluation: Quaternary Research, v. 12, p. 47-82.

Schroeder, R.A. and Bada, J.L., 1976, A review of the geochemical applications of the amino acid racemization reaction: Earth Science Reviews, v. 12, p. 347-391.

Smith, G.G. and Wonnacott, D.M., 1980, The resolution of enantiomeric amino acids by gas chromatography: in Hare, P.E., Hoering, T.C. and King, K., Jr., eds., Biogeochemistry of Amino Acids: N.Y., Wiley, p. 203-214.

Valentine, J.W., 1980, Camalú: A Pleistocene terrace fauna from Baja California: Jour. Paleontology, v. 54, p. 1310-1318.

Wehmiller, J.F., 1977, Amino acid studies of the Del Mar, California, midden site: apparent rate constants, ground temperature models, and chronological implications: Earth and Planetary Science Letters, v. 37, p. 184-196.

__________, 1980, Intergeneric differences in apparent racemization kinetics in mollusks and foraminifera: implications for models of diagenetic racemization, in Hare, P.E., Hoering, T.C and King, K., Jr., eds., Biogeochemistry of Amino Acids: N.Y., Wiley, p. 341-355.

__________, 1981a, Kinetic model options for interpretation of amino acid enantiomeric ratios in Quaternary mollusks: comments on a paper by Kvenvolden *et al.* (1979): Geochimica et Cosmochimica

Acta, v. 45, p. 261-264.

______________, 1981b, Amino acid age estimation of Quaternary mollusks: downtown San Diego and Coronado Island: in Trenching the Rose Canyon Fault Zone, by E. Artim and D. Streiff, Final Tech. Report, U.S. Geological Survey Contract No. 14-08-0001-19118, p. D-2 - D-27.

Wehmiller, J.F. and Hare, P.E., 1971, Racemization of amino acids in marine sediments: Science, v. 173, p. 907-911.

Wehmiller, J.F. and Belknap, D.F., 1978, Alternative kinetic models for the interpretation of amino acid enantiomeric ratios in Pleistocene mollusks: examples from California, Washington and Florida: Quaternary Research, v. 9, p. 330-348.

Wehmiller, J.F. and Emerson, W.K., 1980, Calibration of amino acid racemization in late Pleistocene mollusks: results from Magdalena Bay, Baja California Sur, Mexico, with dating applications and paleoclimatic implications: The Nautilus, v. 94(1), p. 31-36.

Wehmiller, J.F., Hare, P.E. and Kujala, G.A., 1976, Amino acids in fossil corals: racemization (epimerization) reactions and their implications for diagenetic reactions and chronological studies: Geochimica et Cosmochimica Acta, v. 40, 763-776.

Wehmiller, J.F., Lajoie, K.R., Kvenvolden, K.A., Peterson, E., Belknap, D.F., Kennedy, G.L., Addicott, W.O., Vedder, J.G. and Wright, R.W., 1977, Correlation and chronology of Pacific coast marine terraces of continental United States by amino acid stereochemistry - technique evaluation, relative ages, kinetic model ages, and geologic implications: U.S. Geological Survey Open File Report 77-680, p. 1-196.

Wehmiller, J.F., Lajoie, K.R., Sarna-Wojcicki, A.M., Yerkes, R.F., Kennedy, G.L., Stephens, T.A. and Kohl, R.F., 1978a, Amino acid racemization dating of Quaternary mollusks, Pacific coast United States: in Zartman, R.E., ed., Short Papers of the Fourth International Conference, Geochronology, Cosmochronology and Isotope Geology 1978, U.S. Geological Survey Open File Report 78-701, p. 445-448.

Wehmiller, J.F., Lajoie, K.R., Sarna-Wojcicki, A.M. and Yerkes, R.F., 1978b, Unusually high rates of crustal uplift in Ventura County, California, inferred from Quaternary marine terrace chronology: Geol. Soc. Amer. Abstracts with Programs, v. 10(7), p. 513.

Wehmiller, J.F., Lajoie, K.R. and Kennedy, G.L., 1979, Role of thermal history uncertainties in amino-acid racemization age estimation of geological and archaeological samples: Geol. Soc. Amer. Abstracts with Programs, v. 11(7), p. 536.

Williams, K.M. and Smith, G.G., 1977, A critical evaluation of the application of amino acid racemization to geochronology and geothermometry: Origins of Life, v. 8, p. 91-144.

Woods, A.J., 1980, Geomorphology, deformation and chronology of marine terraces along the Pacific coast of central Baja California, Mexico: Quaternary Research, v 13, p. 346-364.

Yerkes, R.F., Greene, H.G., Tinsley, J.C. and Lajoie, K.R., 1980, Seismotectonic setting of the Santa Barbara channel area, southern California: U.S. Geological Survey Open File Report 80-299, p. 1-39; superseded by U.S. Geological Survey Report MF-1169, p. 1-25.

UTILIZING WOOD IN AMINO ACID DATING

N.W. RUTTER and R.J. CRAWFORD

ABSTRACT

D/L ratios of wood have been determined on samples from various Quaternary units in the unglaciated Old Crow area, Yukon and Alaska, most of which, have been subjected to long periods of permafrost conditions. The purpose was to evaluate the usefulness of D/L ratios of amino acids in wood as an aid in correlating equivalent stratigraphic units, determining relative ages, and in a general way, the absolute ages of units. The total amounts of amino acids were determined using a gas chromatograph equipped with a FID Detector and Chirasil-val capillary column (25 m) and controlled by a digital microprocessor terminal which reports peak areas by automatic intergration. D/L ratios of aspartic acid have proved to be useful and reliable.

Results show that D/L ratios of aspartic acid of wood are useful in correlating equivalent units in widely spaced sections, some over 300 km from each other. This is provided the sediments have had similar climatic and environmental histories and that units above and below those being correlated are of widely varying age. It appears that species differentiation is not necessary for gross correlations. D/L ratios of aspartic acid for Holocene sediments vary mostly from approximately 0.01 to 0.08. Late and Mid-Wisconsin 0.14 to 0.24 and early Wisconsin and Sangamon 0.24 to 0.36. Results from samples older than Sangamon are difficult to obtain with accuracy. Although the explanation is not clearly understood, more acceptable ratios are obtained by using a greater amount of sample than is usual in routine analysis. D/L ratios of aspartic acid of 0.50 have so far been obtained.

INTRODUCTION

Many varieties of fossils are now routinely analyzed to determine the D/L ratios of amino acids for relative and absolute dating of sediments. Most investigations have centered on a variety of marine molluscs and mammal bones brought about mainly by the availability of material and success in analyses. Wood, however, has not been widely analyzed (Lee *et al.*, 1976; Engel *et al.*, 1977; Rutter *et al.*, 1980), as it is difficult to prepare samples for accurate analytical results. The procedures outlined below have largely eliminated spurious results.

The objective of this investigation is to evaluate the reliability of D/L ratios of aspartic acid in wood as a tool in stratigraphic correlation and relative age dating by comparing results from the same stratigraphic units of nearby sections and widely spaced sections. This will aid in evaluating over how broad an area reliable correlations can be made in a region that has been subjected to similar paleoclimates and environments. At the same time it will confirm, refine or void parts of the proposed stratigraphy. Another objective is to estimate the absolute age of sediments by comparing D/L ratios of aspartic acid in wood from sediments of known or estimated age.

The present work is based on over 75 analyses of mostly Quaternary age wood from the Old Crow and adjacent basins in the Northern Yukon (Figure 1). This area was selected because, 1) there is an abundance of wood in units of widely varying age and broad distribution, 2) a geological framework has been established for the region, although problems still remain and, 3) the area is in the Continuous Permafrost Zone and little is known about the effects on the rate of racemization of amino acids of fossils subjected to long periods of permafrost conditions.

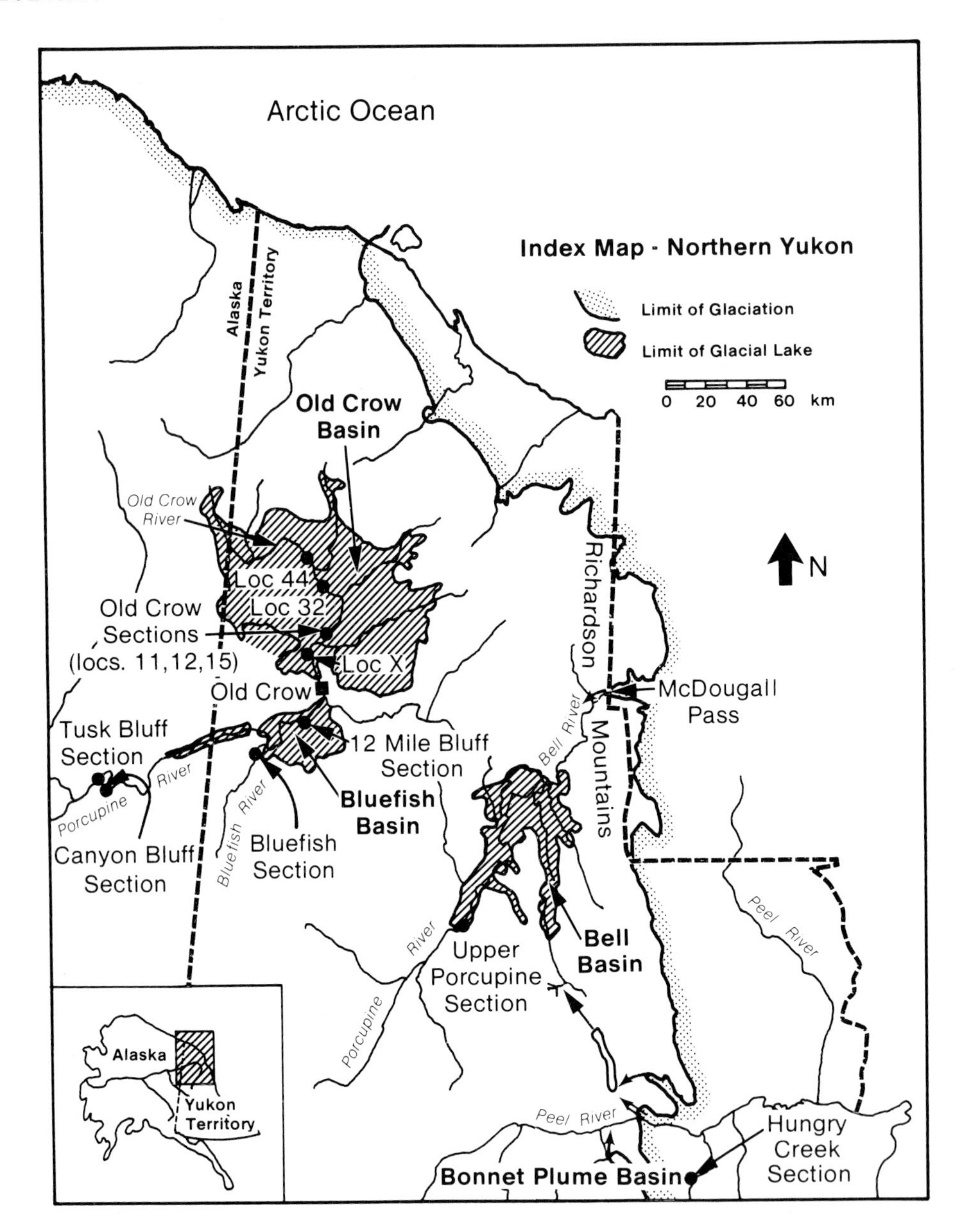

Figure 1 Index map of the Northern Yukon indicating locations of various sections and major basins.

METHODS

Preparation

Volatile enantiomers are prepared in order to record the various amino acids by gas chromatography. Each wood sample is thoroughly cleaned with distilled water, and then air dried on a plastic weighing dish. Next the sample is broken into small fragments and crushed using a mortar and pestle. Periodically the material is sieved through a 20 mesh screen and collected in a plastic weighing boat. The sieved wood particles are washed twice in a plastic disposable centrifuge tube using 2N HCℓ and twice with double distilled water. Between washings, the sample is sonified, centrifuged and decanted. The cleaned particles are transferred to a Buchner funnel connected to a water-vacuum tap and fitted with Whatman glass fibre paper (GF/A - 4.25 cm) and washed several times with double distilled water. The filtrate is discarded and the washed particles are collected in small plastic vials. The vials are covered and placed in a desiccator for vacuum drying. This usually takes several hours.

Derivatization

About 100 mg of washed, dried sample is placed in a glass screw-top culture tube (13 x 100 mm). Added to this is about 6 to 8 ml 5.5N HCℓ (constant boiling). The mixture is allowed to reflux at 108°C for 24 hrs. in a heating block. After heating, the tube is removed and allowed to cool to room temperature. It is then centrifuged after cooling to remove particulate matter. The supernatant liquid is collected using a Pasteur pipette and transferred to a clean culture tube. The sample is then evaporated to dryness in a Speed Vac Concentrator. The residue is dissolved in 1 to 2 ml double distilled water and added to freshly regenerated cation exchange resin (Dowex AG 50W-X8, 50-100 mesh). Next 4 bed volumes of double distilled water are added and the effluent discarded. Two bed volumes of 3N NH_4OH are added to elute the amino acids. About 10 ml of amino acid eluate are collected in a clean 13 x 100 mm screw-top culture tube when the solvent front is about 1.5 to 2.0 cm from the bottom of the column. A substantial amount of heat is emitted from the amino acid band. The excess NH_4OH is evaporated using a Speed Vac Concentrator. It usually takes more than 8 hours to evaporate to dryness.

Esterification is carried out by adding 0.1 ml isopropanol/3.5N HCℓ to the dried eluate. This is sonified until homogeneous, then heated at 100°C for 15 minutes in a oil bath. After evaporation to dryness, about 2 hours in the concentrator, the sample is acylated by adding 0.1 ml PFPA (pentafluoropropionic anhydride) and 0.3 ml distilled $CH_2C\ell_2$ (methylene chloride). The sample is sonified until dissolved and then heated in an oil bath at 100°C for 5 minutes. The excess PFPA and $CH_2C\ell_2$ are cold evaporated on a Büchi rotary evaporator using liquid N_2. Next the sample is washed with 0.5 to 0.1 ml $CH_2C\ell_2$ and after allowing the residue to dissolve completely, it is then cold evaporated to dryness using a rotary evaporator. The sample is then diluted in 0.5 ml $CH_2C\ell_2$ and filtered through a Gelman alpha-200, 0.20 μm metricel filter. The derivative is now ready to be injected into the gas chromatograph. The sample may be diluted with additional $CH_2C\ell_2$ depending upon its concentration upon injection. About 0.2 to 1.0 μl is injected. The gas chromatograph used is a Hewlett-Packard Model 5840A equipped with FID Detector and Chirasil-val capillary column (25 m) and controlled by a digital micro-processor terminal which reports peak areas by automatic integration.

Results

D/L ratios of alanine, valine, leucine, phenylalanine, proline, and aspartic acid are routinely determined. Aspartic acid proved to be the most useful because of the relatively fast rate of racemization and reliability (Kvenvolden, 1980). Only D/L ratios of aspartic acid are reported here.

PHYSICAL ENVIRONMENT

Introduction

The results reported here are from sections in the Old Crow, Bell, Bluefish and Bonnet Plume Basins and from sections on the Porcupine River in Alaska (Figure 1). The Procupine River drains the Bell, Bluefish, and Old Crow basins. It flows northward through the Bell Basin, then westward separating the Old Crow Basin to the north and the Bluefish Basin to the south before continuing southwestward to Alaska. Having escaped glaciation, sediments have been able to accumulate with a minimum amount of erosion and therefore, contain several hundred metres of Quaternary or older material. The Bonnet Plume Basin has been partially glaciated and is drained by the eastward flowing Peel River. Downcutting by major rivers and tributaries within the basins have exposed sections on the order of 50 m, consisting mostly of lacustrine and fluvial sediment containing abundant fossils and organic material. These factors have generated detailed investigations by a number of Quaternary scientists anticipating a nearly complete record of Quaternary environments and events including elucidation of the migration of man into North America (Hughes, 1972; Irving, 1978; Morlan, 1978; Morlan and Matthews, 1978). Major problems in stratigraphic investigations have been in correlating and dating equivalent units between sections. This is hampered by the small number of widespread stratigraphic markers such as volcanic ash, the similarity of units of various ages, lateral facies changes of equivalent units, and that most units are too old to be dated by ^{14}C methods.

OLD CROW BASIN - NEARBY SECTIONS

Stratigraphy

The details of the geology of the Old Crow and adjacent basins are subjects of a series of papers in preparation. Only the elements of the stratigraphy that are necessary to illustrate the utilization of D/L ratios of amino acids in wood will be discussed.

In the Old Crow Basin several major sections are located along a 10 km stretch of the Old Crow River about 30 km north of the village of Old Crow (Figure 1). Three of these have been investigated and sampled in detail (locations 11, 12, and 15). Although problems persist in correlating equivalent units, a basic stratigraphy has been established. A simplified composite section is illustrated in Figure 2. The lower most unit, 1, consists of massive lacustrine clay with silt. The unit extends below water level in all sections so thicknesses are difficult to estimate but 1 to 3 m are commonly exposed at low water level. The upper surface of this unit is irregular, unconformably overlain by reworked lacustrine clay, Unit 2, and containing abundant molluscs and wood. This unit reaches a maximum thickness of only a few metres and is missing in some sections. The overlying unit, Unit 3, with thicknesses on the order of 20 m consists mostly of well-bedded, laminated silt and sand alluvium with minor gravel, containing wood, peat, molluscs and bone. Lateral and vertical facies and structural changes have hindered correlation of beds between sections. Therefore, subdividing this unit has not been possible. In several sections, the upper contact is marked by a well defined unconformity, A, that has been the source of abundant fossils including bones, some of which may be artifacts. Unit 4, consisting mostly of well-bedded silt a few metres thick overlies Unit 3. It is characterized by several horizons marked by cryotubated zones, ice-wedge casts, peat layers, minor unconformities and layers containing wood, molluscs and bone. Unit 4 is overlain by Unit 5, a zone of silt and fine sand. Unit 6 consists of well-bedded glacial lacustrine clay and silt, commonly varved and less than 7 m thick. No wood or other fossils have been found. This unit along with overlying peat forms the upper surface over much of the Old Crow Basin. In places, fluvial silt and sand have been deposited in high level channels.

A series of ^{14}C dates and fission track data from tephra have aided in dating the younger sediments. The tephra, lying just below Unconformity A in Unit 3 has been estimated at between 50 ka and 100 ka (J.A. Westgate, U. of Toronto, pers. commun., 1977). ^{14}C dates from overlying Units 4 and 5 have yielded dates from about 32 ka to >53 ka (GSC 2507, 2574, 2676, 2739) and a 12,460 year old date (GSC 3574) has been obtained from the uppermost silt and sand unit above Unit 6. From these dates and other geological evidence it seems reasonable to place a late Wisconsin age (10 ka to 25 ka) on Unit 6, which was deposited when Continental glaciers covered areas to the east. Units 4 and 5, where most of the dated material has come from is mainly mid-Wisconsin (25 ka to 65 ka) with the stratigraphically lower beds perhaps older

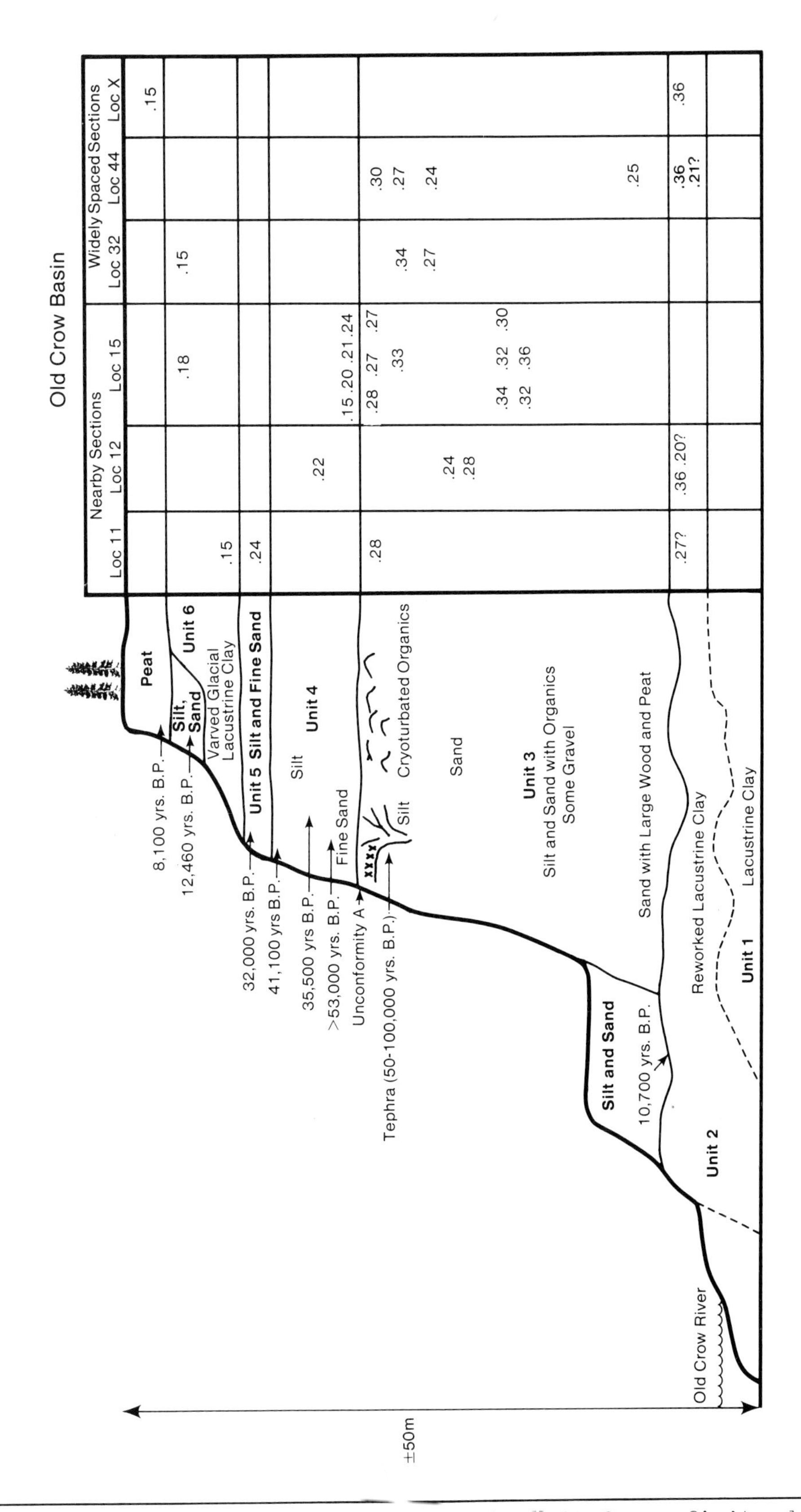

Figure 2 Composite section of the Old Crow Basin showing sample locations and D/L ratios of aspartic acid

than mid-Wisconsin. Below the tephra in Unit 3, no finite dates are available, so the entire unit may be all or in part mid-wisconsin or older. Unit 2 appears to have resulted from the thawing and redeposition of ice-rich sediments derived from Unit 1 below. This, and the presence of abundant wood suggests a relatively warm time perhaps the Sangamon Interglaciation (100 ka). At the present time, little can be said about the age of Unit 1, or its association with glacier activity to the east.

In summary, the sediments of the composite section are waterlain, representing a relatively long period of time, most likely over 100 ka. It seems likely that the sediments have been subjected to long periods of climate similar to, or colder than today and therefore, under permafrost conditions.

Results of Amino Acid Analyses

Figure 2 indicates that most samples for amino acid analyses are from Units 3 and 4 with a few from Units 2, 5 and 6. Two out of three of the results in Unit 2 have lower ratios (0.27, 0.20) than would be expected when compared to results obtained in the overlying unit. We have no satisfactory explanation for this anomaly. It may be due to diagenetic changes that take place in wood with time resulting in a paucity of amino acids that make accurate determinations of D/L ratios difficult. The ratio of 0.36, is probably analytically correct and appears reasonable for Unit 2, if the upper part of Unit 2 is similar in age to the lower part of Unit 3. Stratigraphic data from other sections offer some validity to this interpretation.

Samples from Unit 3 were taken at several stations from three sections at widely spaced intervals within each unit. In general, D/L ratios of aspartic acid increase with increasing age of the sediment, varying from 0.24 to 0.36 with an average of 0.29. Correlation of individual beds between stations or sections cannot be made with any accuracy so D/L ratios obtained from wood at a certain elevation at one station may be at a different stratigraphic position than D/L ratios of wood obtained at about the same elevation at another station. As mentioned above, all that can be said about the age of this unit, is that the upper part is >53 ka. If Unit 2 below contains Sangamon age sediments, then Unit 3 could include part mid-Wisconsin and early Wisconsin age sediments, *i.e.* between approximately 50 to 100-120 ka. The spread of ratios contained in Unit 3 gives an indication of what aspartic acid D/L ratios to expect for this time range in this region. Furthermore, even though Unit 3 is correlated between sections from lithographic evidence, D/L ratios of aspartic acid when compared to D/L ratios of the overlying unit, indicate that correlation of Unit 3 between sections is possible with D/L ratios alone, although problems may arise when separating the older beds of Unit 3 with those of underlying Unit 2.

Wood samples were analysed from Unconformity A at several stations at Location 15 and two from above the unconformity in Location 11 and 12 in Units 4 and 5. D/L ratios of aspartic acid vary between 0.15 and 0.24 with an average of 0.21. Not enough results are available from Units 4 and 5 to reach definite conclusions, but the results suggest that correlations of Units 4 and 5 can be made between stations and nearby sections. The D/L ratio variations obtained along Unconformity A and from Units 4 and 5 are probably representative of what to expect in sediments of ages between approximately 25 ka to 65 ka (mid-Wisconsin) in this region.

Only two D/L ratios of aspartic acid are available from Unit 6. Ratios of 0.15 and 0.18 appear to be typical for deposits between about 10 ka to 25 ka (late Wisconsin).

From the above, it can be concluded that in the Old Crow Basin: 1) there are problems in obtaining accurate D/L ratios of aspartic acid in the older units of sections; 2) D/L ratios of aspartic acid can be used to roughly indicate mid- (and early?) and late Wisconsin sediments and 3) major units can be correlated between stations within a section, and between nearby sections using D/L ratios of aspartic acid if the variations of ratios within a unit are known.

OLD CROW BASIN - WIDELY SPACED SECTIONS

Stratigraphy

Locations 32, 44 and X represent sections that lie at a distance of 20 km, 30 km and 10 km from Location 15 respectively (Figures 1 and 2). The objective of sampling these sections was to test the feasibility of correlating units by D/L ratios of aspartic acid over greater distances than nearby sections but within the same sedimentary basin under similar paleoenvironments. Although, these sections have not been investigated in detail, rough lithologic correlation has been made with units of the composite section given above (Figure 2). There is no problem in correlating Units 1 and 6 but the deposits between Units 1 and 6 contain different characteristics than those of the composite section, but are tentatively correlated with those based on similar stratigraphic position.

Results of Amino Acid Analyses

Unit 2 at Location 44 yielded D/L ratios of aspartic acid of 0.36 and 0.21 (Figure 2). The former ratio is in keeping with what is expected for Unit 2 whereas 0.21 is probably erroneous for the reasons cited in the last section. At location X, a ratio of 0.36 was obtained from what was thought to be the lower part of Unit 3. The ratio of 0.36, however, indicates that perhaps the sample came from beds that are actually equivalent to Unit 2.

Most samples analysed were from Unit 3. Ratios from Locations 32 and 44 vary between 0.24 and 0.34, averaging 0.27. The data available suggest there is no apparent overall increase in D/L ratios with depth, as there was at Location 15. The ratios are similar, however, suggesting that correlation of Unit 3 between Location 15 and Locations 32 and 44 is possible.

D/L ratios of 0.15 were obtained from samples located in the upper part of Unit 6 and just overlying Unit 6 at Locations 32 and X respectively. The ratios are compatible with ratios for late Wisconsin sediments found at Locations 11 and 15.

In conclusion, with the data available it appears that correlation of units, using D/L ratios of aspartic acid is possible between widely spaced sections within the Old Crow Basin.

ADJACENT BASINS

Introduction

Using Location 15 in the Old Crow as a base, D/L ratios of samples from sections in nearby basins and rivers were compared. The objective was to test for long distance correlations of units in regions where paleoclimates may have varied but probably not enough to cause major differences in D/L ratios of samples of about the same age. As seen in Figure 1, two sections are located in the Bluefish Basin, 12 Mile Bluff and Bluefish; one section in the Bonnet Plume Basin at Hungry Creek; and two sections along the Porcupine River, at Tusk Bluff and Canyon Bluff. The furthest section is Hungry Creek about 320 km southeast of Location 15. With few exceptions, equivalent units cannot be correlated by lithologic characteristics in these widely spaced sections, although ^{14}C data, tephra and stratigraphic positions have proved somewhat helpful. D/L ratios were compared and evaluated where stratigraphic positions of units were known and in some cases, were used to determine the stratigraphic position of units based on the success within the Old Crow Basin.

Stratigraphy - 12 Mile Bluff Section (Bluefish Basin)

About 45 km south of Location 15 in the Old Crow Basin, lies 12 Mile Bluff Section in the Bluefish Basin (Figure 1). This section with thicknesses up to 200 m, is exposed on the south bank of the Porcupine River, extending for more than 3 km. The character and thicknesses of units vary from location to location. A simplified cross-section is illustrated in Figure 3. Units that have been correlated by lithology, ^{14}C dates, tephra and stratigraphic position, with units within the Old Crow Basin are designated by the same unit number.

The lower units consist of sand, silt and grit beds with abundant tree trunks, roots, branches, twigs and cones, ice wedge casts and cryoturbated zones. Although the age is uncertain, the plant remains suggest that the units are pre-Quaternary. The overlying clay, silt unit is equated to Unit 1 in the Old Crow Basin. Overlying units may be equivalent to Unit 3 in the Old Crow Basin, but direct correlation is questionable. Near the upper part of the unit, however, the same tephra that crops out in the Old Crow Basin is present so that a rough correlation can be made with the upper part of Unit 3. Although the prominent unconformity (Unconformity A) above the tephra is not identified at 12 Mile Bluff, the characteristics of the overlying deposits are similar to those of the Old Crow sections that these may well be equivalent at least in part to Units 4 and 5. The overlying deposit consists of well bedded, partly varved, glacial lacustrine clay and silt, most likely equivalent to Unit 6 in the Old Crow Basin.

Results of Amino Acid Analyses - 12 Mile Bluff Section

Figure 3 shows D/L ratios of aspartic acid determined from units at two locations in the Big Bluff section. Only one reliable D/L ratio from the lower units was obtained. This reached 0.50 making it the highest ratio obtained in the entire region, suggesting that the lower units are older than any units found in the Old Crow Basin. This age difference was suspected from other stratigraphic criteria and was confirmed by D/L ratios.

A suite of ratios was obtained from sediments above and below the tephra in Unit 3 (?). The results should be roughly the same as those found in about the same stratigraphic position at Location 15. Tephra is found just below Unconformity A at Location 15 whereas in the 12 Mile Bluff section, the unconformity has not been identified. At Location 15 the ratios above the unconformity are lower than those found below. About the same ratios (between 0.27 and 0.29) are found in the sediments in roughly equivalent beds at Station 2 in the 12 Mile Bluff section, whereas in Station 1 the ratios are lower (between 0.17 and 0.20) just above, at, and below the ash and about the same (0.28) 3 m below the ash. Therefore, the only real discrepancies in the ratios are for those at, or just below the ash at 12 Mile Bluff; they should be higher to be compatible with those at Location 15. Analytically the ratios are believed to be correct. The discrepancy can only be explained by different racemization rates under different conditions.

Stratigraphy - Bluefish Section (Bluefish Basin)

The other section sampled in the Bluefish Basin is named the Bluefish Section located about 60 km southwest of Location 15 in the Old Crow Basin on the west bank of the Bluefish River (Figure 1). The section is about 25 m thick (Figure 4). Above the lowermost unit, Tertiary sediment with coal, lies a unit of alluvial silt and sand, with gravel near the base, containing wood pieces throughout. Overlying this unit unconformably, is alluvial gravel and sand that underlies lacustrine silt, the uppermost unit. The lacustrine silt is correlated with Unit 6 of the Old Crow composite section based on similar lithology and stratigraphic position. The units below, however, cannot be correlated with any assurance. The age of the silt, sand unit is older than 53 ka based upon a ^{14}C date (GSC 2373-3) on wood from the overlying gravel, sand unit. Samples for amino acid analyses were taken systematically throughout the silt, sand unit.

(Results of Amino Acid Analyses - Bluefish Section (Bluefish Basin)

The amino acid analyses from the silt, sand unit were reported earlier but are discussed here in relation to long distance correlations (Rutter *et al.*, 1980). All samples analysed were *Picea*.

As can be seen in Figure 4, results vary between 0.17 and 0.31, averaging 0.24. Although the sediments cannot be correlated lithologically with sediments at 12 Mile Bluff or those in the Old Crow Basin, the D/L ratios, with few exceptions are similar to those obtained for samples from the upper part of Unit 3 (?) at 12 Mile Bluff and the upper

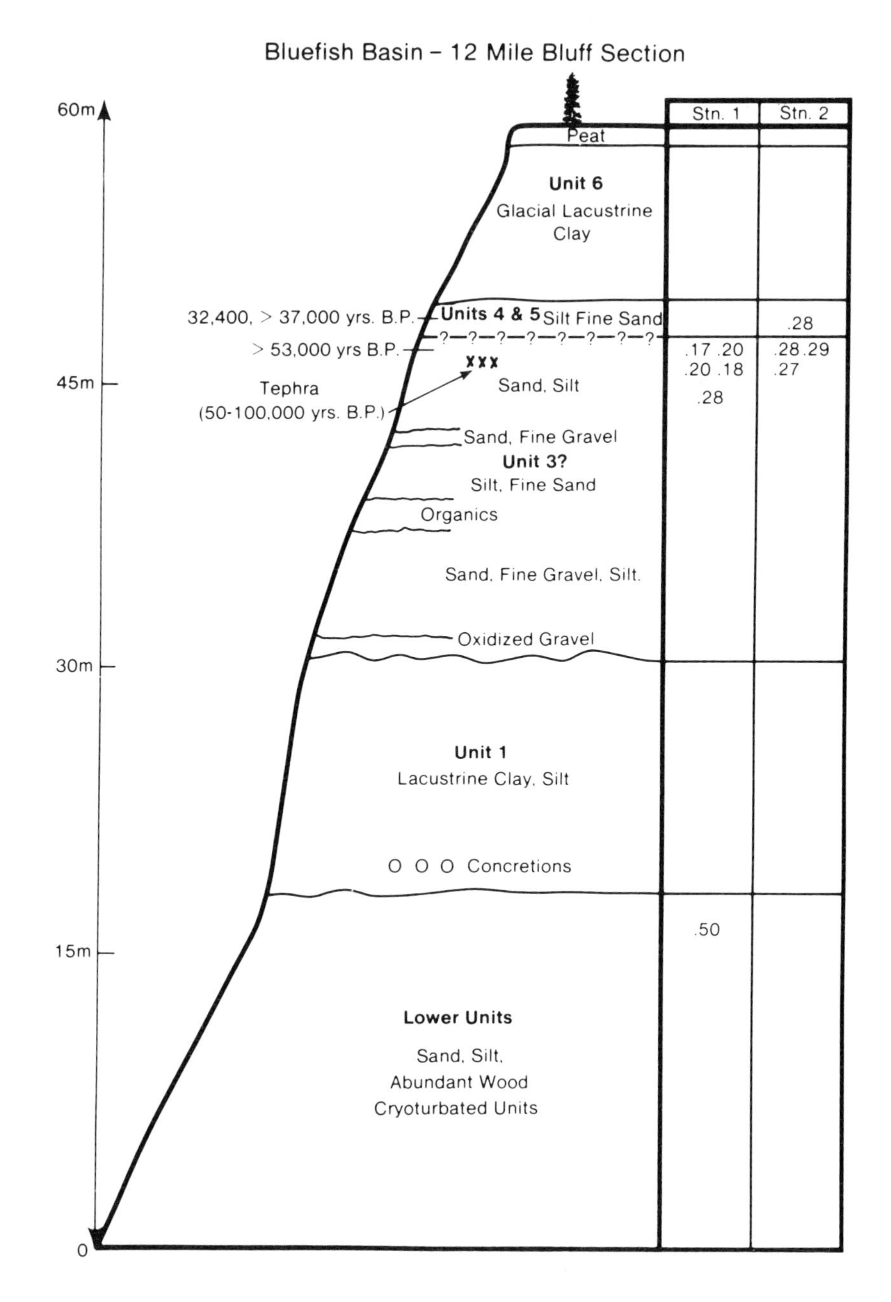

Figure 3 12 Mile Bluff composite section showing sample locations and D/L ratios of aspartic acid.

part of Unit 3 at Location 15 that are between about 50 and 100 ka.

Stratigraphy - Hungry Creek Section (Bonnet Plume Basin)

The Hungry Creek section in the Bonnet Plume Basin is located about 320 km southeast of Location 15 in the Old Crow Basin, a much greater distance in relation to other sections discussed (Figure 1). No attempt has been made to lithologically correlate units for the Old Crow Basin but ^{14}C dates have aided in identifying equivalent deposits and verifying the amino acid results obtained.

The lowest unit consists of a gravel unit more than 4 m thick (Figure 5). This is overlain by about 11 m of silt, clay and sand with abundant organic layers including wood which underlies about 3 m of till. The till is derived from glaciers that flowed from the east terminating not far west of this section. Near the top of the silt, clay and sand unit, a small rounded fragment of wood has been dated at 36,900 ± 300 BP (GSC 2422). This date is especially important because

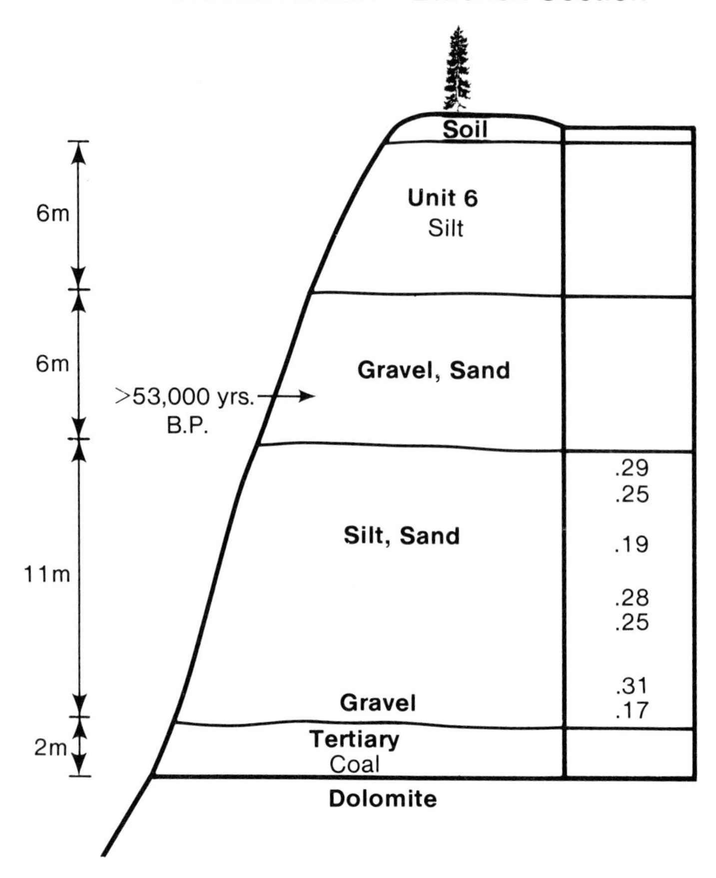

Figure 4 Bluefish Section showing sample locations and D/L ratios of aspartic acid.

it gives an indication of the maximum age of the last glaciation to effect this area. The validity of this date was questioned on stratigraphic grounds and the possibility that the wood had been redeposited (Hughes *et al.*, 1981). Above the till is about 2.5 m of sandy silt overlain by peat.

Results of Amino Acid Analyses - Hungry Creek Section (Bonnet Plume Basin)

To aid in evaluating the validity of the ^{14}C date, several wood samples were taken from below the dated sample but within the same unit for amino acid analysis. The D/L ratios of aspartic acid derived from this unit vary between 0.12 and 0.21 with an average of 0.16 (Figure 5). These ratios are compatible with those in the Old Crow Basin for deposits within the range of traditional ^{14}C dating older than Holocene. It is suggested then that the ^{14}C date is correct. Therefore, the unit is roughly equivalent to the upper part of Unit 4, and 5 in the Old Crow Basin, and the till is late Wisconsin and equivalent to Unit 6 in the Old Crow Basin.

PORCUPINE RIVER - ALASKA

Introduction

In eastern Alaska, on the banks of the Porcupine River a number of thick Quaternary and Late Tertiary sections crop out (Figure 1). Two of these sections have been described by R. Thorsen (pers. comm., 1980) and later sampled for amino acid analysis. The purpose was to aid Thorsen in his stratigraphic studies and again to test long distance correlation while trying to correlate equivalent units with those of the Old Crow Region.

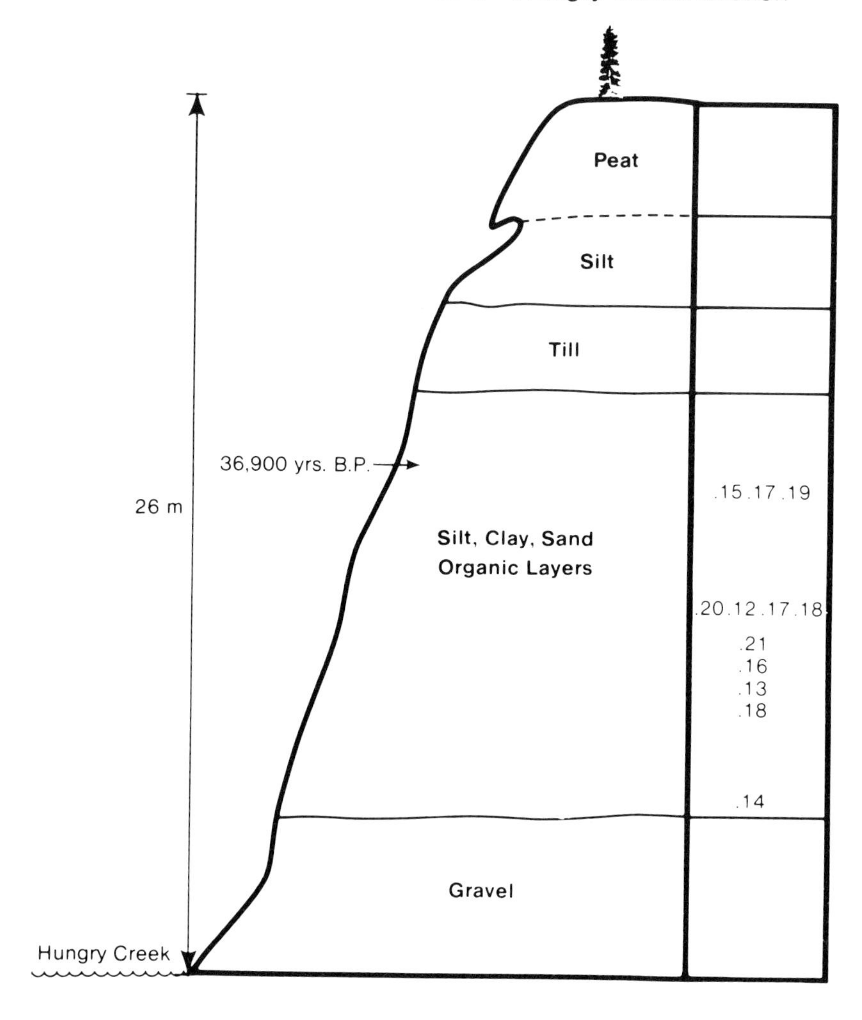

Figure 5 Hungry Creek section showing sample locations and D/L ratios of aspartic acid.

Stratigraphy - Canyon Bluff Section (Porcupine River)

The Canyon Bluff Section is located about 130 km southwest of Location 15 in the Old Crow Basin on the south bank of the Porcupine River (Figure 1). The section consists of about 25 m of Tertiary (?) sands underlying about 12 m of Quaternary deposits. The lower Quaternary deposit consists of about 7 m of gravel, with some sand and silt, commonly cryoturbated. The overlying unit consists of about 1 m of silt. Above the silt unit is about 2 m of cryoturbated gravel with increasing silt content toward the top. This is overlain by about 1 to 2 m of fluvial gravel which in turn is overlain by a thin blanket of silt containing a cryoturbated soil horizon.

Results of Amino Acid Analyses - Canyon Bluff Section (Porcupine River)

Wood samples for amino acid analysis were taken from three widely spaced horizons within the gravel unit (Figure 6).

Relatively high D/L ratios of aspartic acid of 0.37, 0.38 and 0.42 were obtained. This unit appears to be relatively old and may be equivalent to the lower part of Unit 3 and Unit 2 in the Old Crow Basin. A ratio of 0.30 was obtained from the overlying silt unit indicating a relatively old age but younger than the underlying unit.

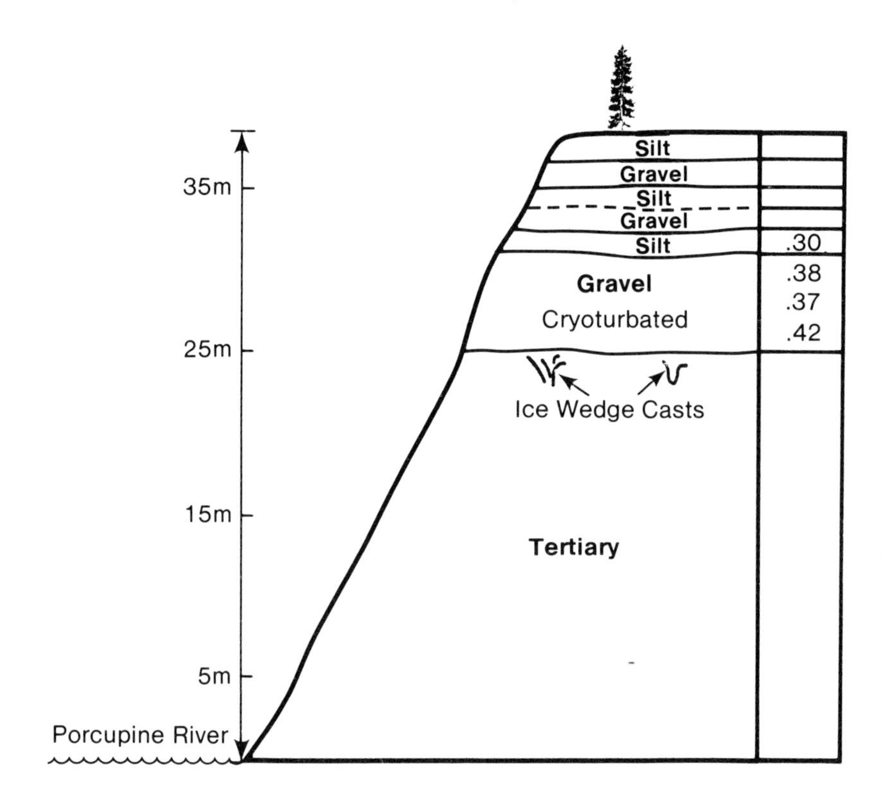

Figure 6 Canyon Bluff section showing sample locations and D/L ratios of aspartic acid.

Stratigraphy - Tusk Bluff Section (Porcupine River)

The second section sampled on the Porcupine River in Alaska is named Tusk Bluff. It is located about 130 km from Location 15 in the Old Crow Basin and about 5 km downstream from the Canyon Bluff Section (Figure 1). The Quaternary geology is considerably different from Canyon Bluff and is considered by Thorsen (pers. commun., 1980) to be relatively younger. As with Canyon Bluff the Quaternary deposits are underlain by Tertiary (?) sediments (Figure 7). About 4 m of Tertiary (?) sediments are exposed above water level overlain by about 21 m of interbedded sand, silt, pebbly sand and gravel. Near the top of the unit a ^{14}C bone collagen date of about 29 ka was obtained. Overlying this unit is about 7 m of cryoturbated silt.

Results of Amino Acid Analyses - Tusk Bluff Section (Porcupine River)

D/L ratios of aspartic acid were determined from samples collected from the Tertiary (?) sediments. Ratios of 0.30 and 0.38 demonstrate that this deposit is old, but not Tertiary. These sediments are probably equivalent to the gravel unit at Canyon Bluff and may be equivalent to the lower part of Unit 3 and Unit 2 in the Old Crow Basin. D/L ratios of aspartic acid of 0.19 and 0.18 were determined from wood about 1 m below the 29 ka date in the pebbly sand subunit. These ratios fit in well with ratios determined for finite dated sediments, slightly older than late Wisconsin at the Hungry Creek section and, even though there is not similar data to draw on, for similar aged sediments in the Old Crow Basin.

Stratigraphy - Upper Porcupine Section (Bell Basin)

The Upper Porcupine Section is located in the Bell Basin about 150 km southeast of Location 15 in the Old Crow Basin (Figure 1). Samples

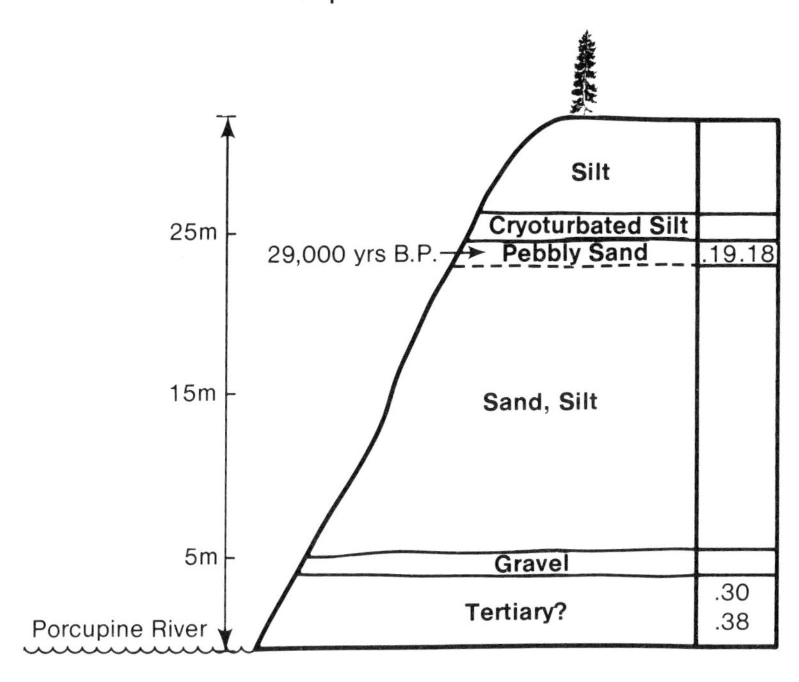

Figure 7 Tusk Bluff section showing sample locations and D/L ratios of aspartic acid.

were collected from a Holocene section that forms a terrace located beside the Upper Porcupine River. Results of this investigation were reported earlier (Rutter *et al.*, 1980), but are repeated here to compare Holocene D/L ratios of aspartic acid with those obtained from wood in older deposits. The section consists of 9 m of fluvial sand and silt with abundant wood fragments throughout, overlain by about 3 m of peat (Figure 8). The only ^{14}C date available, 9190 years BP (GSC 2461), is derived from wood taken from some of the oldest sediments within the unit. It is a maximum age for the initiation of deposition of the material that forms the terrace. The paucity of trees in the region today compared with the abundance of wood found throughout the section (*Populus* and *Salix*), suggests these sediments were deposited during a period warmer than today.

Results of Amino Acid Analyses - Upper Porcupine Section (Bell Basin)

Samples were taken at various intervals throughout the entire section all stratigraphically above the oldest dated beds. Ratios vary between 0.01 and 0.08 with an average of 0.06, but are probably all about the same age, because the terrace sediments represent large scale pointbar deposits, and therefore, deposited relatively fast, say in a few hundred years. The ratios seem reasonable for Holocene wood, when compared with wood from older sediments.

DISCUSSION AND CONCLUSIONS

All in all, determining D/L ratios of the aspartic acid in wood has been useful as an aid in correlating equivalent depositional units, determining relative ages, and in a general way, the absolute age of units. Long range correlation has been successful providing the sediments have had similar climatic and environmental histories and that units above and below those being correlated are of widely varying age. D/L ratios of aspartic acid from Holocene sediments vary mostly from approximately 0.01 to 0.08, late and mid-Wisconsin 0.14 to 0.24, and early Wisconsin and Sangamon 0.24 to 0.36. Results from samples older than Sangamon are difficult to obtain with accuracy. Although the explanation is not clearly understood, more acceptable ratios are obtained

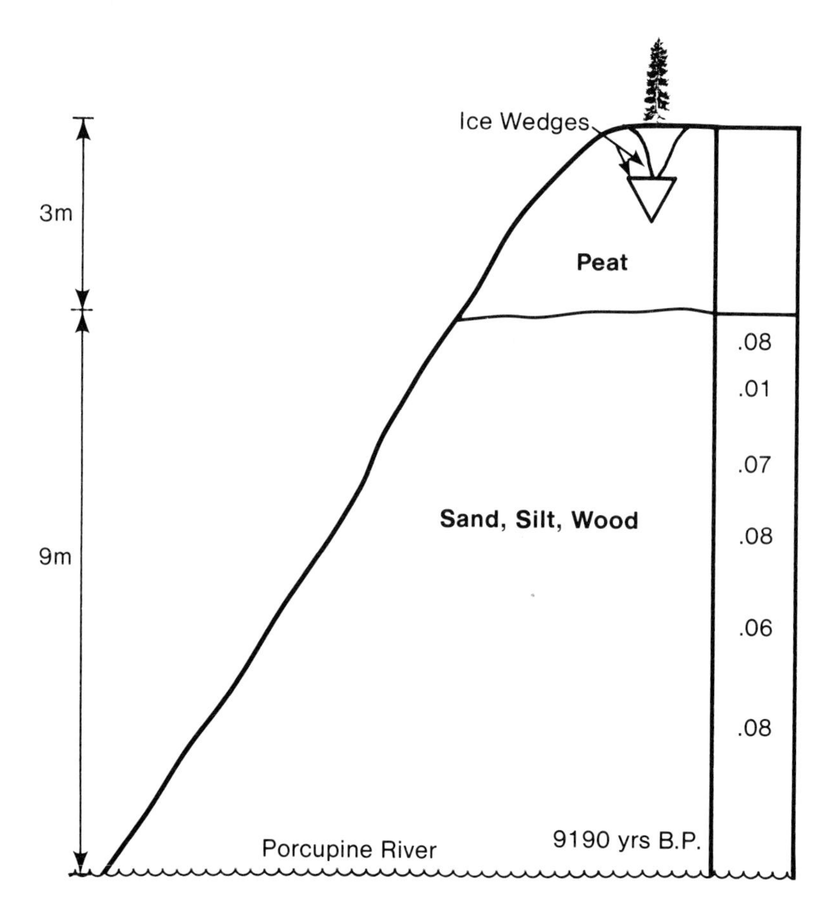

Figure 8 Upper Porcupine section showing sample locations and D/L ratios of aspartic acid.

(that is, on the order of 0.50) by using a greater amount of original sample than usual in routine analysis.

Differences in D/L ratios in wood of the same age can be caused by diagenetic changes before and after burial, contamination, variations in analytical procedures, variation in paleoenvironments (including temperature and climate), and variation in the rate at which various species racemize. It is difficult to comment on what diagenetic changes may have taken place before or after burial, or what samples may have been contaminated. However, comments can be made on other aspects.

We believe the results of our analytical procedures are accurate to within 5 to 7%. This is based on over 1000 samples analyzed in a variety of materials. Accuracy and precision are checked by frequently determining amino acids of standards, analyzing modern wood and by analyzing the same sample more than once. In addition, we have made interlaboratory comparisons by analyzing the same sample.

Paleoenvironmental factors, particularly temperature variations have probably had the greatest effect on D/L ratios of specimens of the same age. The older the samples, the more likely they are to show variation. Long term temperature fluctuations in the Quaternary are well known. In addition, local factors related to the presence or absence of permafrost may cause variation. For instance, how long have the specimens been under permafrost conditions, and at what temperatures? What specimens have been subjected to active layer melting and what specimens have not? Although difficult to answer, it is safe to say that most specimens have been subjected to long periods of permafrost conditions. Therefore, we assume racemization has been retarded.

Although racemization rates are species specific for shell, bone, and some other substances, this does not appear to be the case in our studies of wood. The racemization rate depends upon the position of the aspartic acid residue in the polypeptide chain (Smith and Silva de Sol, 1980). It is also dependent on the nature of the adjacent peptide units (R. Moir, pers. commun., 1982), thus it is possible that the protein material in wood is similar from one species to another whereas the amino acid sequence varies from one species to another species in say, shells. It is also known that the matrix in which protein is intercalated can cause a variation in the amount of racemization. In the wood samples, the aliquots are from structural material and do not have a possible functional variation as in non-plant material.

ACKNOWLEDGEMENTS

We would like to thank the team of the Yukon Refugium Project for hours of discussion on the stratigraphy of the Old Crow Region. These include Drs. O.L. Hughes and J.V. Matthews Jr., (Geological Survey of Canada); R.E. Morlan and C.R. Harington (National Museums of Canada); and C.E. Schweger (University of Alberta). The present work has been financed by grants from the Natural Science and Engineering Research Council and the Geological Survey of Canada.

REFERENCES CITED

Engel, M.H., Zumberge, J.E. and Nagy, B., 1977, Kinetics of amino acid racemization in *Sequioadendron giganteum* heartwood: Anal. Biochem., v. 82, p. 415-422.

Hughes, O.L., 1972, Surficial geology of the northern Yukon Territory and northwestern District of Mackenzie, Northwest Territories: Geological Survey of Canada Paper 69-36.

Hughes, O.L., Harington, C.R., Janssens, J.A., Matthews, J.V., Jr., Morlan, R.E., Rutter, N.W. and Schweger, C.E., 1981, Upper Pleistocene stratigraphy, paleoecology and archaeology of the northern Yukon interior, eastern Beringia 1. Bonnet Plume Basin: Arctic, v. 34, p. 329-365.

Irving, W.H., 1978, Pleistocene archaeology in eastern Beringia, in Bryan, A.L., ed., Early Man in America - from a Circum-Pacific Perspective: Archaeological Researches International, Edmonton, p. 96-101.

Kvenvolden, K.A., 1980, Interlaboratory comparison of amino acid racemization in a Pleistocene mollusk; *Saxidomus giganteus*, in Hare, P.E., ed., Biogeochemistry of Amino Acids, N.Y., Wiley, p. 223-232.

Lee, C., Bada, J.L. and Person, E., 1976, Amino acids in modern and fossil woods: Nature, v. 259, p. 183-186.

Morlan, R.E., 1978, Early man in northern Yukon Territory: perspectives as of 1977, in Bryan, A.L., ed., Early Man in America - from a Circum-Pacific Perspective: Archaeological Researches International, Edmonton, p. 78-95.

Morlan, R.E. and Matthews, J.V., Jr., 1978, New dates for early man: GEOS, 2-5, Winter 1978.

Rutter, N.W., Crawford, R.J. and Hamilton, R., 1980, Correlation and relative age dating of Quaternary strata in the continuous permafrost zone of northern Yukon with D/L ratios of aspartic acid of wood, freshwater molluscs, and bone, in Hare, P.E., ed., Biogeochemistry of Amino Acids, N.Y., Wiley, p. 463-475.

Smith, G.G. and Silva de Sol, B., 1980, Racemization of amino acids in dipeptides shows $COOH-NH_2$ for non-sterically hindered residues: Science, v. 207, p. 765-767.

TREE-RING DATING IN CANADA AND THE NORTHWESTERN U.S.

M.L. PARKER, L.A. JOZSA, SANDRA G. JOHNSON and PAUL A. BRAMHALL

ABSTRACT

Dendrochronology has been used recently in Canada and the northwestern United States to date archaeological sites, driftwood accumulations of a lake created by a surging glacier, driftwood on raised beaches, glacial moraines, rates of alluviation, flooding, ice jamming and forest fires. Archaeological tree-ring samples have been dated from a caribou-trap site in the northern Yukon, Ozette Village on the Olympic Peninsula, Kitwanga National Historic Site, British Columbia and the Bell Site near Lillooet, British Columbia. Dates also were obtained from log cabins at Silver City, a gold-rush ghost-town in the Yukon.

Coastal, high-elevation of high-latitude sites have generally been considered to be the ones that produce tree-ring material of poorer dendrochronological quality than do the semi-arid sites. However, it is now possible to obtain dates from the poorer-quality samples from these sites because of the development of several new techniques. X-ray densitometry is used to measure ring density as well as ring width. Maximum ring density has been used to provide crossdating on the coastal, high-latitude and high-elevation wood samples that could not be dated using ring width alone. Computer crossdating has been used to match tree-ring samples that have been difficult to date by other matching techniques. Using these computer crossdating techniques, tree-ring patterns have been matched over distances greater than 500 kilometers. Some living trees in Washington and British Columbia exceed 1300 years in age. There is a potential for building tree-ring chronologies, using both living and dead tree material, back in time for thousands of years.

INTRODUCTION AND BACKGROUND

The recorded observation of the relationship between time and the diameter growth of trees dates back at least two millenia to Theophrastus, the student and successor of Aristotle. A number of botanists, foresters and astronomers in Europe and America in the 18th and 19th Centuries understood that rings in trees are annual in nature and some even understood the principle of crossdating between tree-ring series, Studhalter, 1959. However, it was not until the work of A.E. Douglass (1919, 1929, 1935, 1937) in the 20th Century that dendrochronology developed as a recognized and continuing field of scientific investigation. In 1901, the idea occurred to Douglass that the growth of the pines and junipers in northern Arizona should depend on the year's, always limited, moisture supply and that this might be reflected in the width of the annual rings (Giddings, 1962).

In 1904 Douglass tested this idea by matching the pattern of wide and narrow annual rings formed in the cross sections of stems of different trees growing in the same area. In the years that followed,

Douglass developed techniques to facilitate crossdating and chronology building (the summarizing and temporal extension of annual ring-width records from different trees into a single "tree-ring chronology").

The potential for tree-ring dating was recognized early by the southwestern archaeologists. In 1916, Douglass first received samples of prehistoric wood. Crossdating of wood and charcoal samples from a number of different southwestern Indian ruins proceeded until by the 1920's a number of "floating chronologies" existed. These floating chronologies provided crossdating between samples of the same site and between sites but were not tied down to calendar years.

This 585-year-long floating chronology, that Douglass called the "Relative Dating Series", was compiled from beams from Aztec, Pueblo Bonito, Cliff Palace, Betatakin and other Indian ruins from the southwest. Haury (1962) has recorded the dramatic moment, when in 1929, a charcoal beam from a ruin in Showlow, Arizona, provided the ring sequence required to crossdate the Relative Dating Series with the tree-ring chronology that Douglass had compiled from living trees from the area. This event marked the climax of years of technique development, crossdating and chronology building by Douglass and his associates. At this one moment in time, the calendar-year dating of many major ruins in the southwest was accomplished.

This story has been repeated many times, *i.e.*, using the crossdating technique, building living-tree chronologies and chronologies of undated samples, and finally, matching the two series to provide calendar dates of previously undated samples (Figure 1). A good example of this is when Giddings (1952) built a floating chronology from tree-ring samples from prehistoric sites on the Kobuk River in Alaska and then dated these sites by matching this floating chronology with a living-tree chronology.

The work begun by Douglass has continued and has been greatly expanded upon by E. Schulman, B. Bannister, C.W. Ferguson, H.C. Fritts, C.W. Stockton, J. Dean, W. Robinson, T.P. Harlan, M.A. Stokes, V.C. Lamarche and many others at the Laboratory of Tree-ring Research in Tucson, Arizona (Stokes and Smiley 1968; Fritts 1976; see also issues of the Tree-Ring Bulletin). Tree-ring laboratories have been established in many parts of the world using the techniques and principles developed in Arizona.

The successful tree-ring dating reported in this paper has resulted, to a large extent, from applying the techniques developed by Douglass and his students. In all cases, however, this dating was achieved by applying two additional and relatively new techniques: (1) X-ray densitometry, and (2) computer crossdating.

All trees are not of equal quality for crossdating purposes. There is a wide range in dendrochronological quality of ring series from those that display good ring-to-ring variation and provide good dating of chronologies, to those that are of such poor quality that they cannot be dated. Trees growing in the semi-arid regions of western North America, from the Fraser River Valley in British Columbia to Mexico, have a limited moisture supply and usually produce good-quality tree-ring records. These ring series that have marked ring-to-ring variation are known as "sensitive". Conversely, trees growing in coastal, high-elevation or high-latitude environments usually produce ring series with little ring-to-ring width variation, or "complacent" tree-ring chronologies (Figure 2).

This paper deals with the use of the methods of X-ray densitometry and computer crossdating (augmenting the techniques developed by Douglass) to date the poorer quality tree-ring samples from Canada and the northwestern United States.

TWO METHODS THAT IMPROVE CROSSDATING POTENTIAL

For years, dendrochronological research has been concentrated

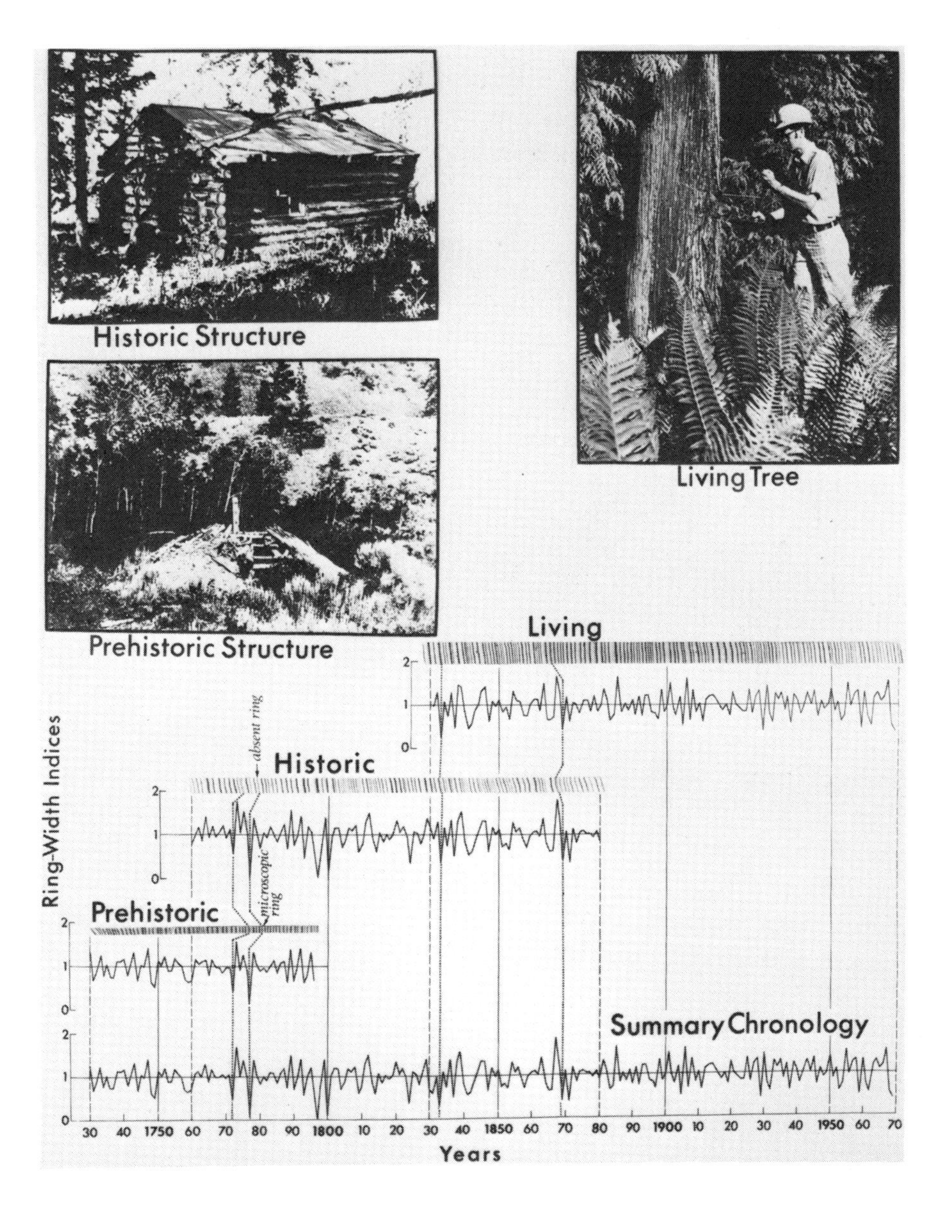

Figure 1 Historic and prehistoric wood structures can be dated by matching their ring patterns against those found in living trees of known age.

mainly on those trees from the semi-arid regions that produce good-quality tree-ring width series. However, extensive and spectacular tree-ring material is available from more mesic sites such as Ozette Village on the Olympic Peninsula. We now know that by applying several new techniques, it is possible to obtain tree-ring dates from these sites. We can look forward to repeating in many new areas, the scenario followed by Douglass years ago in the southwest, of cross-dating, chronology building and then the calendar-year dating of tree-ring samples.

X-Ray Densitometry

Various radiation, light and mechanical techniques have been developed to study the within-ring density variations in wood samples (Cameron *et al.*, 1959; Marian and Stumbo, 1960; Green, 1964, 1965; Green and Worrall, 1964; Harris, 1969; Kawaguchi, 1969); but perhaps the most successful method, X-ray densitometry, was pioneered by Polge (1963, 1965 and 1966) during the 1960's. The techniques in X-ray densitometry used in the research described in this paper were developed in Ottawa, Ontario, with the Geological Survey of Canada

Figure 2 Complacent ring series from trees on the moist Brtish Columbia coast have rings that are more uniform in width than are the rings of the sensitive chronologies of trees from the province's dry interior or the Northwest Territories.

between 1967 and 1970 (Parker 1969a, 1969b, 1970a, 1970b; Parker and Henoch 1969 and 1971; Jones and Parker 1970; Parker and Meleskie 1970) and in Vancouver, British Columbia with the University of British Columbia; the Western Forest Products Laboratory, Canadian Forestry Service; and Forintek Canada Corp. between 1970 and the present (Parker 1972 and 1976; Parker and Jozsa 1973a, 1973b, 1977a and 1977b; Parker and Kennedy 1973; Parker, Jozsa and Brue. 1973; Parker, Schoorlemmer and Carver, 1973; Parker, Barton and Jozsa 1974; Parker, Bunce and Smith 1974; Parker *et al.*, 1976, 1977, 1979, 1980 and 1981; Heger *et al.*, 1974; Johnson, 1982).

The X-ray densitometry system currently in use in Vancouver uses an in-motion X-ray technique to produce radiographs of wood and charcoal samples and an on-line computerized scanning densitometer to collect and process data from the radiographs (Parker *et al.*, 1980). Detailed data of earlywood and latewood components of both width and density are rapidly and accurately collected and stored on magnetic tape. The on-line computer and peripheral devices greatly facilitate processing, printing, plotting and dating of tree-ring series.

If we compare this method of recording tree-ring data with those methods used by Douglass in the early years, we can note that preparing a given sample by the X-ray densitometry method is much more time consuming. In many cases, a dendrochronologist working in the southwestern U.S. can break a piece of archaeological charcoal by hand, examine it with a hand lens or low-power microscope and date it by the memory method (Douglass, 1946) in a few minutes. If the memory method does not produce results, the sample may be dated by the skeleton-plot technique (Stokes and Smiley, 1968). This method also is a fairly rapid one. Even in the southwest, however, where good-quality tree-ring samples exist, many samples cannot be dated because of their poor quality.

In most of Canada and the northwestern U.S., the very good quality tree-ring width samples are rare. Most ring-width series are complacent. It has been demonstrated, however, that the coastal high-elevation and high-latitude specimens may be sensitive with respect to ring-to-ring density variation and are, therefore, of good quality for dating or for climatic studies if density is used (Parker and Henoch, 1971; Parker 1976; Parker *et al.*, 1981, 1982). Year-to-year variations in temperature are reflected especially in annual ring maximum density (Figure 3). Traditional methods of ring-width analysis are not adequate for dating most of these wood samples. For this reason X-ray densitometry is the most appropriate existing method to be used to date tree-ring samples from Canada and the northwestern U.S.

Computer Crossdating

The computer crossdating program in use at Forintek, the Shifting Unit Dating Program (SUDP), was developed by Parker at the Laboratory of Tree-Ring Research (Parker, 1967), modified at the Geological Survey of Canada (Parker, 1970a and 1970b), and modified again at the Western Forest Products Laboratory with the assistance of Paul Bramhall.

This program used computer techniques to crossdate tree-ring series. A portion (unit) of an undated tree-ring series is correlated with a dated master tree-ring series in all possible positions. The positions and correlations of the three best matches between the unit and the master series are recorded. Successive units (of designated length and increment) are correlated with the master of all possible locations until the entire undated series has been matched with the entire dated series. The validity of the crossdating is evaluated by the sequential placement of the units and the values of the correlation coefficients. This technique has been used successfully to crossdate tree-ring series over distances greater than 500 km (Figure 4) for samples from: (1) California and the Four Corners area (Parker, 1967), (2) Ontario and Nova Scotia (Parker, 1970b), and (3) several areas along the British Columbia coast (Forintek Canada Corp., no date).

Computer crossdating was used to date the tree-ring material, described in a following section, from geological, archaeological and historic sites in the Yukon Territory, Northwest Territories, British Columbia, Manitoba and Washington.

METHOD USED FOR STANDARDIZING AND SUMMARIZING TREE-RING DATA

A major problem that dendrochronologists face in dealing with most tree-ring data is the removal of non-climatic trends, such as the growth trend, and averaging these data in some standard form without inadvertently removing features that should be retained. Objectives are to produce tree-ring chronologies that are of good quality for dating purposes and that accurately reflect the influence of climate.

The "raw data" produced by the X-ray densitometry system described here consist of: (1) ring-width values in 0.01 mm units, and (2)

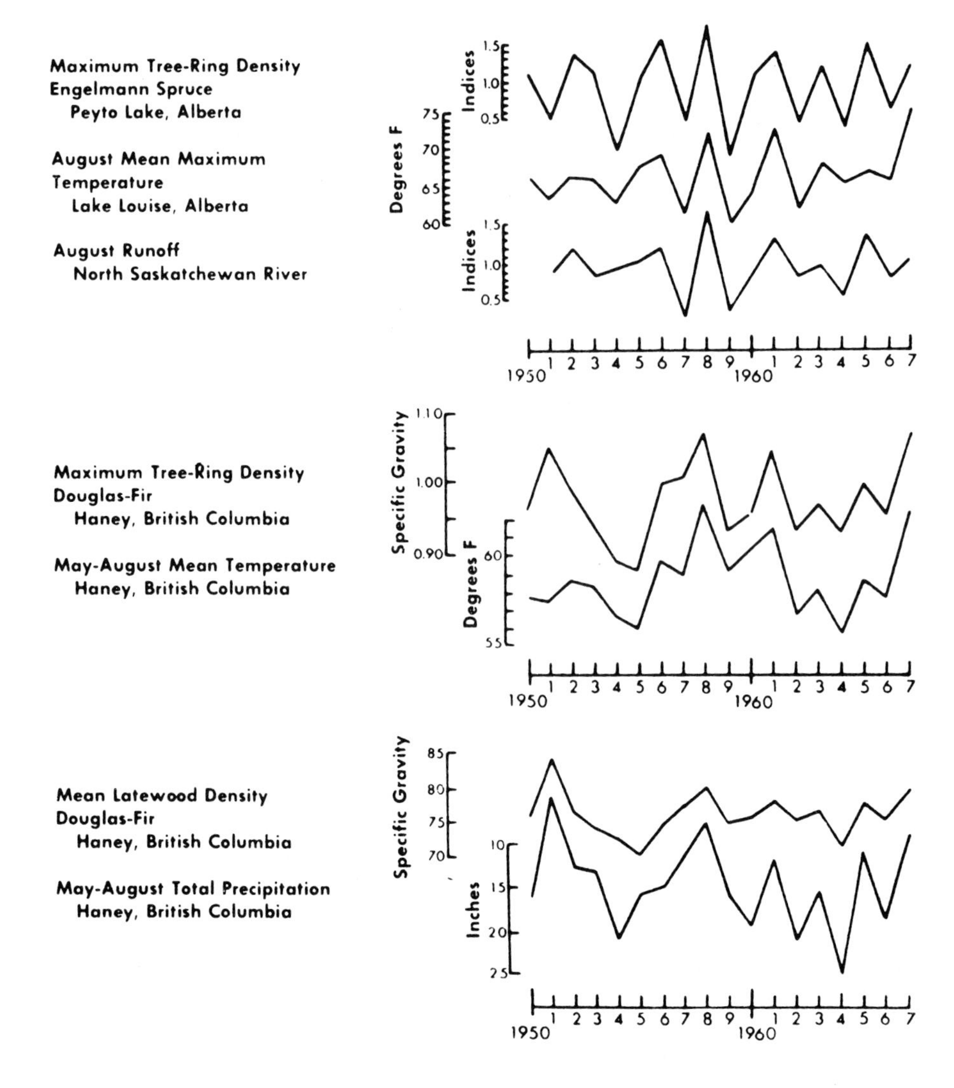

Figure 3 Tree-ring density of the high-elevation spruce is shown to be closely related to August temperature and river runoff. The two lower graphs show the close relationship of May through August temperatures and precipitation to tree-ring density in the British Columbia coastal Douglas-fir.

maximum ring-density values in g/cm^3 measured at 0.01 mm increments along the scanned portion of a tree-ring sample. These raw data are "standardized" by converting them into indices (ratio of observed value to fitted trend, yielding values with a mean of 1.00) by removing the growth trend (and sometimes other trends or fluctuations). The standardized data (indices) for all samples are then "summarized", or averaged, to produce a summary or "master" chronology for the site.

This procedure is illustrated in Figure 5. In a previous publication (Parker *et al.*, 1981) the "A", "B" and "C" components have been described: "A" defined as the growth trend; "B", the short-term fluctuations greater than 10 years in length; and "C", the year-to-year variations. Various data processing programs are used to produce an "A", a "B", a "C" and a "B & C" chronology from the raw data of each tree-ring sample. These data are then averaged with data from other tree-ring sample series to produce the summary chronologies.

If tree-ring data are averaged without removing the growth trend, non-climatic fluctuations related to the age of the tree will be included in the chronology. Therefore, it is essential that the growth trend be removed. However, variation due to climate may be removed

Figure 4 Living trees up to 4 metres in diameter have been sampled with a power-driven increment borer. Using computer crossdating techniques, tree-ring patterns have been matched over distances greater than 500 kilometers.

inadvertently, if proper techniques are not used in this procedure.

One method for removing the growth trend (the "A" component) from tree-ring series, utilized a digital-filter with a length of 60 years (Parker, 1970a). This method is designed to remove the growth trend and the non-climatic surges in growth, such as tree growth release after a forest fire. Tree-ring indices (in the form of our "B & C" chronology) are produced by calculating deviations from a growth-trend line through the raw data values.

The "B and C" chronologies are obtained from the tree-ring indices (the "B & C" chronology). This is done by using a digital-filter technique with a length of 13 years (Parker, 1970a). The "B" chronology is the estimated curve which is a weighted running mean. Deviations from this line are the year-to-year fluctuations, or the "C" chronology. The "C" chronology type has proven to be the form most useful for dating purposes.

SUCCESSFUL CROSSDATING IN NORTHERN AND WESTERN NORTH AMERICA

X-ray densitometry and computer crossdating were used in all cases of successful dating of archaeological or driftwood tree-ring material from Canada or the northwestern U.S. In nearly all cases maximum ring

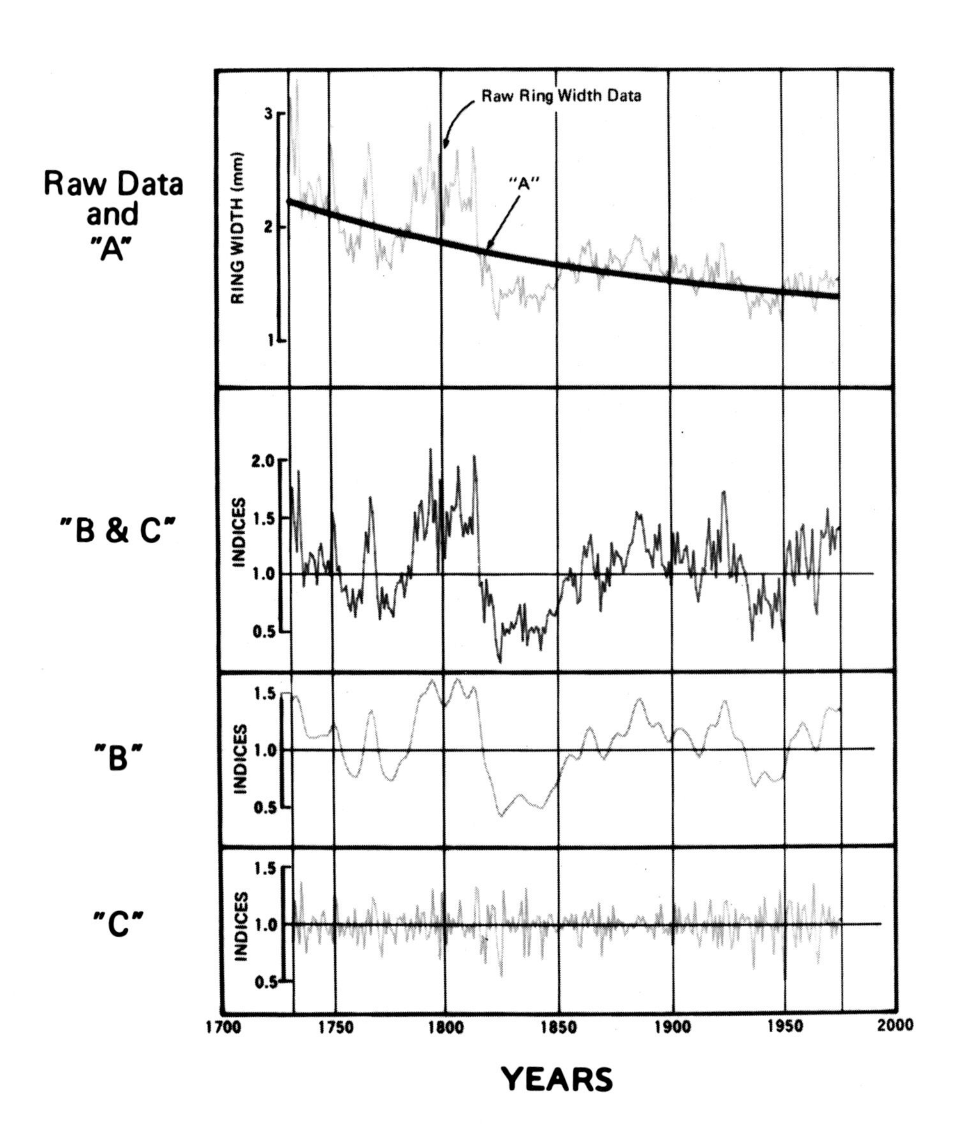

Figure 5 The "A", "B", and "C" components of a tree-ring series. Data used in this example are ring width values of a white spruce tree from Cri Lake, Quebec.

density was the parameter used.

Northern Yukon Adze-cut Stump

The first successful dating of an archaeological tree-ring sample in Canada was the dating of an adze-cut stump from a caribou trap located in the Old Crow area in the northern Yukon Territory. This sample was submitted to the Geological Survey of Canada by William Irving. The SUDP was used to match the ring series of this sample against ring series from a single white spruce (*Picea glauca* (Moench) Voss) tree from near Inuvik, NWT, a distance of about 250 km. A date of 1647 (very variable outside) was obtained from both ring width and maximum ring density.

The significance of this is that it demonstrates that tree-ring dates can be obtained in this northern region by crossdating samples derived from areas long distances apart and even by using single-tree chronologies.

Bell Site near Lillooet, British Columbia

Wood and charcoal samples of Douglas-fir *(Pseudotsuga menziesii* (Mirb.) Franco), lodgepole pine *(Pinus contorta* Dougl.) and ponderosa pine *(Pinus ponderosa* Laws.) from the Bell Site were submitted to the Western Forest Products Laboratory by A.H. Stryd for tree-ring analysis

(Stryd, 1979). The Bell Site is a prehistoric pithouse village site in the British Columbia interior near Lillooet.

A number of tree-ring samples crossdated with one another and a composite chronology derived from 9 samples was constructed. A date of 1854 was obtained for the most recent ring-year on this composite by using SUDP to match this chronology against a master chronology constructed from Douglas-fir trees collected near Pavilion Lake by H.C. Fritts in 1966. The successful match was obtained by using ring width rather than maximum ring density. This suggests that although maximum ring density is the better parameter to use with samples from the wetter and colder sites, ring width is better for crossdating samples from the dry interior sites.

Additional crossdating was obtained between other samples from different houses at the site and between the Bell Site samples and another site in this region, but no calendar-year dates were determined probably because the material is too old to be matched with the existing master chronology.

Kitwanga National Historic Site, British Columbia

Wood and charcoal samples from the Kitwanga National Historic Site, Battle Hill, were submitted to Forintek Canada Corp. for tree-ring analysis (Jozsa, Parker, Bramhall, Kellogg and Rowe, 1980). Tree-ring master chronologies were built from recently-felled western hemlock (*Tsuga heterophylla* (Raf.) Sarg.) and western red cedar (*Thuja plicata* Donn) trees from the Kitwanga area.

Although a large proportion of the forest in that area consists of western hemlock, none of the eight species represented in the archaeological tree-ring collection was hemlock. The tree ring samples from the Battle Hill site were generally of very poor dendrochronological quality - complacent, short and fragmentary. However, one date (1749) was obtained, using SUDP, for a western red cedar sample.

Ozette Village Site, Washington

A very large number of tree-ring samples have been removed from the prehistoric village of Ozette located on the west coast of the Olympic Peninsula in Washington. A pilot study was undertaken by Forintek Canada Corp. for the Washington Archaeological Research Center (WARC) to determine if tree-ring dates could be derived from this material (Jozsa, Parker and Bramhall, 1980).

Sixteen archaeological tree-ring samples were analyzed and one date (1613) was obtained by using SUDP to crossdate a western red cedar sample with a composite chronology built from two recently-felled western red cedar trees.

Under a more recent contract between Forintek and WARC, better quality master chronologies have been built for Douglas-fir and western red cedar trees from the Olympic Peninsula. An additional tree-ring date was obtained (1719vv) by matching an archaeological sample with the western red cedar master chronology.

There are living western red cedar trees in the area that exceed 1300 years in age, however, some of these are very complacent and lack circuit uniformity. These trees are of too poor quality to be used for dendrochronological studies even if X-ray densitometry is employed.

However, the good quality of some tree-ring series, the age of the trees and the extensiveness of the archaeological material, make this a potentially fruitful area for tree-ring research.

Silver City, Yukon Territory

In 1979, eleven tree-ring samples were collected from spruce logs used in the construction of cabins in the gold-rush town of Silver City in the Yukon (Jozsa *et al.*, 1981; Clague *et al.*, 1982). X-ray densitometry was used to analyze these samples and the tree-ring data were matched against tree-ring chronologies derived from living white spruce trees growing in the vicinity.

Dates were produced for all eleven samples indicating that the log-cabin trees were cut in 1905 or shortly thereafter. The significance of this is that the tree-ring series of trees in that region are of excellent quality for dating purposes and it should be possible to obtain dates from the many historic structures in this northern area.

Neoglacial Lake Alsek, Yukon Territory

The intermittent damming of the Alsek River by Lowell Glacier at various times in the past has resulted in the formation of numerous beaches. Some of these beaches contain driftwood strandlines. Driftwood samples were collected in 1979 from beaches near Haines Junction, Yukon Territory (Jozsa *et al.*, 1981, Clague *et al.*, 1982).

Sixteen tree-ring dates were obtained from driftwood samples from several of the more recent beaches. These dates ranged from 1675 to 1848. All dates were obtained through the use of X-ray densitometry and computer crossdating.

Raised Beaches on The Hudson Bay Coast

In 1981, driftwood samples were collected from raised beaches near the mouth of the Owl River which flows into Hudson Bay midway between Churchill and York Factory, Manitoba (Parker *et al.*, 1982). X-ray densitometry and computer crossdating were used to date twelve of these tree-ring samples. The dates range from 1849 to 1961.

Figure 6 Tree rings can be used to date events such as landslides by changes in growth patterns and compression wood.

Almost all of the samples processed from the youngest beaches were dated. This demonstrates that it is possible to date driftwood in the Hudson Bay area. Some driftwood in the region is thousands of years old, making it a good area for future dendrochronological research.

Other Tree-Ring Dating in Canada

In addition to the above examples, dendrochronology has been used in a number of minor projects such as the dating of a bonfire used to consume the body of a murder victim, the dating of a driftwood sample from the Horton River in the northern Northwest Territories, the dating of forest fires and a number of landslides (Figure 6).

Dating methods other than X-ray densitometry have sometimes been used. These methods include ring counts the use of compression wood, scars, traumatic resin canals or other techniques. These methods are simple but useful for dating such events as flooding or ice jamming (Figure 7) (Parker, Jozsa and Bruce, 1973).

CONCLUSIONS

There are living trees in the northwestern United States and Canada that exceed 1300 years in age. There are tree-ring samples in geological and archaeological contexts that are thousands of years old. The recent development of new techniques, particularly X-ray densitometry, and computer crossdating, have been used to augment older techniques to obtain many tree-ring dates during the last several years. There are economically significant applications of dendrochronological

Figure 7 The scars on these three tree sections indicate the year in which flooding or ice jamming occurred along the river where the trees grew.

research to climatic, wood quality and environmental studies. The combination of the availability of good-quality tree-ring material to study, the development of new techniques and the existence of many useful applications, makes Canada and the northwestern United States logical regions in which to conduct tree-ring research.

REFERENCES CITED

Cameron, J.F., Berry, P.F. and Phillips, E.W.J., 1959, The determination of wood density using beta rays: Holzforschung. v. 13, no. 3, p. 78-84.

Clague, J., Jozsa, L.A. and Parker, M.L., 1982, Dendrochronological dating of glacier dammed lakes; an example from Yukon Territory, Canada: Arctic and Alpine Research, v. 14, in press.

Douglass, A.E., 1919, Climatic Cycles and Tree Growth, Vol. I. Carnegie Inst. Wash. Publ. 289.

_______________, 1929, The secret of southwest solved by talkative tree rings: Nat. Geogr. Mag., v. 5,6, no. 6, p. 736-770.

_______________, 1935, Dating Pueblo Bonito and other ruins of the southwest: Nat. Geogr. Soc. Contrib. Tech. Pap., Pueblo Bonito Ser. 1.

_______________, 1937, Tree rings and chronology: Univ. Ariz. Bull., v. 8, no. 4, Phys. Sci. Ser. 1.

_______________, 1946, Precision of Ring Dating in Tree-Ring Chronologies: Laboratory of Tree-Ring Research Bulletin, No. 3. Tucson.

Forintek Canada Corp. (No date), Talkative Tree-Rings. News For You, SS-104, 2 pp.

Fritts, H.C., 1976, Tree Rings and Climate: London, Academic Press, 567 pp.

Giddings, J.L., Jr., 1952, The Arctic Woodland Culture of the Kobuk River: Museum Monographs, The University Museum, University of Pennsylvania, Philadelphia, 143 pp.

___________, 1962, Development of tree-ring dating as an archaeological aid, in Kozlowski, T.T., ed., Tree Growth: N.Y., Ronald Press, p. 119-132.

Green, H.V., 1964, Supplementary details of construction of the stage and drive assembly of the scanning microphotometer, Pulp Pap. Res. Inst. Can., Res. Note 41, 7 p.

____________, 1965, Wood characteristics IV: The study of wood characteristics by means of a photometric technique, Pulp Pap. Res. Inst. Can., Tech. Rep. 419, 17 pp.

Green, H.V. and Worrall, J., 1964, Wood quality studies I: A scanning microphotometer for automatically measuring and recording certain wood characteristics: Tappi, v. 47, no. 7, p. 419-427.

Harris, J.M., 1969, The use of beta rays in determining wood properties Part 1 - 5: N.Z.J. Sci., v. 12, no. 2, p. 396-451.

Haury, E.W., 1962, HH-39: Recollections of a dramatic moment in southwestern archaeology: Tree-Ring Bull., v. 24, no. 3-4, p. 11-14.

Heger, L., Parker, M.L. and Kennedy, R.W., 1974, X-ray densitometry: A technique and an example of application: Wood Science, v. 7, no. 2, p. 140-148.

Johnson, L.W., 1982, Tree rings that tell stories: British Columbia Lumberman, January 1982, 2 pp.

Jones, F.W. and Parker, M.L., 1970, G.S.C. tree-ring scanning densitometer and data acquisition system: Tree-Ring Bulletin, v. 30, no. 1-4, p. 23-31.

Jozsa, L.A., Parker, M.L. and Bramhall, P.A., 1980, Feasibility of Tree-Ring Dating at Ozette, Contract report (80-543) for the Washington Archaeological Research Center, Forintek Canada Corp., Unpublished manuscript, 22 pp.

Jozsa, L.A., Parker, M.L., Bramhall, P.A., Kellogg, R.M. and Rowe, S., 1980, Wood and Charcoal Sample Analysis for the Kitwanga National Historic Site, Contract report (WR 156-79) for Parks Canada, Western Region, Forintek Canada Corp., Unpublished manuscript 22 pp.

Jozsa, L.A., Parker, M.L., Johnson, S.G. and Bramhall, P.A., 1981, Dendrochronological Dating of Strandlines from Neoglacial Lake Alsek, Contract report (80-68-534) for the Geological Survey of Canada. Forintek Canada Corp., Unpublished manuscript, 47 pp.

Kawaguchi, M., 1969, A Fourier analysis of growth ring photo-electric analysis curves: J. Jap. Wood Res. Soc., v. 15, no. 1, p. 6-10.

Marian, J.E. and Stumbo, D.A., 1960, A new method of growth ring analysis and the determination of density by surface texture measurements: For. Sci., v. 6, no. 3, p. 276-291.

Parker, M.L., 1967, Dendrochronology of Point of Pines (MA Thesis): Department of Anthropology, University of Arizona, 168 pp.

____________, 1969a, Tree-ring chronology building in eastern Canada and Alberta, Report of Activities, Part A: April to October, 1968: Geological Survey of Canada Paper 69-1, Part A, p. 121-2.

____________, 1969b, Dendrochronological investigations in Canada, Report of Activities, Part B: November 1968 to March 1969: Geological Survey of Canada Paper 69-1, Part B, p. 67-8.

____________, 1970a, Dendrochronological techniques used by the Geological Survey of Canada, in Smith, J.H.G. and Worrall, J., eds., Tree-Ring Analysis with Special Reference to Northwest America: The University of British Columbia, Faculty of Forestry, Bulletin No. 7, p. 55-66. (Also published in 1971 as Geological Survey of Canada Paper 71-25, 30 pp.)

____________, 1970b, Some new techniques used in dendrochronological investigations in Canada, Report of Activities, Part B: November 1969 to March 1970, Geological Survey of Canada Paper 70-1, Part B. 71-4.

____________, 1972, Techniques in X-Ray Densitometry of Tree-Ring Samples, Paper presented at the 45th Annual Meeting of the Northwest Scientific Association, Forestry Section, Western Washington State College, Bellingham, 11 pp.

____________, 1976, Improving tree ring dating in northern Canada by X-ray densitometry, SYESIS No. 9, 163-172.

Parker, M.L. and Henoch, W., 1969, Preliminary Report on Dendrochronological Investigations at Peyto Glacier, Alberta, Paper presented at the North Saskatchewan Headwaters Meeting, December 5, 1969, Ottawa, 2 pp.

____________, 1971, The use of Engelmann spruce latewood density for dendrochronological purposes: Canadian Journal Forest Research, v. 1, no. 2, p. 90-98.

Parker, M.L. and Jozsa, L.A., 1973a, Dendrochronological investigations along the Mackenzie, Liard and South Nahanni Rivers, Northwest Territories - Part I: Using tree damage to date landslides, ice-jamming, and flooding, Technical Report 10, Hydrologic Aspects of Northern Pipeline Development, Environmental-Social Committee, Northern Pipelines, Task Force on Northern Oil Development, Report No. 73-3, p. 313-464.

____________, 1973b, X-ray scanning machine for tree-ring width and density analysis: Wood and Fiber, v. 5, no. 3, p. 192-197.

____________, 1977a, Use of the on-line computer-densitometer system to rapidly produce summary density profiles: Bi-Monthly Research Notes, v. 33, no. 2, p. 13.

____________, 1977b, What Tree Rings Tell Us. Forest Fact Sheet, Canadian Forestry Service, 4 pp.

Parker, M.L. and Kennedy, R.W., 1973, The status of radiation densitometry for measurement of wood specific gravity. Proc. International Union of Forest Research Organizations (IUFRO), Division 5 meetings in Cape Town and Pretoria, South Africa, September and October, 1973, 17 pp.

Parker, M.L. and Meleskie, K.R., 1970, Preparation of X-ray negatives of tree-ring specimens for dendrochronological analysis: Tree-Ring Bulletin, v. 30, no. 1-4, p. 11-22.

Parker, M.L., Jozsa, L.A. and Bruce, R.D., 1973, Dendrochronological Investigations along the Mackenzie, Liard, and South Nahanni Rivers, Northwest Territories - Part II: Using Tree-Ring Analysis to Reconstruct Geomorphic and Climatic History, Technical Report to Glaciology Division, Water Resources Branch, Department of the Environment, under the Environmental-Social Program. Northern Pipelines, 104 pp.

Parker, M.L., Schoorlemmer and Carver, L.J., 1973, A computerized scanning densitometer for automatic recording of tree-ring width and density data from X-ray negatives: Wood and Fiber, v. 10, no. 2, p. 120-130.

Parker, M.L., Barton, G.M. and Jozsa, L.A., 1974, Detection of crystalline lignans in western hemlock by radiography: Wood Science and Technology, v. 8, no. 3, p. 229-232.

Parker, M.L., Bunce, H.W.F. and Smith, J.H.G., 1974, The use of X-ray densitometry to measure the effects of air pollution on tree growth near Kitimat, British Columbia: Proc. International Conference on Air Pollution and Forestry: Marianské Lazne, Czechoslovakia, 15 pp.

Parker, M.L., Hunt, K., Warren, W.G. and Kennedy, R.W., 1976, Effect of Thinning and Fertilization on Intra-Ring Characteristics and Kraft Pulp Yield of Douglas-Fir, Applied Polymer Symposium, no. 28, 1075-1086.

Parker, M.L., Bruce, R.D. and Jozsa, L.A., 1977, Calibration, Data Acquisition and Processing Procedures used with an Online Tree-Ring Scanning Densitometer, Presented at IUFRO Group P4.01.05, Instruments Meeting, Corvallis, Oregon, Sept.8-9, 1977, 20 pp.

____________, 1980, X-Ray Densitometry of Wood at the WFPL Technical Report No. 10, Forintek Canada Corp., 18 pp.

Parker, M.L., Jozsa, L.A., Johnson, S.G. and Bramhall, P.A., 1981, Dendrochronological studies on the coasts of James Bay and Hudson Bay, Climatic Change in Canada -2. Syllogeus, No. 33, p. 129-188.

Parker, M.L., Bramhall, P.A. and Johnson, S.G., 1982, Tree-Ring Dating of Driftwood from Raised Beaches on the Hudson Bay Coast, Contract report for the National Museums of Canada (KL229-1-4102) M.L. Parker Co., Inc. and Forintek Canada Corp., Manuscript submitted for publication in Climatic Change in Canada-3, Syllogeus, 71 pp.

Polge, H., 1963, L'analyse densitométrique de clichés radiographiques: une nouvelle méthode de détermination de la texture du bois: Ann. Ec. Natl. Eaux Forêts, St. Rech. Expér., v. 20, no. 4, p. 530-581.

__________, 1965, Study of wood density variations by densitometric analysis of X-ray negatives of samples taken with a Pressler auger: Proc. IUFRO, Sect. 41, Melbourne, Aust., 19 pp.

__________, 1966, Etablissement des courbes de variation de la densité du bois par exploration densitométrique de radiographies d'échantillons préléves à la tarière sur des arbres vivants: Ann. Sci. For., Nancy, v. 23, no. 1, p. 1-206.

Studhalter, R.A., 1956, Early history of crossdating: Tree-Ring Bull., v. 21, no. 1-4, p. 31-35.

Stokes, M.A. and Smiley, T.L., 1968, An Introduction to Tree-Ring Dating: Chicago, University of Chicago Press, 73 p.

Stryd, A.H., 1979, The Lillooet archaeological project laboratory analysis program (Part 2): A statement to the Social Sciences and Humanities Research Council for Canada for research grant #S75-1241. Cariboo College, Kamloops, B.C., p. 21-24.

DEVELOPMENT AND APPLICATION OF A LICHENOMETRIC DATING CURVE, BROOKS RANGE, ALASKA

PARKER E. CALKIN and JAMES M. ELLIS

ABSTRACT

Lichenometry is a dating technique that uses lichens to obtain estimates of relative or absolute ages of rock-bearing substrates. It is based on the general assumptions that: a) stabilization and succeeding colonization of lichen-free rocks occur shortly after deposition and b) subsequent lichen growth occurs with a predictable pattern as a function of time. Species often reach ages of several thousand years, perhaps even to ~9000 yr BP.

Searches of bouldery Holocene moraines fronting small cirque glaciers in the arctic and alpine terrain of the central Brooks Range reveal patterns of selected lichen species consistent on the basis of maximum thallus diameters. These patterns show close parallelism to ridges marking successive ice-marginal positions. They have been mapped along with lichen trimline and density data as isophyses or graphed as frequency distributions to yield relative ages and modes of movement for over 50 glaciers.

Absolute ages have been derived through development of a lichen growth curve based largely on the commonly used species *Rhizocarpon geographicum s.l.*, and the fast-growing *Alectoria minuscula/pubescens*. The *R. geographicum* curve has a "great (rapid) period" of growth ~200 years. This part of the curve is controlled indirectly by historic and dendrochronologic data as well as by direct measurements of the *Alectoria* species which grows seven times faster. The subsequent linear (slower) growth phase of *R. geographicum* approximates 3 mm per century and is calibrated to 1300 yr BP by radiocarbon dates. This *lichen factor* of "3" is comparable to those from southern Alaska, Baffin Island, Swedish Lapland, and the Colorado Front Range. A colonization period of ~30 years following substrate stabilization is built into the curve.

We have computed surface ice velocity by measuring lichens along 1200 m of a supraglacial boulder train. Lichenometric maps of moraines have been used to reconstruct former glaciers. Lichenometry and radiocarbon analysis of vegetation emerging from beneath a retreating ice margin has provided a chronology of glacial advance and retreat spanning the last 2500 years. A regional synthesis of data suggests a complex glacial chronology involving seven periods of expansion during the past 4500 yr BP. In addition to glacial studies, we have made considerable use of lichens to indicate substrate stability and to estimate age of rock glaciers as well as other periglacial landforms.

INTRODUCTION

Purpose

The intent of this paper is to review the basic concepts and procedures of the technique of lichenometry in the context of their practical application to dating Holocene glacial features of the central

Brooks Range. While we cannot cover all of the aspects of lichenometry that may prove useful in individual cases, we hope to provide an introduction for those unfamiliar with lichenometry and/or for those who have yet to apply the technique. For a more comprehensive, single text coverage of lichenometry, we refer readers to Locke *et al.* (1979) or to a series of articles in Arctic and Alpine Research, 5(4), 1973.

Need in the Brooks Range

Lichens have world-wide distribution (Brown *et al.*, 1976) but they are particularly useful as a Holocene dating tool in polar and alpine regions. Here materials for other methods such as radiocarbon or tree ring dating are sparse, historical records are short or fragmentary, and long-living species particularly useful for dating are widespread on rock surfaces. Such is the case along the Continental Divide of the east-west trending Brooks Range (Figure 1).

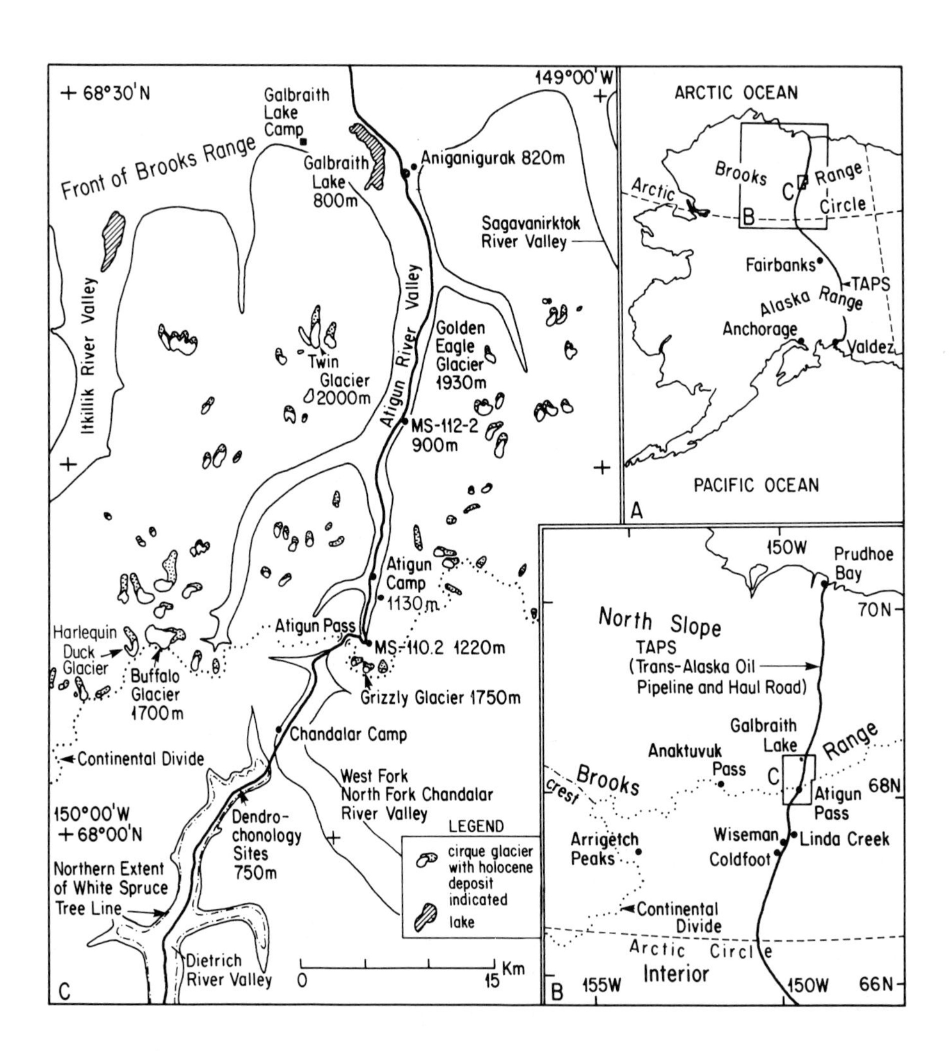

Figure 1 Location map showing the Brooks Range and Atigun Pass field area.

Our initial studies of Holocene glaciation undertaken in 1977 in the Atigun Pass area of the east-central Brooks Range (Figure 1; Ellis and Calkin, 1979) revealed that high bouldery moraines of the small cirque glaciers bore a lichen cover with selected species surprisingly consistent in maximum thallus diameters along individual ridges. In addition, these lichens were markedly smaller and more consistently sized than those on boulders of the older, Pleistocene, drift sheets extending beyond downvalley. Therefore, we used lichenometry to date these deposits and a variety of associated forms (Calkin and Ellis,

1980). We have extended our applications from the Atigun pass area into the west-central Brooks Range (Ellis *et al.*, 1981) and in 1981 eastward 300 km to the high glacierized areas of the eastern Brooks Range.

THE STUDY AREA

Geologic Setting

The studies referred to in this paper were undertaken principally between elevations of 1000 m and 2000 m flanking the Continental Divide as it rises gently northeastward from the spectacular, granite Arrigetch Peaks in the west, through the sedimentary terrains of Anaktuvuk Pass and Atigun Pass. Repeated glaciation since at least early Pleistocene time (Hamilton and Porter, 1975) has left a rugged alpine topography with sharp peaks and col divides that reach 1000 to 1500 m above the floors of intervening U-shaped valleys. Scattered north-oriented cirque glaciers less than two kilometers long, remain below peaks north of the Continental Divide.

Climate

Lichenometry was most successful above and north of the spruce tree line within the zone of continuous permafrost (Ferrians, 1965). Mean temperatures rise above freezing in May through September (Haugen, 1979). In Atigun Pass, the main area of lichen study, the mean annual temperature at 1450 m altitude is about -14°C. Annual precipitation at Atigun Pass ranges between 400 and 700 mm of which about 50 percent is snow (Calkin and Ellis, 1980, Figure 2). Areas near or below the cirque glaciers are generally snow-free from late June through August. Precipitation may be of similar magnitude to the southwest in the Arrigetch Peaks although winter snowfall may be heavier there (Ellis *et al.*, 1981; Hamilton, 1981a).

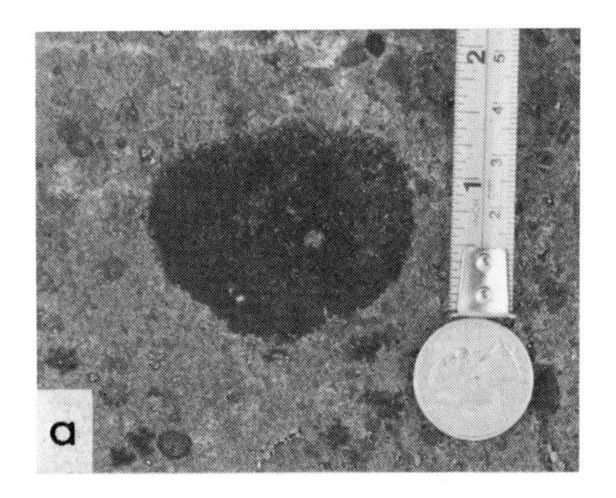

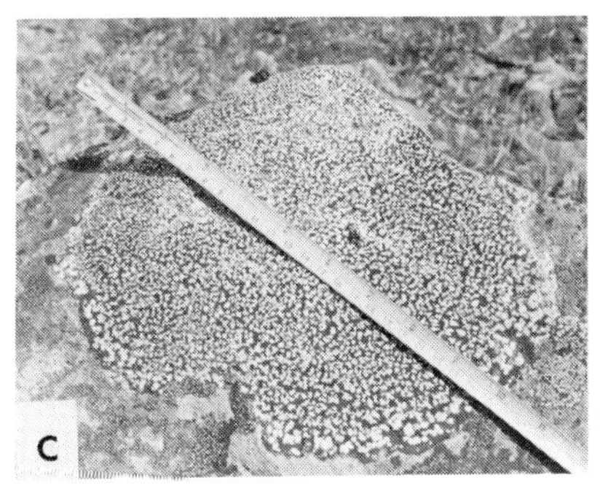

Figure 2 Photographs of major lichens used in the central Brooks Range study, including a) *Alectoria minuscula*; b) *Rhizocarpon geographicum* s.l.; and c) *Rhizocarpon eupetraeoides/inarense*. See Calkin and Ellis (1980, Figure 3) for voucher specimens.

COLLECTION AND ORGANIZATION OF LICHEN DATA

Development and Basic Assumptions of Lichenometry

The basic development of lichenometry through its acceptance as a valid, relative and absolute dating tool may be ascribed to Roland Beschel (1950, 1961), although the technique was reported as early as 1933 (see Locke *et al.*, 1979). Its basic concepts, limitations, and its application to glacial and climatic studies are particularly well presented in specialized studies by Andrews and Webber (1964), Benedict (1967, 1968), Orwin (1970), Worsley (1973), Jochimsen (1973), and Karlén (1973). In Alaska, our investigations in the Brooks Range (Calkin and Ellis, 1980) were preceded by lichenometric work in the Alaska Range by Reger and Péwé (1969) and that in the St. Elias and Wrangell Mountains farther south by Denton and Karlén (1973).

Lichenometry may be defined as a dating method that used lichens to indicate relative and absolute ages of rock-bearing substrates. It is based on assumptions that: a) rock debris is free of lichens when deposited; b) colonization occurs shortly after surface stabilization; c) subsequent lichen growth (as expressed by increase of weight, or most commonly diameter or axes) occurs with a predictable pattern; and d) within an area of similar climate, growth is a function of time passed since colonization (Beschel, 1961; Karlén 1973). Therefore, the diameter of the largest lichen should be proportional to the age of the substrate in question. The latter presupposes that the lifespan of the species measured is greater than the age of the substrate surface.

Beschel (1957) noted that the actual growth of lichens proceeds in small steps. Annual increases in thallus size of slow-growing species are subject to fluctuations which are smoothed out over periods of several years. Furthermore, the frequency distribution of lichen size classes on any one surface shows a regular relationship; recent work suggests this may be log normal (Anderson and Sollid, 1971).

Lichen Species and Problems of Identification

Lichens consist of algae and fungi growing in close association. Those species used in lichenometry have a plant body (thallus) that attaches itself to rock substrates and tends toward a near-circular growth pattern that can be readily measured. At least 35 species or subspecies have been used to date (Locke *et al.*, 1979). Details of morphology and help in field identification of these may be obtained from various manuals (Hale, 1969). However, for the average field worker it is desirable to follow preliminary field identification by collecting voucher specimens. These can be verified under laboratory conditions by a lichenologist.

We have measured six lichen species in our studies in the Brooks Range (see Calkin and Ellis, 1980 for amplification). All are widespread in this area on siliceous clastic, crystalline, and tough metasediment substrates. The first four, as identified below, were among species studied on Baffin Island since 1950 (Andrews and Webber, 1964; 1969; Miller and Andrews, 1972; G.H. Miller, 1973; Andrews and Barnett, 1979).

Alectoria minuscula (Figure 2a) is a black fibrous lichen used in conjunction with the similar fruticose *Alectoria pubescens*. Specimens reach diameters greater than 140 mm; these often are deteriorated in the center, perhaps indicating the onset of senescence (end of life). *Umbilicaria proboscidea* is a dark gray to black foliose lichen with delicate leaves radiating from a central structure which is attached to the substrate. These three species are fast growing, have similar growth rates, and are common on fresh debris surfaces.

We have used three yellow-green species of the crustose genera *Rhizocarpon,* one of the most abundantly represented lichen genera in the Arctic (Thomson, 1967, 1979). These are all slow-growing species and include *R. geographicum* (Figure 2b), the species most commonly reported in lichenometric studies. This is a rather variable species which is described by many different names and even subdivided as a group (Thomson, 1967, 1979). Difficulty in field identification of these variants has led us and other field workers to use the species name in the general sense as *Rhizocarpon geographicum sensu lato (s.l.)*. *R. superficiale* is very similar in appearance and may have been included in some of our measurements (Calkin and Ellis, 1980, Figure 3c).

Rhizocarpon eupetraeoides (Nyl.) Blomberg and Forssell and *R. inarense* (Vainio) *vainio* are particularly noticeable on late Pleistocene and early Holocene substrates in the central Brooks Range. They may only be clearly distinguished from each other by a chemical iodine test on the medulla (interwoven fungal thread between the algal layer) (Thomson, 1967, p. 441). Both belong to the *Rhizocarpon* group Alpicola (Runem.) which include *R. alpicola* found in Europe and Asia and used extensively in lichenometric studies in Scandinavia by Denton and Karlén (1973, Figure 7). *R. eupetraeoides/inarense* (Figure 2c) displays more vivid yellow-green aureoles (tiny button-like reproductive parts at the thallus surface) than *R. geographicum* (Figure 2b).

There are difficulties and frustrations in the field identification of the various species of *Rhizocarpon*. For example, we found them difficult to distinguish when less than 5 or 6 mm in diameter. In addition, *R. geographicum* lichens were difficult to distinguish, if present at all, on surfaces where thalli resembling *R. eupetraeoides/inarense* in the 20 to 40 mm size range were common. Our references to *R. geographicum* may generally be applied to *R. eupetraeoides/inarense* for thallus diameters to 150 mm. Above 150 mm, *R. geographicum* thalli were often found to coalesce or disintegrate, while the *R. eupetraeoides/inarense* thalli were more circular and distinct. We were able to measure distinct *R. geographicum* thalli to 250 mm and *R. eupetraeoides/inarense* to 450 mm. These maxima are very similar respectively to those measured for *R. geographicum* on Baffin Island (Andrews and Barnett, 1979) and for *R. alpicola* in Swedish Lapland (Denton and Karlén, 1973) where they are believed to reach ages on the order of 8,000 to 9,000 years.

Measurement and Sampling

General

For the purposes of this study, we followed the experience of Webber (Andrews and Webber, 1964; personal communication, 1977), Benedict (1968) and Karlén (1973) who adapted the fundamental assumption of Beschel (1950). That is, only the maximum diameter (longest axis) of the largest lichen thallus is an indicator of substrate age, because it is assumed to be the oldest and to represent the optimum growth rate for the site studies. Some variants of this procedure are common in the literature. For example, Locke *et al.* (1979) advocate measurement of the shortest axis (largest inscribed circle) of the largest lichens. Some workers have used dry weight, particularly when measurement was more difficult in the case of some foliose or fruticose species. The measurement of lichen area is tedious, but this additional technique is particularly useful when recording increases in thallus size during direct measurements of lichen growth rate.

Many studies now average a group of the largest lichens, for example, the largest five (Karlén, 1979). Size-frequency measurements (Benedict, 1967; Lindsay, 1973) take considerable time but are desirable when surface age is a goal and help to reveal "the effects of climatic change, substrate disturbance, lichen senility, and competition" (Locke *et al.*, 1979, p. 14) as well as to exclude anomalous-sized thalli. We excluded anomalous thalli by discounting those where maximum diameters were about 20 percent or more above the next largest diameter or cluster of diameters. Only lichens with distinct thallus margins were used, and of these, only those slightly elliptical to circular-shaped. We recorded the minimum as well as maximum diameters to the nearest 1 mm along with the lichen species, orientation, and its substrate inclination and lithology.

During our lichenometric mapping we apportioned our time to cover the area as thoroughly as possible, looking at boulders and cobbles for the largest measurable lichen. Measurements of the 10 to 15 largest lichens were recorded in small areas of 5 to 10 m^2 or continuously along linear landforms. Valid results for dating of the Neoglacial moraines usually required mapping areas in excess of 500 m^2 Karlén (1973).

Fixed area searches, short selective traverses across the terrain, or time limits on sampling (Birkeland, 1973; Rampton, 1970) have been used by other workers to provide data for statistical analysis.

Some Environmental Considerations

Environmental factors play a part in sampling techniques and their relative importance is variable from region to region. The availability of sunlight, moisture, and stability of the substratum were critical factors in the central Alaska Range, 500 km to the south of the Atigun Pass (Reger and Péwé, 1969). Rock surfaces bearing the largest lichen thalli in this area display a dominant southwest orientation; however, in the Front Range of Colorado (Birkeland, 1973) or the Swiss Alps (Haeberli *et al.*, 1979) *Rhizocarpon geographicum* does not occur on south-facing surfaces. We found no strong relationships between lichen growth and orientation or height above the moist ground surface in our Brooks Range studies. On moister slopes and over local areas of abundant moisture, such as the Arrigetch Peaks, red-colored *Trentepholia iolithus* algae were ubiquitous on new rock surfaces. This appeared to inhibit lichen cover as it precedes lichen colonization (D. Cooper, personal communication, 1980).

Substrate stability severely controls lichen growth and creep or tumbling of bouldery rubble is a persistent problem in the Brooks Range as in other alpine areas and is commonly related to melting of subsurface ice masses; this in turn, is more prevalent in the topographically-shaded areas such as the steep granite-walled cirques of the Arrigetch Peaks (Ellis *et al.*, 1981) than in more open areas. While it was impossible to completely avoid unstable deposits in such areas, we were careful to make lichenometric comparisons between areas or forms of maximum stability.

Moraine ridge crests bore the most consistent lichen diameters on stable surfaces (Ellis and Calkin, 1979). On unstable substrates, the lower areas yielded the more reliable lichenometric evidence. However, at the same time, special efforts were taken to sample well-drained topographic areas where snow did not linger. Several recent workers (Billings, 1974; Flock, 1978; Haeberli *et al.*, 1979; Koerner, 1980) have emphasized that lichen growth is strongly inhibited by longer lasting, but still annual snow fields. These often produce sharp trimlines.

Variations in substrate lithology affect lichen growth but details of this control are not clearly understood. Therefore we and others have attempted to minimize these unknowns by measuring lichens on rocks of similar composition. Lithology is a first-order control of topography and because our studies were principally concerned with the high resistant uplands, we were able to work with quite uniform and resistant rock substrates. These consisted of resistant siliceous conglomerates, and sandstones or quartzites throughout most of the central Brooks Range and uniform granitic rocks in the Arrigetch Peaks. Our measurements show no clear relationship of growth with silica content. The major limitation of lichen growth with these rocks is their resistance to weathering. For example, *R. geographicum* only rarely reached diameters of greater than 25 mm on easily-weathered shales and phyllites.

The species used in this study do not grow on limestones and this must be taken into account when planning lichenometric studies (Osborne and Taylor, 1975). Below outcrops of limestone we searched for the siliceous exotic stones or measured lichen thalli on chert nodules which occurred within the limestones.

Interspecific Ratios

We measured the ratios of thallus diameters for a range of sizes among four of the lichen species utilized in this study. Such ratios are invaluable for transfer of potential diameters and age relationships from one species to another. In our study it has allowed comparison to work on Baffin Island (Andrews and Webber, 1964; 1969; Miller and Andrews, 1972; G.H. Miller, 1973). Furthermore, a field concept of the size ratios between the species provided a handy tool to check for anomalous thallus diameter measurements. Only the maximum diameter lichens characteristic of the age of a landform were used in our ratio determinations.

The size ratio of *A. minuscula* to *A. pubescens* was 1.04 and *A. minuscula* was 1.06; therefore, we used these species interchangeably in lichenometric dating. The ratio of *A. minuscula/pubescens* to *R. geographicum* was in the range of 6.5 to 7.0 (Figure 3; Calkin and Ellis, 1980, Table 1). The ratios are similar to those attained on Baffin

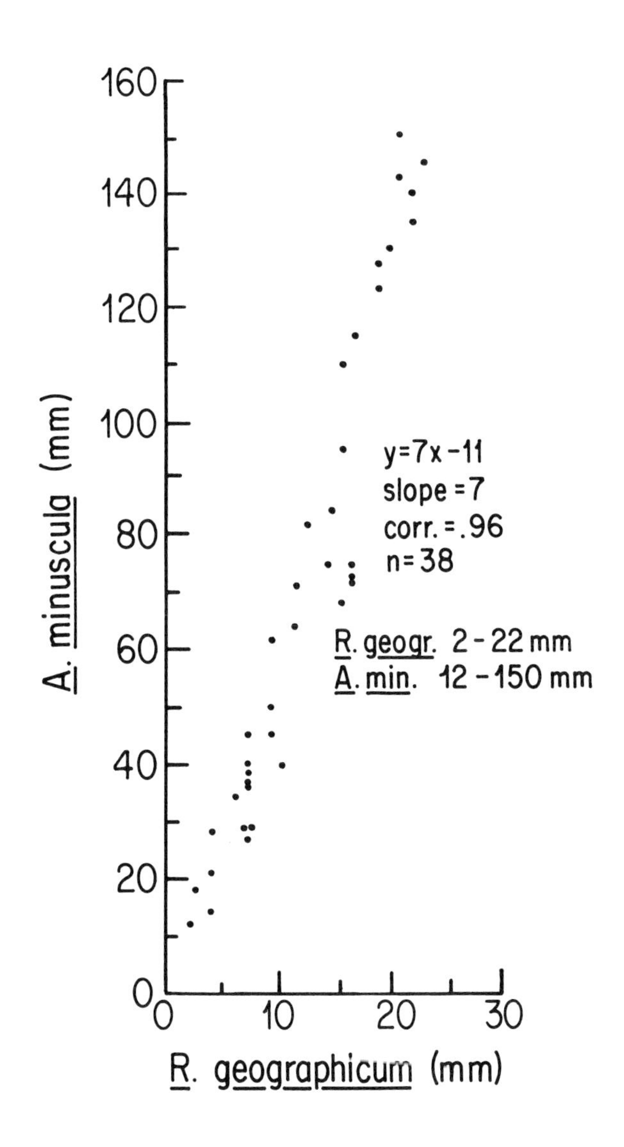

Figure 3 Sample of growth rate relationship between *Alectoria minuscula* and *Rhizocarpon geographicum s.l.* Taken from measurements at five glaciers near Atigun Pass.

Island (Locke *et al.*, 1979, Table 2). However, data of Andrews and Webber (1969), Miller and Andrews (1972) and G.H. Miller (1973, Figure 2) who used direct measuring methods have shown that the interspecific ratio of *A. minuscula* to *R. geographicum* changes continuously with time. A frequency curve of such ratios displays a broad, although symmetrical pattern (see Locke *et al.*, 1979, Figure 3).

Table 1. Sample of direct growth measurement data for *Alectoria minuscula* and *Alectoria pubescens* extrapolated to *Rhizocarpon geographicum*.

A. minuscula and *A. pubescens*		*R. geographicum*	
Maximum diameters (mm) 1977	Planimetrically-determined diameter change (mm/yr)[a] 1977 - 1979	Extrapolated growth rate (mm/100 yr)[b]	For effective diameter range (mm)
30.0	+1.8		
31.5	+1.7	25	5.0 - 5.5
35.0	+1.2		
35.5	+1.9		
42.5	+1.6	24	6.0 - 6.7
47.3	+1.8		

[a]Diameter = 2.0 $(\text{area}/\pi)^{1/2}$ from tracings made in early August of 1977 and 1979.

[b]An interspecific ratio of 7:1 for *A. minuscula/pubescens:R. geographicum* is assumed (see Calkin and Ellis, 1980, Tables 1 and 2).

One problem in making our measurements was the potential difference in colonization period between species. Our studies suggest that *R. geographicum* and *A. minuscula* can colonize or become visible to the unaided eye on surfaces contemporaneously along with *A. pubescens* and *Umbilicaria proboscidea*; this is in contrast to results of work by Andrews and Webber (1964), Hale (1967), and G.H. Miller (1973).

RELATIVE DATING FROM LICHEN MEASUREMENTS

Lichens may be used to provide relative ages either by direct comparison of thallus diameters or by percentage of lichen cover. Our measurements of the percentage of lichen cover in the central Brooks Range have so far yielded only very crude relative age data. Some examples of this use are given in Figures 4 and 5. The variation of lichen-cover percentage is not only time dependent but is affected in alpine glaciated areas such as the Brooks Range by a varying degree of restriction of each of the various fast- or slow-growing species to certain favorable sites. These in turn, vary irregularly in space and time because of topography, snow cover, boulder size, and instability derived from ice-cores. Frequently the degree of cover can be 2nd or 3rd degree recycling. Therefore, age alone, was difficult to derive in the short mapping period available for each deposit. Given the field time, these problems might be approached in the manner of Haeberli *et al.* (1979). They applied lichen cover studies to an active rock glacier after first mapping the ecological variables controlling lichen distribution. Benedict (1967, 1968) as well as Birkeland (1973) have been more successful in using the percentage of total lichen cover in Colorado where instability and possibly ecological changes are less of a problem. Birkeland (1973; Burke and Birkeland, 1979) used an estimate of the percentage of lichen cover on 50 boulders in the Mt. Sopris area as a mapping technique, but found that the mean cover for these had a linear relationship with the maximum cover on one boulder, a much more rapid observation to make.

Figure 4 Photographs showing typical substrate lichen cover at Harlequin Duck Glacier with a) Early Holocene-aged rock glacier section near terminus; and b) upper portion of debris lobe in transition zone where moraines are ~400 lichen years old. Note that while the overall lichen cover is nearly similar in the two areas, the percent of crustose (lighter tone) lichen is greater on the older surface and weathering and soil development much further advanced. See Figure 5 for locations and further detail of lichen cover.

GENERATION OF A LICHEN GROWTH CURVE

Introduction

In order to assign absolute ages to surfaces on which lichen thalli sizes are recorded, it is necessary to construct growth curves for the major lichen species used in the region. This is not always a simple task because the environmental factors which influence growth rate of a species may have changed somewhat through the lichen age. Even under constant environmental conditions, the growth rate varies over the lichen lifespan (Benedict, 1967). Initial growth of the species used in this and most similar studies is rapid. For a contrasting trend see Armstrong (1976). This "great period" of growth (Beschel, 1950) lasts from 40 to 500 yr during which rates decelerate until a generally linear, long-term period of growth is attained (Figure 6). In the case of *Rhizocarpon geographicum* or related species, this may extend over almost 9,000 yrs. (Denton and Karlén, 1973; Andrews and Barnett, 1979).

Lichen factors (growth rates expressed in mm/100 yr) compiled from 29 *R. geographicum* curves by Webber and Andrews (1973; Locke *et al.*, 1979) show that environmental factors of precipitation, growing season, and high temperature vary directly with lichen growth rate during the great growth period (Porter, 1981). Beschel (1961) and Ten Brink (1973) made this particularly clear by direct measurements over a 12-year period in west Greenland when growth rates were "highly sensitive to

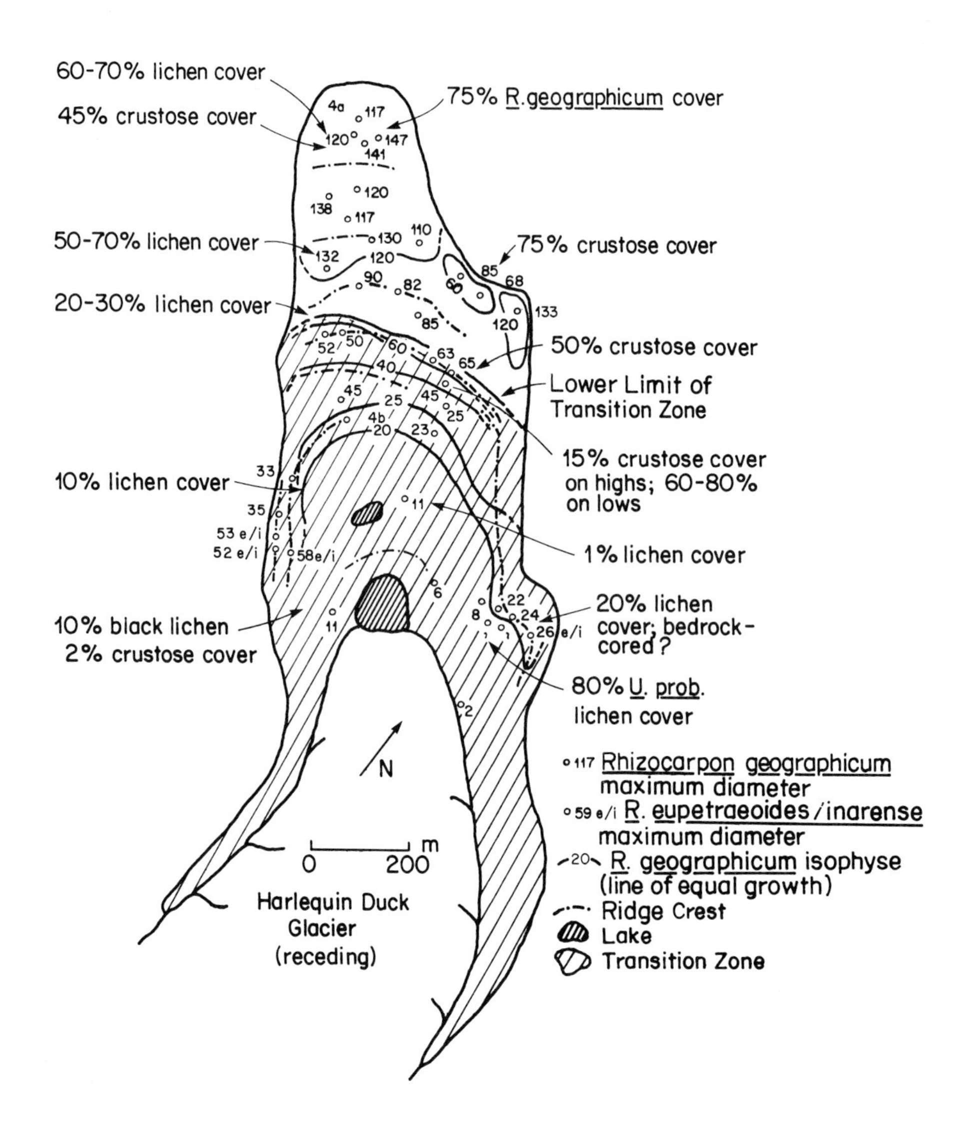

Figure 5 Lichenometric map of glacier-cored deposits of Harlequin Duck Glacier (Figure 1). In the near-glacier area of stable, Neoglacial moraines (called the "transition zone" and shown by diagonal pattern) isophyses connect points of equal lichen diameter. The predominantly, gravity - mobilized rock deposits downslope of the transition zone, show irregular lichen ages. Crustose lichen which usually includes *R. geographicum*, is distinguished from general lichen cover which may include many fast-growing fruticose or foliose taxa. Locations of Figure 4a and 4b are shown.

local climate and inversely proportional to continentality" (Ten Brink, 1973, p. 323). Similarly, G.H. Miller (1973) found elevationally-related climatic control important to growth on eastern Baffin Island.

Climatic differences may have less effect on the long-term (linear?) growth rate of *R. geographicum* although tightening of the absolute control for this part of the curve may eventually prove otherwise. Curves derived from Baffin Island (Miller and Andrews, 1972), the Colorado Front Range (Benedict, 1967), southeastern Alaska (Denton and Karlén, 1973) and the Brooks Range (Calkin and Ellis, 1980 and this paper) all yield growth rates (lichen factors) of approximately 3mm/100 yr for intervals of over 1000 yr (Figure 6).

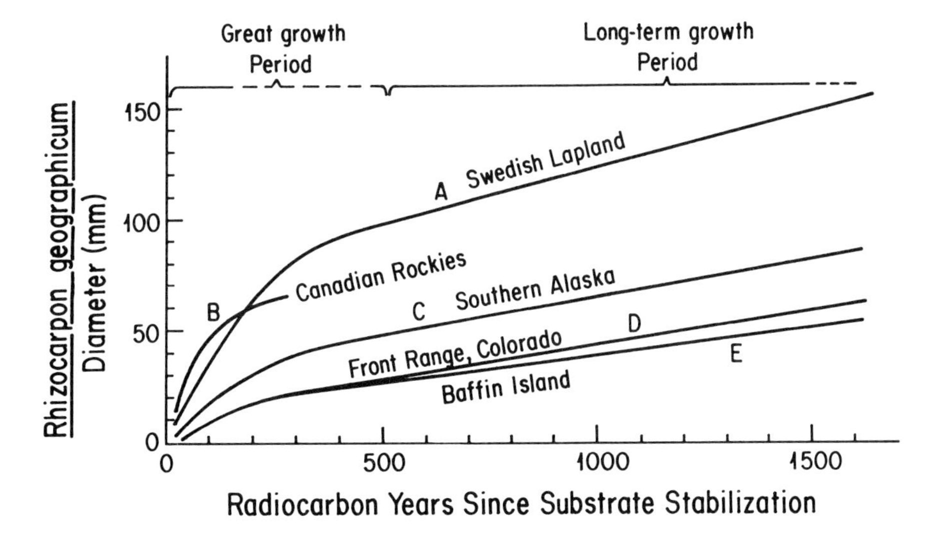

Figure 6 Growth curves for *Rhizocarpon geographicum/alpicola* (A. Denton and Karlén, 1973) and *Rhizocarpon geographicum* (B. Luckman, 1977; C. Denton and Karlén, 1973; D. Benedict, 1967; and G. Miller and Andrews, 1972). In general decreasing precipitation, temperature, and length of growing season result in a slower early growth rate. After Calkin and Ellis, 1980.

Direct Measurements of Growth Rate

The growth rates of *Alectoria minuscula* and *A. pubescens* between the diameters of 30 mm and 72 mm have been measured directly in the Atigun Pass area by both direct tracing (Hale, 1959) and a more accurate photographic method of Hooker and Brown, (1977) (Table 1). In the latter method, we photographed the thalli from a perpendicular position using a 35 mm camera with a standard 50 mm lens. A coin served as scale. G.H. Miller (1973) and Locke *et al.* (1979) have made detailed recommendations for measurement that will allow for more uniform future duplication by other workers. Areas in both enlarged tracings and photographic prints were determined by planimeter. These in turn were converted to theoretical diameters ($2\ (\text{area}/\pi)^{\frac{1}{2}}$) assuming circular thalli to determine growth between times of field recordings.

Measurements of lichen thalli were made on bouldery Neoglacial drift associated with Grizzly Glacier (1680 m) and Buffalo Glacier (1615 m) (Figure 1) over a composite period of three growing seasons (1977-79). Records and measurements were made on well-drained, representative sites, and during times when the lichens were dry to minimize swelling or shrinkage (G.H. Miller, 1973). These data (Calkin and Ellis, 1980, Table 1) were used to help construct an *Alectoria minuscula/pubescens* growth curve and then extrapolated to the indirectly-controlled growth curve of the slow-growing *Rhizocarpon geographicum* by using the interspecific ratio of 7:1 (Figure 7). As diameters of *R. geographicum* increased from 5.0 to 10.1 mm. the growth rate decreased from about 25 mm/100 yr to 14 mm/100 yr. The range in growth rates of the two *Alectoria* species ranged between 1 and 1.9 mm/yr in the Atigun Pass area; this is on the same order of magnitude as rates determined by G.H. Miller (1973, Table 2) on Baffin Island.

The variability in radial growth rates between some fast-growing, foliose lichens of equal size may be high, even under standard growth conditions (Armstrong, 1976). However, experience and the literature suggest that environmental factors on various scales form the major limitations of lichen curves generated by direct measurement. The climatic conditions including those of annual snowcover under which lichen growth are measured, may be atypical of those controlling long-term growth. In addition, direct measurements are usually made on a

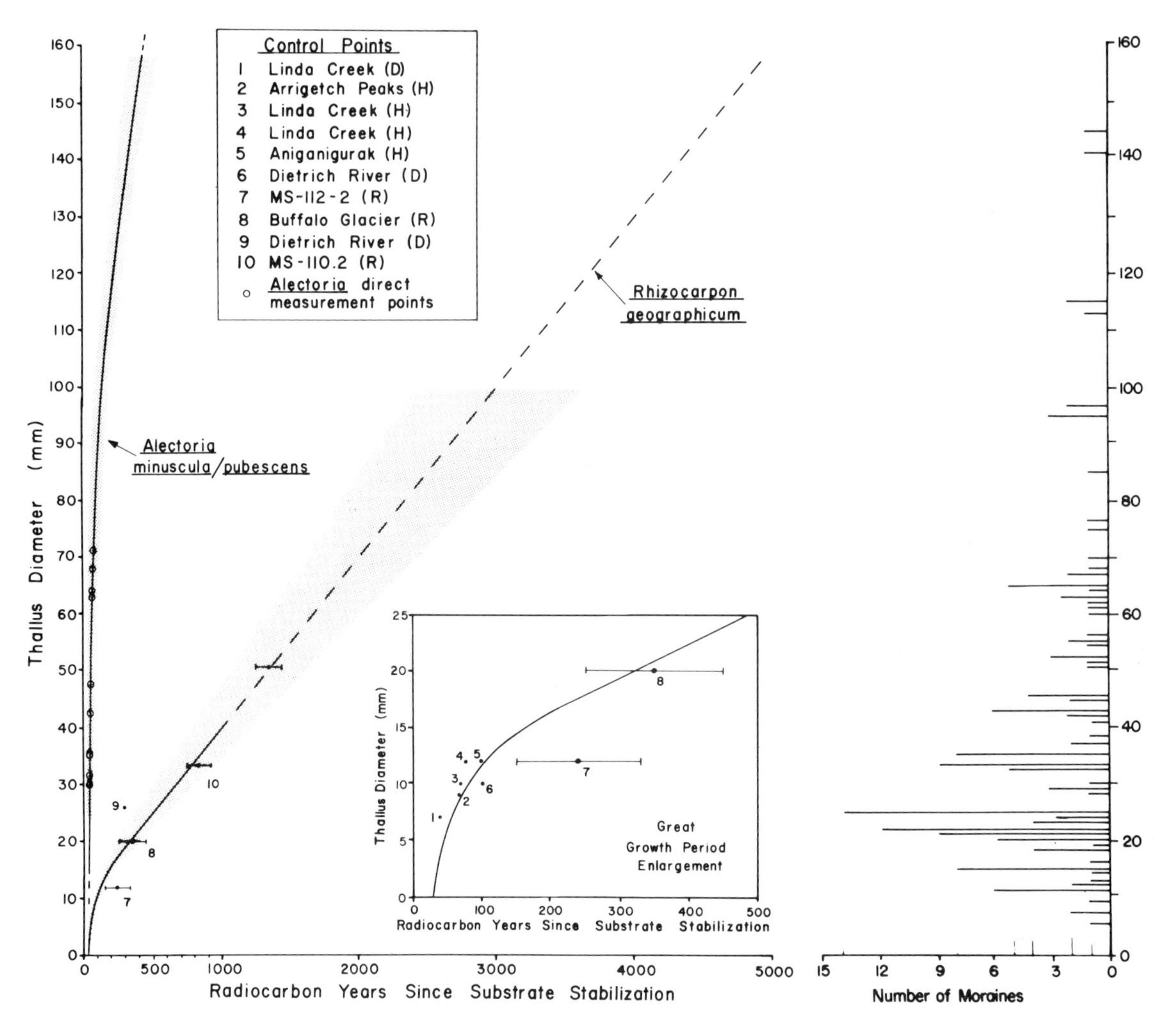

Figure 7 *Rhizocarpon geographicum* and *Alectoria minuscula/pubscens* growth curves for the central Brooks Range. Data on *A. minuscula/pubscens* curve show direct measurements. Control points 1-10 for *R. geographicum* are indicated as dendrochronologic (D), historic (H), or radiocarbon (R) and are located on Figure 1 with elevations. The curve has not been adjusted since placement of Atigun Camp radiocarbon point 11 of 1300 yr/50mm. Shaded zone indicates qualitative ±20 percent age reliability. Curve is not fixed to any given year; to convert ^{14}C given on abscissa to ^{14}C yr BP (before AD 1950), subtract AD 1950 from current AD date and deduct this value from graph's ^{14}C age. Frequency histogram at right shows number of moraines characterized by *R. geographicum* of indicated maximum thallus diameter. The graph suggest 7 to 8 clusters of moraines that may mark withdrawal from advanced glacial positions (glacial advances?).

variety of lichen thalli that have lower average growth rates than that largest lichen used for dating and perhaps a product of a more favorable growth environment. Therefore, the curve yields estimates of the maximum surface ages (Locke *et al.*, 1979).

INDIRECT CONTROL OF GROWTH RATE

Curves from Other Areas

Since climate is a primary control of lichen growth rate, particularly during the "great period" we assumed that a first approximation of a *R. geographicum* curve for the central Brooks Range area should resemble the relatively well-controlled curve of central Baffin Island (Figure 6). There, mean annual temperature is -10°C and annual precipitation is about 350 to 400 mm (Andrews and Webber, 1964).

Radiocarbon Dates

Three radiocarbon dates have been used as control points in construction of both growth curves (Figure 7) and a fourth point (11) has been plotted. These control points (located on Figure 1) are derived from organic materials underlying uniform, stable surfaces bearing boulders with a well developed cover of *R. geographicum*. Representative maximum diameters were determined from this cover.

The most useful and reliable date was 320 ± 100 yr BP (BGS-522) for control point 8 obtained on woody vegetation buried by deposition of an outermost Holocene lateral moraine at Buffalo Glacier (Calkin and Ellis, 1980, Figure 6). This date was the first direct control of Neoglacial moraine deposition obtained in the Brooks Range. Control points 7 (210 ± 90 yr BP (BGS-547) of Alyeska site MS-112.2) and 10 (800 ± 100 yr BP (BGS-547) at MS-110.2) as well as point 11 (1300 ± 100 yr BP (BGS-670) near Atigun Camp) were obtained from peat layers which occurred within sand and gravel alluvium 0.65, .95 and .85 m respectively below present vegetated surfaces. The maximum diameters of lichen characterizing these surfaces (Figure 6) must represent minimum figures. This is because of the time factor for deposition, erosion, or non-deposition which may have intervened between burial of peat and lichen colonization on the overlying substrates.

Dendrochronologic Data

Tree ring counts taken at the white spruce tree line 15 km south and 750 m below Atigun Pass (Figure 1), afford approximate ages of the largest *R. geographicum* lichens on alluvial fans and meltwater channel surfaces (control points 1, 6 and 9). At the gold mining area of Linda Creek (Figure 1) a 39 yr old white spruce sapling growing out of a tailing pile characterized by maximum 7 mm *R. geographicum* provided control point 1.

Historic Controls

The operations of hand-worked gold placers at the Linda Creek gold mining center south and 1000 m below Atigun Pass are very well documented (Maddren, 1913; Cobb and Kachadoorian, 1961; and Cobb, 1976). Therefore, the abandonment of tailing piles or gravel clearings may be dated to the year as in control points 3 and 4 (Calkin and Ellis, 1980, Figure 5). Lichen dating at this site is subject only to the unknown period of colonization. Since the summer's growing season is warmer here than in the main area of application (Haugen, 1979), the diameters of *R. geographicum* measured in this mining area are considered maximum values for the lichen growth curve.

Rifle cartridges lying about the abandoned Nunamiut Eskimo village of Aniganigurak (control point 5; Figure 1) indicate occupation no earlier than 1873 and abandonment between 1880 and 1890 (Corbin, 1971). A very limited population of measurable *R. geographicum* on scattered sandstone erratics showed evidence of being overturned during this occupation.

Sequential photographs were taken on the retreating margins of selected cirque glaciers in the Arrigetch Peaks in 1911, 1962 (Hamilton, 1965), and in 1979 (Ellis *et al.*, 1981). These have provided minimum periods of colonization and growth rates. No visible lichen colonization of *R. geographicum* was found in an ice-marginal zone exposed during the 17 yr since 1962. However, there was a general down-valley increase of *R. geographicum* thalli diameters to about 9 mm within the zone exposed by retreat in the total 51 yr between 1911 and 1962. Boulders deposited directly on bedrock and apparently at about the time of the 1911 photographic survey, displayed *R. geographicum* up to 11 mm, *R. eupetraeoides/inarense* to 10 mm and *Alectoria pubescens* to 62 mm.

Construction of the Lichen Curve

We have used direct measurement extrapolated from *A. minuscula/ pubsecens,* the lichen curves from areas of similar climate, radiocarbon dates, tree ring counts, ages derived from archeological sites, historic records of mining areas, and sequential photographs to construct a *Rhizocarpon geographicum* growth curve for the central Brooks Range. All dates from these control points are adjusted to AD 1979 and are presented in radiocarbon years BP on Figure 7. Adjustment to radiocarbon years is a common procedure because the critical older sections of growth curves are based on ^{14}C dates which do not convert accurately to calendar years (Locke *et al.*, 1979). The curve is drawn through the various lichen control points so as to incorporate estimates of the time lag separating death or burial of the carbon-dated vegetation, the stabilization of overlying surface and lichen colonization.

Period of Colonization

Evidence from direct measurement cited in the literature suggest that the time lag between substrate stabilization and initial lichen colonization for the species used in this study range between 10 and 50 yr under similar climatic regimes (Andrews and Webber, 1964; Benedict, 1968; Rampton, 1970; Miller and Andrews, 1972; Beschel and Weidick, 1973; Karlén, 1973). In the Arrigetch Peaks area, *Alectoria minuscula* and *A. pubescens* as well as *R. geographicum* appeared between 17 and 68 yr following deglaciation (see Historic Control). One hundred kilometers northwest of Arrigetch Peaks, W.P. Brosge (personal communication, 1980) photographed *R. geographicum* lichens to 2 mm diameter on a granitic rock surface scraped clean by him 29 years before. At the receding Chamberlin Glacier in the northeastern Brooks Range comparison of a 1958/59 photograph (Holmes and Lewis, 1961) and the 1981 ice-marginal position demonstrated no lichen colonization after ~23 years of substrate exposure. Based on these data, we have assumed a colonization period of 30 yr which is incorporated in the growth curves of Figure 7.

Great Growth and Linear Growth Periods

The exact patterns and period of deceleration during the "great period" of growth varies but our curve is similar in shape to others (Figure 6) to the extent that it shows an interval of very rapid growth followed by a tapering off where an exponentially decreasing growth rate is approximated (see Porter, 1981). This extends through the first 200 yr; at this point, the Brooks Range curve, (and others) appears to become linear with growth at approximately 3 mm/100 yr. The *Rhizocarpon geographicum* curve for the Central Brooks Range is linearly extrapolated beyond a 1,300 ± 100 yr BP date to 5000 yr BP based on measurement of *R. geographicum* in other alpine areas of the world (Benedict, 1967; Miller and Andrews, 1972; Denton and Karlén, 1973; Andrews and Barnett, 1979).

There is some evidence from the Brooks Range supporting one point on the curve extension. This is based on a correlation of Holocene construction of moraines and stream terraces that allows maximum *R. geographicum* thallus diameters of 145 mm to be tied to dates of ~4000 yr BP, respectively (Calkin and Ellis, 1980; Hamilton, 1981b).

A qualitative ± 20 percent age reliability is shown by the shaded zone surrounding the growth curve of Figure 7. There is no statistical basis for this; however, it is inserted following the procedure of Miller and Andrews (1972, p. 1137) as "a subjective interpretation of the reliability of the curve at present."

The *Rhizocarpon geographicum* growth curve can be applied with great reservation to thallus diameters of *R. eupetraeoides/inarense* which are less than 150 mm in diameter. The *A. minuscula* growth curve (Figure 7) generated entirely by direct or indirect measurements on this species also represents *A. pubescens* and can be used for dating deposits to 400 yr BP. *Umbilicaria proboscidea* is incorporated in this latter curve but the species is only useful for deposits less than 100 yr old.

APPLICATIONS AND PROBLEMS OF DATING WITH LICHEN

Most of our applications of lichenometry so far in the Brooks Range have been directed toward reconstructing and dating of glaciers or rock glacier positions during the Holocene. Cirque glaciers in the central Brooks Range are presently wasting away with equilibrium lines within 100 m of the glacier heads (Ellis and Calkin, 1979). Lichenometric maps of Holocene moraines have been used to reconstruct former glaciers and their equilibrium-line altitudes. The use of lichen trimlines for these reconstructions is difficult because care has to be taken to avoid confusion of true ice-trimlines with zones containing depleted lichen populations. These latter may only reflect contemporary balance with lingering, annual snow cover (see Haeberli *et al.*, 1979; Koerner, 1980).

Lichens frequently survive supraglacial transport, particularly on large boulders. At several of more than 50 glaciers mapped, we observed supraglacial to ice-cored situations where medial and lateral moraines were populated by taxa representing 10 to 20 lichenometric years of growth. On Twin Glacier, north of Atigun Pass (Figure 1), we recorded the progressive increase of *R. geographicum* thalli diameters on boulders from their colonization below the cirque headwall to a point 1200 m downglacier along an active flowline. From this we derived a mean velocity of glacier flow of 6 m/yr, a figure compatible with measured flow rates elsewhere in the Brooks Range (Wendler *et al.*, 1975).

Some of the lichens of supraglacial debris survive deposition, particularly on shrinking ice masses (Matthews, 1973). This rarely occurs during glacial advances in the Brooks Range.

Preservation of lichen under active glaciers is much less common than this surface transport; however, it has been reported from the Canadian Arctic Islands as well as Greenland (see Koerner, 1980), and now in the central Brooks Range (Calkin and Ellis, 1981a). Conditions favorable for preservation of lichens beneath glaciers are apparently related to the frozen basal conditions of these polar or subpolar glaciers and, of course, the lack of burial by glacial drift. Recession of Golden Eagle Glacier (Figure 1) from a Neoglacial maximum marked by *Rhizocarpon geographicum* thalli of up to 19 mm, is currently exposing an 800 m^2 surface with undisturbed lichen covered boulders surrounded by dead moss and unsorted patterned ground.

R. geographicum up to 72 mm diameter, brightly colored and visually indistinguishable from living forms, occur near the terminus of Golden Eagle Glacier but become progressively bleached and deteriorated 40 m beyond the glacier toe. A sprinkling of boulders which have melted out from supraglacial positions are distinguished by their more angular and relatively lichen-free character. Radiocarbon analysis of the dead emergent moss surrounding the boulders dates a Neoglacial advance across the site at 1120 ± 180 yr BP. Maximum diameters of the preserved lichens indicate that a minimum ice-free interval of 1500 to 2500 yr preceded this glacial expansion.

Consistent patterns of lichen diameters are used to distinguish stable, ice-cemented or ice-cored moraines of open terrain from unstable, glacier-cored deposits (Ellis and Calkin, 1979). Furthermore, they help substantiate the reliability of the lichenometric method in this area. The stable, looped moraines of Holocene age which front the small cirque glaciers of the central Brooks Range, are characterized by: a) a general lack of lichen cover on recently deposited ice marginal ridges; b) lichen thalli that become successively larger in diameter outward from the margin corresponding to successively greater substrate age as based on geometry and independent geologic criteria; and c) a lichen cover bearing a substantial number of thalli close to the maximum measured diameter on individual ridges. Variation of the maximum diameter often falls within a 2 to 3 mm range along ridge crests up to 1 km long.

The results of this sampling have been recorded on maps which display areas segmented on the basis of lichen ages (see also Denton and Karlén, 1973) and isophyses which connect points of equal lichen diameters (Figure 5; Calkin and Ellis, 1980, Figure 8). Data from these maps may be presented as histograms plotting the frequency of individual moraine ridges bearing diameters of mapped lichens at intervals. These in turn may be correlated with the lichen curve as in Figure 7.

We have constructed a Holocene glacial chronology based largely on such lichenometric data together with five ^{14}C dates directly associated with moraines and relative weathering and soil criteria (Ellis and Calkin, 1981). This chronology is far more complicated than formerly envisioned (Detterman *et al.*, 1958; Porter and Denton, 1967) and includes seven major expansions of similar magnitude extending from 4500 yr BP to 350 yr BP (Calkin and Ellis, 1981b).

A less distinct but complimentary chronology is defined by lichenometric analysis of rock glacier deposits in this area (Ellis and Calkin, 1979). Gravity-controlled movement of these deposits disrupts or precludes the continuity of lichen thalli diameters at successive distances from glacier heads (Figure 5). However, maximum lichen diameters near the snouts of ice-cored rock glacier tongues (Figure 5) are indicative of glacial development and surface movement in early Holocene time. "Transition zones" (Foster and Holmes, 1965) of distinct morainal ridges occur between downslope rock glacier tongues and upslope small cirque glacier remnants (Figure 5). These carry a lichen cover limited to the late Holocene (see Figure 4b) but reflect a chronology, similar to that of stable moraine ridges unassociated with rock glaciers. Details of lichenometric studies of rock glacier deposits are discussed by C.D. Miller (1973), Luckman and Osborne (1979), Haeberli *et al.* (1979), and Johnson (1980).

ACKNOWLEDGEMENTS

This study was supported by NSF grants from the Department of Polar Programs through the Research Foundation of the State University of New York. Photographic work was partly funded by Sigma Xi. Botanical identifications were provided by W.A. Webber, P.J. Webber, J.W. Thomson, D. Copper, and R. Zander. We are grateful to T.D. Hamilton and the USGS, Placid Oil Co., and ALASCOM for logistical support. Field assistants were M. Bruen, S. Walti, and T. Lowell.

REFERENCES CITED

Anderson, J.T. and Sollid, J.L., 1971, Glacial chronology and glacial geomorphology in the margin zones of the glaciers, Midtdalsbreen and Nigardsbreen, south Norway: Norsk Geogr. Tidsskr., v. 25, p. 1-38.

Andrews, J.T. and Barnett, D.M., 1979, Holocene (Neoglacial) moraine and proglacial lake chronology, Barnes Ice Cap, Canada: Boreas, v. 8, p. 341-358.

Andrews, J.T. and Webber, P.J., 1964, A lichenometrical study of the northwestern margin of the Barnes Ice Cap: A geomorphological technique: Geogra. Bull., No. 22, p. 80-104.

_______________, 1969, Lichenometry to evaluate changes in glacial mass budgets: as illustrated from north-central Baffin Island, N.W.T.: Arctic and Alpine Research, v. 1, p. 181-194.

Armstrong, R.A., 1976, Studies on the growth rates of lichens, in Brown, D.H., Hawksworth, D.L., and Bailey, R.H., eds., Lichenology: Progress and Problems, The Systematics Assoc., Sp. Vol. 8: London, Academic Press, p. 309-322.

Benedict, J.B., 1967, Recent glacial history of an alpine area in the Colorado Front Range, U.S.A. I. Establishing a lichen-growth curve: Jour. Glaciology, v. 6, p. 817-832.

______________, 1968, Recent glacial history of an alpine area in the Colorado Front Range, U.S.A. II. Dating the glacial deposits: Jour. Glaciology, v. 7, p. 77-87.

Beschel, R.E., 1950, Flechten aus Altersmasstab rezenter Moränen (Lichens as a yardstick of age of late-glacial moraines): Zeitschrift für Gletscherkunde und Glazial Geologic, v. 1, p. 152-161.

______________, 1957, A project to use lichens as indicators of climate and time: Arctic, v. 10, p. 60.

______________, 1961, Dating rock surfaces by lichen growth and its application to glaciology and physiography (lichenometry), in Raasch, G.O., ed., Geology of the Arctic, Vol. II, Toronto, Univ. of Toronto Press, p. 1044-1062.

Beschel, R.E. and Weidick, A., 1973, Geobotanical and geomorphological reconnaissance in West Greenland, 1961: Arctic and Alpine Research, v. 5, p. 311-319.

Billings, W.D., 1974, Arctic and Alpine vegetation: plant adaptations to cold summer climates, in Ives, J.D. and Barry R.G., eds., Arctic and Alpine Environments, London, Methuen, p. 403-444.

Birkeland, P.W., 1973, Use of relative age-dating methods in a stratigraphic study of rock-glacier deposits, Mt. Sopris, Colorado: Arctic and Alpine Research, v. 5, p. 401-416.

Brown, D.H., Hawksworth, D.L. and Bailey, R.H., eds., 1976, Lichenology: Progress and Problems (The Systematics Assoc., Spec. Vol. 7): London, Academic Press, 551 p.

Burke, R.M. and Birkeland, P.W., 1979, Reevaluation of multi-parameter relative dating techniques and their application to the glacial sequence along the eastern escarpment of the Sierra Nevada, California: Quaternary Research, v. 11, p. 21-51.

Calkin, P.E. and Ellis, J.M., 1980, A lichenometric dating curve and its application to Holocene glacier studies in the central Brooks Range, Alaska: Arctic and Alpine Research, v. 12, p. 245-264.

______________, 1981a, A cirque glacier chronology based on emergent lichens and mosses: Jour. Glaciology, v. 27 (in press).

______________, 1981b, Holocene glacial history of the central Brooks Range, Alaska: Geol. Soc. Amer. Abs. with Programs, v. 13(7) (in press).

Cobb, E.H., 1976, Summary of references to mineral occurrences (other than fuel and construction materials) in the Chandalar and Wiseman quadrangles, Alaska: U.S. Geol. Survey Open File Rept., 76-340, 200 p.

Cobb, E.H. and Kachadoorian, R., 1961, Index of metallic and non-metallic mineral deposits of Alaska compiled from published reports of Federal and state agencies through 1959: U.S. Geol. Survey Bull., 1139, 363 p.

Corbin, J.E., 1971, Aniganigurak (S-67): A contact period Nunamiut Eskimo village in the Brooks Range, in Cook, J.P., ed., Final Report of the Archeological Survey and Excavations Along the Alyeska Pipeline Service Company Pipeline Route: Dept. Anthropology, Univ. Alaska, p. 272-296.

Denton, G.H. and Karlén, W., 1973, Lichenometry: Its application to Holocene moraine studies in southern Alaska and Swedish Lapland: Arctic and Alpine Research, v. 5, p. 347-272.

Detterman, R.L., Bowsher, A.L. and Dutro, J.T., Jr., 1958, Glaciation on the Arctic Slope of the Brooks Range, northern Alaska: Arctic, v. 11, p. 43-61.

Ellis, J.M. and Calkin, P.E., 1979, Nature and distribution of glaciers, Neoglacial moraines, and rock glaciers, east-central Brooks Range, Alaska: Arctic and Alpine Research, v. 11, p. 403-420.

________________, 1981, Environments and soils of Holocene moraines and rock glaciers, central Brooks Range, Alaska (abstract), in Evenson, E., ed., INQUA Symposium on the Genesis and Lithology of Quaternary Deposits, August 20-30, 1981, Jackson, Wyoming.

Ellis, J.M., Hamilton, T.D. and Calkin, P.E., 1981, Holocene glaciation of the Arrigetch Peaks, Brooks Range, Alaska: Arctic, v. 34, p. 158-168.

Ferrians, O.J., Jr., 1965, Permafrost map of Alaska: U.S. Geol. Survey Misc. Field Studies Map I-445, Scale 1:2,500,000.

Flock, J.W., 1978, Lichen-Bryophyte distribution along a snow-cover-soil-moisture gradient, Niwot Ridge, Colorado: Arctic and Alpine Research, v. 10, p. 31-47.

Foster, H.L. and Holmes, G.W., 1965, A large transitional rock glacier in the Johnson River area, Alaska Range: U.S. Geol. Survey Prof. Pap., 525-B, p. B112-B116.

Haeberli, W., King, L. and Flotron, W., 1979, Surface movement and lichen-cover studies at the active rock glacier near the Grubengletscher, Wallis, Swiss Alps: Arctic and Alpine Research, v. 11, p. 421-441.

Hale, M.E., 1959, Studies on lichen growth rate and succession: Bull. Torrey Botanical Club, v. 86, p. 126-129.

__________, 1967, The Biology of Lichens: London, Edward Arnold, p. 97.

__________, 1969, How to Know the Lichens: Dubuque, Iowa, Brown, 226 p.

Hamilton, T.D., 1965, Alaskan temperature fluctuation and trends: an analysis of recorded data: Arctic, v. 18, p. 105-117.

______________, 1981a, Multiple moisture sources and the Brooks Range glacial record (abstract), in Tenth Annual Arctic Workshop, March 12, 13, 14, 1981, Boulder, Colorado: Inst. of Arctic and Alpine Research, Univ. Colorado, Boulder, p. 16-18.

______________, 1981b, Episodic Holocene alluviation in the central Brooks Range: chronology, correlations. and climatic implications, in United States Geological Survey in Alaska: Accomplishments during 1979: U.S. Geolog. Survey Circ., 823-B, p. B21-B21.

Hamilton, T.D. and Porter, S.C., 1975, Itkillik glaciation in the Brooks Range, northern Alaska: Quaternary Research, v. 5, p. 471-497.

Haugen, R.K., 1979, Climatic investigations along the Yukon River to Prudhoe Bay haul road, Alaska, 1975-1978. Informal extract from Final Federal Highway Administrative Contract Report Environmental Engineering Investigations along the Yukon River Prudhoe Bay Haul Road, Alaska: Hanover, New Hampshire, Cold Regions Research and Engineering Laboratory, 23 p.

Holmes, G.W. and Lewis, C.R., 1965, Quaternary geology of the Mount Chamberlin area, Brooks Range, Alaska: U.S. Geol. Survey Bull.,

1201-B, 32 p.

Hooker, T.N. and Brown, D.H., 1977, A photographic method for accurately measuring the growth of crustose and foliose saxicolous lichens: Lichenologist, v. 9, p. 65-75.

Jochimson, M., 1973, Does the size of lichen thalli really constitute a valid measure for dating relict glacial deposits?: Arctic and Alpine Research, v. 5, p. 417-424.

Johnson, P.G., 1980, Comments on glacier-rock glacier transition in the southwest Yukon Territory: Arctic and Alpine Research, v. 12, p. 195-204.

Karlén, W., 1973, Holocene glacier and climatic variations, Kebnekaise Mountains, Swedish Lapland: Geografiska Annaler, v. 55A, p. 29-63.

__________, 1979, Glacier variations in the Svartisen area, northern Norway: Geografiska Annaler, v. 61A, p. 11-28.

Koerner, R.M., 1980, The problem of lichen-free zones in arctic Canada: Arctic and Alpine Research, v. 12, p. 87-94.

Lindsay, D.C., 1973, Estimates of lichen growth rates in the maritime antarctic: Arctic and Alpine Research, v. 5, p. 341-346

Locke, W.W., III, Andrews, J.T. and Webber, P.J., 1979, A Manual for Lichenometry (British Geomorphological Research Group, Technical Bulletin No. 26): London, British Geomorphological Research Group, Inst. British Geographers, 47 p.

Luckman, B.H., 1977, Lichenometric dating of Holocene moraines at Mount Edith Cavell, Jasper, Alberta: Canadian Jour. Earth Sci., v. 14, p. 1809-1822.

Luckman, B.H. and Osborn, G.D., 1979, Holocene glacier fluctuations in the middle Canadian Rocky Mountains: Quaternary Research, v. 11, p. 52-77.

Maddren, A.G., 1913, The Koyukuk-Chandalar region, Alaska: U.S. Geol. Survey Bull. 532, 119 p.

Matthews, J.A., 1973, Lichen growth on an active medial moraine, Jotunheiman, Norway: Jour. Glaciology, v. 12, p. 305-313.

Miller, C.D., 1973, Chronology of neoglacial deposits in the northern Sawatch Range, Colorado: Arctic and Alpine Research, v. 5, p. 385-400.

Miller, G.H., 1973, Variations in lichen growth from direct measurements: Preliminary curves for *Alectoria minuscula* from eastern Baffin Island, N.W.T., Canada: Arctic and Alpine Research, v. 5, p. 333-339.

Miller, G.H. and Andrews, J.T., 1972, Quaternary history of northern Cumberland Peninsula, east Baffin Island, N.W.T., Canada. Part IV: Preliminary lichen growth, curve for *Rhizocarpon geographicum*: Geol. Society Amer. Bull., v. 83, p. 1133-1138.

Orwin, J., 1970, Lichen succession on recently deposited rock surfaces: New Zealand Jour. Botany, v. 8, p. 452-477.

Osborn, G. and Taylor, J., 1975, Lichenometry of calcareous substrates in the Canadian Rockies: Quaternary Research, v. 6, p. 111-120.

Porter, S.C., 1981, Lichenometric studies in the Cascade Range of Washington: establishment of *Rhizocarpon geographicum* growth curves at Mount Rainier: Arctic and Alpine Research, v. 13, p. 11-23.

Porter, S.C. and Denton, G.H., 1967, Chronology of Neoglaciation in the North American Cordillera: Amer. Jour. Sci., v. 265, p. 177-210.

Rampton, V., 1970, Neoglacial fluctuations of the Nataxhat and Klutlan glaciers, Yukon Territory, Canada: Canadian Jour. Earth Sci., v. 5, p. 1236-1263.

Reger, R.D. and Péwé, T.L., 1969, Lichenometric dating in the central Alaska Range, in Péwé, T.L., ed., The Periglacial Environment: Montreal, McGill-Queens Univ. Press, p. 223-247.

Ten Brink, N.W., 1973, Lichen growth rates in West Greenland: Arctic and Alpine Research, v. 5, p. 323-331.

Thomson, J.W., 1967, Notes on *Rhizocarpon* in the Arctic: Nova Hedwigia, v. 14, p. 421-481.

______________, 1979, Lichens of the Alaskan arctic slope: Buffalo, Univ. Toronto Press, 314 p.

Webber, P.J. and Andrews, J.T., 1973, Lichenometry: A commentary: Arctic and Alpine Research, v. 5, p. 295-302.

Wendler, G., Benson, C., Fahl, C., Ishikawa, N., Trabant, D. and Weller, G., 1975, Glacio-Meteorological studies of McCall Glacier, in Weller, G. and Bowling, S.A., eds., Climate of the Arctic (24th Alaska Science Conference, 1973). Geophysical Institute, University of Alaska, p. 334 338.

Worsley, P., 1973, An evaluation of the attempt to date the recession of Tundsbergdalsbreen, southern Norway, by lichenometry: Geografiska Annaler, v. 54A, p. 137-141.

LANDFORMS AND LANDSCAPES AS MEASURES OF RELATIVE TIME

DONALD R. COATES

ABSTRACT

W.M. Davis developed the first relative age terrane models with his geographical cycles of youth, maturity, and old age for fluvial landscapes in humid and arid regions. Later workers followed his lead delineating relative age phases for coastal, karst, and periglacial environments. These were descriptive models, but the newer tools of quantitative geomorphology can provide in some cases a numerical base for determining relative age of landscapes....such as drainage density and hypsometric integrals.

Relative age comparisons are only valid when landform and landscape sets have the same geology, hydrology, climate, and topographic framework. Physiographic boundaries cannot be crossed for purposes of contrasting age. Caution is also needed because of such constraints as the principle of equifinality. Since similar-appearing landforms can be created by different processes, a knowledge of landform genesis is crucial. Furthermore landforms that may have been initiated at the same time, may progress differently if magnitude and frequency of acting processes are not similar.

Relative dating rules of cross-cutting features, superposition, and degree of complexity can provide guidelines for developing a chronology of events. Also when appropriately analyzed the geomorphic indicators of landform size, hillslope steepness and shape, and landscape fabric can provide useful age determination.

Each geologic process that influences the earth's surface, whether endogenic or exogenic, has its own characteristic fingerprint that aids in revealing the sequence of processes. Fluvial, glacial, and faulted terranes are used to illustrate these basic principles. In fluvial systems, barbed tributaries, wind gaps, point bars, terraces, alluvial fans, and deltas can provide important information in dating relative sequences of erosion or deposition. Glacial landforms such as preglacial strandlines, umlaufbergs, erosional sluiceways, and moraines may contain special features that assist in diagnosing their relative age. Faulted areas may contain many landforms that indicate relative development, such as degree of spur and facet modification, water table readjustment, and stream channel profile changes. Landform analysis in the Basin and Range province has been successfully used to place relative chronologies on events that range from about 100,000 to 1 million years ago.

INTRODUCTION

Geomorphology is concerned with research on the age of materials and surfaces. Since the late 1960's, however, there has been an even greater emphasis on time because of renewed interest in environmental affairs. Historically a wide variety of dating methods have been used to develop chronologies useful in determining rates and types of changes to landforms and the processes that create them. The new urgency in calculating accurate time frames stems from geologic hazards on human activities as well as human impacts to natural systems. The use of "hindcasting" techniques to predict and forecast future events has become a necessity in land use management, and is a standard function in the United States for preparing environmental impact statements under the National Environmental Policy Act of 1970. Thus it has become important to assess the time factor for determining recurrence intervals of such phenomena as earthquakes, flooding, landsliding, and volcanic activity. Although this paper is centered around relative time, many of the approaches that are discussed here have relevance to the broader issues of geologic time.

A basic tenet of geomorphology has been that if both the exogenic and endogenic forces and conditions remain constant a characteristic system of landforms will develop. Such an idea was first enunciated by Gilbert (1877) and later restated by Hack (1960, 1975) as the concept of dynamic equilibrium. Time was viewed as only incidental and emphasis was placed on the stability of landforms so long as the variables remained constant. An alternate model of landscape development was introduced by Davis in 1899 whereby time was ranked, along with structure and process variables, as being in the trinity of factors that influence terrain character.

William Morris Davis deserves credit for introducing the most popular system for the relative dating of landscapes. He undoubtedly reasoned that if Darwin's model for evolution in the organic world was valid, there must be a similar developmental model of inorganic features for the evolution of topography with the passage of time. Davis (1899) first used the model in the description of changes that progress in a humid climate, and assigned the classes of change in the anthropomorphic images of youth, maturity, and old age (Figure 1). The premise that resulted in this value judgment was that a landmass was rapidly uplifted during an initial phase, and that thereafter base level was static. With such postulates fluvial topography evolved through time from conditions of large expanses of undissected uplands in youth, to a time of maturity when all terrain was in maximum slope. The "geographical cycle" was completed during old age when relief was greatly subdued to a "peneplain" condition. Davis, and also his followers (Cotton, 1947) expanded on the theme to include both climatic and volcanic "accidents" that cause interruptions to the "normal" cycle. For example, base level changes would lead to renewed erosion, so that a polycyclic landscape forms containing elements of both destroyed or partially destroyed earlier surfaces. The relict features were then used to diagnose a relative sequence of events.

The orderly progression of landforms as propounded by Davis seemed to the majority of observers to be so probable and logical that it was widely adopted by most geomorphologists. Davis (1905) extended his ideas to include an evolution of features during an arid cycle, and other workers extrapolated the thesis to cover nearly all geomorphic processes and climates. Beede (1911) and Cvijic (1918) developed the erosion cycle for karst terrane. Johnson (1919) instituted coastal erosion cycles for shorelines that were either undergoing emergence or submergence. Peltier (1950) adopted the theme for periglacial regions. This process approach assigned to each environmental system its own individual assemblage of landforms that gradually changed. The wide spectrum of landforms from diverse terranes and climates occupies niches in a continuum of features that evolve through time. Each terrane is viewed as having unique signals that mark it as having developed during a particular time framework in the total hierarchy of landscapes.

The relative time cycle for landscape development has received several important challenges, and these are summarized by Flemal (1971). For example, Penck (1953, although first written in German in 1924) and King (1953) challenged the Davisian idea of slopes becoming gentler with time and the proposition that denudation is proportional to relief. The Davisian cycles were also attacked by such arguments as (1) the earth is too mobile with too many climate and base level fluctuations

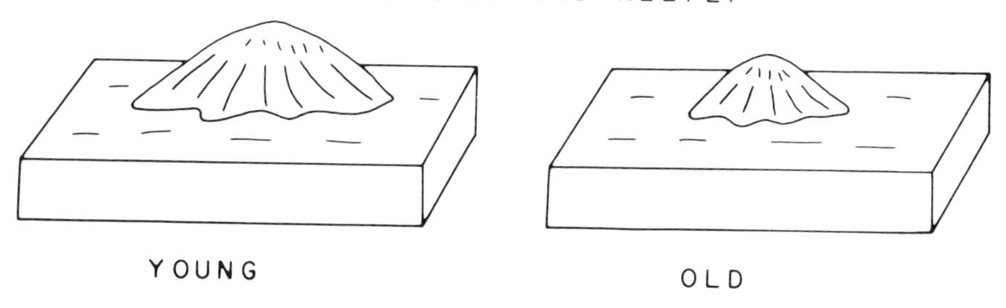

Figure 1 If environmental and geologic conditions are constant, a larger landform will be a reflection of a younger age.

to form standard landscapes, (2) landforms cannot be ascribed to creation by a single mechanism, (3) natural systems are open and not closed, (4) threshold events are not sufficiently accounted for with uniform evolution of features, and (5) dynamic equilibrium can occur within shorter intervals than the mature phase of Davis' cycles.

In an attempt to resolve the question of cycles and time, Schumm and Lichty (1965) designed a system that contained three different scales of time. Cyclic time was operative for periods $>10^6$ to 10^2 range, and steady time occurred from 10^2 to 10^{-2}. With such a system, the independent and dependent variables that interact in landscape development can change with the passage of time. For example, in steady time, the relief or the volume within a landmass above base level will be an independent variable, but with cyclic time such factors will be dependent.

The stage is now set for the evaluation of those features that may form by processes operating through time, and to provide some guidelines that bear on this type of an inventory. It must be made clear that our present quest is for the elucidation of phenomena that are pertinent to the development of relative time indicators as revealed by the topography. It would be erroneous to assume this is the actual and only way geomorphologists analyze features. Instead the "earth detective" will use all available dating techniques at his disposal in order to unravel the complete, and when possible in absolute terms, chronology of events that have affected the earth's surface form and materials.

TIME AND GEOMORPHOLOGY

A principal objective for the geomorphologist is the analysis of the sculpturing of the earth's surface, which produces landforms and landscapes. Bates and Jackson (1980) define a landform as "Any physical, recognizable form or feature of the Earth's surface, having a characteristic shape, and produced by natural causes...." (p. 349). In similar fashion, landscape is defined as "The distinct association of landforms, especially as modified by geologic forces that can be seen in a single view, *e.g.* glacial landscape" (p. 349). Thus a hill is a landform, and a valley is a landform, and the manner in which they are related can constitute a landscape. It is the slope of the terrain that gives it individuality. Thus a basic component of geomorphology is the analysis of slope....length, steepness, shape, direction, and integrated fabric....that constitutes the building blocks of the topography. Furthermore the genesis and rate of change of slope comprise the time frame that is necessary for a complete understanding of the geomorphological personality of the area being investigated. For this comprehension it is important to establish both absolute and relative time indices, but this paper includes only the latter which can still reveal an evolutionary sequence of features.

Systems and Theory

The evaluation of time necessarily involves a systems approach because for any given input of energy at a specific time there is a resulting output whereby the energy transformation is manifest in work accomplished. In natural systems the processes of feedback and linkage become intermixed so that when the time dimension is added the output from one system may become the input to another system. The complexity of these interrelationships have been shown in the classic paper by Schumm and Lichty (1965).

The adoption of a strategy to evaluate the relation of time with space and form becomes both a philosophical and theoretical exercise. Landforms might be interpreted as tending toward a variety of semi-equilibrium states, such as static, metastable, or steady-state (dynamic equilibrium). Another approach would be to select a model that most closely fits one's perception of the landscape system (Thornes and Brunsden, 1977). The deterministic model might be used by the pragmatist because temporal changes then become closely related to process and response, input to output, and cause to effect. However, one might argue this is not realistic because there is so much variability and uncertainty in natural systems that only a stochastic model can establish "truth". Such models incorporate degrees of randomness among the variables, and provide results in terms of statistical solutions or probabilities. The stochastic model can be designed to be simple or complex, and the accuracy becomes a function of the precision of the data input.

Boundary Conditions

It is now important to establish some guidelines that will aid us to steer a course for the determination of time and terrain change. Kirkby (1980) set some rules for landform change and developed the idea of dominance domains. The character of these determine whether a landform, such as the stream head, will undergo uniform progressive change or be subjected to a threshold condition, whereby an extraordinary transformation is produced. One measure of geomorphic effectiveness that can be useful in calculating dominance domains for different processes is sediment transport. This acts in an analogous manner to landform evolution, much as the flow of money being an indicator of economic conditions.

The next set of parameters that need inclusion for terrain assessment involve the magnitude, frequency, and sequence of environmental events that are required to produce landscape evolution. These domains of geomorphic activity possess a set of forces that change over space and time. Their interaction with materials of contrasting resistance produce measurable sediment yield responses. Differences in terrain character may contain memory features that reveal past events. Identification of these memory features (special elements in the landscape) comprise the tools that aid to unravel the time dimension. Topographies that contain only short memory systems are more likely to obey rules of the stochastic model, whereas those topographies with long memories are deterministic.

The search for conditions that favor change, and thereby create the fingerprints used for differentiating relative time sequences, ultimately rests with forces and forms, and the production of inequalities. The keys to this analysis involve the concepts of thresholds and relaxation time (Coates and Vitek, 1980). There is a wide range of different threshold types, but my principal concern in this paper is the combination of events that have triggered a memorable response within the geomorphic system. The responses within this system to drastically change impulses are subject to feedback mechanisms that either dampen (negative) or enhance (positive) the change. The amount and direction of landscape modification is a reflection of the total environment situation.. ..rock type and structure, landmass size and shape, relief and base level, climate and vegetation *etc*. Brunsden (1980) has characterized the relations in the form of two equations.

The landscape change factor (F_{LC})

$$F_{LC} = \frac{\text{magnitude of barriers to change}}{\text{magnitude of disturbing forces}}$$

There are two end members in this system: (a) mobile states with high sensitivity to change such as beaches, and (b) slowly responding states which are largely insensitive to short-range change such as plateaus.

The transient form ratio (TFr)

$$TFr = \frac{\text{mean relaxation time of the system}}{\text{mean recurrence time of events causing change}}$$

Within this context relaxation time is the time necessary for a system to achieve a new equilibrium following an impulse that has changed the original character of the system. When the ratio is <1.0 the impulses of change are insufficient to prohibit rapid recovery to a characteristic state of the system. When the ratio is >1.0 the impulses arrive so rapidly the system cannot adjust.

Thus analysis of landscape evolution and chronology must recognize the sequence in which adjustments have occurred. The degree of success of this undertaking depends upon the sensitivity of the terrain to possess a memory. Some landforms are so insensitive as to almost transcend time, such as major escarpments and inselbergs, and thus are useless as Quaternary time indicators. Other landforms may be so fragile as to be erased with nearly every passing event, such as beach cusps. It is the middle ground that is most helpful to the Quarternarist....those situations where changes are caused and/or preserved within the 10^{-1} to 3×10^{6} range.

Timescales

In the most recent book on this topic (Cullingford, Davidson and Lewin, 1980) geomorphic time is divided into three orders: short 10^{1} to 10^{2}, medium 10^{3} to 10^{4}, and long 10^{4} to 10^{5}. Cambers (1976) applied the Schumm and Lichty model for coastal landforms and determined that cliffs in soft rock constituted steady time because they attained equilibrium in 2 to 100 yr. For more resistant coastal features Kirk (1977) calculated it took shore platforms 10^{2} to 10^{3} yr to evolve to a stable, or steady time form. The recent study by Brunsden and Jones (1980) of coastal cliffs developed on soft rocks at West Dorset, England, indicated a time period of 10^{2} yr for integration of process and form to establish a dynamic equilibrium system. The mass movement phenomena they investigated showed there were diagnostic landform assemblages that could indicate stages of development in an overall sequence of events.

Wolman and Gerson (1978) have attempted to quantify landform time changes by the determination of effectiveness of an event for creating change as related to the relaxation time, or return period. In mountain areas they calculated large ranges for the transient form ratio (TFr), with a ratio of 5.0 for Tanasawa Mountains, Japan, and of 0.25 for the Appalachians. The inference of these numbers is that low ratios are associated with relatively long geologic time spans. For example, the Dorset sea cliffs have a 2.5 ratio when composed of clay, and a 10 ratio in sand.

Timescales for relative dating of the Quaternary can be developed from: (1) direct field measurements of process and form, but great care must be exercised that the forms being studied were made by the process being measured, (2) use of historical data, (3) comparative qualitative analysis of earth features, and (4) comparative quantitative analysis using field and map numbered data.

RELATIVE AGE DETERMINANTS

There are several fundamental principles that can be used as indicators of the relative age of the terrain. The validity for some of these guidelines is increased when the origin of features being compared is known, and environmental conditions have been similar throughout their development.

Stratigraphic Indicators

Some of the concepts from stratigraphic geology can be borrowed and can be equally useful for deciphering relative form sequences in the topography of an area.

1. Law of superposition.....that younger materials rest on older materials. Thus, any landform that can be seen to rest on a landform that is now partly hidden by the topmost landform is younger than the one that is partly obscured. Alluvial fans, deltas, and moraines provide good examples where this technique can be used.

2. Law of cross-cutting relations.....that materials had to be in place and predate any events that intrude or cut them. Wind gaps, barbed tributaries, and intrusive troughs (as in the Finger Lakes region, N.Y.) provide examples of this concept.

3. Law of complexity.....that those rock sequences which contain the greatest number of diverse features and structures are older than less complicated rock masses. This rule can be especially applied to those landscapes that have undergone multiple cycles of development. Remnant features in polygenetic terrane are particularly helpful when using this law. Again the Finger Lakes provide a good example because there are vestiges of glacial and fluvial events, and of glacial episodes, interglacial events, postglacial, and later Holocene features.

Geomorphic Indicators

When used with care, there are several suggestive terrain signatures that can provide strong evidence for the development of sequential landscape chronologies. Each of the following statements should be prefaced with the disclaimer that "other things being equal such features may be an important indication of relative time."

1. Landform size. As in all other measures, if two different landforms were initially the same height and mass, then a diminution in these variables should be a reflection of age difference (Figure 1).

2. Hillslope steepness. Although the amount of slope is dependent on many factors, many studies have shown that hillslopes trend toward an equilibrium steepness through time. Davis contended that hillslopes were steepest during the "mature stage" of the geographical cycle. My studies, and that of many colleagues, have shown there is a maximum slope of equilibrium for each lithologic type. Deviations from this steepness provide a signal of relative age for the feature.

3. Hillslope shape. The subject of hillslope shape is much too involved to be discussed adequately, but a few generalizations are relevant. Convex hillslopes may be considered as "gravity controlled", whereas concave slopes are "wash, or water controlled". Prior to inception of fluvial activity, the principal denudational process would be gravity movement, thus producing convex slopes. With continued development, slopes undergo a transition to concave shapes as stream networks work headward. In this scenario only the transportation and erosional parts of slopes have been considered for it is assumed that waste products are being completely evacuated from hillslope locations.

4. Landscape fabric. The total array of features that constitute

the landscape can prove helpful in revealing relative ages of diverse elements. This is especially true when multi-generation events have occurred and been indelibly inscribed into the terrain. It is the analysis of these landscape memory systems that provides the relative dating clues. Barbed tributaries are one such feature. With conventional networks stream junction angles are acute and point downstream. However, when a competing system draining the opposite direction captures the headwaters of a disadvantaged system, the uppermost tributaries are incorporated into the newer system. The resulting pattern displays piracy because the junction angles are now "barbed" and the original acute angle no longer points in the direction of valley lowering. Such a geometry thus reveals reversal and shows the tributaries antedated the master stream (Figure 2).

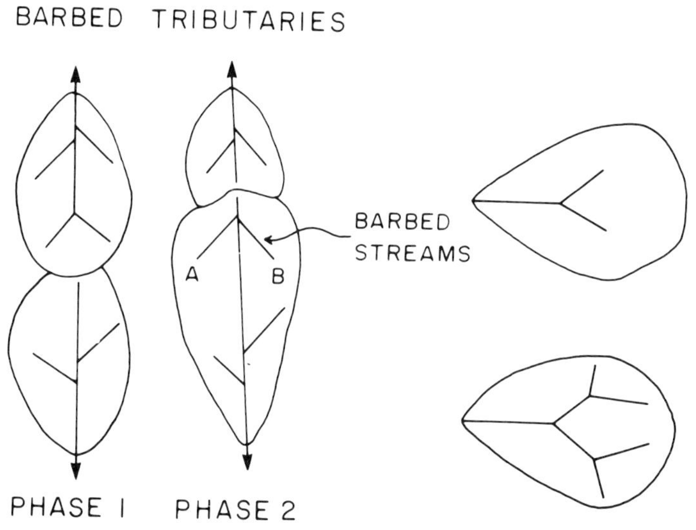

Figure 2 The use of barbed tributaries as an age criterion. Two competing and oppositely flowing drainage systems are shown in Phase 1. In Phase 2 the south-flowing master stream has captured part of the previously north-flowing drainage. Stream segments A and B are now "barbed" because they point in opposite direction to flow of master stream.

Water gaps and wind gaps are other examples that may indicate a poly-phased landscape. Thus, wind gaps may reveal a prior history that is no longer part of the current fluvial environment. Such features that are not linked to present conditions provide evidence for their greater antiquity. Misfit stream valleys and other out-of-adjustment forms give clues to their earlier ancestry. In some cases, as described by Dury (1964) several rivers in England now flow in valleys whose form developed by the greatly enlarged flows during Pleistocene glacial episodes. The Binghamton, N.Y. area contains many such valleys that are similarly underfit. In addition some stream valleys are overfit, such as the Tioughnioga River Valley. Although certain sections are wider and represent the usual topographic flavor of the region, other parts contain narrow, incised valleys, and are "overfit" in relation to the total system. The overall appearance of such "beaded valleys" indicates different ages of development. Coates and Kirkland (1974) have therefore described such topography as a multicycle sluiceway.

Quantitative Morphology

A consistent theme in the geographical cycle of Davis and others who followed his leadership is the absence of quantitative indices or numbers that signal different developmental stages (and thereby relative ages). It is important to note that Davis deliberately chose this approach, in spite of his ability to handle numbers....as evidenced by his early works in astronomy and meteorology. He simply preferred to use his exceptional command of prose to present ideas in an elegant descriptive manner.

Although there were excursions into numbers by various geomorphologists prior to 1945, it was largely because of the pioneer work of Horton (1945) that the "quantitative revolution" was initiated in geomorphology. His earliest disciples were A.N. Strahler and the Columbia "school of geomorphology". Several landform analysis parameters that were initiated during the 1945-52 period are applicable for use in assigning numbers to the aging process in fluvial terranes.

Drainage density (the ratio of total stream lengths to area) was one of the earliest methods to be successfully applied to an age-related problem. Ruhe (1952) showed that the location of different age till sheets corresponded with appropriate dissimilar drainage densities. The Kansan, or oldest drift, has a 7.9 to 10.2 drainage density, whereas three successively younger Wisconsinan drifts have drainage densities of 6.1 to 7.7, 4.7 to 5.4, and 1.9 to 2.1 (Figure 3).

Hypsometric analysis (the relation of remaining landmass above base level to a total volume with the same area and height) has been used by Strahler (1952) and others as a technique for determining relative age. The hypsometric integral can be viewed as a percent of landmass remaining. Strahler equated high numbers to an "inequilibrium stage (youth), medium numbers around 50 percent as the "equilibrium stage", and low numbers as the "monadnock phase".

Several other quantitative methods will also yield numbers that can be placed into a system for ordering the relative age of a landscape. Stream bifurcation ratios increase through time to an equilibrium state. For example, typical bifurcation ratios (number of tributary branchings per master stream of next highest order) in dendritic systems usually fall within the 3.5 to 4.5 range. Thus if the bifurcation ratio is considerably less it most probably indicates a basin that is not fully extended, and thus immature in development. However exceptions may occur in extremely ancient and pre-Pleistocene terranes.

The degree of stream straightness, or its sinuosity ratio (channel length *vs* valley straightline length), is another index that can provide insight into the relative age of a stream. If structural control is not a factor, most streams increase curvature with time. Anyone who has channelized a stream knows this rule because after channelization the stream invariably attempts to re-establish its original meandering course. Straight streams endure only when erosional forces expand all energy in downcutting and there is complete evacuation of valley debris. Ultimately, however, valley widening ensues whereby the channel is widened by lateral migration of the thalweg with accompanying development of a floodplain. Sinuosity can then be a measure of relative aging in this process. Low numbers reflect earlier developmental stages than higher numbers.

CONSTRAINTS TO A SYSTEM FOR RELATIVE DATING

There are two realms of constraints that restrict the operation of a universal system whereby relative dates can be established for all terranes....primary constraints and secondary constraints. The principal rule is to compare topographies that have the most environmental factors in common. For example it would be erroneous to attempt correlation of relative ages of the Scotland and West Virginia highlands. There are too many dissimilarities so the comparison would be invalid.

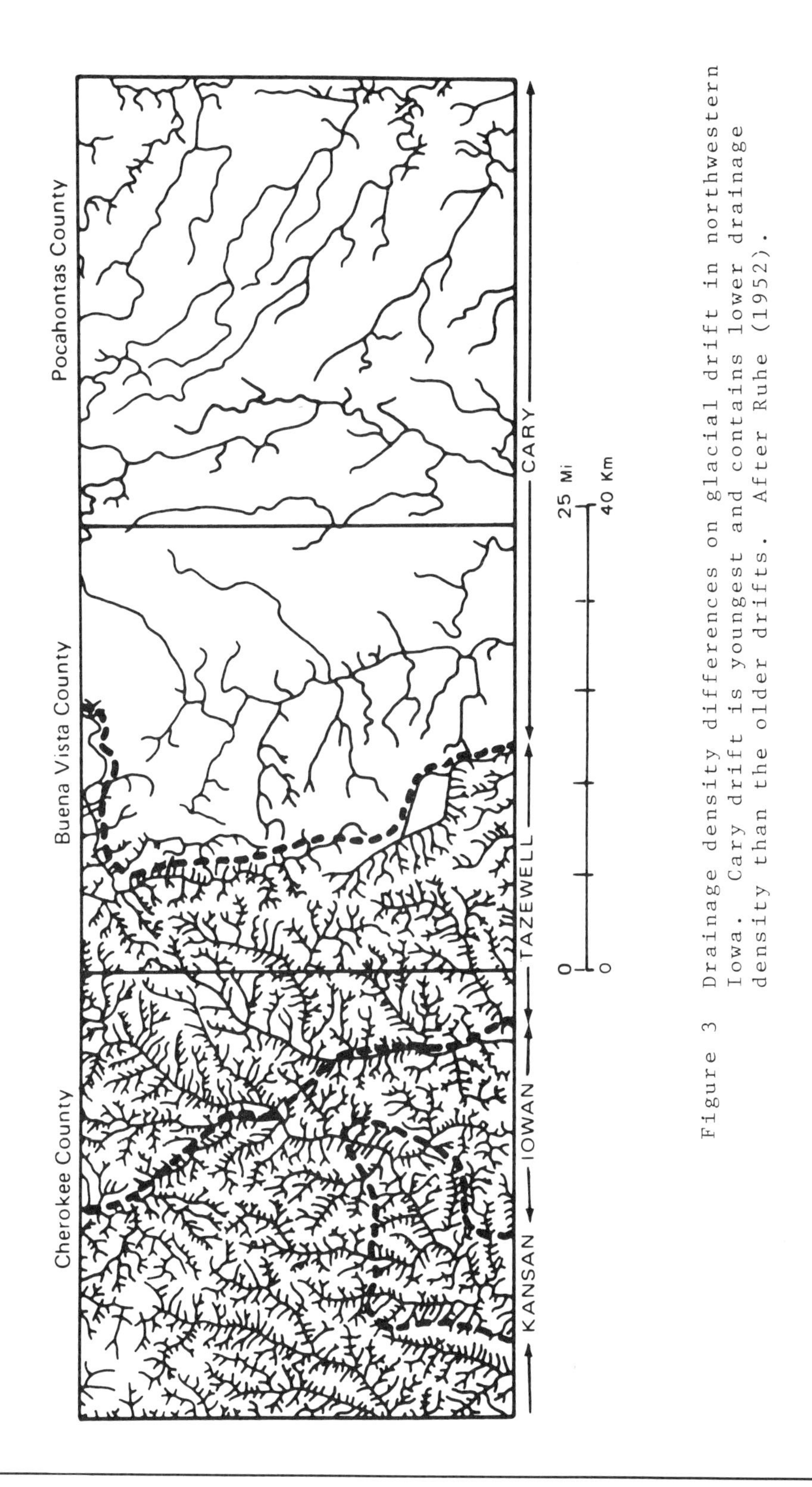

Figure 3 Drainage density differences on glacial drift in northwestern Iowa. Cary drift is youngest and contains lower drainage density than the older drifts. After Ruhe (1952).

Primary Constraints

1. Lithology and soils. The composition of materials determines the effectiveness of denudational processes. They also influence type and amount of vegetation, infiltration and runoff properties of precipitation, and character of weathering *etc.*

2. Structure. Here we include both the gross relations such as the nature of folding and faulting, and the smaller scale features of bedding, foliation, jointing, cleavage, *etc.* A highly fractured

terrane will progress differently through time than others with fewer discontinuities.

3. Geologic processes. Not only do the endogenic forces of volcanism and diastrophism produce different terranes, but the exogenic forces also differ, among themselves.....running water, groundwater, gravity movements, shoreline activity, glaciers, and wind.

4. Landform magnitude. For best comparisons of relative time whenever possible only those features that started their development with similar landmass volume and relief should be used. The progression of features in a large landmass may differ from those of a smaller one.

5. Climate. The idea of homologous landscape development in different climates as proposed by King (1953, 1967) is unacceptable. The concept of morphogenetic regions as expressed by Peltier (1950) is more applicable. Even Davis (1905) recognized that different criteria must be employed to evaluate the aging process in humid areas as contrasted with arid areas. Climate influences not only water availability and vegetation, but some rock types, such as limestone have inverse topographic relations (valley-forming in humid terrane, and cliff-forming under dry conditions).

6. Physiographic province. When all factors are summed, their blend is the production of distinctive physiographic provinces. Such distinctive regions should not be crossed for comparative purposes. Thus landforms as developed in the Folded Appalachians may bear a different developmental history than those in the Appalachian Plateau.

Secondary Constraints

The secondary limits that should be considered when evaluating topographic relative time sequences are not so pervasive and universal as primary constraints, but yet can create problems if neglected.

1. Principle of equifinality. As applied to landforms, the meaning of this concept is that similar appearing features may have formed by different processes. Thus, there may be no unique solution for some cases. A 50 m high hill with 20° slopes may be created by many different processes, and its evolution may have taken entirely different time periods. Therefore when determining relative age, a primary requirement is knowledge of the genesis of the landform.

2. Polygenetic terranes. Many landscapes represent hybrid systems that developed from the interaction of several different processes. Although in some regions this may create special problems (especially when the differences are not recognized), in other areas such complexity may prove a blessing. In south-central N.Y. the sequence of glacial episodes, and interglacial periods have each left sufficiently unique landform arrays that the total landscape can be placed into a relative-dating chronology.

3. Relict terranes. One should not have a blind faith in the law of uniformitarianism. It is easy for the field observer to look at and measure terrain features, and to assume present conditions have created the landscape personality he now witnesses. However the magnitude and frequency of activity may have changed and the processes may have altered. Thus, if the slopes were an inheritance from a vastly different environment such as "hindcasting" procedure will provide a false foundation for forecasting future development of the slopes.

4. Single cycle landforms. The most difficult features to assess are those with uninterrupted histories. The absence of intervening markers reduces the number of yardsticks that can be used for comparison purposes.

5. Topographic features that reverse with time. Although restricted to Quaternary time in this discussion, under certain conditions

landform reversals can occur within this time span. Hillslopes may begin with convex profiles (Figure 4), then change to concave, and reverse to convex again at advanced stage....especially when depositional materials are involved as in the pediment slopes near Phoenix, Arizona (Moss, 1977). Hillslopes may be initially gentle, steepen with age, and again decline in angle of profile at some subsequent age. Of course this was a basic component of Davis' humid geographic cycle. Many quantitative studies of hillslopes in different stream order systems, show that first and second order slopes are more gentle than third and fourth order slopes.

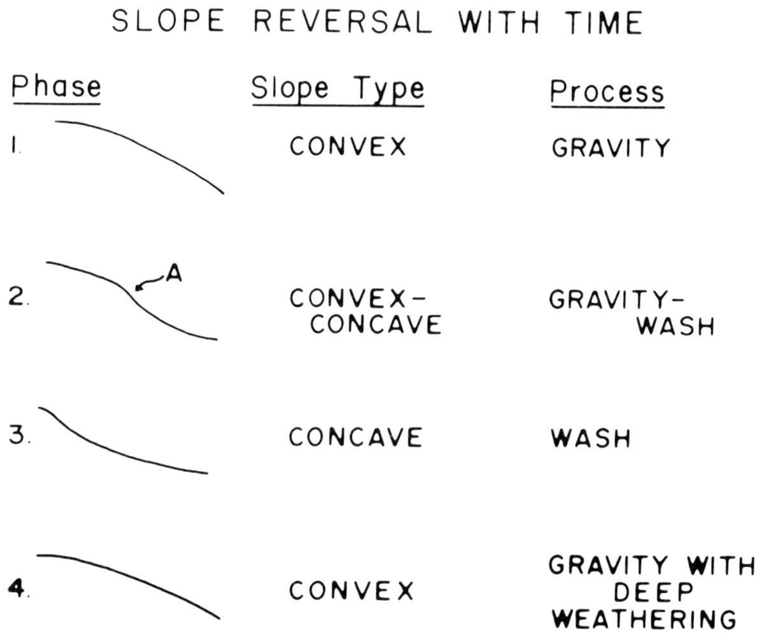

SLOPE REVERSAL WITH TIME

Phase	Slope Type	Process
1.	CONVEX	GRAVITY
2. (A)	CONVEX-CONCAVE	GRAVITY-WASH
3.	CONCAVE	WASH
4.	CONVEX	GRAVITY WITH DEEP WEATHERING

Figure 4 Slope reversal with time. During initial Phase 1 the hillside has a convex slope prior to establishment of running water development. With increasingly greater influence of water the slope becomes increasing concave in Phases 2 and 3. A deep weathering Phase 4 stage caused slope to again revert to a convex form.

6. Thresholds events. Conditions before and after thesholds have occurred may reveal exceptionally diverse terrain characteristics. Failure to recognize the hierarchy can lead to inaccurate landscape analysis.

7. Rate of change of variables. Phases in landform development reflect the magnitude and rapidity of forces that produce change. Therefore, the end products may appear similar, but greater force acting more often will create the same form in a fraction of the time it will take a weaker process.

8. The abbreviated Quaternary length of time. Some landscapes may be so insensitive that they fail to reveal significant changes within the 3 my of Quaternary time.

9. Human landscape impacts. The full scale and character of topographic changes that can be attributed to human activity raises important questions. In some cases, human impacts may have been so severe as to blur natural changes in the land-water ecosystem.

RELATIVE DATING OF FLUVIAL LANDFORMS

Although every geologic process that interacts with the earth's surface has its own set of relative age landform indicators, examples of three of these processes will be emphasized....fluvial, glacial, and faulted terrane.

Stream Channel Development

The relative age of a channel can be determined from both the deposits and the channel changes in meandering rivers. When a bend progressively migrates across a floodplain a series of point bars may be left in its wake that indicates both relative time and position of formation. In addition, during flooding the channel constriction of a meander bend can be severed with intrusion of a new channel through the older deposits (Figure 5). Hickin (1974) and Hickin and Nanson (1975) have used such criteria to determine channel sequences for meandering rivers.

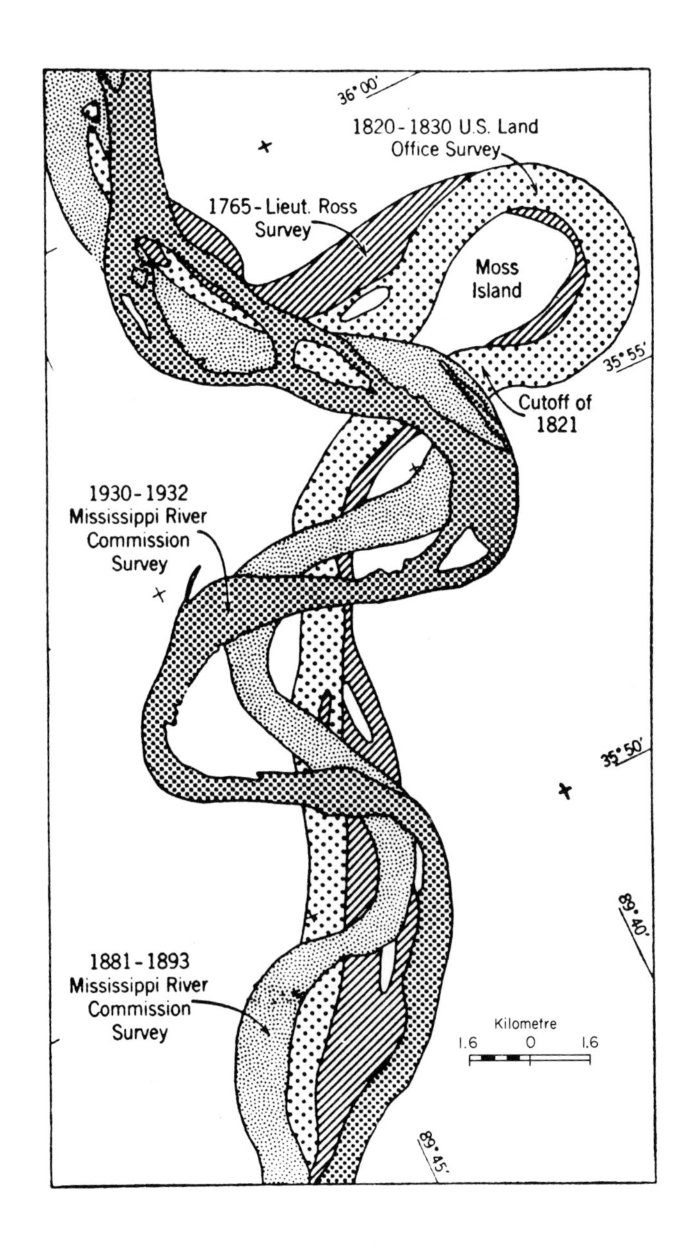

Figure 5

Principle of cross cutting relations as revealed by four surveys of the Mississippi River. Younger surveys show the river has cut through channels formerly occupied by earlier river positions. After U.S. Army Corps of Engineers and A.N. Strahler.

Stream Terraces

Terraces carved in valley fill sediments provide a systematic technique to determine relative age. Davis (1902) in his classic study of New England terraces, and Fairbridge (1968) outline the use of such landforms. If there is no upset in normal terrace evolution, the higher the terrace the older its formation (Figure 6). Fluvial history chronologies have thus been analyzed by numerous authors such as Jenkins (1964), Everitt (1968), and Womack and Schumm (1977). Glacio-fluvial terraces have also been studied by Peltier (1949) and Brunnacker (1975).

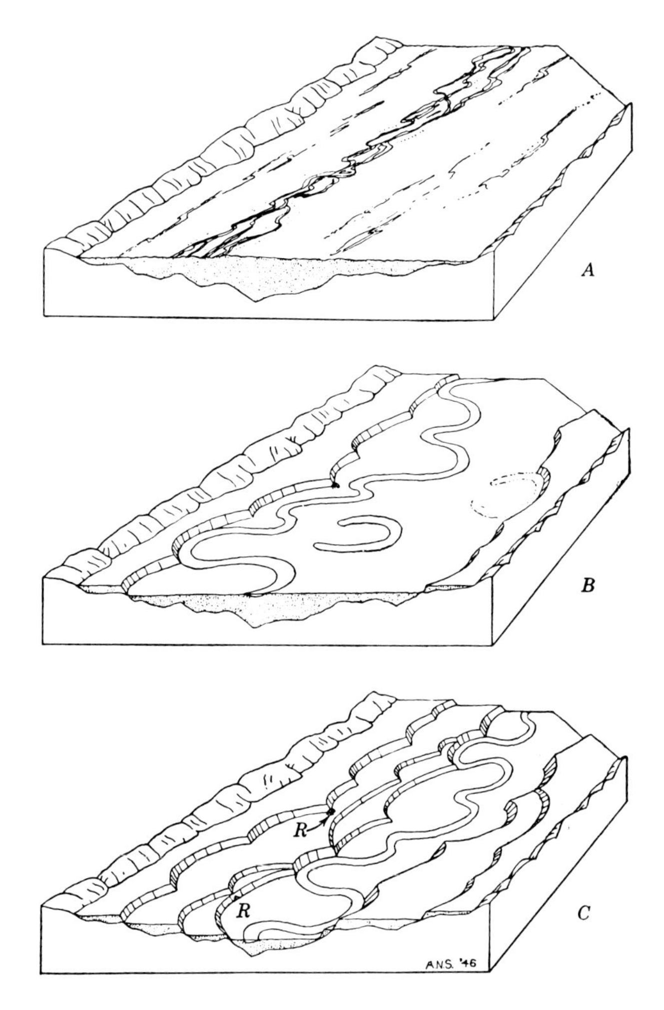

Figure 6 Relative age of alluvial terraces can be inferred by their position in the valley. Older terraces are higher than younger terraces. (After Strahler, 1951).

Alluvial Fans and Deltas

These stream depositional forms have many common elements. They both form at a drainage terminus and are composed of sediments discharged from an initiating channel. The sediment regime changes reflect the channel shifting over the landform, leaving a legacy that reveals the sequence of formation (Gole and Chitale, 1966). The youngest depositing event is the one that has covered the greatest number of different prior depositional sites (Figure 7). In similar manner the youngest channel is the one that has incised through the most former channel sites. These criteria have been used in the alluvial fan studies of Bull (1964), Hunt and Mabey (1966), and Ruhe (1967). Kolb and Van Lopik (1966) provide classic work for dating of the Mississippi River delta, and Coleman (1968) synthesizes the geomorphic evolutionary development in deltaic environments (Figure 8).

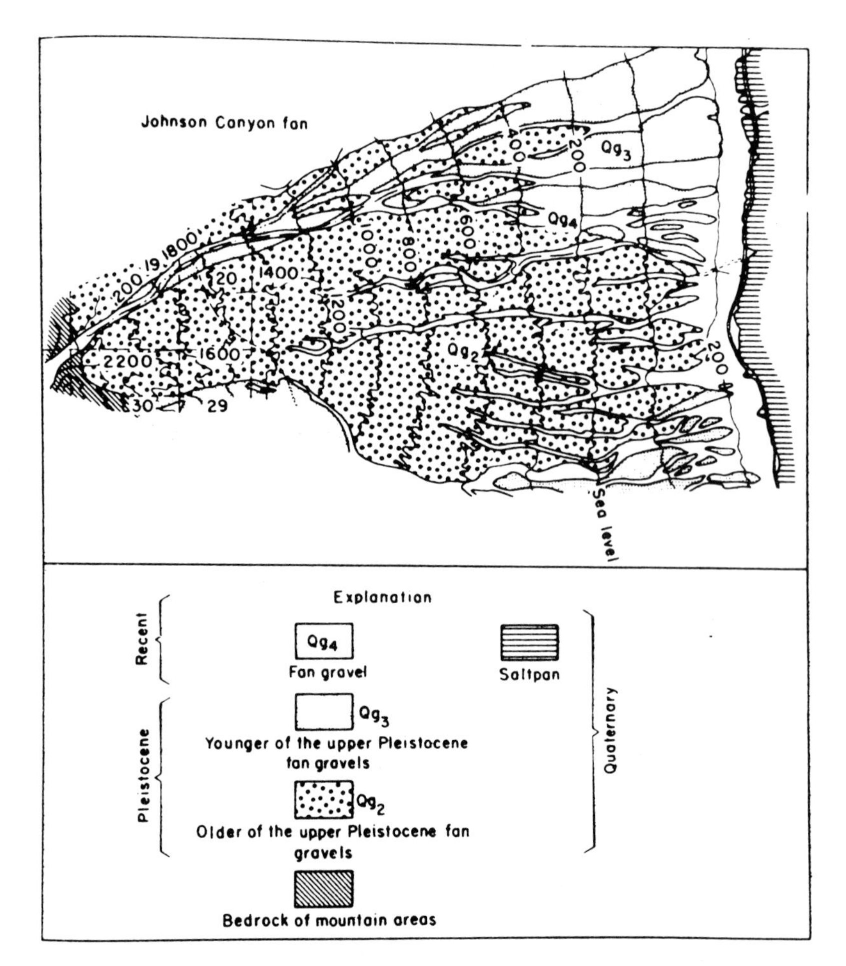

Figure 7 Map view and explanation of sediments comprising Johnson Canyon alluvial fan in Death Valley, California (after Hunt and Mabey, 1966). The younger fan gravels occur superposed on older gravels.

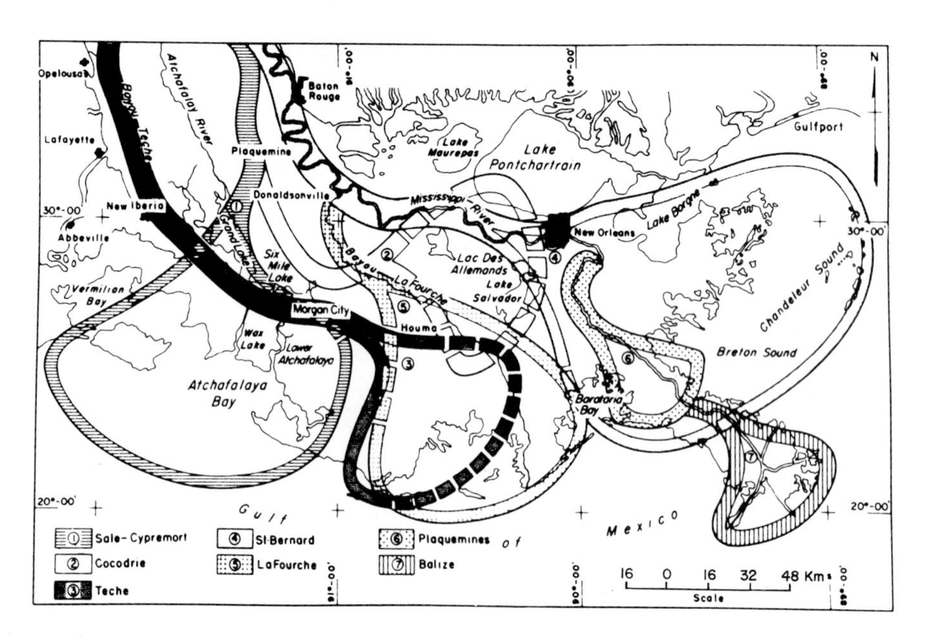

Figure 8 Mississippi River delta time sequences (after Kolb and Van Lopik, 1966). Respective ages for deltas are Sale-Cypremort >4600 yBP; Cocodrie 4600-3500 yBP; Teche 3500-2800 yBP; St. Bernard 2800-1000 yBP; LaFourche 1000-300 yBP; Plaquemine 550-500 yBP; Balize <550 yrs.

RELATIVE DATING OF GLACIAL LANDFORMS

Size and form of terrain features created by glacial action are particularly vexing in their inability to contain relative age indicators. Such landforms as eskers, kames, drumlins, cirques, U-shaped valleys, *etc.* largely reflect the intensity and duration of the process that formed them. However, some glacial landforms are useful for relative time chronologies.

Strandlines

These surfaces represent the land-water interface of proglacial lakes and can be used (Elson, 1967) as time-line indicators. Post-glacial rebound produces a tilting of the former horizontal strandlines, and the greater the distortion of the strandlines the older the formation of that surface (Figure 9).

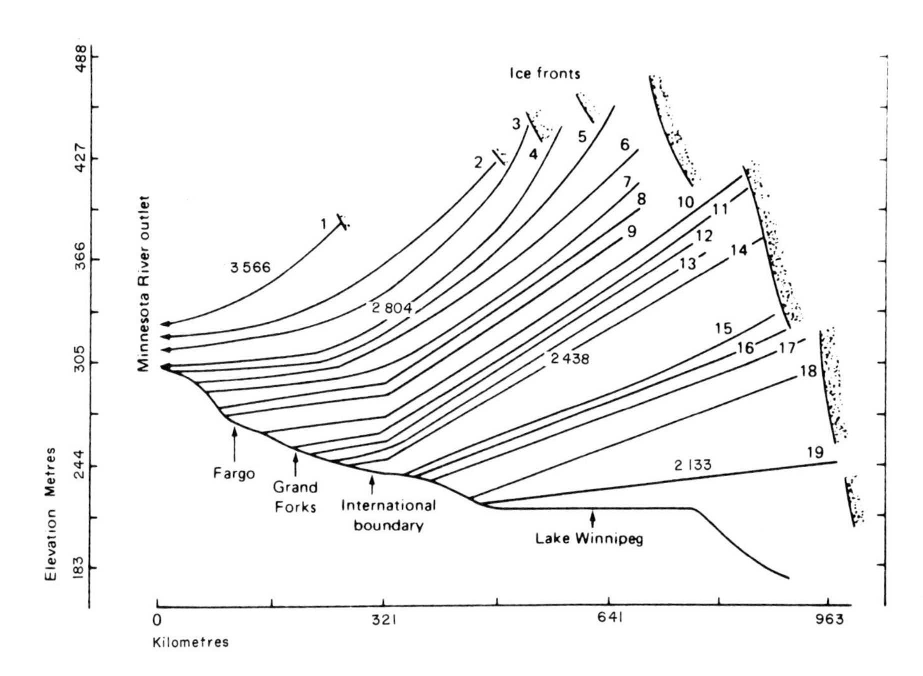

Figure 9 Glacial rebound produced these upwarped strandlines of glacial Lake Agassiz. Four of the former shorelines were radiocarbon dated as being 11,700 yBP, 9200 yBP, 8000 yBP, and 7000 yBP. Note that the older strandlines contain the steepest gradients. (After Elson, 1967).

Umlaufbergs

Coates (1974) provided a brief description of these forms in the glaciated Appalachian Plateau, and indicated that differences in their development might reveal the number of cycles they experienced (Figure 10). In this region, umlaufbergs develop when ice blocks meltwater drainage pathways, forcing water to incise a new channel which severs the hill from its original upland setting. Features that can be used to decipher the cycles of evolution include: (1) The presence of tributaries. If there are no tributaries, the umlaufberg would be first cycle because insufficient postglacial time has not allowed stream development. (2) Height of umlaufberg compared with upland height. The greater the height disparity, the older the umlaufberg, and (3) Size of valleys. Younger umlaufbergs generally have large differences in width of the two valleys. For example with repeated glaciations or with longer time spans, isolation becomes increased for the bedrock outlier.

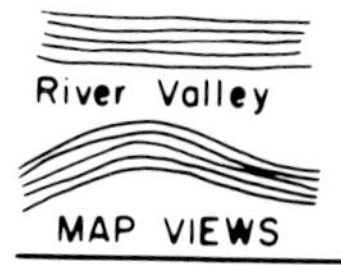

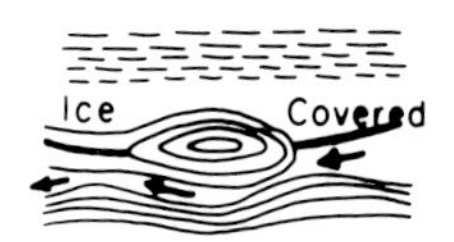

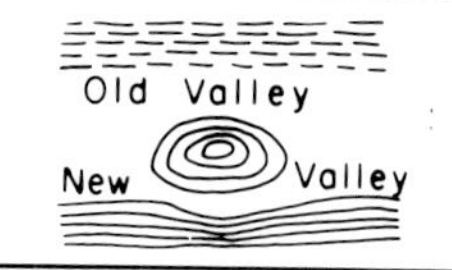

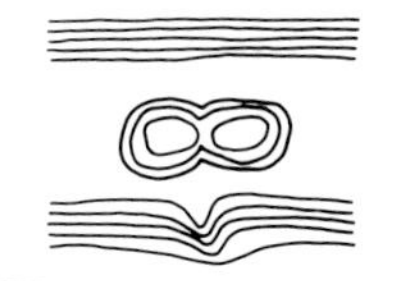

Figure 10 Umlaufberg development.

Sluiceways

Sluiceways, as used in this paper, are major courses for glacial meltwater rivers. In addition, I have restricted discussion to those with pronounced erosion. They differ from overflow channels or spillways by virtue of denudation to the general base level position of fluvial topography. When each glacial and interglacial episode leaves its imprint, then a chronology of sluiceway cycles can be calculated.

First cycle sluiceways develop when meltwater carves a new valley or combines elements from pre-existing valleys in a new flow pattern. This stage can be recognized by the absence of well integrated tributaries and of till in the valley.

Second cycle sluiceways contain better developed tributaries because during interglacial episodes sufficient time has elapsed for their creation. If ice covered the area, till may have been left in parts of the valley.

Multicycle sluiceways invariably contain total valley systems that are beaded as already described for the Tioughnioga River. Parts of the system will contain wide valleys, others will be narrow. Parts will have fully integrated tributary development, other parts will have a paucity of stream junctions.

Moraines

Several other glacial landforms may offer possibilities for relative dating purposes. For example, within a series of recessional moraines, as in midwestern U.S., the overlap of ridges can determine their sequence. In correlating moraines, the straightness of the ridge may offer dating clues. Such straightness may be a function of ice thickness, or indicate the effectiveness of postglacial erosion. Thus, straight moraines would have been formed by ice that was thicker, therefore older, whereas serrate ridges would represent either thinner ice, or landforms that have undergone considerable erosion.

RELATIVE DATING OF ENDOGENIC PROCESSES

The earth's interior forces of volcanism and diastrophism can produce some landforms that reveal the relative age of the process of formation. Lava flows can be placed into a chronologic sequence because they obey the rules of superposition. The breaching of craters, as in cinder cones can provide insight into the effectiveness, and perhaps length of time, of the denudational process. Cross-cutting of joint sets can determine their sequence, and many workers have shown the influence of joints in terrain development under certain environmental conditions. Intrusion by salt domes can develop diagnostic features regarding their relative age, and whether they are still active or not. This becomes a practical problem, because salt domes have been viewed as a possible recepticle for nuclear wastes. The salt domes in Texas have been shown to contain younger features near the coast and to be relatively older farther inland.

Faults

Knowledge of possible faulting, and the recurrence interval of seismic activity, is crucial in land planning of unstable regions. The siting of nuclear power plants is especially critical, and regulations that govern their location require knowledge of the fault age. The following list indicates some of the types of geomorphic evidence that can be used to assess the relative dates of faults: (1) Cross cutting relations; (2) Degree of change of spurs and facets (Figure 11); (3) Retreat of escarpment from original fault position; (4) Degree of re-establishment of water table that may have been cut by vertical fault movement; (5) Degree of adjustment in re-establishment of stream equilibrium profiles.

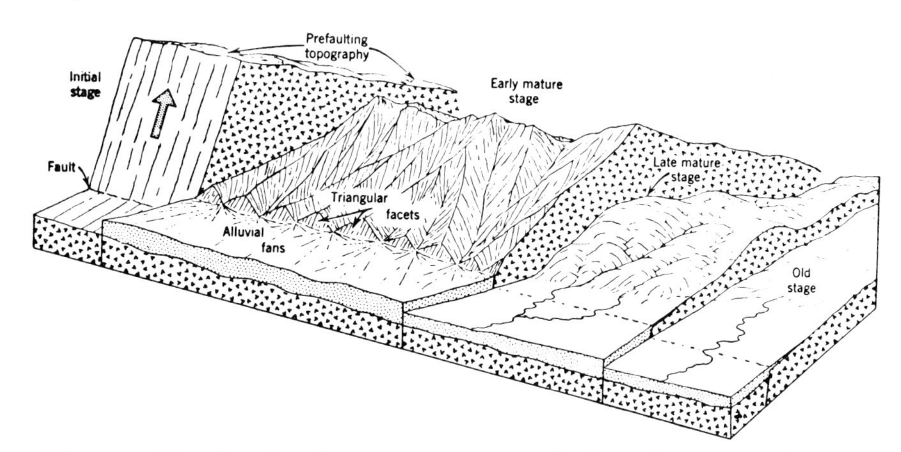

Figure 11 Evolutionary erosional development showing relative age differences in topographic sequences of a tilted fault block. (After W.M. Davis and A.N. Strahler).

Bull (1977, 1978) and Hamblin, Damon, and Bull (1981) have provided some guidelines in the rapidly growing discipline of tectonic geomorphology. They show how a straight mountain front might be indicative of more recent and active fault movement. Residual knobs of bedrock protruding above the erosional plain of a pediment is generally associated with a more sinuous mountain front, and both elements are suggestive of relative tectonic quiescence.

In the San Gabriel Mountains, California, Bull (1978) used stream terraces to determine the frequency and magnitude of local faulting episodes. Where terraces cross fault zones, the youngest terraces were generally not faulted, whereas older terraces revealed progressively more ground rupture. At a specific point on a stream that position acts as a local base level, affecting slopes and streams upstream and downstream. When an event occurs to change elevation of that point the base level conditions become changed. A combination of techniques were used including characteristics of terraces, soil profiles, and alluvial fan deposits that enabled Bull to determine relative ages of faulting and magnitudes during the 10^5 to 10^6 yr period. In the Mojave Desert, Bull (1977) was particularly interested in determining which faults had been active during the period from 500,000 to 1 million yr ago. His landscape studies showed that the eastern and western parts had been tectonically inactive whereas the central area had received repeated fault movements.

Along the eastern margins of the Basin and Range province, Hamblin, Damon, and Bull (1981) combined K-Ar dating of Cenozoic basalts and landform analysis to determine the deformation rate, and its variance in time and space. Radiometric dating made possible the estimation for rate of regional and local uplift by comparing profiles of ancient streams preserved beneath the isotopically dated lava flows with profiles of present streams. Aggradation and degradation processes directly reflect the tectonics of the locale where a lava flow is extruded. If the stream channel is lowering the lava becomes buried in alluvium, but if the stream is stable the displaced channel approaches its previous profile. Topographic inversion upstream from a fault will occur when the displacement causes accelerated erosion. These studies showed that the Grand Wash area is rising 26 m/my, the area between the Grand Wash and Hurricane faults is rising 90 m/my, and the block east of the Hurricane fault is rising 390 m/my.

CONCLUSIONS

Although relative dating does not have the precision of absolute dating, it can still provide important and necessary insights into the chronology of events that shape the earth's surface. The techniques used to establish the relative dates of events are being increasingly used because of the necessity to establish an environmental data base for land use planning.

REFERENCES CITED

Bates, R.L. and Jackson, J.A., eds., 1980, Glossary of Geology, 2nd ed.: Falls Church, Va., Amer. Geol. Inst., 749 p.

Beede, J.W., 1911, The cycle of subterranean drainage as illustrated in the Bloomington, Indiana, Quadrangle: Proc. Indiana Acad. Sci., v. 20, p. 81-111.

Brunnacker, K., 1975, The mid-Pleistocene of the Rhine Basin, in Butzer, K.W. and Isaac, G.L., eds., After the Australopithecines: The Hague, Mouton, p. 189-224.

Brunsden, D., 1980, Applicable models of long term landform evolution: Zeit. Geomorph. Suppl. Bd., 36, p. 16-26.

Brunsden, D. and Jones, D.K.C., 1980, Relative time scales and formative events in coastal landslide systems: Zeit. Geomorph. Suppl. Bd., 34, p. 1-19.

Bull, W.B., 1964, Geomorphology of segmented alluvial fans in western Fresno County, California: U.S. Geol. Survey Prof. Paper 352-E, p. 89-129.

__________, 1977, Tectonic geomorphology of the Mojave Desert, California, Contract report for the Office of Earthquake Studies, U.S. Geol. Survey, Menlo Park, California, 188 p.

__________, 1978, South front of the San Gabriel Mountains, southern California, Contract report for the Office of Earthquake Studies, U.S. Geol. Survey, Menlo Park, California, 100 p.

Cambers, G., 1976, Temporal scales in coastal erosion systems: Trans. Inst. Brit. Geog.: New Series, v. 1, p. 246-256.

Coates, D.R., 1974, Reappraisal of the Glaciated Appalachian Plateau, in Coates, D.R., ed., Glacial Geomorphology: Binghamton, State Univ. of New York, p. 205-243.

Coates, D.R. and Kirkland, J.T., 1974, Application of glacial models for large-scale terrain derangements: in Mahaney, W.C., ed., Quaternary Environments: Proc. of a Symposium: York Univ., no. 5, p. 99-136.

Coates, D.R. and Vitek, J.D., eds., 1980, Thresholds in Geomorphology: London, George Allen & Unwin, 498 p.

Coleman, J.M., 1968, Deltaic evolution: in Fairbridge, R.W., ed., Encyclopedia of Geomorphology: N.Y., Reinhold, p. 255-260.

Cotton, C.A., 1947, Climatic Accidents in Landscape-Making: Wellington, Whitcombe and Tombs, Ltd., 354 p.

Cullingford, R.A., Davidson, D.A., and Lewin, J. eds., 1980, Time-scales in Geomorphology: N.Y., Wiley, 360 p.

Cvijic, J., 1918, Hydrographie souterraine et evolution morphologique du karst: Rec. Trav. Insts. Geog. Alpine (Grenoble), v. 6, no. 4, 56 p.

Davis, W.M., 1899, The geographical cycle: Geog. Jour., v. 14, p. 481-504.

__________, 1902, River terraces in New England: Mus. of Comparative Zoology Bull., v. 38, p. 281-346.

__________, 1905, The geographical cycle in an arid climate: Jour. Geol., v. 13, p. 381-407.

Dury, G.H., 1964, Principles of underfit streams: U.S. Geol. Survey Prof. Paper 452-A, 67 p.

Elson, J.A., 1967, Geology of Glacial Lake Agassiz, in Mayer-Oakes, W.J., ed., Life, Land, and Water: Winnipeg, Univ. of Manitoba Press, p. 37-95.

Everitt, B.L., 1968, Use of the Cottonwood in an investigation of the recent history of a flood plain: Amer. Jour. Sci., v. 266, p. 417-439.

Fairbridge, R.W., 1968, Terraces, fluvial - environmental control: in Fairbridge, R.W., ed., Encyclopedia of Geomorphology: N.Y., Reinhold, p. 1124-1138.

Flemal, R.C., 1971, The attack on the Davisian system of geomorphology: a synopsis: Jour. Geol. Ed., v. 19, p. 3-13.

Gilbert, G.K., 1877, Report on the Geology of the Henry Mountains (Utah): U.S. Geog. and Geol. Survey of the Rocky Mountain Region, Washington, D.C., U.S. Govt. Printing Office, 160 p.

Gole, C.V. and Chitale, S.V., 1966, Inland delta building activity of Kosi River: Amer. Soc. Civil Eng. Jour. Hydraul: Div. HY-2,

p. 111-126.

Hack, J.T., 1960, Interpretation of erosional topography in humid temperate regions: Amer. Jour. Sci. (Bradley Vol.) 258-A, p. 80-97.

__________, 1975, Dynamic equilibrium and landscape evolution: in Melhorn, M.N. and Flemal, R.C., eds., Theories of Landform Development: Binghamton, State Univ. of New York, p. 87-102.

Hamblin, W.K., Damon, P.E., and Bull, W.B., 1981, Estimates of vertical crustal strain rates along the western margins of the Colorado Plateau: Geology, v. 9, p. 293-298.

Hickin, E.J., 1974, The development of meanders in natural river channels: Amer. Jour. Sci., v. 274, p. 4]4-442.

Hickin, E.J. and Nanson, G.C., 1975, The character of channel migration on the Beatton River, northeast British Columbia, Canada: Geol. Soc. Amer. Bull., v. 86, p. 487-494.

Horton, R.E., 1945, Erosional development of streams and their drainage basins: hydrophysical approach to quantitative morphology: Geol. Soc. Amer. Bull., v. 56, p. 275-370.

Hunt, C.B. and Mabey, D.R., 1966, Stratigraphy and structure, Death Valley, California: U.S. Geol. Survey Prof. Paper 494-A, 162 p.

Jenkins, O.P., 1964, Geology of placer deposits: Calif. Div. Mines Bull. 135, p. 146-216.

Johnson, D.W., 1919, Shore Processes and Shoreline Development: N.Y., Wiley, 584 p.

Kirk, R.M., 1977, Rates and forms of erosion on intertidal platforms at Kaikoura Peninsula, South Island, New Zealand: N.Z. Geol. and Geophys., v. 20, no. 3, p. 571-613.

King, L.C., 1953, Canons of landscape evolution: Geol. Soc. Amer. Bull., v. 64, p. 721-752.

__________, 1967, The Morphology of the Earth, 2nd ed.: N.Y., Hafner, 726 p.

Kirkby, M.J., 1980, The stream head as a significant geomorphic threshold: in Coates, D.R. and Vitek, J.D., eds., Thresholds in Geomorphology: London, George Allen & Unwin, p. 53-73.

Kolb, C.R. and Van Lopik, R.R., 1966, Depositional environments of the Mississippi River and deltaic plain southeastern Louisiana, in Shirley, M.L. and Ragsdale, J.A., eds., Deltas and their Geologic Framework, Houston Geol. Soc., p. 17-61.

Moss, J.H., 1977, The formation of pediments: scarp backwearing or surface downwasting, in Doehring, D.O., ed., Geomorphology of Arid Regions, Binghamton, State Univ. of New York, p. 51-78.

Penck, W., 1953, Morphological analysis of land forms: Czech, H., and Boswell, K.C., (eds. and trans.): London, Macmillan. 429 p.

Peltier, L.C., 1949, Pleistocene terraces of the Susquehanna River, Pennsylvania: Pa. Geol. Survey Fourth Series Bull. G 23, 158 p.

__________, 1950, The geographic cycle in periglacial regions as it is related to climatic geomorphology: Ann. Assoc. Amer. Geog., v. 40, p. 214-236.

Ruhe, R.V., 1952, Topographic discontinuities of the Des Moines lobe: Amer. Jour. Sci., v. 250, p. 46-56.

__________, 1967, Geomorphic surfaces and surficial deposits in southern New Mexico: N.M. Bureau of Mines and Mineral Resources Memoir No. 18.

Schumm, S.A. and Lichty, R.W., 1965, Time, space and causality in geomorphology: Amer. Jour. Sci., v. 263, p. 110-119.

Strahler, A.N., 1951, Physical Geography: N.Y., Wiley, 442 p.

______________, 1952, Hypsometric (area-altitude) analysis of erosional topography, Geol. Soc. Amer. Bull., v. 63, p. 1117-1142.

Thornes, J.B. and Brunsden, D., 1977, Geomorphology and Time: N.Y., 208 p.

Wolman, M.G. and Gerson, R., 1978, Relative scales of time and effectiveness in watershed geomorphology: Earth Surf. Proc., v. 3, p. 189-208.

Womack, W.R. and Schumm, S.A., 1977, Terraces of Douglas Creek, northwestern Colorado: an example of episodic erosion: Geology, v. 5, p. 72-76.

RELATIVE DATING OF SOILS AND PALEOSOLS

W.J. VREEKEN

ABSTRACT

Relative dating of a soil includes establishing the succession and duration of pedogenic processes involved in the principal stages of its development, as much as the duration of various environmental and geomorphic conditions that may have successively prevailed in the soil-landscape, apart from determining the total duration of the formation of that soil. Practices of relative dating include dating of geomorphic surfaces by stratigraphical means, and the use of pedogenic indices, whether as direct measures of soil age, or indirectly, when interpreting soils as environmental indicators. These practices are reviewed and numerous principles, bearing on them, are formulated. Application of the stratigraphical approach towards pedochronological analysis of functional soil-landscape units is illustrated with examples from enclosed and open hill-slope systems.

INTRODUCTION

Relative dating of an object amounts to reconstructing conditions, processes and events thought to bear on its nature. Relative dating of a soil places it in an historical framework of geologic, geomorphic, climatic and biotic conditions, processes and events, some of which may have been anthropogenically influenced. Relative soil dating involves simultaneous consideration of soil age and soil genesis.

A soil (Ruhe, 1972) is a naturally occurring three-dimensional body with morphology and properties resulting from effects of climate, flora and fauna, parent rock materials and topography imparted through time (Figure 1A). A soil occupies a portion of the land surface, is mappable, and commonly is composed of horizons that parallel the ground surface. A vertical section downward through all the horizons of the soil is called a soil profile.

A paleosol is a soil that began forming on a landscape in the past (Ruhe,1965). Three basic kinds of paleosols are relict, buried and exhumed soils. Relict soils (Figure 1B) began forming on landscapes in the past that escaped destruction or burial and persist as relict geomorphic surfaces within the present-day topography (Thornbury, 1969). Because relict soils remained subaerially exposed during and after changes in geomorphic regimen, pedogenesis continued and younger pedogenic features were superprinted on their older, relict features. A soil that developed under various climatic and biotic, but essentially unchanged geomorphic, regimens (Figure 1C), also may be regarded as a relict soil, because it may have retained imprints from past pedogenic regimens. Normally however, such a soil is called a polygenetic soil. Actually, most soils and paleosols are polygenetic because climatic and biotic factors have changed during their development.

A buried soil (Figure 1D) formed on a preexisting landscape and was subsequently buried by younger sediment or rock. An exhumed or

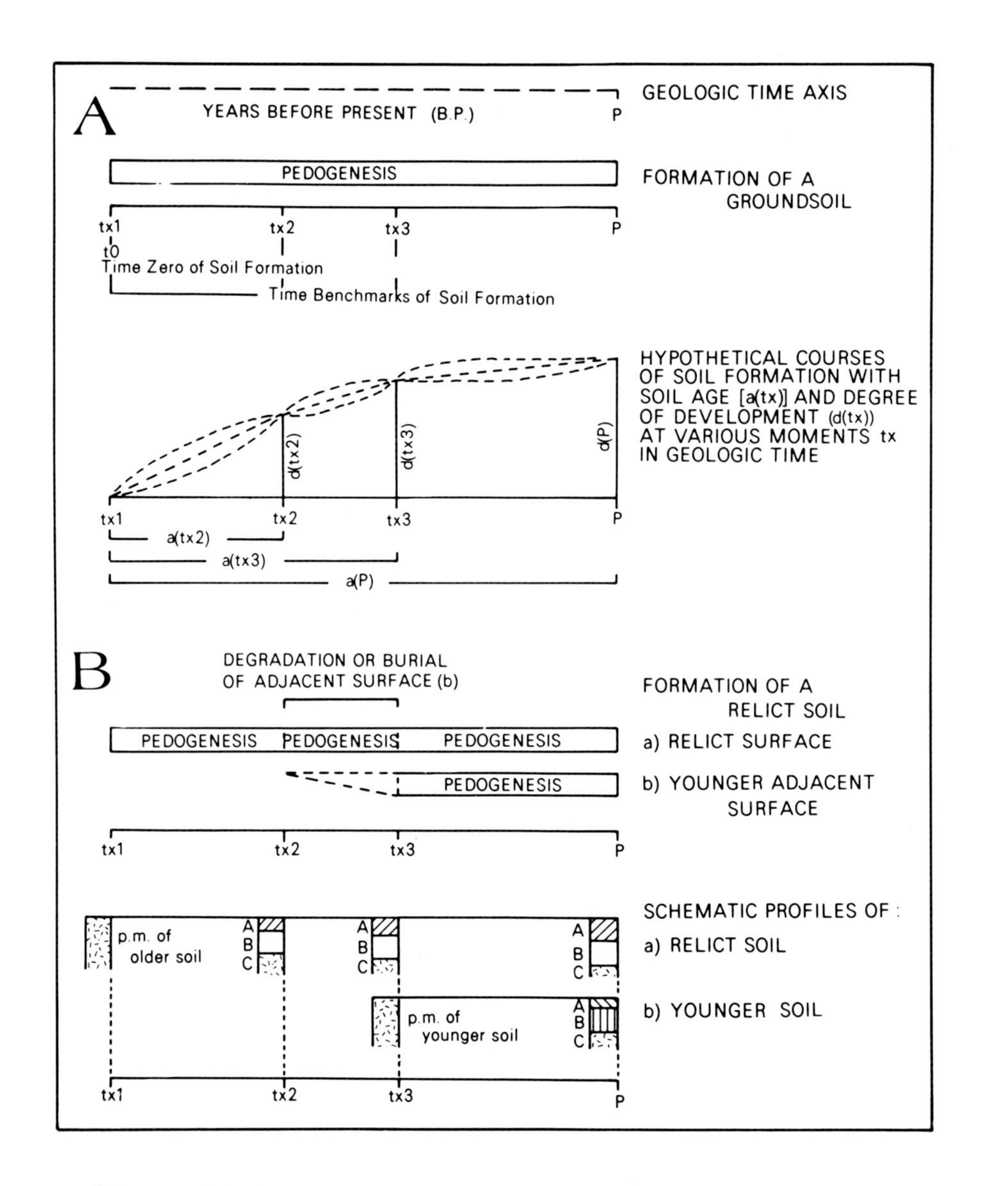

Figure 1A-1B Schematic illustrations of the Formation of Groundsoils (A) and Relict Soils (B).

resurrected soil (Figure 1E) is a soil that, once buried, has been re-exposed on the present-day topography by erosion of the covering mantle. A soil, once buried under thin overburden, can later be subjected to postburial soil formation, extending through that overburden, whether the latter has been partly eroded or not (Figure 1F). The result is called a welded profile (Ruhe and Olson, 1980). Soil formation continuing during gradual or intermittent burial processes gives rise to a cumulic profile. A buried soil that has not been subjected to soil formation after final burial may be called a fossil soil.

The age of a soil corresponds to the duration of its exposure to the soil-forming factors, *i.e.* the length of its pedogenic interval. For a nonburied soil, this is defined by distance on the time axis between the point marking incipience (time zero) of soil formation and the point that marks the present (Figure 1A,B,C). The age of a buried soil is defined by the distance between the points that mark incipience and cessation (by burial) of soil formation (Figure 1D). Changes incurred by a soil while not exposed to the soil-forming factors are called diagenic (Figure 1D,E.F.). The age of an exhumed soil is the sum of the age attained before burial and the duration of exposure since exhumation (Figure 1E). Similarly, the age of the buried portion of a welded profile is a composite age (Figure 1F).

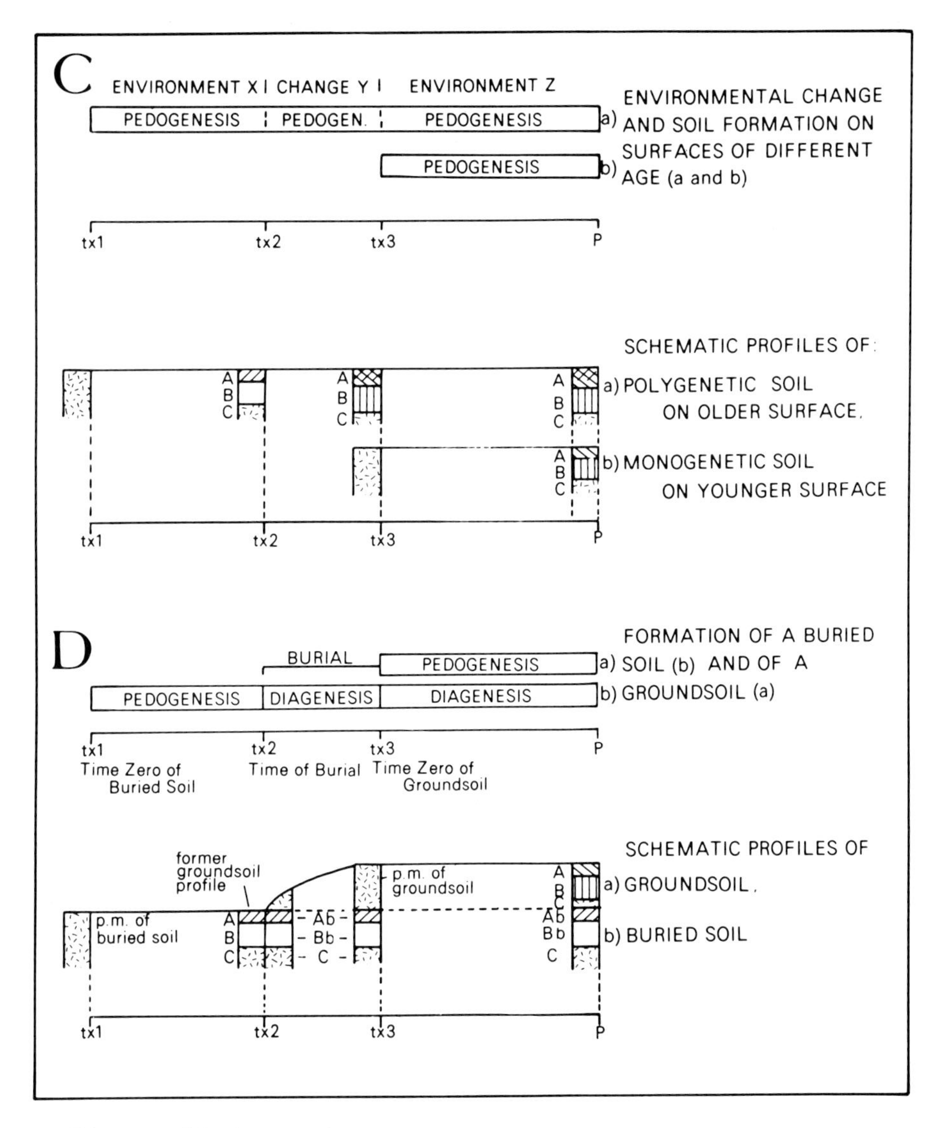

Figure 1C-1D Schematic illustrations of the Formation of Polygenetic Soils (C), and Buried Soils (D).

More time benchmarks are needed than those delimiting pedogenic intervals, to answer the basic pedochronological questions (Ruellan, 1971):

1. What were the principal stages in the formation of a given soil?

2. Did the soil evolve in environments or physical landscapes that differed from those prevailing now?

3. What was the succession of processes responsible for profile differentiation observable now?

Pedochronological analysis serves to resolve soil characteristics, time-transgressively acquired and superprinted on one another as they are, in terms of the historical environmental record that is contained in coeval sedimentary sequences. Firm chronological control on such correlation is provided only by physically tracing the soil towards those sediments. Such linking requires stratigraphical analysis of the broader soil-landscape.

Soil-sediment continua in the landscape are hydrophysical and hydrochemical in nature. Destruction of soil profiles by erosion and mass-wasting provides detritus for accumulation of clastics in colluvia

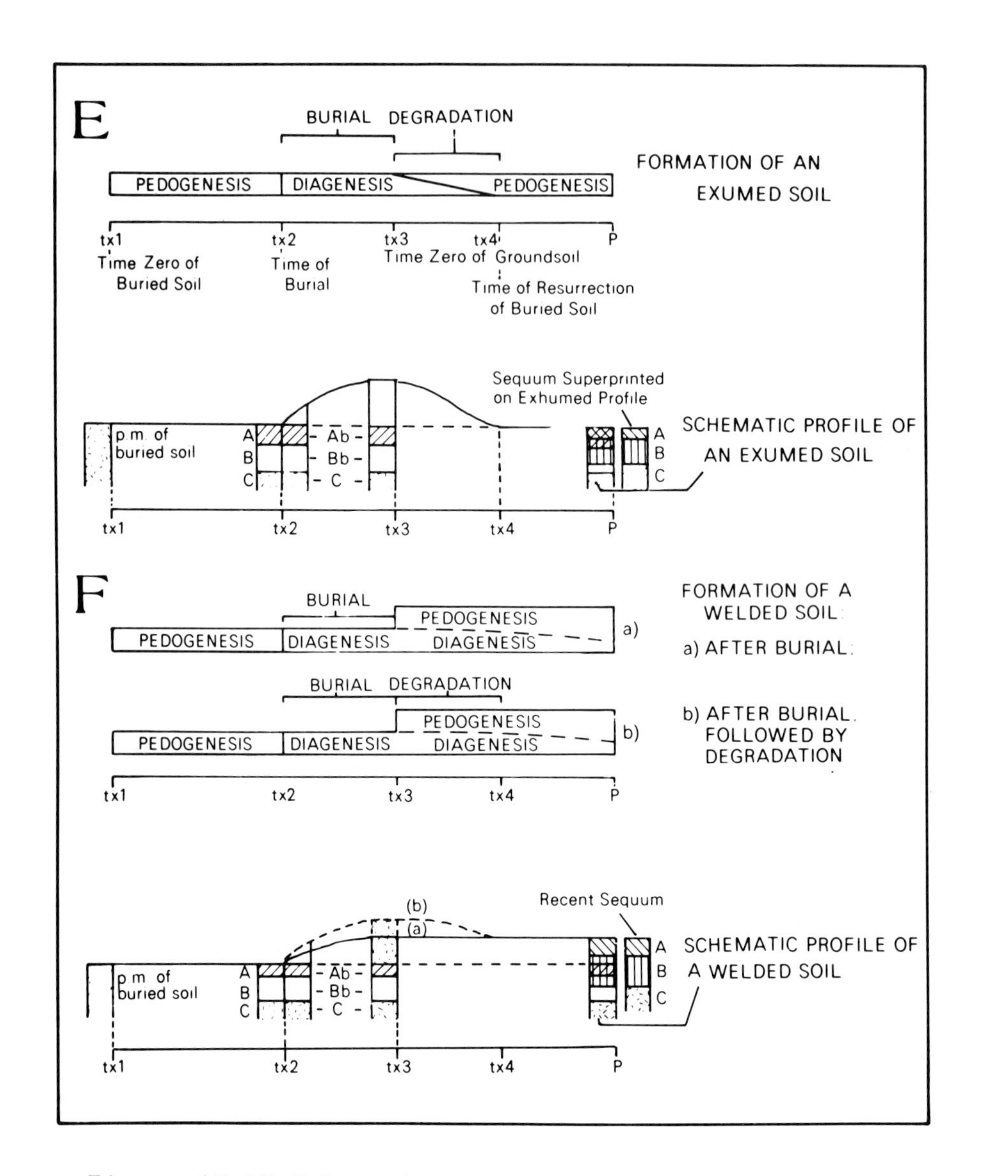

Figure 1E-1F Schematic illustrations of the Formation of Exhumed Soils (E), and Welded Soils (F).

and alluvia. The hydrophysical soil-sediment continuum ranges from *in-situ* soil, through transported soil material in which the soil fabric is retained, to raw sediment (Gerasimov, 1971). Similarly, chemical leaching and precipitation of constituents from and within members of the hydrophysical continuum provide for a hydrochemical continuum ranging from *in-situ* profiles including the soil solution, to organo-chemical sediments that may accumulate in suitable depositories. Continuity of these processes results in superprinting of their impacts in the soil-landscape. This complexity is accounted for in Milne's catena:

> Soil catena refers to the sequence of soils encountered between the crest of a low hill and the floor of the adjacent swamp or thalweg, the soil profile changing from point to point of this traverse in accordance with conditions of drainage and past history of the land surface. Two variants of the catena can be distinguished in the field. In one, the topography was modeled, by denudation or other process, from a formation originally similar in lithologic character. Soil differences were then brought about by drainage conditions, differential transport of eroded material, and leaching, translocation, and redeposition of mobile chemical constituents. In the other variant, the topography was carved out of two or more superposed formations which

differ lithologically (Milne, 1936).

Pedogenic hysteresis and pedogenic inertia are historical factors that compound the progressively cumulative and therefore time-transgressive way in which soil features are acquired. Pedogenic hysteresis means that "a particular factor in soil development may be operating at a changed value, or may have been eliminated altogether, but its effects are erased only slowly, or may persist indefinitely if they were of an irreversible or destructive kind" (Milne, 1936). Pedogenic inertia (Bryan and Teakle, 1946) states "that specialized soil-forming processes, once established, tend to continue (or persist) in spite of changes in original conditions of formation. Once past certain physical and chemical thresholds, the pedogenic regimen continues by virtue of its inertia" (Finkl, 1980). These considerations are related to the fundamental pedochronological questions. They must be kept in mind as underscoring complexities that may be involved both in soil dating and in using soils as age criteria in the Quaternary.

DATING SOILS BY DATING SURFACES

A soil is maximally as old as the geomorphic surface on which it began forming, and minimally as old as the base of its overburden. Soil changes incurred after burial remain under diagenesis as long as pedogenesis imposed on the overburden does not extend beneath its base.

A geomorphic surface, unit, or complex of surfaces

1. is a part of the land that is specifically defined in space and time and may include many landforms;

2. is a mappable feature whose geographic distribution is portrayed on maps or aerial photographs and whose geometric dimensions are specified and analyzed;

3. has an association with other geomorphic units (surfaces) which is defined in order to place it in a space and time sequence;

4. has an association with rock or sediment below it, or on it, which must be specified;

5. is dated by relative or absolute means, which must be specified;

6. is labeled with a geographic name (Ruhe and Vreeken, 1976).

Time zero of soil formation on a given surface reflects the nature of that surface. When a surface, whether constructional or erosional, stabilizes at some point in time, time zero of soil formation is marked by the onset of surface stabilization. If the surface remains instable, erosion or deposition may dominate weathering and soil-forming processes, and time zero of soil formation cannot be assigned to a specific moment. Also, if pedogenic processes dominate, but erosional or depositional processes continue, time zero cannot be pinpointed because it transgresses time.

Cessation of soil formation again reflects the nature of geomorphic processes responsible for burial, and may be abrupt or gradual. Abrupt cessation, suggested by a sharp upper boundary of the buried soil, often was preceded by partial truncation. In such a case, the burial event provides a minimum date for the end of soil development and there may have been a considerable pedogenic hiatus. Gradual cessation, suggested by pedogenic features within the basal increments of the overburden, still may have been preceded by partial truncation of the pre-burial soil profile. Here, the beginning of burial provides again a minimum date for the end of development of the older soil, and it provides a maximum date for the time-transgressive time zero of the subsequent cumulic soil.

Relative dating of geomorphic surfaces involves combined use of the Principle of Superposition, *i.e.* younger beds are on top of older beds providing they have not been overturned, and of the Principle of Cross-Cutting Relationships, *i.e.* in a cross-cutting relationship, the feature that is cut is older than the feature that cuts across it. Applied to the landscape, this results in the following Principles for Dating of Geomorphic Surfaces:

1. A surface is younger than the youngest material that it cuts;

2. A surface is younger than any structure that it bevels;

3. A surface is younger than any material of which there are distinguishable fragments or fossils in alluvial deposits that are now on top of that surface;

4. A surface is contemporaneous with alluvial deposits that lie on it, and it is the same age as, or older than other terrestrial deposits lying on it;

5. A surface is older than valleys cut below it;

6. A surface is younger than erosion remnants above it;

7. A surface is older than deposits in valleys cut below it;

8. A surface is younger than any erosion surface standing at a higher level;

9. A surface is older than any erosion surface at a lower level (Ruhe, 1969a).

These principles are found illustrated in numerous publications and have been summarized by Ruhe (1969a) and by Daniels *et al.* (1971).

DATING SOILS BY WAY OF SOIL PROPERTIES

Indices of soil development have been and continue being used as measures of relative soil age. The approach is as follows. One or more properties from a series of soils differing in age, are keyed to relative age values to produce a reference chronosequence. Alternatively, complete soil profiles are arranged into development sequences, in which soils are designated as fully mature, half-mature, *etc.* (Jenny, 1941), or as infantile, juvenile, virile and senile (Mohr and Van Baren, 1954). Subsequently, non-dated soils are assigned a certain age upon comparison with the reference sequence, while applying the Principle of Uniformitarianism, *i.e.* "The same processes and laws that operate today, operated throughout geologic time, although not necessarily with the same intensity as now" (Thornbury, 1969). Principal chronosequences were discussed by Vreeken (1975). Bockheim (1980) gave the most recent review of soil chronofunctions.

There are two variants in the application of this approach. The first may be called Dating by Pedogenic Index. Here, the reference sequence is used under the assumption that soils differing in terms of this index "represent the consecutive stages of some soil forming process" and that "these soils, occurring as they do at one end of the evolution series, have in the past gone through stages characterized by some preceding younger member of the same series, *i.e.* they have been endowed earlier with features which are now the property of soils that stand at the beginning of the evolutionary series" (Rode, 1961). Thus, environmental constancy is implicit. Two subvariants may be identified, *i.e.* Dating by Average Pedogenic Index and Dating by Variable Pedogenic Index.

The second variant may be called Dating by Paleopedogenic Index. Here dating is based again on a reference sequence, but on the assumption that certain properties of the older soil members could not have developed under the present environment, such anomalous properties are then ascribed to paleopedogenesis (Butler, 1967). The anomalous properties, when encountered in undated soils, become a criterion for their age. Thus, environmental change is emphasized.

Dating by Average Pedogenic Index

In this method, the soil is viewed as a result of the relative effective age (duration) of weathering of its parent material (Ruhe, 1969b). Average rate of soil change is estimated by dividing the difference between the value of a property for a soil and that of its presumed parent material, through an appropriate absolute or relative time measure. Relative effective age of another soil is then determined by dividing the value of its index property through that rate value. Index properties used include depth of carbonate leaching (Smith, 1942), solum thickness (Hutton, 1951), and molar ratios of CaO/ZrO_2 (Beavers *et al.*, 1963). Such relative effective age values, which actually are development ratios, have been subsequently used to explain between-profile differences in other soil properties. Implicit assumptions are those of monogenetic soil development, environmental constancy, and that all types of soil development have been replicated throughout geologic time.

Soil monogenesis implies that the soil-forming factors remained constant during the course of soil formation. Using this premise, functional analysis of modern soils, introduced by Jenny (1941) and updated by Birkeland (1974), has abundantly illustrated the limiting effectiveness on soil development of factors other than time. Soils of similar age tend to differ in accordance with climatic, vegetational... *etc.* setting. When adding to this the more realistic notion that soil-forming factors tend to interact and change through time, one arrives at the view of soil as the product of a functional synthesis of the soil-forming factors (Stephens, 1947). This view implies *a priori* assumption of potential soil polygenesis and can be combined with conclusions reached from functional analysis into the Principle of Limiting Pedogenic Factors

> The degree of development of a soil profile may reflect the limiting effectiveness of one, or of more than one soil-forming factor, but soil development must be evaluated with a view on the integrated action of the soil-forming factors, changeable as these may be through time.

This Principle implies that differences in expression of soil-forming factors can partly substitute for one another and yet effect similar pedogenic imprints. That is the element of equifinality that limits the diagnostic value of pedogenic indices as measures for soil age. In soil formation, the significance of the factor time as a simple measure of process duration is probably outstripped by its role as a time-space or frame for successive environmental conditions and processes. Thus, the very effort of constructing average rate measures of long term soil change is questionable, let alone their use for the purpose of soil dating.

Finally, when transplanting rates of processes backwards into time for purposes of dating soils from the more distant past by this method, the assumption of environmental and pedogenic replication throughout time clashes with the Principle of Environmental Uniqueness

> No environment in geological history is ever exactly duplicated. Especially progressive organic evolution, continental drift, and plant succession during climatic change are not replicated (After Nairn, 1965).

In conclusion, caution expressed by Ruhe (1969b) against Dating by Average Pedogenic Index is formulated here as the Principle of Fallibility of Pedogenic Indices as Estimators of Soil Age:

> Using a property of soil to construct a time measure that, in turn, is used to show that soils differ because of age, involves a circular argument, unless some independent criterion of environmental

history relevant to soils is brought in.

Dating by Variable Pedogenic Index

This method acknowledges that a newly exposed surficial material may adjust by internal change, at variable rate, to attain a state that is more in equilibrium with the prevailing environment. An initially rapid rate of adjustment would progressively decrease as a hypothetical equilibrium state would be approached, if soil development were comparable to a thermodynamic system adjusting to change (Lavkulich, 1969). This then could apply to pedogenesis under environmental conditions permitting monogenesis. If environmental conditions were to change during pedogenesis, rates of pedogenic adjustment could change again (Yaalon, 1971), to follow a new trend, a shifted pedogenic course.

According to this concept, reference chronosequences could be envisaged, to typify rate variations of monogenetic processes under constant and specified environmental conditions. Age benchmarks to peg the sequences would necessarily have to be obtained using non-pedogenic criteria. Then, the age of an undated soil that developed under the same conditions, could be picked off the reference curve, on the strength of its pedogenic index property. Provided, of course, that the latter value would not plot on the flat, *i.e.* time-independent portion of the curve depicting the chronofunction, and provided that the curve was ascending only. Sags in the reference curve would allow for multiple age options, if a single index were to be used. Sags in such a reference curve could reflect, for example, the depletion of nutrients, *e.g.* P, necessary to maintain levels of biotic productivity and of related pedogenic indices (Walker, 1965). Rate variations of processes under changing environmental conditions would be again more variable. The corresponding compounded reference chronosequences would be less useful for purposes of dating unknown soils.

Pedologists have constructed chronosequences ever since Dokuchaev's first attempt in 1883 (Dickson and Crocker, 1954). Their primary objective was pedochronologically oriented. The traditional approach has been to employ soils of different age, present in the groundsoil continuum of regions with multiple geomorphic surfaces, and to interpret them as a time series. These so-called post-incisive chronosequences (Vreeken, 1975) constitute the vast majority of the sequences, reviewed by Bockheim (1980), and represent a basis for contemporary pedological models.

Regrettably, the pedochronological significance of these post-incisive sequences is highly overrated. Basic limitations are imposed by the kind of soil evidence selected in the field (Vreeken, 1975). Also, their traditional interpretation is confused by monogenetic bias and circular logic, oblivious of cautions expressed by Rode (1961) and in Nairn's (1965) Principle of Environmental Uniqueness. Unless a post-incisive soil chronosequence can be proven to reflect monogenesis, and until someone can prove that soil history repeats itself, this sequence does not provide unambiguous information on rate variations of pedogenetic processes through time, because it does not provide time-lapse information. These traditionally used chronosequences are unfit for the testing of thermodynamics-inspired models. Neither do they provide the pedochronological information ascribed to them.

The secondary objective of chronosequence studies is to use them for purposes of dating formations and surfaces of unknown age. To this end, the same approach is followed as outlined before, but the reference sequences are used in a purely empirical manner. Examples can be found in Birkeland (1974, 1978) and Birkeland *et al.* (1979, 1980). The main problem that may arise is the problem of equifinality, *i.e.* whereby soil-forming factors can partly substitute for one another or operate differently through time and yet effect similar pedogenetic imprints. This includes the potential overprinting of impacts from pedogenic hysteresis and inertia.

Clearly, such empirical reference sequences are not likely to be exportable far beyond the relatively small area in which they were established, because environmental factors and their history may spatially differ. On grounds of similar pedogenic indices, two soils might be equated that are actually quite apart in age. In addition, site selection is crucially important, both when establishing the reference sequence and when sampling the undated soil. Site stability, with reference to geomorphic processes, is one prerequisite and would dictate the use of level upland or interfluve sites, thus restricting the technique to only one hillslope element from the catenary soil-landscape (Ruhe, 1960). Also, the dated as well as undated soils should always have occupied this comparable hillslope position. If not, one risks comparing profiles that differ both in age and catenary position, hence process setting. These two basic approaches towards explaining soil-topography relationships were already (in 1930) reviewed by Norton and Smith. Actually they complement one another (Ruhe and Walker, 1968; Vreeken, 1973) according to the original catena concept (Milne, 1936). The restriction to geomorphically stable sites implies that the soils to be used are almost automatically polygenetic, if not even relict paleosols. Polymorphic (Simonson, 1978) and bisequal profiles may characterize the older soils, adding complexity to comparisons between profiles of different age.

Indices measuring degree of soil development are being used as indices for relative age in the absence of reference chronosequences (Birkeland, 1974). This is done in areas where alternative age criteria are absent or have not been explored as yet. Qualitative as the results of this approach may be at the best, soils should not be used as age criteria, to the exclusion of other relative dating methods. *Ibidem* Pedogenic indices remain fallible as estimators of soil age.

Dating by Paleopedogenic Index

Most soil-landscapes are polycyclic, having developed under various geomorphic, climatic and/or biotic regimens. Thus paleopedogenic imprints are normal phenomena, whether as relict or polygenetic soil features, or as buried soils in restricted landscape positions. This applies both to soil-landscapes that are largely subaerially exposed, and to those that have been buried *in toto*. Such imprints have been used as relative age indices for the geomorphic surfaces on which they are found. However, prerequisites must be met.

Diagnosis of soil features as paleopedogenic is the first prerequisite. This requires criteria to distinguish between features originating from successive pedogenetic regimens, and is normally complicated by the differential persistence of older features when subjected to further pedogenesis or diagenesis. It also requires distinction between soil, weathered zone and sediment. Analytical and micromorphological criteria can be helpful but, taken by themselves, are of limited diagnostic value when referring to single profiles or vertical sections only. The genetic soil concept requires soil to exhibit lateral variability within the three-dimensional landscape, in accordance with differences in the soil-forming factors. Catenary trends of variability provide the primary distinction between soil and not-soil. Criteria for the identification of buried soils were reviewed by Valentine and Dalrymple (1976), but no review is available for relict soils. In both categories, there is a shortage of information on hydrophysical and hydrochemical soil-sediment continua. This is not pronounced for aquatic environments not traditionally studied by pedologists (Buurman, 1975). These include paludal, lacustrine, deltaic, estuarine, tidal flat and other marine environments. These gaps may cause much paleopedogenic evidence to go undetected.

Buried Soil-Landscapes

Buried soil-landscapes are significant markers in terrestrial stratigraphical studies and serve as macrocriteria for the relative age of their overburdens and substrates. Reviews are in Birkeland (1974), Mahaney and Fahey (1976) and Finkl (1980). Prerequisite is that these soil-landscapes have been assessed in terms of the geomorphic surfaces contained, and in terms of catenary trends of variability within these surfaces, as well as the differences in these trends between adjacent surfaces, both on the local and regional scale. Because such prerequisites often are not met, telecorrelation of buried soils on the basis of relative degree of profile development alone, is advised against (Birkeland, 1974).

Isolated buried soil profiles or, conversely, their local absence should not form a basis for regional landscape inferences. For example the type locality for the Yarmouth Interglacial paleosol, 82 years later, was shown to contain a pre-Illinoian age soil at the type level. The Yarmouth soil was stratigraphically lower (Hallberg *et al.*, 1980). On the other hand, absence of paleosols from certain buried geomorphic surfaces in northeast Iowa, while attributed to the Iowan Glaciation during more than 70 years, was finally shown to reflect a subaerial erosion cycle (Ruhe *et al.*, 1968). The uniformitarian application of what is known about the groundsoil continuum entails the Principle of Regional Paleogeomorphic Significance of Paleopedogenic Evidence *per se*.

> As much as soil formation, erosion and deposition are simultaneously active within present day landscapes, paleopedogenic evidence from a limited number of sites is insufficient to infer regional/landscape stability. Absence of paleopedogenic evidence from limited sites is insufficient to infer regional instability. Detailed soil-landscape analyses are a prerequisite.

Subaerially Exposed Soil-Landscapes

Polycyclic soil-landscapes contain various geomorphic surfaces and, correspondingly, an age mosaic of soils, including relict paleosols. The paleopedogenic index method of relative dating presumes that environment of soil formation can be read from the constitution of soils themselves: Assuming that morphology and properties of relatively young soils should reflect the currently operative environmental framework, these soils serve as an environmental norm to gauge the features of other soils in the same area. If latter features are dissimilar to the norm, they are regarded as anomalous and interpreted to originate from a different environmental framework. Thus, they are interpreted as paleopedogenic, and indicative of greater age.

The first question regarding this method is: What soils unambiguously reflect the currently operative environmental framework, and could serve as environmental norms? Interpretation of soils as environmental indicators is rooted in the Climatic Soil Zonality Concept. But, most soils showing a zonal distribution resembling that of bioclimatic divisions are polygenetic, as much as distribution and expression of latter divisions are resultants of change and succession. Therefore, as long as those soils, used as environmental norms, are not pedochronologically and genetically fully understood, their normative value for paleo-environmental reconstruction, with all attendant inferences on age, is doubtful. This reservation may be termed the Non-Axiom of Climatic Soil Zonality:

> The principal differences between soils, and certainly between paleosols, are not necessarily climatically zonal, but present climate alone does not control and influence the biological, chemical and most of the physical processes in soil formation. Differences due to other soil-forming factors, including the variable impact of all soil-forming factors through time may be more important.

A second question is, whether one really could interpret the character of an old soil in terms of present-day environment, using soil evidence alone. The Principle of Environmental Uniqueness suggests that such interpretations can be tentative only. Thus, support from independent, non-pedological paleo-environmental criteria will be required (Butler, 1967). Although Bryan and Albritton (1943) reconstructed environment from buried paleosols in North America, Ruhe (1965) drew attention to equifinal pedogenic features when substituting time and various other soil-forming factors, thus underscoring the Principle of Limiting Pedogenetic Factors. Butler (1967) warns against reconstructing environmental histories from soil-derived evidence, where applied to non-buried soils in Australia.

CONCLUSION

In conclusion, as much as pedogenic criteria are fallible as measures of soil age, they are fallible as paleoenvironmental and corresponding age indicators. Stratigraphically supported criteria for the environmental significance of soil features must be established first. This involves soil-landscape analysis and the delineation of geomorphic surfaces.

REFERENCES CITED

Beavers, A.H., Fehrenbacher, J.B., Johnson, P.R. and Jones, R.L., 1963, CaO-ZrO_2 Molar Ratios as an Index of Weathering: Soil Sci. Soc. Amer. Proc., v. 27, p. 408-412.

Birkeland, P.W., 1974, Pedology, Weathering, and Geomorphological Research: Oxford University Press, Oxford.

Birkeland, P.W., 1978, Soil Development as an Indication of Relative Age of Quaternary Deposits, Baffin Island, N.W.T., Canada: Arctic and Alpine Research, v. 10, no. 4, p. 733-747.

Birkeland, P.W., Colman, S.M., Burke, R.M., Shroba, R.R. and Meierding, T.C., 1979, Nomenclature of Alpine Glacial Deposits, or, What's in a Name?: Geology, v. 7, p. 532-536.

Birkeland, P.W., Burke, R.M. and Walker, A.L., 1980, Soils and Subsurface Rock-Weathering Features of Sherwin and pre-Sherwin Glacial Deposits, Eastern Sierra Nevada, California: Geol. Soc. Amer. Bull., Part I, v. 91, p. 238-244.

Bockheim, J.G., 1980, Solution and Use of Chronofunctions in Studying Soil Development: Geoderma, v. 24, no. 1, p. 71-85.

Bryan, K. and Albritton, C.C., 1943, Soil Phenomena as Evidence of Climatic Changes: Amer. Jour. Sci., v. 241, p. 469-490.

Bryan, W.H. and Teakle, L.J.H., 1946, Pedogenic Inertia - a Concept in Soil Science: Nature, v. 164, p. 969.

Butler, B.E., 1967, Soil Periodicity in Relation to Landform Development in Southeastern Australia: in Jennings, J.N. and Mabbutt, J.A. (eds.), Landform Studies from Australia and New Guinea, Austr. Nat. Univ. Press, Canberra, p. 231-255.

Buurman, P., 1975, Possibilities of Paleopedology: Sedimentology, v. 22, p. 289-298.

Daniels, R.B., Gamble, E.E. and Cady, J.G., 1971, The Relation between Geomorphology and Soil Morphology and Genesis: Advances in Agronomy, v. 23, p. 51-88.

Dickson, B.A. and Crocker, R.L., 1954, A Chronosequence of Soils and Vegetation near Mt. Shasta, California, I. Definition of the Ecosystem Investigated and Features of the Plant Succession; J. Soil Sci., v. 4, no. 2, p. 123-141.

Finkl, C.W., 1980, Stratigraphic Principles and Practices as Related to Soil Mantles: Catena, v. 7, p. 169-194.

Gerasimov, I.P., 1971, Nature and Originality of Paleosols: in Yaalon, D.H. (ed.), Paleopedology, I.S.S.S./U.N.E.S.C.O., Jerusalem.

Hallberg, G.R., Fenton, T.E., Kemmis, T.J. and Miller, G.A., 1980, Yarmouth Revisited: Guidebook for the 27th Field Conference, Mid West Friends of the Pleistocene. Iowa Geol. Survey, Iowa City, and Iowa State University, Ames, Iowa.

Hutton, C.E., 1951, Studies of the Chemical and Physical Characteristics of a Chrono-Litho-Sequence of Loess-derived Prairie Soils of Southwestern Iowa: Soil Sci. Soc. Amer. Proc., v. 15, p. 318-324.

Jenny, H., 1941, Factors of Soil Formation: McGraw-Hill Co., New York.

Lavkulich, L.M., 1969, Soil Dynamics in the Interpretation of Paleosols: p. 25-37 in: Pawluk, S. (ed.) Pedology and Quaternary Research, Univ. Alberta Print. Dept., Edmonton.

Mahaney, W.C. and Fahey, B.D., 1976, Quaternary Soil Stratigraphy of the Front Range, Colorado, in Mahaney, W.C., ed., Quaternary Stratigraphy of North America, Dowden, Hutchison and Ross, Stroudsburg, Pa, p. 319-352.

Milne, G., 1936, A Provisional Soil Map of East Africa: East African Agr. Res. Sta., Amani Memoirs.

Mohr, E.C.J. and Van Baren, F.A., 1954, Tropical Soils: Van Hoeve, The Hague.

Nairn, A.E.M., 1965, Uniformitarianism and Environment: Palaeogeogr., Palaeoclim., Palaeoecol., v. 1, p. 5-11.

Norton, E.A. and Smith, R.S., 1930, The Influence of Topography on Soil Profile Character: Jour. Amer. Soc. Agronomy, v. 22, p. 251-262.

Rode, A.A. 1961, The Soil Forming Process and Soil Evolution, Israel Program for Sci. Transl., Jerusalem.

Ruellan, A., 1971, The History of Soils: Some Problems of Definition and Interpretation, p. 3-13 in Yaalon, D.H. (ed.), Paleopedology, I.S.S.S./U.N.E.S.C.O., Jerusalem.

Ruhe, R.V.,1960, Elements of the Soil-Landscape, Trans. 7th Int. Cong. Soil Sci., v. 4, p. 165-170, Madison.

__________, 1965, Quaternary Paleopedology, in Wright, H.E. and Frey, D.G. (eds.), The Quaternary of the United States, p. 744-764, Princeton University Press, Princeton.

__________, 1969a, Quaternary Landscapes in Iowa, The Iowa State Univ. Press, Ames.

__________, 1969b, Application of Pedology to Quaternary Research, p. 1-23, in Pawluk, S. (ed.), Pedology and Quaternary Research, Univ. Alberta Print. Dept., Edmonton.

__________, 1972, Pedology (Soil Science), p. 911-917, in Fairbridge, R.W. (ed.), Encyclopedia of Geochemistry and Environmental Sciences, Van Nostrand Reinhold Co., New York.

Ruhe, R.V., Dietz, W.P., Fenton, T.E. and Hall, G.F., 1968, Iowan Drift Problem, Northeastern Iowa: Iowa Geol. Survey Rept. Invest. 7, Iowa City.

Ruhe, R.V. and Walker, P.H., 1968, Hillslope Models and Soil Formation. I. Open Systems, in Trans. 9th Int. Congr. Soil Sci., Adelaide, S.A., v. 4, p. 551-560.

Ruhe, R.V. and Vreeken, W.J., 1976, Soil-Geomorphology Reconnaissance in the Darien Project Area: Final Report (mimeo) to Org. Amer. States, Washington.

Ruhe, R.V. and Olson, C.G., 1980, Soil Welding: Soil Sci., v. 130, n. 3, p. 132-139.

Simonson, R.W., 1978, A Multiple-Process Model of Soil Genesis, in Mahaney, W.C. (ed.) Quaternary Soils, Geo Abstracts, Norwich, England.

Smith, G.D., 1942, Illinois Loess - Variations in its Properties and Distribution - A Pedologic Interpretation: Ill. Agric. Exp. Sta. Bull. 490, p. 139-184.

Stephens, C.G., 1947, Functional Synthesis in Pedogenesis: Trans. Roy. Soc. S. Austr., v. 71, p. 168-181.

Thornbury, W.C., 1969, Principles of Geomorphology: Wiley and Sons, Inc., New York.

Valentine, K.W.G. and Dalrymple, J.B., 1976, Quaternary Buried Paleosols: A Critical Review: Quat. Res., v. 6, p. 209-222.

Vreeken, W.J., 1973, Soil Variability in Small Loess Watersheds: Clay and Organic Carbon Content, Catena, v. 1, p. 181-196.

_____________, 1975, Principal Kinds of Chronosequences and their Significance in Soil History: Jour. Soil Sci., v. 26, no. 4, p. 378-394.

Walker, T.W., 1965, The Significance of Phosphorus in Pedogenesis, in Hallsworth, E.G. and Crawford, D.V., eds., Experimental Pedology, Butterworths, London, p. 295-315.

Yaalon, D.H., 1971, Soil-Forming Processes in Time and Space, in Yaalon, D.H. (ed.), Paleopedology, I.S.S.S./U.N.E.S.C.O., Jerusalem, p. 29-40.

DATING WITH POLLEN: METHODOLOGY, APPLICATIONS, LIMITATIONS

A.M. DAVIS

ABSTRACT

Although the dating function of pollen has been largely superceded by radiocarbon, the former is still valuable particularly for dating beyond the range of ^{14}C, in determining the chronology of recent and short-term events, and in sediments where ^{14}C cannot be used. Pollen stratigraphies may be useful as checks on the accuracy of radiocarbon dating.

The basis for dating with pollen is the recognition of assemblages or zones, usually from the percentages of the taxa, but sometimes from pollen influx. The determination of zones is influenced by a number of site-related variables and by preparation and analytical constraints. Pollen spectra vary with the environment of deposition. Pollen records from lake sediments tend to represent regional inputs, while those from peats are usually dominated by local producers. There may be considerable internal variability that in lakes reflects size and shape, and the position of the sequence within the basin. Pollen spectra from soils, caves, inshore, and deepwater marine deposits may all be distinctive and different. All can be zones, but the taxa on which zonation is based may vary and the zones may not be synchronous.

The pollen spectra may also be influenced by differential preservation and by redeposition from older sediments. The record may be distorted by misidentification, particularly a problem in the early days of pollen analysis. The validity of zonation is in part conditioned by the size and composition of the pollen sum.

Zonation of pollen sequences can be determined subjectively or objectively by several numerical methods: clustering and divisive techniques, principal components analysis, sequential correlation, *etc*. The latter methods allow division into zone and sub-zones by identifying the locations and 'powers' of groupings and/or boundaries. Subjective and objective approaches seldom yield dissimilar results.

The broad synchroneity of pollen zones in Holocene sequences from eastern North America and northwest Europe illustrate the value of zonation as a dating device. However, effectiveness and synchroneity may be much less obvious in sequences from sedimentary environments other than lakes. Zonation may be extended as a relative dating device beyond the range of ^{14}C. Four-or five-phase models of glacial/interglacial pollen spectra have been applied in Europe and North America. The recognition of these depends largely on the assumption of the ecological stability of the taxa on which zonation is based. Pollen influx has been used to estimate the length of interglacials.

Pollen analysis is valuable in determining the chronology of recent or short-term events outside of the resolution of ^{14}C.

INTRODUCTION

Within the range of radiocarbon, pollen stratigraphies are generally supported by ^{14}C dates. However, for most sequences, the number of dates is small, and due to environmental and procedural constraints many of the chronologies are unreliable. Ogden (1977) estimated that there were then approximately 300 sediment cores from eastern North America. Less than 30 of these had more than three radiocarbon dates. Terasmae (this symposium) indicates that less than 50% of ^{14}C dates are accepted by those submitting the material. Some of the rejections are presumably due to collection or processing problems; others to the unsuitability of materials (Ohlsson, 1979). Lake sediments often yield unduly old dates because of reservoir effects (Ohlsson and Florin, 1980). Shotton (1972) has demonstrated the effect of 'hard water' error on dating. In a sediment composed of an algal mud matrix with inclusions of wood and small twigs, the latter dated 1700 years younger than the matrix. The dates from wood and twigs were in agreement with the pollen stratigraphy reliably dated elsewhere. The 'old' date on the algal mud was the result of synthesis of carbonate by algae.

Dates from cave sediments and archaeological sites are often equivocal. The accuracy of dates from peat depends much on the care used to eliminate contamination by the deep roots of plants younger than the matrix.

Radiocarbon is routinely utilized to *ca.* 40,000 BP. Beyond that, pollen stratigraphies may provide crude relative dates, *i.e.* glacial *vs.* interglacial, or position within the glacial/interglacial cycle. Modern or short-duration events cannot be adequately dated by radiocarbon. Pollen analysis may allow determination of absolute age within a standard deviation of ^{14}C. Pollen influx may provide a 'floating' absolute chronology within the range of ^{14}C or beyond it (Dabrowski, 1971; Mehringer *et al.*, 1977).

ZONATION AND DATING

The most common approach to dating with pollen is through the recognition of biostratigraphic units: pollen zones. Prior to the development of radiocarbon dating, this was the only method. It remains important despite the availability of ^{14}C. Zonation of pollen diagrams should be determined entirely on the observed pollen assemblages. However, other criteria have been used. Birks and Birks (1979) suggest that pollen zones have been defined as units of inferred past vegetation, units of inferred past climate, as units of sediment lithology, and as units of inferred time.

The association of zones with time-bound climate and vegetation changes predates the use of ^{14}C. The Blytt-Sernander scheme of late-glacial and Holocene climatic change was based on peat stratigraphy and emphasized moisture as a control. Inevitably, pollen analysis developed by von Post after 1916 was also climatically-constrained. Von Post, recognizing temperature as a major control, proposed a tripartite Holocene with an episode of increasing warmth, followed by a climatic optimum, which was replaced by an episode of decreasing warmth. Sears (1935) attempted to apply the European pollen/climate sequence to North America. He recognized four zones (Boreal, Atlantic, Subboreal, Subatlantic). These zones were considered to be biostratigraphic units with time-parallel boundaries.

Increasing use of ^{14}C to date pollen sequences has shown that the assumption of time-parallel zones is doubtful. The pollen records from New England show a lack of synchroneity in the appearance of major taxa. The climatic optimum, or hypsithermal appears to be time-transgressive (Wright, 1976). Richard (1977) has shown that the tundra zone at the base of pollen sequences in southern Quebec dates at 11,400 BP at Mount Shefford and at 7200 BP in the southern Laurentides. Smith and Pilcher (1973) have demonstrated that the Alnus rise in Britain, assumed to mark the start of the Atlantic at *ca.* 7500 BP transgresses some 2000 years. The difficulties of identifying zones with regional application have become obvious as the number of pollen sequences has increased. In Britain, the zonation scheme proposed by Godwin (1940) is still used, but has frequently been replaced or complemented by local schemes. In North America, zonation schemes often have a regional application, but interregional comparison may be difficult (Figure 1).

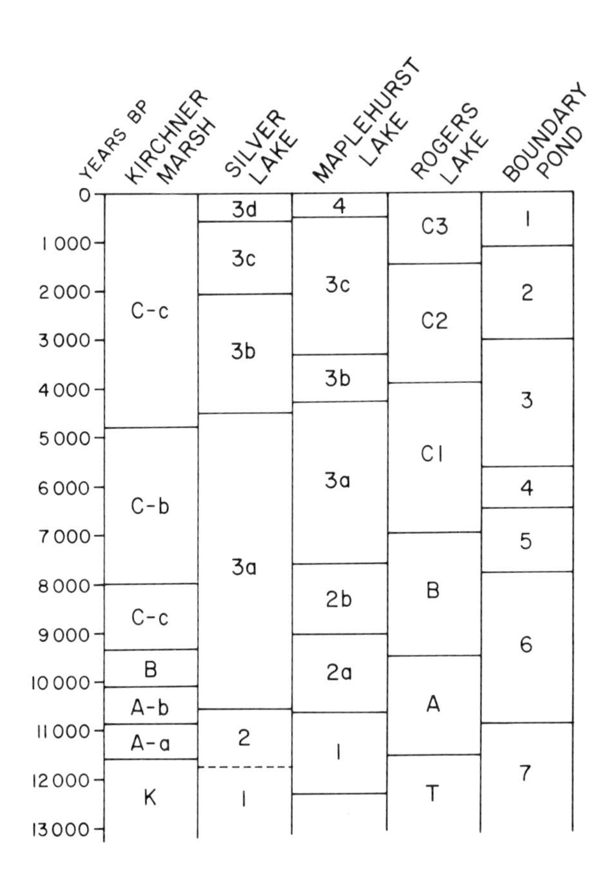

Figure 1 Comparison of pollen zonation schemes from eastern North America. Some similarities are evident, despite the lack of a common zonation nomenclature. K.M. (Wright *et al.*, 1963); S.L. (Ogden, 1966); M.L. (Mott and Farley Gill, 1978); L.P. (Davis, 1967; B.P. (Mott, 1977).

Pollen Behaviour and Site Variables

Clearly, the derivation of any zonation scheme is dependent on the variations in the pollen spectra. The pollen record from any site is determined by many pollen behaviour and site variables. These include pollen productivity, dispersal, deposition/redeposition, pollen preservation, site size and character. Analytical constraints, particularly the number of grains counted, and the composition of the pollen sum, influence representation of the pollen assemblage, hence zonation. A brief review of these variables is presented below. Birks and Birks (1979) presented a more detailed synthesis.

The pollen rain at any location is largely determined by productivity and dispersal. Janssen (1973) recognized four components: local, extralocal, regional, and extraregional pollen. Most pollen has a short dispersal distance, consequently local components, *i.e.* those growing at the site or within a few meters of it, should dominate the spectra. Such domination is usually obvious in records from peatlands and small lakes. Local input is small in the centre of large lakes and also in environments where local productivity is small. In tundra areas, the pollen rain contains a large extraregional component due to the low inputs from other components of the pollen rain.

The incorporation of pollen grains into sediments may involve a number of physical and chemical processes each of which will contribute to the character of the pollen spectra and hence, to the distinctiveness of zonation. Tauber (1965) has identified three dispersal elements in the pollen rain of a forested area: trunk space pollen, above canopy pollen, and rain-out pollen. The relative contribution of each varies with the size of the basin of deposition, above canopy and rain-out components becoming more important as basin size increases. In lake sediments pollen spectra may show considerable horizontal variation (Davis *et al.*, 1971; Davis and Brubaker, 1973). Mixing by burrowers may homogenize pollen through 15 cm of sediment (Davis, 1974). This may supress fluctuations in the pollen record and influence the determination of pollen zones.

Pollen spectra may be modified by the effects of differential decay and by contamination from older sediments. Dimbleby (1976) concluded that after two years almost no pollen remained in soils with pH > 6.0, but Bryant (1969) has recovered pollen in soils of pH 8.9 although identification was difficult because of corrosion. Corrosion can severely skew a pollen spectrum. Resistant types become over-represented and the number of species represented may be reduced to produce a spectrum that may be unlike that originally deposited.

Conversely, pollen spectra may be augmented by redeposited pollen. Sometimes the redeposited component may be morphologically distinguishable from contemporary pollen *e.g.* pre-Tertiary or Tertiary pollen in late Quaternary sediments. The alien grains may be corroded. However, there may be no differences in degradation. Apparent ecological incompatibility may not be a valid criterion.

Numerical Constraints and Objective Zonation

The character of a pollen record is in part determined by numerical constraints. The reliability of the sample as representative of the population is dependent on the number of grains counted. Normally, 300 to 500 grains is sufficient, but numerical stability is influenced by the numbers of taxa and the frequency of each taxon. Stability is achieved rapidly when the taxon is common, more slowly if the taxon is rare (Figure 2).

Determination of confidence limits can provide an expression of reliability. Mosimann (1965) produced formulae for the necessary calculations for taxa 'inside' and 'outside' the pollen sum. The values are easily determined by computer or programmable calculator although Maher (1972) has produced nomograms for the purpose. In the same paper, he illustrates the technique with a pollen record from Molas Lake, Colorado. The presence of confidence limits on a pollen diagram helps to filter signal from noise and may be useful aid to subjective zonation.

The character of a pollen diagram is dependent not only on the size of the pollen sum but on the constitution of that sum. Zonation is almost always based on percentage values of component taxa. Those percentages are determined from a pollen sum which is usually the sum of the frequencies of less than the total number of taxa present. Sometimes only tree and shrub taxa will provide the sum. Commonly, non-arboreal (herb) taxa will be included, but aquatics and spores are usually excluded (Wright and Patten, 1963). The choice of the constituents of the pollen sum will determine the percentages, the shape of the pollen diagram, and hence subjective zonation of that diagram.

The bias likely in subjective zonation may be eliminated by the numerical zonation techniques (provided all taxa are included). Several of these objective methods are presently being used. They are summarized in Table 1 and by Birks and Birks (1979). Those most frequently used are the single-link clustering technique, CONSLINK, developed by Gordon and Birks (1972), and their two divisive techniques SPLITINF, based on total information content, and SPLITLSQ, which divides *via* the sum of least squares. CONSLINK has been proposed as the standard zonation technique for pollen data used in the International Geological Correlation Programme (IGCP). The results of the three techniques above

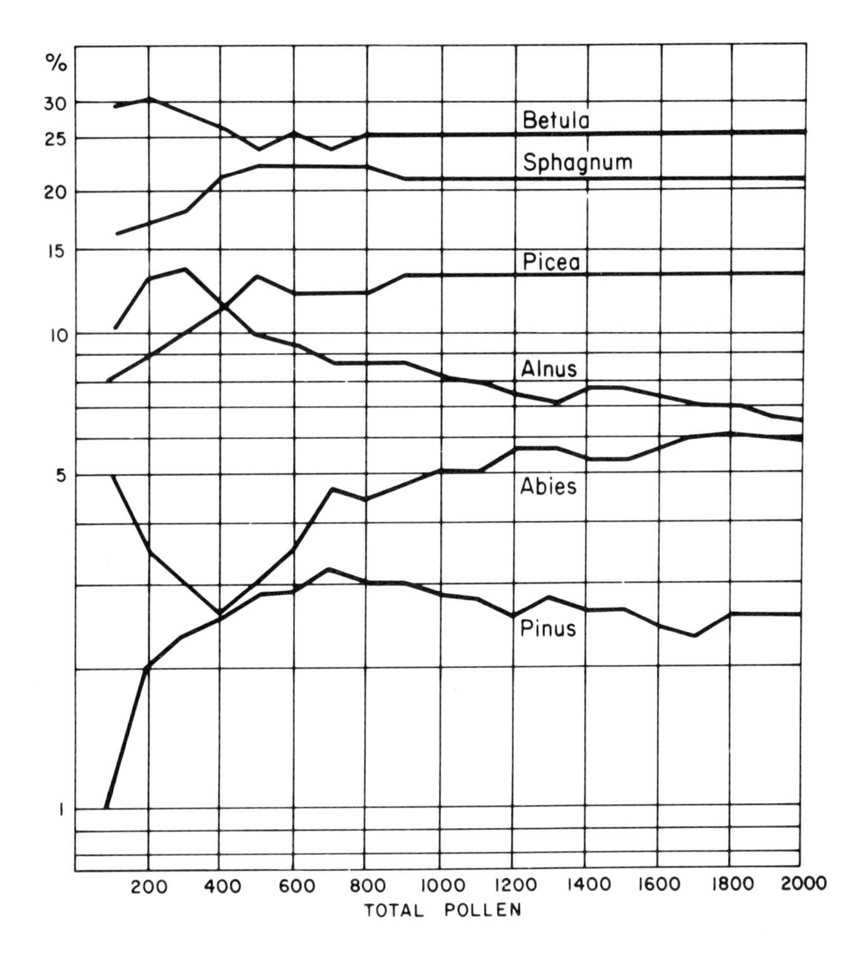

Figure 2 Variation in percentages in one sample as total count increases. Stability is achieved earliest in those taxa that are common (after Birks and Birks, 1979).

******

Table 1. Some approaches to objective zonation and comparison of pollen diagrams

ORDINATION	A. Non-metric multi-dimensional scaling (Gordon & Birks, 1974).
	B. Principal components analysis (Birks, 1974; Adam, 1974).
CLASSIFICATION	A. Divisive techniques using total information content (SPLITINF) and sum of least squares deviations (SPLITLSQ).
	B. Constrained single-link clustering method, CONSLINK (Gordon & Birks, 1972).
	C. Sequential correlation (Yarranton & Ritchie, 1972).
COMPARISON OF DIAGRAMS	A. Ordination techniques used above, MDSCAL, PCA.
	B. Sequence slotting, SLOTSEQ (Gordon & Birks, 1974).

can be included in a pollen diagram as dendrograms. Figure 3 illustrates the application of these techniques and principal components analysis to the diagram for Scaleby Moss, England (Birks, 1974). Objective techniques can be compared to each other, and to the local and standard British zonations.

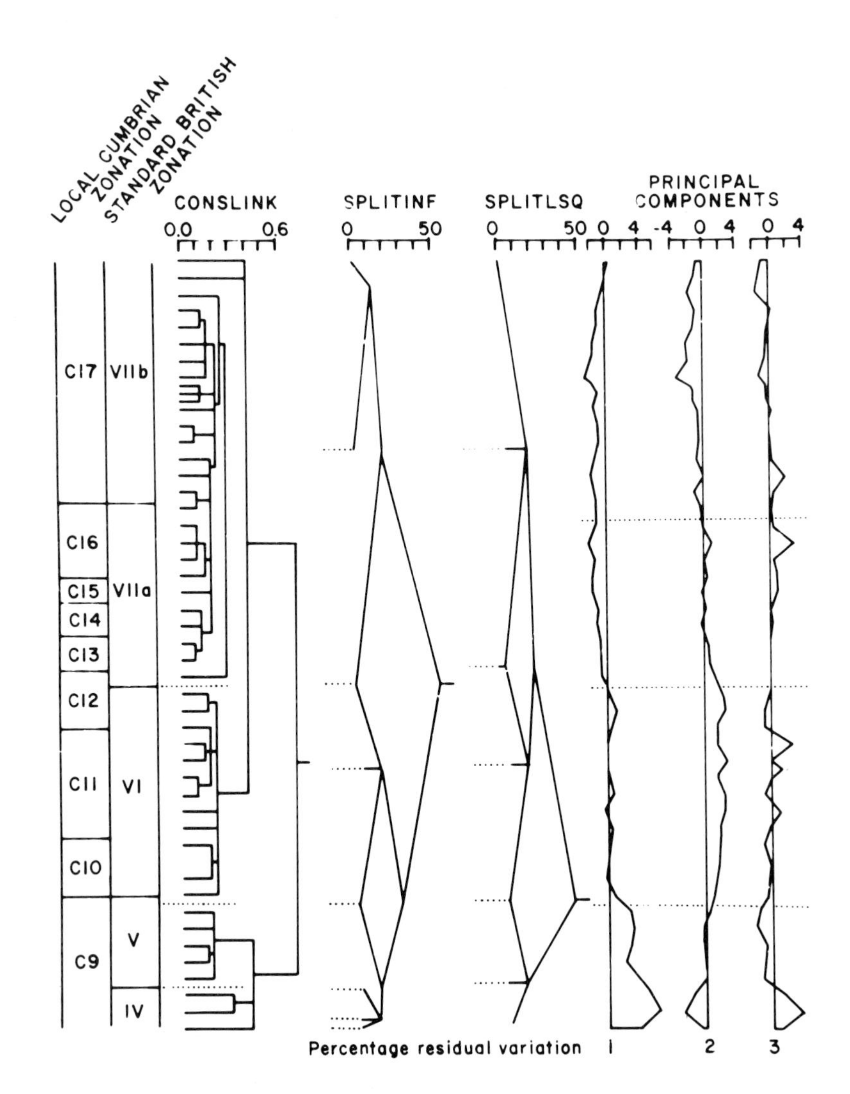

Figure 3 Results of the application of some objective zonation techniques to a diagram from Scaleby Moss, England. Note general concordance of boundaries (after Birks, 1974).

Walker and Wilson (1978) have proposed an alternative to these multivariate methods that is founded on the basic statistics (mean, variance, *etc.*) of individual taxa. Their technique involves influx not percentage values and therefore requires reliable absolute dating of the sequence.

Objective comparison of diagrams, useful to the recognition of regional zones is possible through principal components analysis, multi-dimensional scaling, and by a sequence slotting technique, SLOTSEQ (Gordon and Birks, 1974). The latter is suggested as regular procedure for IGCP projects (Birks, 1979).

Objective zonation generally confirms the schemes derived subjectively. However, the techniques provide a hierarchy of division that allows assessment of the importance of each division and the distinctiveness of each zone. They are particularly useful for the identification of sub-zones.

The section above outlines the variables that influence the character of a pollen diagram. The constraints suggest perhaps that zonation is inappropriate as a dating technique. However, where regional pollen assemblage zones are well-established normally with ^{14}C control, sequences can be dated *via* their pollen stratigraphies. The comparability is best where some of the site variables can be considered constant. Faegri and Iversen (1975) suggest that lakes of *ca.* 5000 m^2 are optimum sites for the study of regional pollen inputs. Digerfeldt (1979) has outlined criteria for the selection of lakes. The regional assemblage zones for northwestern Minnesota are well established (McAndrews, 1966) as are those for southern Ontario (McAndrews, 1981) and Quebec (Richard, 1980). In England, the zonation scheme originally proposed by Godwin (1940) has been confirmed by more recent work, despite the recognition of many local variants.

It is important, however, that zonation be based entirely on the assemblage present and not manipulated to fit a pre-existing scheme. Objective zonation should allay this temptation.

Pollen zones tend to occupy broad time periods. Consequently they provide only crude dating. Boundaries between zones may be well-dated but may be metachronous. The presence of large local producers may mask the regional pollen rain enough to produce assemblages that may be misinterpreted and thus incorrectly dated. The spruce zone at the base of a sequence from Blue Mounds Creek, Wisconsin, is well-defined (Figure 4). Its demise is rapid as it is elsewhere in the Midwest. However, the disappearance of boreal woodland occurred at *ca.* 11,000 BP in this area, not at 9500 BP as it is dated at Blue Mounds Creek (Bernabo and Webb, 1977; Davis, 1977). The spruce zone here is a product of black spruce growing on the site long after the general northward migration of boreal forest. The explanation is confirmed by macrofossil analysis.

Non-climatic controls such as soil development, differential migration and disturbance complicate the pollen spectra and zonation. These controls may be progressive (time-bound?) or episodic, local or regional (Iversen, 1973). Brubaker (1975) has demonstrated that substrate differences cause persistent modification of pollen spectra. Local disturbances caused by human activity have been largely responsible for the difficulties in recognizing regional pollen assemblage zones in Europe after 5000 BP (Birks and Birks, 1979).

Dating beyond the Range of Radiocarbon

Zonation of pollen sequences is important for dating sequences beyond the range of ^{14}C, although it normally provides only a crude relative dating within the glacial/interglacial cycle. In Europe, there has long been recognition of a single progressive to retrogressive cycle (Firbas, 1949; Iversen, 1958; Andersen, 1966). More recently the concept has been applied to North American sequences (Wright, 1972; Kapp, 1977). The length of interglacials and glacials based on varve-counts is variable. The full cycle appears to require 50-100,000 years. Interglacials, the shorter of the two components, seem to vary in length from 10-20,000 years (Davis, 1976).

Iversen (1958) has summarized the glacial/interglacial cycle in terms of climate, substrate, vegetation structure, composition and succession (Figure 5). Turner and West (1968) have suggested a four zone interglacial sequence. Zone 1, or Pretemperate phase, is marked by increasing forest closure and by boreal taxa. Zone II, Early-temperate or Mesocratic phase, has increasing amounts of deciduous trees. Zone III, the Late-temperate or Oligocratic phase, has late-arriving deciduous trees. In Zone IV, the Post-temperate phase, there is a return to boreal taxa and increasingly open conditions.

In North America, Wright (1972) has applied the model to Pleistocene interglacials and the Holocene. Kapp (1977) has suggested the addition of Zone V, Full glacial phase, for unglaciated areas. Figure 6 illustrates the application of the model to the Sangamon/Wisconsin

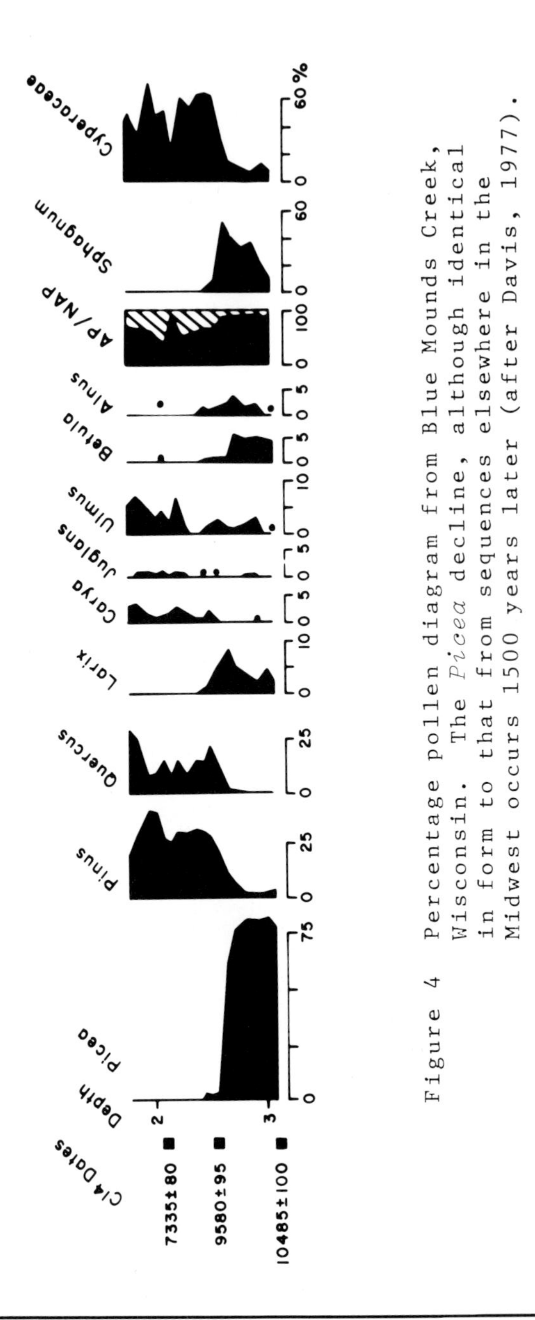

Figure 4 Percentage pollen diagram from Blue Mounds Creek, Wisconsin. The *Picea* decline, although identical in form to that from sequences elsewhere in the Midwest occurs 1500 years later (after Davis, 1977).

age sequence from the Pittsburg Basin, Illinois (Grüger, 1972).

The zonation scheme for the glacial/interglacial sequence is simple when compared to those derived for Holocene sequences. Its simplicity is both advantageous and misleading. Undated and partial sequences may be 'slotted' into one of the phases in the cycle and a crude relative date assigned. Assignment to the appropriate cycle requires some absolute or stratigraphic date control. Complications arise from the palynological dissimilarities between successive interglacials. It cannot be assumed that the pollen spectra at any phase of a cycle will be identical to those in the same phase of other cycles. In East Anglia, Zones II and III of successive interglacials show sometimes pronounced differences in their pollen spectra (Figure 7). With some taxa, the differences are variations in frequency and timing. Some species are absent from some interglacials.

The length of interglacials is best approximated by varve-counts, although these provide only floating absolute chronologies. The partially varved sequence at Marks Tey, England, indicates that Zone II and III in the Hoxnian Interglacial lasted about 4500 years (Turner, 1970).

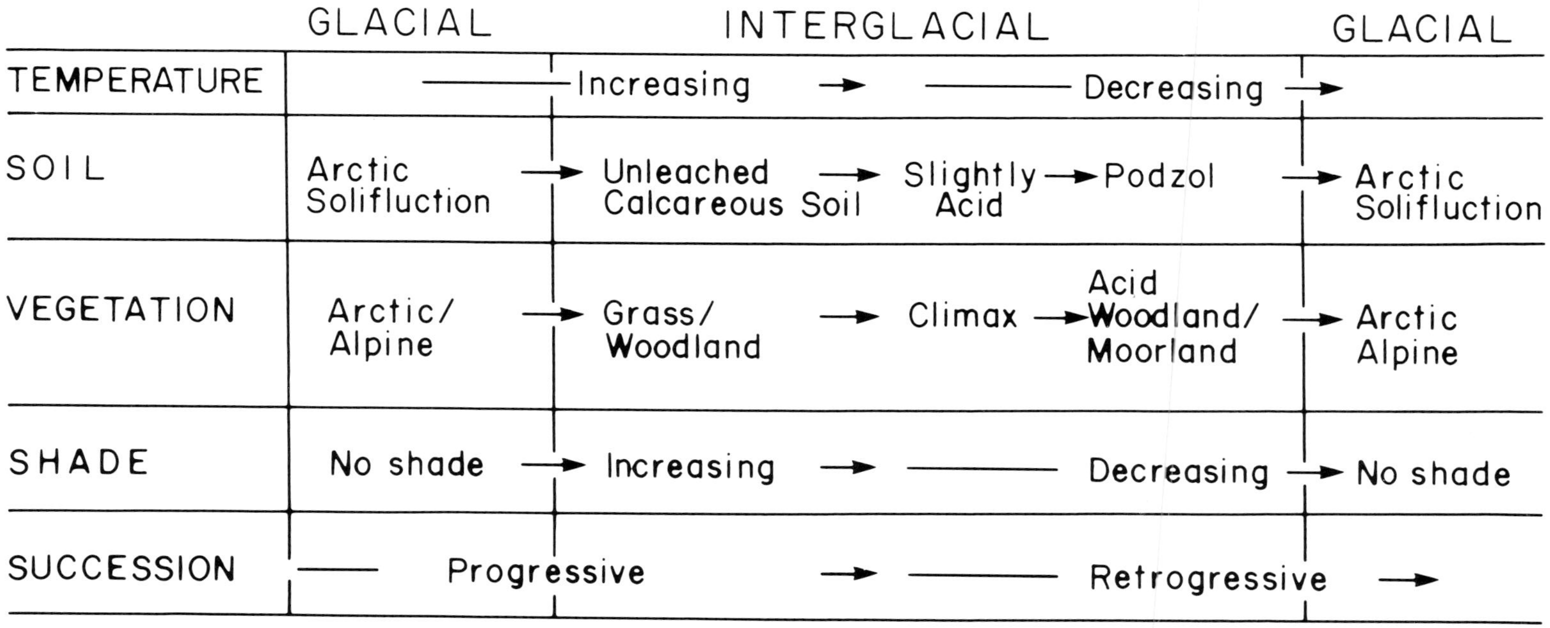

	GLACIAL	INTERGLACIAL	GLACIAL
TEMPERATURE	——	Increasing → —— Decreasing	→
SOIL	Arctic Solifluction →	Unleached Calcareous Soil → Slightly Acid → Podzol →	Arctic Solifluction
VEGETATION	Arctic/ Alpine →	Grass/ Woodland → Climax → Acid Woodland/ Moorland →	Arctic Alpine
SHADE	No shade →	Increasing → —— Decreasing →	No shade
SUCCESSION	—— Progressive	→ —— Retrogressive	→

Figure 5 Generalised changes associated with the glacial/interglacial cycle (after Iversen, 1958).

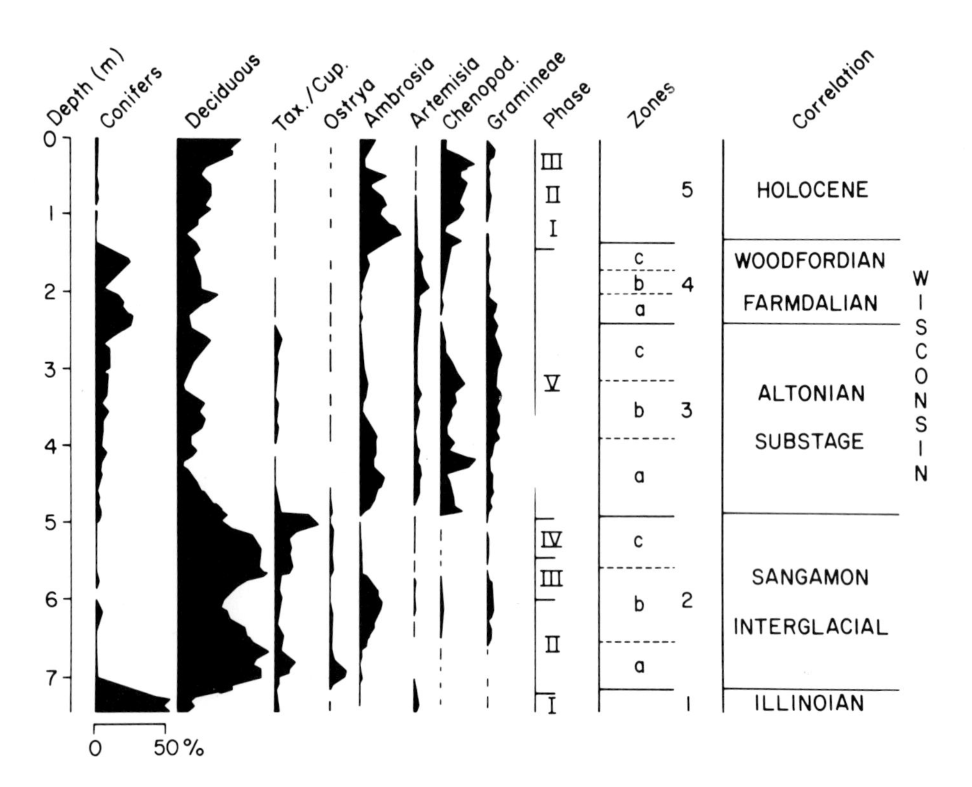

Figure 6 Application of glacial/interglacial pollen zonation to the sequence from Pittsburg Basin, Illinois (after Grüger, 1972, and Kapp, 1977).

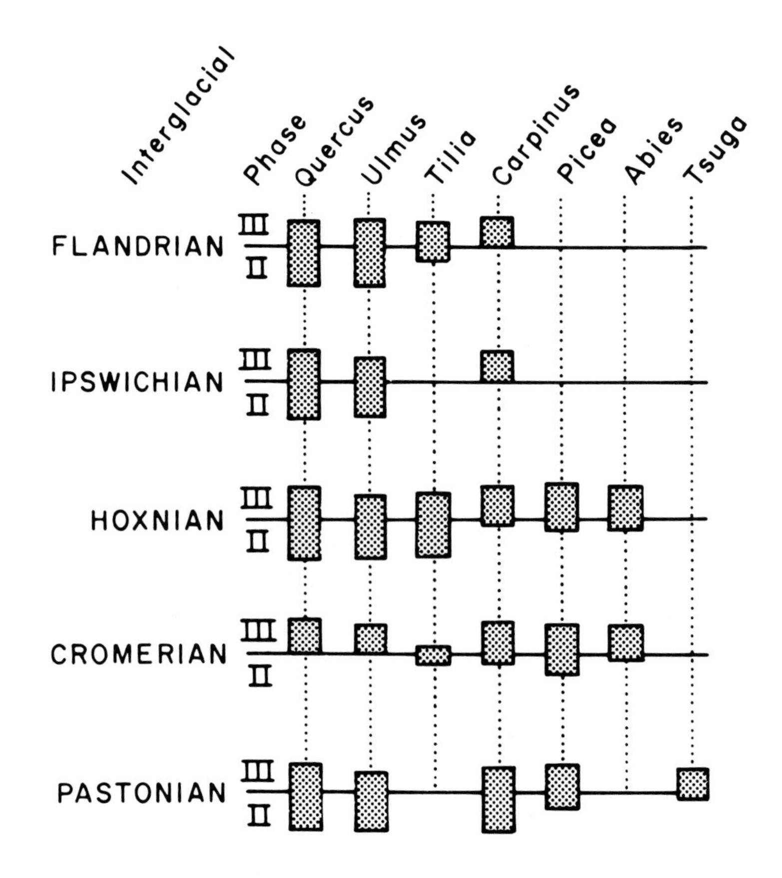

Figure 7 Variations in pollen spectra in Zones II and III through successive interglacials, East Anglia (after West, 1970).

Dabrowski (1971) attempted to use pollen influx to determine the length of the Eemian Interglacial. Modern influx measurements provided surrogates for Eemian rates. His calculations of *ca.* 18,000 palynochrones (a palynochrone = approx. 1 year) is considerably longer than Müller's (1974) estimate of 11,000 years based on varves.

SHORT DURATION AND RECENT EVENTS

Pollen analysis may be useful in dating recent events and those of short duration that fall within the standard deviation of ^{14}C. The recent events may be dated historically or by varve-counting. If the pollen stratigraphies are distinctive such events can provide well-defined dated marker horizons. In North America, settlement by immigrants is most often indicated by a rapid rise in *Ambrosia*. This increase in ragweed can usually be dated in most areas to an accuracy of a decade. In the varved sequence at Crawford Lake, Ontario, ragweed increases rapidly at about 1850 (Boyko, 1973). Oldfield (1969) has recognized the impact of the dissolution of the monasteries (*ca.* 1540 AD), intensification of agriculture at the start of the Napoleonic wars (*ca.* 1800 AD), and more recent forestation in pollen records from the southern Lake District, England.

Disease may quickly decimate an important forest species. Chestnut blight was first reported in New York in 1906. By 1930, chestnut had been almost eliminated from the forests of eastern North America. The demise is rapid and clearly marked in pollen records (Brugam, 1978).

Tephra units often provide distinctive marker horizons in peat and lake sediments in the northwestern U.S. and southwestern Canada. Within the range of radiocarbon these are usually dated by bracketing each unit with ^{14}C dates. However, the length of time involved in the deposition of the unit cannot be determined in this way. Mehringer *et al.* (1977) have illustrated the value of pollen influx in estimating the length of those usually short-term depositional events. At Lost Trail Pass Bog, Bitterroot Mountains, Montana, Glacier Peak ash (*ca.* 11,500 BP) and Mazama ash (*ca.* 6700 BP) are present in the sediments. Radiocarbon dates throughout the sequence allowed determination of sedimentation rates and hence pollen influx. Mean influx rates were compared to those derived from the ash. It was determined that the Glacier Peak ash was made up of two falls separated by 10-25 years. The Mazama ash, 7.3 at Lost Trail Pass Bog, was deposited over several years (Figure 8). The bulk, 4.5 cm (unit A), fell between fall and spring, based on very low pollen content and the lack of pollen from spring and summer pollinating trees. Another 1 cm, unit B, fell the following year, and the remaining 1.7 cm, unit C, the year after, although there were probably light, sporadic falls for up to three years.

CONCLUSIONS

Although ^{14}C dating has become routine on pollen sequences, it has not replaced the use of pollen stratigraphy for age determination. The two can be considered complementary. Many pollen sequences have only one or two ^{14}C dates on them. Analyses done prior to the development of the radiocarbon dating technique have only the pollen stratigraphy on which assumptions about age can be made. Despite the constraints outlined above, zonation remains important in the determination of a chronology. The recognition of regional zonation schemes and inter-regional relationships between these schemes is critical. These are probably best established *via* the objective techniques now available, although subjective, ecologically-based zonation is unlikely to be significantly different.

The complication evident in Holocene sequences tends to be ignored when zonation of older materials is attempted. The models for the glacial/interglacial cycle are almost invariably more simple than those for the latest, the late Pleistocene/Holocene. Without absolute control, any chronology from the glacial/interglacial cycle is relative. The slotting of partial sequences must be tentative.

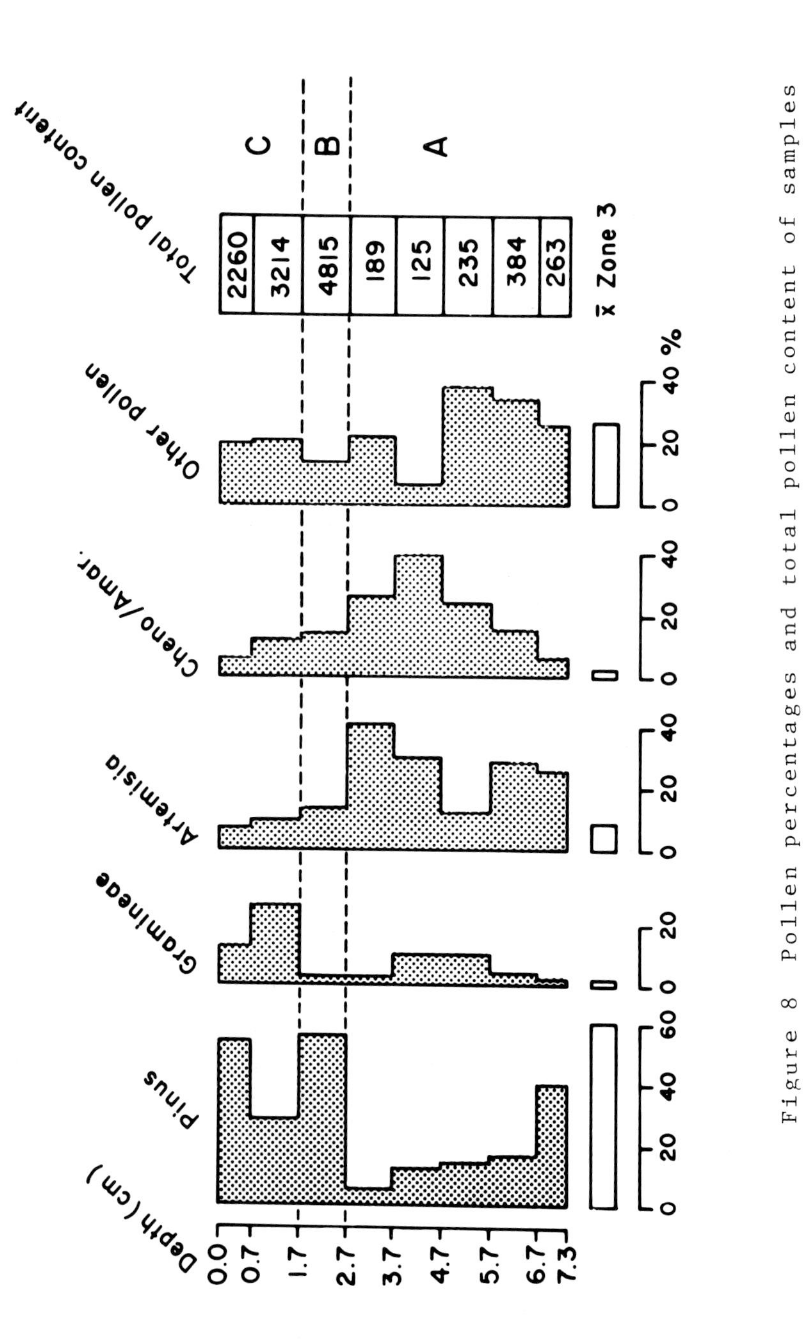

Figure 8 Pollen percentages and total pollen content of samples of Mazama Ash from Lost Trail Pass Bog, Montana (after Mehringer *et al.*, 1977).

Pollen influx data can be useful to provide floating absolute chronologies for short duration events, or some estimate of age within the range of radiocarbon in materials unsuitable for ^{14}C dating. The influx standards can be either modern surrogates or fossil ones established with radiocarbon or varves.

REFERENCES CITED

Adam, D.P., 1974, Palynological applications of principal component and cluster analysis: J. Res. U.S. Geol. Survey, 2, p. 727-741.

Andersen, S.T., 1966, Interglacial vegetational succession and lake development in Denmark: Palaeobotanist, v. 15, p. 117-127.

Bernabo, J.C. and Webb, T., III, 1977, Changing patterns in the Holocene pollen record of northeastern North America: a mapped summary: Quaternary Research, v. 8, p. 64-96.

Birks, H.J.B., 1974, Numerical zonations of Flandrian pollen data: New Phytologist, v. 73, p. 351-358.

__________, 1979, Numerical methods for the zonation and of correlation of biostratigraphic data; in Berglund, B., ed., Paleohydrologic Changes in the Temperate Zone in the Last 15000 Years: sub-project B, v. 1, p. 99-123.

Birks, H.J.B. and Birks, H.H., 1979, Quaternary Palaeoecology: Arnold, 289 p.

Boyko, M., 1973, European impact on vegetation around Crawford Lake in southern Ontario: M.Sc., Botany Dept., Univ. Toronto, Toronto.

Brubaker, L.B., 1975, Post-glacial forest patterns associated with till and outwash in northcentral Upper Michigan: Quaternary Research, v. 5, p. 499-527.

Brugam, R.B., 1978, The human disturbance history of Linsley Pond, North Branford, Connecticut: Ecology, v. 59, p. 19-36.

Bryant, V.M., Jr., 1978, Palynology: a useful method for determining paleoenvironment: Texas J. Science, v. 30, p. 25-42.

Dabrowski, M.J., 1971, Palynochronological materials - Eemian interglacial: Bull. L'Acad. Pol. Sciences, v. 19, p. 29-36.

Davis, M.B., 1967, Pollen accumulation rates at Rogers Lake, Connecticut, during late and postglacial time: Rev. Paleobot. Palynol., v. 2, p. 219-230.

__________, 1976, Pleistocene biogeography of temperate deciduous forests: Geoscience and Man, v. 13, p. 13-26.

Davis, M.B., Brubaker, L.B. and Beiswenger, J.M., 1971, Pollen grains in lake sediments: pollen percentages in surface sediments from southern Michigan: Quaternary Research, v. 1, p. 450-467.

Davis, M.B. and Brubaker, L.B., 1973, Differential sedimentation of pollen grains in lakes: Limnol. Oceanog., v. 18, p. 635-646.

Davis, R.B., 1974, Stratigraphical effects of tubificids in profundal lake sediments: Limnol. Oceanog., v. 19, p. 466-488.

Digerfeldt, G., 1979, Guidelines for lake investigations, in Berglund, B., ed., Paleohydrologic Changes in the Temperate Zone in the Last 15000 Years: Sub-project B, v. 1, p. 31-48.

Dimbleby, G.W., 1976, A review of pollen analysis of archaeological deposits, in Davidson, D.A. and Shackley, M.C., eds., Geoarchaeology, Duckworth, p. 347-354.

Faegri, K. and Iversen, J., 1945, Textbook of Pollen Analyses: Copenhagen, Munksgaard, 295 p.

Firbas, F., 1949, Spät-und nacheiszeitliche Waldgeschichte Mitteleuropas nördlich der Alpen. Bd. 1. Allgmeine Waldgeschichte: Fisher Verlag, 480 p.

Godwin, H., 1940, Pollen analysis and forest history of England and Wales: New Phytologist, v. 39, 370-400.

Gordon, A.D. and Birks, H.J.B., 1972, Numerical methods in Quaternary palaeoecology. I. Zonation of pollen diagrams: New Phytologist, v. 71, 961-979.

__________, 1974, Numerical methods in Quaternary palaeoecology. II. Comparison of pollen diagrams: New Phytologist, v. 73, 221-249.

Grüger, E., 1972, Late-Quaternary vegetation development in south-central Illinois: Quaternary Research, v. 2, p. 217-231.

Iversen, J., 1958, The bearing of glacial and interglacial epochs on the formation and extinction of plant taxa: Uppsala Univ. Arsskr. 6, p. 210-215.

__________, 1973, The development of Denmark's nature since the last glacial: Dan Geol. Unders. Ser. 5, #7c.

Janssen, C.R., 1973, Local and regional pollen deposition, in Birks, H.J.B. and West, R.G., eds., Quaternary Plant Ecology, Blackwells, p. 31-42.

Kapp, R.O., 1977, Late Pleistocene and postglacial plant communities of the Great Lakes Region; in Romans, R.C., ed., Geobotany, Plenum, p. 1-27.

Maher, L.J., Jr., 1972, Nomograms for computing 0.95 confidence limits of pollen data: Rev. Palaeobot. Palynol., v. 13, p. 85-93.

McAndrews, J.H., 1966, Postglacial history of prairie, savanna and forest in northwestern Minnesota: Mem. Torrey Bot. Club, 22, 72 p.

__________, 1981, Late Quaternary climate of Ontario: temperature trends from the fossil pollen record, in Mahaney, W.C., ed., Quaternary Paleoclimate, Geo. Abstracts, Norwich, p. 319-333.

Mehringer, P.J., Jr., Blinman, E. and Petersen, K.L., 1977, Pollen influx and volcanic ash: Science: v. 198, p. 257-261.

Mosimann, J.E., 1965, Statistical methods for the pollen analyst. Multinominal and negative multinomial techniques, in Kummel, B.G. and Raup, D.M., eds., Handbook of Palaeontological Techniques, Freeman, p. 636-673.

Mott, R.J., 1977, Late-Pleistocene and Holocene palynology in southeastern Québec: Géogr. phys. Quat., v. 31, p. 139-149.

Mott, R.J. and Farley-Gill, L.D., 1978, A late-Quaternary pollen diagram from Woodstock, Ontario: Can. J. Bot., v. 15, p. 1101-1111.

Müller, H., 1974, Pollenanalytische Untersuchungen and Jahresschichtenzählugen an der eem-zeitlichen Keiselgur von Bispingen/Luhe: Geol. Jb., A21, p. 149-169.

Ogden, J.G., III, 1966, Forest history of Ohio. Radiocarbon dates and pollen stratigraphy of Silver Lake, Logan County, Ohio: Ohio J. Science, v. 66, p. 387-400.

__________, 1977, Pollen analysis: state of the art: Géogs. phys. Quat., v. 31, p. 151-159.

Ohlsson, I.U., 1979, A warning against radiocarbon dating of samples containing little carbon: Boreas, v. 8, p. 203-207.

Ohlsson, I.U. and Florin, M-B., 1980, Radiocarbon dating of dy and peat in Getsjö area, Kolmarden, Sweden, to determine the rational limit of Picea: Boreas, v. 9, p. 289-305.

Oldfield, F., 1969, Pollen analysis and the history of land use: Advmnt. Sci., v. 25, p. 298-311.

__________, 1977, Végétation tardiglaciare au Québec méridional et implications paléoclimatiques: Géogr. phys. Quat., v. 31, p. 161-176.

Richard, P., 1980, Paléophytogéographie post-wisconsinienne de Québec-Labrador: bilan et perspectives: Notes et Documents, 80-01, Dépt. Géog. Univ. Montréal, 30 p.

Sears, P.B., 1935, Glacial and postglacial vegetation: Botanical Rev. v. 1, p. 37-51.

Shotton, F.W., 1972, An example of hard-water error in radiocarbon dating of vegetable matter: Nature, v. 240, p. 460-461.

Smith, A.G. and Pilcher, J.R., 1973, Radiocarbon dating and the vegetational history of the British Isles: New Phytologist, v. 72, p. 903-914.

Tauber, H., 1965, Differential pollen deposition and the interpretation of pollen diagrams: Dan. Geol. Unders. Sev. 2, 89, 69 p.

Turner, C., 1970, The Middle Pleistocene deposits at Marks Tey, Essex: Phil. Trans. Royal Soc. London, B, v. 257, p. 373-440.

Turner, C. and West, R.G., 1968, The sub-division and zonation of interglacial periods: Eiszeitalter und Gegenwart, v. 19, p. 93-101.

Walker, D. and Wilson, S.R., 1978, A statistical alternative to the zoning of pollen diagrams: J. Biogeog., v. 5, p. 1-21.

West, R.G., 1970, Pleistocene history of the British flora, in Walker, D. and West, R.G., eds., Studies in the Vegetational History of the British Isles, Cambridge Univ. Press, p. 1-11.

Wright, H.E., Jr., 1972, Interglacial and postglacial climates: the pollen record: Quaternary Research, v. 2, p. 274-282.

__________________, 1976, The dynamic nature of Holocene vegetation. A problem in paleoclimatology, biogeography and stratigraphic nomenclature: Quaternary Research, v. 6, p. 581-596.

Wright, H.E., Jr. and Patten, H.C., 1963, The pollen sum: Pollen et Spores, v. 5, p. 445-50.

Wright, H.E., Jr., Winter, T.C. and Patten, H.L., 1963, Two pollen diagrams from southeastern Minnesota: Bull. Geol. Soc. Amer., v. 74, p. 1371-1396.

Yarranton, G.A. and Ritchie, J.C., 1972, Sequential correlation as an aid in placing pollen zone boundaries: Pollen et Spores, v. 14, p. 213-223.

MAMMOTHS, BISON AND TIME IN NORTH AMERICA

C.R. HARINGTON

ABSTRACT

North American Land Mammal Ages are valuable in approximating geological time in the Quaternary, particularly where other more precise dating methods (such as radiocarbon, tephra and paleomagnetic) are lacking. Mammoths and bison are particularly important fossil indicators, for the earliest mammoth remains in North America mark the beginning of the Irvingtonian Land Mammal Age (about 1.8 million years ago), and the earliest bison remains mark the beginning of the Rancholabrean Land Mammal Age (about 1.2 million years ago). It is worth noting that the Rancholabrean may have occurred earlier in northern North America than in southern North America, because bison seem to have been delayed in penetrating the southern plains.

The main morphological changes seen in mammoths and bison as time proceeded were, respectively: (a) an increase in number and compression of enamel plates, and a thinning of enamel in molar teeth; and (b) at least during the last 12,000 years, a tendency to progressive reduction in size of horncores.

More specifically, southern and woolly mammoths seem to be useful indicators of early (Nebraskan to Kansan) and late (mainly Wisconsin) Pleistocene deposits, respectively. Small-horned bison are generally useful time-guides for the last 12,000 years, whereas steppe bison usually indicate deposits older than 12,000 BP.

INTRODUCTION

Mammoths and bison, two immigrants from Eurasia, are particularly important because they have been used by Savage (1951) in establishing Irvingtonian and Rancholabrean North American Land Mammal Ages. This is a major division of the Quaternary (approximately the last 2 million years of geological time) based on fossil vertebrate assemblages (Figure 1). Savage emphasized that the terms Irvingtonian and Rancholabrean are based on faunal names and on animals within those faunas from California, and that they are entirely separate from rock formational names. He indicated that the earliest mammoth remains mark the beginning of the Rancholabrean. Ironically, the latter age almost came to an end about 1890 due to human overhunting of bison (Roe, 1970; Banfield, 1974).

The exact time demarcation of these ages may, perhaps, only be established on a regional basis and in the future, when there is better geochronological control over middle Pleistocene land mammal faunas. Presently, I suggest that mammoths first entered North America about 1.8 million years ago during the Nebraskan Glaciation and at the beginning of the Quaternary, and that bison first entered North America during the Kansan Glaciation about 1.2 million years ago (Figure 1). Of course, the earliest times of entry of mammoths and bison into northern North America would precede the times they became established in the south.

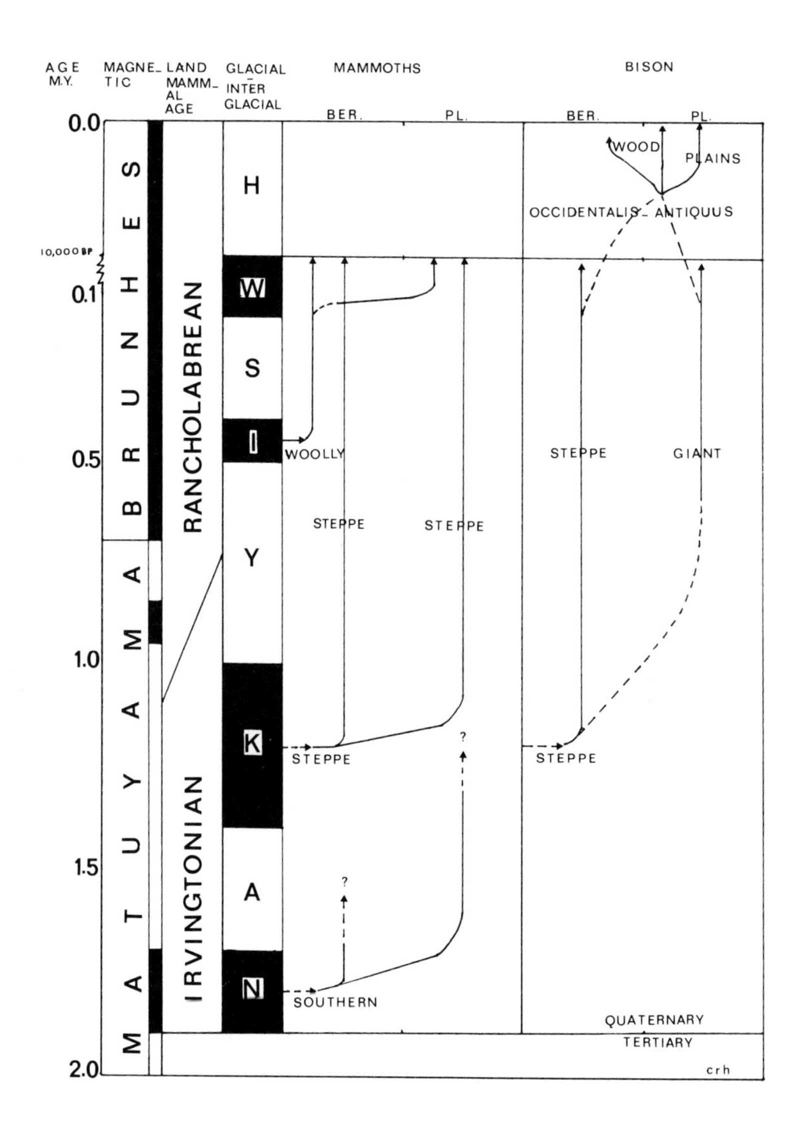

Figure 1 Suggested chronology and regional distribution of mammoths and bison in North America during the Quaternary. Chronological data in three columns on left are approximate, (modified from Ericson and Wollin (1968), Maglio (1973) and Kurten and Anderson (1980)), and phylogenies are highly simplified. For example, I have not dealt with the origin of the dwarf mammoths of Santa Rosa Island, California (Stock and Furlong 1928). Also, I have avoided the issue of the presence of steppe-like bison on the plains until more is known about their geological age and affinities. Perhaps steppe bison (*e.g. Bison crassicornis* from Massachusetts radiocarbon dated at approximately 21,200 BP (Romer 1951)) penetrated the plains a second time - before the peak of the last glaciation. Some of these steppe-like forms may also represent smaller-horned derivatives of giant bison stock. Magnetic: Normal polarity (black) from top to bottom - Brunhes, Jaramillo, Olduvai; reversed polarity (white). Glacials (black) and Interglacials: N (Nebraskan); A (Aftonian); K (Kansan); Y (Yarmouth); I (Illinoian); S (Sangamon); W (Wisconsin); H (Holocene - last 10,000 years). Note break in scale at W/H boundary.

Each of the two columns on the right dealing with mammoths and bison are subdivided into Beringia ("BER." - unglaciated parts of eastern Siberia, Bering Isthmus, Alaska and Yukon) on the left, and Plains ("PL." - heartland of the continent) on the right, in order to show possible relationships between populations in those regions. Dashed lines indicate points of greatest uncertainty. Upright arrowheads mark evident times of extinctions.

MAMMOTHS

Mammoth remains, particularly complete molar teeth, are sometimes useful biostratigraphic indicators, for they are commonly widespread geographically and they tend to preserve well. The most primitive species known from North America, the southern mammoth *(Mammuthus meridionalis)* probably entered this continent from Eurasia near the beginning of the Quaternary, when Repenning recognizes an invasion of rodents including *Pliomys* and *Synaptomys* (Repenning, 1980; Harington, 1980a). Several molars from the northern Yukon resemble those of the most advanced southern mammoth stock (Bacton Stage) or Eurasia, and I suspect that they represent the earliest North American mammoths (Harington 1977; Figure 2). This species also appears to be present in the Wellsch Valley fauna in Saskatchewan, where it may have lived nearly 1.7 million years ago (A. MacS. Stalker, pers. commun., 1976), and at Bruneau, Idaho about 1.36 million years ago (Maglio, 1973).

North American steppe mammoths *(Mammuthus columbi* and/or *Mammuthus armeniacus)* are the least valuable biostratigraphically and their origins present problems (Harington 1980b). It is now widely accepted that the Columbian mammoth *(Mammuthus columbi*, under which I include the imperial mammoth, *Mammuthus imperator,* following Kurtén and Anderson 1980) arose from the southern mammoth in North America (*e.g.* Maglio, 1973). As an alternative, I propose that steppe mammoths reached North America from Eurasia, replacing southern mammoths during the Kansan Glaciation about 1.2 million years ago. I think that they invaded with primitive muskoxen like *Soergelia* and *Euceratherium* and perhaps *Phenacomys, Clethrionomys, Neodon* and *Pitymys* (Repenning's (1980) "Irvingtonian II" microtines).

From the evidence I have seen, I can only stress the similarities of Eurasian and North American steppe mammoths. Some specimens I have referred to *Mammuthus cf. armeniacus* from the northern Yukon may record the earliest North American entry of steppe mammoths from Eurasia (Harington, 1977). Also, Savage's (1951) description and measurements of *Mammuthus columbi* molars from Irvingtonian localities fit well with those of the Eurasian steppe mammoths now included in *Mammuthus armeniacus*. Further, Henry Fairfield Osborn (1942), a paleontologist with great experience in studying elephants, derived Columbian mammoths from Eurasian steppe mammoths now included in *Mammuthus armeniacus* by Maglio (1973). If, indeed, Eurasian and North American steppe mammoths are not synonymous, we need a way of differentiating them -- preferably on molar characteristics. Although Maglio (1973) states that the North American lineage leading from southern to Columbian mammoths has broader molars, most steppe mammoth molars I have measured from the northern Yukon are slightly narrower than the average of a sample from Eurasia (Harington, 1977).

Steppe mammoths (Figure 3) seem to have occupied North America for more than 1 million years. The most advanced type (sometimes considered as a separate species, *Mammuthus jeffersonii)* survived on the plains until about 11,000 years ago (Kurtén and Anderson, 1980). Evidently some steppe mammoths lived in eastern Beringia during the peak of the Wisconsin Glaciation approximately 20,000 years ago (Harington, 1980b).

Figure 1 Continued

Mammoths: southern *(Mammuthus meridionalis)*; steppe *(M. armeniacus* and/ or *M. columbi)*; woolly *(M. primigenius)*.

Bison: steppe (earliest(?) - *Bison alaskensis*, and latest - *B. crassicornis:* sometimes they are grouped under the name *B. priscus)*; giant *(B. latifrons)*; *"occidentalis-antiquus* complex" *(B. bison occidentalis* - evidently derived from steppe bison stock in Beringia; *B. bison antiquus* - evidently derived from giant bison stock in the plains); wood *(B. bison athabascae* - evidently derived mainly from *B. bison occidentalis* stock on the northern and western flanks of the Canadian plains); plains *(B. bison bison* - evidently derived from *"occidentalis-antiquus"* stock on the plains).

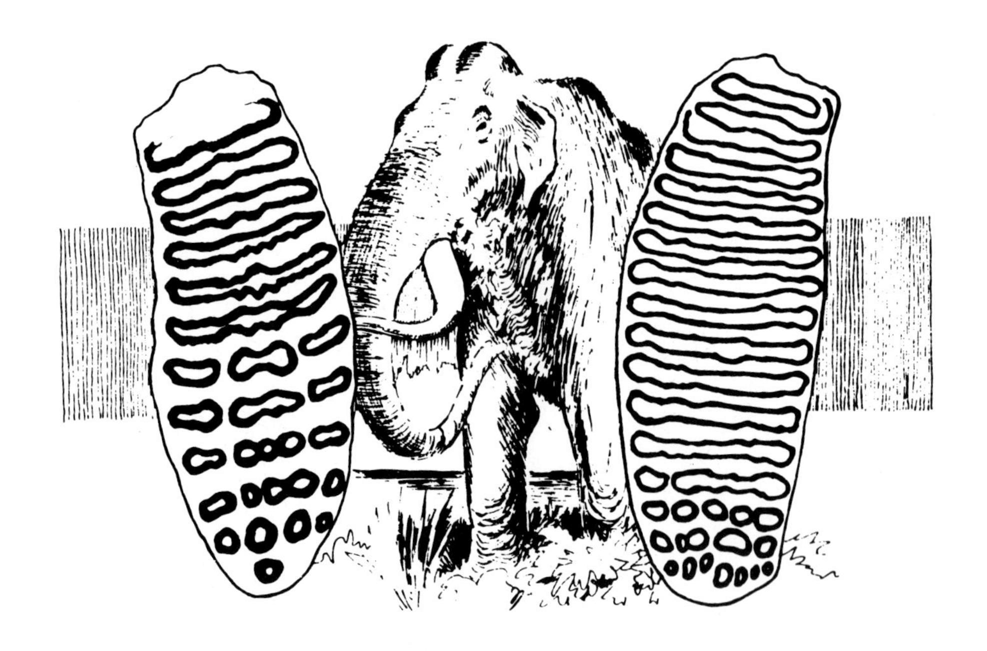

Figure 2 Mammoth teeth are often useful indicators of geological age and past environment. The lower third molar on the left (grinding surface shown with enamel plates in black) is from a southern mammoth *(Mammuthus meridionalis)* from Old Crow Basin, Yukon Territory, and shows the "primitive" condition of relatively few, thick, widely-spaced enamel plates. The lower third molar on the right, from a woolly mammoth *(Mammuthus primigenius)* from the same region of the Yukon, shows the "advanced" condition involving more numerous, thinner, more compressed enamel plates. Ink sketch by Charles Douglas.

However woolly mammoths *(Mammuthus primigenius)* were dominant there during that period. Conceivably steppe and woolly mammoths may have shared their range both north and south of the Wisconsin ice sheets.

Woolly mammoths (Figure 4) migrated from Eurasia to North America in late Illinoian or early Wisconsin time (Harington, 1980a). In the source area, northeastern Siberia, an early type of woolly mammoth is known from the Utka Beds (probably late Illinoian), and the advanced type was established there by early Wisconsin time (Iedoma Suite) (Sher, 1971). Evidently woolly mammoths first reached the northern plains (Medicine Hat, Alberta) during the early Wisconsin (Stalker and Churcher, 1970). The species spread to eastern Canada during the mid-Wisconsin about 45,000 years ago (Churcher, 1968a). It seems to have occupied discontinuous tundra-like range south of the Wisconsin ice sheet, extending from southern British Columbia to southern Ontario, and perhaps farther eastward. Scattered fossils of other tundra-adapted mammals including lemmings, tundra muskoxen and caribou also occur in this zone (Harington and Ashworth, in preparation). Woolly mammoths died out in North America about 13,000 years ago. I consider this species to be an indicator of Wisconsin-age deposits in southern Canada and northern United States, at least.

Figure 3 Restoration of a North American steppe mammoth *(Mammuthus columbi)*. They seem to have been the most common and widespread of North American mammoths. Ink sketch by Charles Douglas.

Figure 4 Restoration of a woolly mammoth *(Mammuthus primigenius)*. They were adapted to tundra-like range and were well adapted to cold conditions. Ink sketch by Charles Douglas.

BISON

The study of North American fossil bison is fraught with problems. Experts disagree on many points: namely, geological age beyond 40,000 BP, phylogeny and distributional history. I think that clarification of many of these problems depends upon gaining a better perspective on the history of Eurasian fossil bison. To this end, D.M. Shackleton and M.L. Weston, sponsored by the National Museum of Natural Sciences, began a study in 1979. They are processing biostratigraphic and metric data on western European fossils. Data from 185 skulls, 16 mandibles and 168 postcranial elements were collected from bison fossils in Soviet institutions in 1982.

Bison had entered central Alaska by Kansan (?) time according to remains from the Fox Gravel unit near Fairbanks. The age estimate rests on Péwé's (1975a, b) stratigraphic interpretation of an exposure at Eva Creek. Other early specimens of bison from Alaska are from deposits of possible Yarmouth Interglacial age near Baldwin Peninsula; Illinoian age near Fairbanks (a fission-track age of approximately 450,000 years from the Easter Ash Bed suggests that *Bison* remains from this part of the Gold Hill Loess may be Illinoian in age (T.L. Péwé, pers. commun., 1982)); and Sangamon Interglacial age near Tofty. As no published descriptions of these fossils are available, it is difficult to know whether they should be treated merely as *Bison* sp., or whether - as seems likely - they represent steppe bison (*sensu* Kurtén and Anderson, 1980).

In the Yukon, I have collected steppe bison specimens with unusually large horncores (*Bison alaskensis*), one of which yielded a radiocarbon date of >39,000 BP (I-5405). I consider it to be a suitable ancestor for later, smaller-horned steppe bison (*Bison crassicornis*) and the giant bison (*Bison latifrons*) of southern North America, and to be closely related to, if not synonymous with, *Bison priscus gigas* (Harington and Clulow, 1973; Harington, 1980a). The latter bison was widespread in Eurasia during the middle Pleistocene (equivalent in age from Yarmouth to Sangamon time in Flerov's (1972) terminology).

The earliest occurrences of bison on the plains appear to be of Yarmouth age in Canada (Stalker and Churcher, 1970) and of post- Kansan to early Illinoian age in the United States (Schultz and Hillerud, 1977). Giant bison are first definitely recorded there in late Illinoian time and seem to have survived until late Wisconsin time (Kurtén and Anderson, 1980).

Northern small-horned bison, or western bison (*Bison bison occidentalis*) evidently arose from steppe bison (*Bison crassicornis*; Figure 5) stock in Beringia toward the close of the last glaciation. Probably warmer, moister conditions occurring in Beringia then, and the resulting more heavily-wooded terrain contributed to the demise of steppe bison and the rise of western bison (Harington, 1977). These bison spread rapidly southward, possibly mixing with southern small-horned bison (*Bison bison antiquus*) that had evolved from larger-horned stock south of the Wisconsin ice sheets (Guthrie, 1970). It is difficult to untie this knot. For example, I have examined skulls with horncores that represent nearly every possible combination of characters thought to be diagnostic of *Bison bison occidentalis* and *Bison bison antiquus*. For utilitarian reasons, I call this mosaic "the *occidentalis-antiquus* complex". In fact, such small-horned bison are generally useful indicators of 12,000-6,000 BP deposits.

It is interesting to speculate on the evolution of modern bison from their small-horned bison ancestors. As a hypothesis for testing, I suggest that the Hypsithermal (about 7,000-5,000 BP) placed relatively great and sudden stress on small-horned bison herds that were best adapted to woodland or parkland conditions, and that some were able to adapt to life on the arid grasslands (plains bison, *Bison bison bison*) while others withdrew northward and westward following the retreating margins of the boreal and subalpine forest, or remained in such areas (wood bison, *Bison bison athabascae*). Adaptation of plains bison

Figure 5 Restoration of steppe bison *(Bison crassicornis)* bulls fighting. This species was most common in unglaciated areas of Alaska and the Yukon during the last glaciation. Based on skeletal material from a late-Wisconsin site near Dawson, Yukon Territory. Ink sketch by Bonnie Dalzell.

probably involved substantial morphological and behavioural changes, whereas I suspect that wood bison differ little (except for smaller horns) from *Bison bison occidentalis* (Harington, 1977).

A bison skull from Banff, Alberta yielding a radiocarbon date of 3,240 ± 90 BP (I-11654) indicates that some western bison survived this period of crisis in intermontane pockets (Appendix; Figures 6, 7). This is in contrast to large samples of skulls best referred to modern bison of similar geological age from sites farther east near Fort Saskatchewan, Alberta, where I have been collecting, and near Bottrell, Alberta where L.V. Hills and M. Wilson (pers. commun., 1981) have been collecting.

CONCLUSION

In summary, division of the North American Quaternary into Irvingtonian and Rancholabrean Land Mammal Ages is generally useful - particularly where more precise physical dating techniques (*e.g.* radiocarbon, tephra, paleomagnetic) are not applicable, and where microtine (*e.g.* the *Ondatra idahoensis* → *annectens* → *nebracensis* → *zibethicus* lineage extending from early Irvingtonian to Recent (Nelson and Semken, 1970)), and other useful indicators, such as *Soergelia* (Harington, 1977), constitute supplementary fossil evidence of geological age. Remains of southern and woolly mammoths seem to be useful indicators of early (Nebraskan to Kansan) and late (mainly Wisconsin) Pleistocene deposits, respectively. Steppe mammoth remains, alone, are not very useful in this regard. Small-horned bison are generally useful time-guides for the last 12,000 years, whereas steppe bison are indicative of deposits older than 12,000 BP. Giant bison are not as useful in this respect as they seemed to be a decade ago, for they evidently survived from the Yarmouth Interglacial to late Wisconsin time.

The main morphological changes seen in molars of the relatively "primitive" southern mammoth to the relatively "advanced" woolly mammoth are: an increase in number and compression of enamel plates; and a thinning of enamel as time proceeded. In bison, at least during the last 12,000 years, there is tendency to progressive reduction in size of horncores (Wilson, 1980).

Finally, I wish to emphasize the possibility that the Rancholabrean may have occurred earlier in northern North America (Kansan?) than in southern North America (Yarmouth?), because bison seem to have been delayed in penetrating the southern plains.

ACKNOWLEDGEMENTS

I am grateful to: Edward J. Hart (Archives of the Canadian Rockies) for the loan of the bison skull from Banff, Alberta; Harry Foster (National Museums of Canada) for photographs of that skull; Charles H. Douglas (National Museum of Natural Sciences), and Bonnie Dalzell for the ink sketches; and Mrs. Gail Rice for typing the manuscript.

APPENDIX

The Banff Archives Bison

In the late summer of 1967, Mrs. Catharine Whyte collected a damaged posterior cranial fragment of an adult bison (Figures 6, 7) uncovered by a bulldozer operator excavating the foundation of the Archives of the Canadian Rockies building in Banff, Alberta. The specimen came from a depth of approximately seven feet (2.1 m) in "glacial silt" (buff sandy silt matrix was found in cavities in the cranial bone). Radiocarbon analysis of bone collagen from the base of the cranium yielded a date of 3,240 ± 90 BP (I-11638), which approximately dates the stream terrace from which the specimen was derived. It is noteworthy that N. Rutter considered the terraces in this area unlikely to be older than 3,500 BP (Wilson, 1975), before the specimen was dated. This Banff terrace is evidently correlative with the late T2 terrace deposits in the Bow Valley farther east (M. Wilson, pers. commun., 1981).

On the basis of the relatively long (compared to modern bison) estimated spread of the horncores (approximately 780 mm), the rather narrow (compared to *Bison bison antiquus)* frontal region between horncores and orbits (301 mm), and the backswept horncores (Figure 6; Table 1), I consider that the specimen is best referred to the western bison *(Bison bison occidentalis)*. Therefore, some bison with marked characteristics of western bison, survived in at least one pocket near the eastern flank of the Rocky Mountains until about 3,000 years ago, whereas contemporaneous bison from two sites farther east (see text) in Alberta are best referred to modern bison. Since western bison occurred near Cochrane as early as 11,000 BP (Churcher, 1968b, 1975), the species occupied this part of western Alberta for some 8,000 years.

Peculiar features of the Banff specimen are the pronounced depressions on the right (medial to the right orbit) and left (medial to the left hornbase) frontals: apparently these depressions are not due to erosion during transport in a stream bed, but resulted from damage (by fighting or possibly disease?) in life.

***

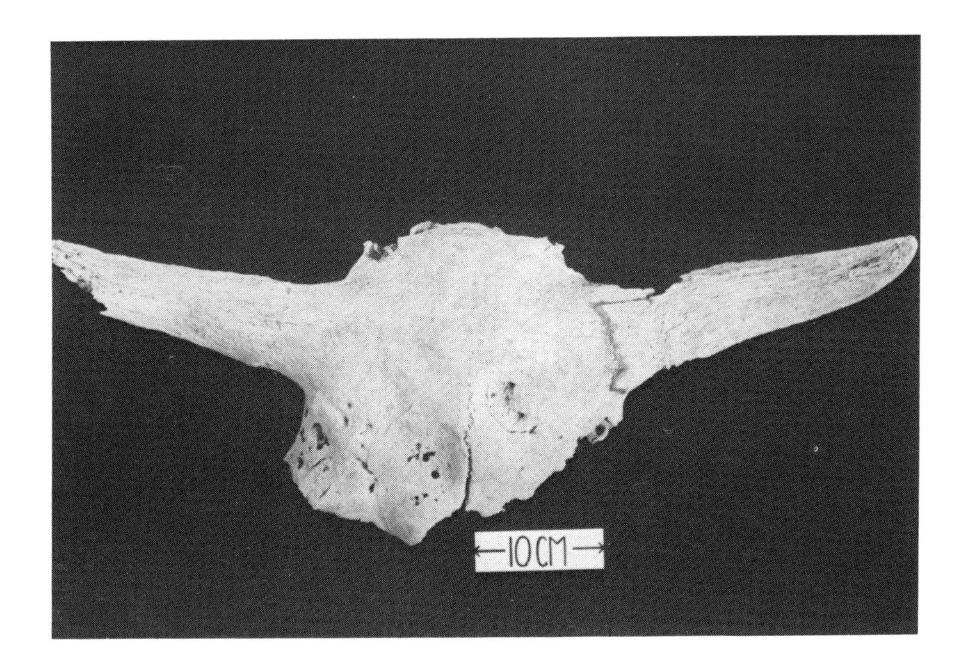

Figure 6 Dorsal view of a bison skull fragment from a low terrace at Banff, Alberta. Bone from this specimen has been radiocarbon dated at about 3,240 BP indicating that there were some late-surviving western bison *(Bison bison occidentalis)* in pockets near the moutain front.

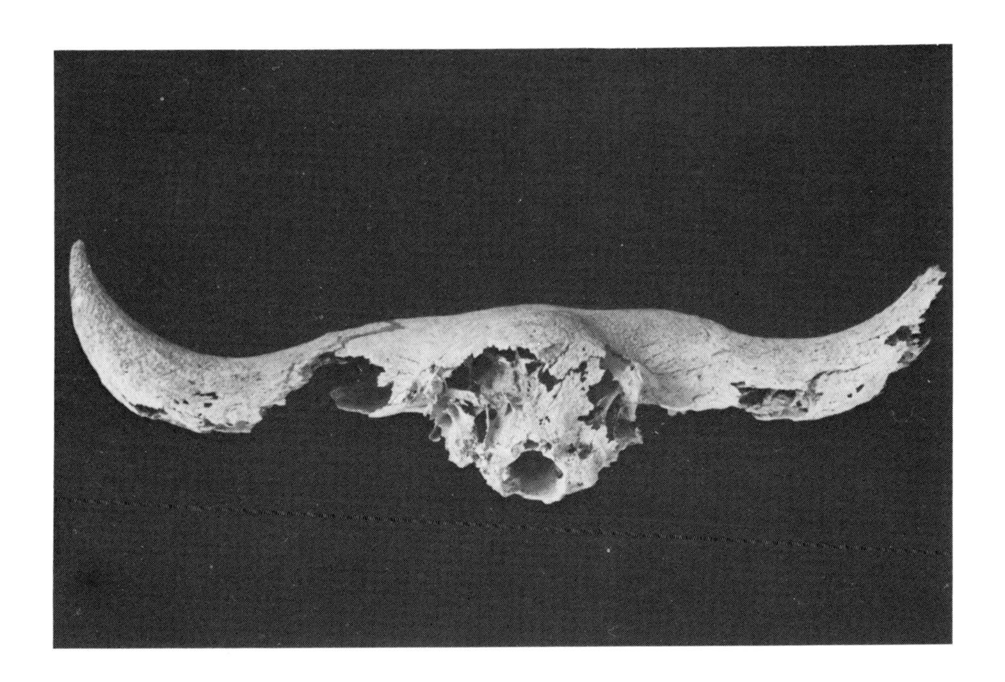

Figure 7 Posterior view of the western bison cranial fragment shown in Figure 6.

REFERENCES CITED

Banfield, A.W.F., 1974, The Mammals of Canada: University of Toronto Press, 438 p.

Churcher, C.S., 1968a, Mammoth from the middle Wisconsin of Woodbridge, Ontario: Canadian Jour. Zoology, v. 46, p. 219-221.

_______________, 1968b, Pleistocene ungulates from the Bow River gravels at Cochrane, Alberta: Canadian Jour. Earth Sciences, v. 5, p. 1467-1488.

_______________, 1975, Additional evidence of Pleistocene ungulates from the Bow River gravels at Cochrane, Alberta: Canadian Jour. Earth Sciences, v. 12, p. 68-76.

Ericson, D.B. and Wollin, G., 1968, Pleistocene climates and chronology in deep-sea sediments: Science, v. 162, p. 1227-1234.

Flerov, K.K., 1972, The earliest forms and history of the genus *Bison*: Akademia Nauk S.S.S.R., Siberian Division, Amalgamated Science Council Biological Science, Theriology, v. 1, p. 81-86. (In Russian).

Guthrie, R.D., 1970, Bison evolution and zoogeography in North America during the Pleistocene: Quarterly Review of Biology, v. 45, no. 1, p. 1-15.

Harington, C.R., 1977, Pleistocene mammals of the Yukon Territory (Ph.D. Dissert.): Edmonton, University of Alberta, 1060 p.

_______________, 1980a, Faunal exchanges between Siberia and North America: evidence from Quaternary land mammal remains in Siberia, Alaska and the Yukon Territory: Canadian Jour. Anthropology, v. 1, no. 1, p. 45-49.

_______________, 1980b, Radiocarbon dates on some Quaternary mammals and artifacts from northern North America: Arctic, v. 33, no. 4, p. 815-832.

Harington, C.R. and Ashworth, A.C., (In preparation), A mammoth *(mammuthus primigenius)* tooth from late Wisconsin deposits in Cass County, North Dakota: p. 1-16 (MS).

Harington, C.R. and Clulow, F.V., 1973, Pleistocene mammals from Gold Run Creek, Yukon Territory: Canadian Jour. Earth Sciences, v. 10, no. 5, p. 697-759.

Kurtén, B. and Anderson, E., 1980, Pleistocene Mammals of North America: N.Y., Columbia University Press, 442 p.

Maglio, V.J., 1973, Origin and evolution of the Elephantidae: Trans. Amer. Phil. Soc. (New Series), v. 63, pt. 3, p. 1-149.

Nelson, R.S. and Semken, H.A., 1970, Paleoecological and stratigraphic significance of the muskrat in Pleistocene deposits: Geol. Soc. Amer. Bull., v. 81, p. 3733-3738.

Osborn, H.F., 1942, Proboscidea, Vol. II: N.Y., American Museum Press, p. 805-1675.

Péwé, T.L., 1975a, Quaternary geology of Alaska: U.S. Geol. Survey Prof. Paper 835, p. 1-145.

__________, 1975b, Quaternary stratigraphic nomenclature in unglaciated central Alaska: U.S. Geol. Survey Prof. Paper 862, p. 1-32.

Repenning, C.A., 1980, Faunal exchanges between Siberia and North America: Canadian Jour. Anthropology, v. 1, no. 1, p. 37-44.

Roe, F.G., 1970, The North American Buffalo: a Critical Study of the Species in its Wild State, Second Edition: University of Toronto Press, 991 p.

Romer, A.S., 1951, *Bison crassicornis* in the late Pleistocene of New England: Journal of Mammalogy, v. 32, p. 230-231.

Savage, D.E., 1951, Late Cenozoic vertebrates of the San Francisco Bay region: University of California Publications, Bulletin Department of Geological Sciences, v. 28, no. 10, p. 215-314.

Schultz, C.B. and Hillerud, J.M., 1977, The antiquity of *Bison latifrons* (Harlan) in the Great Plains of North America: Trans. Nebraska Academy of Sciences, v. 4, p. 103-116.

Sher, A.V., 1971, Mammals and Stratigraphy of the Pleistocene of the Extreme Northeast of the U.S.S.R. and North America: Moscow, Nauka. (English translation published in the International Geology Review, v. 16, p. 1-284, 1974).

Stalker, A. MacS. and Churcher, C.S., 1970, Deposits near Medicine Hat, Alberta, Canada: Chart published by the Surveys and Mapping Branch, Department of Energy, Mines and Resources, Ottawa.

Stock, E. and Furlong, E.L., 1928, The Pleistocene elephants of Santa Rosa Island, California: Science, (New Series), v. 58, p. 140-141.

Wilson, M., 1975, Holocene fossil bison from Wyoming and adjacent areas (M.A. Thesis): Laramie, University of Wyoming, 276 p.

__________, 1980, Morphological dating of late Quaternary bison on the northern plains: Canadian Jour. Anthropology, v. 1, no. 1, p. 81-85.

Table 1. Measurements of a postglacial western bison *(Bison bison occidentalis)* cranial fragment from Banff, Alberta

Specimen	Measurements (mm)[a]								
	1	2	3	4	5	6	7	8	9
Banff Archives site, Alberta	780[b]	800[b]	265	380[b]	82[b]	85	260[b]	325	301

[a] 1. Spread of horncores (tip to tip), 2. Greatest spread of horncores (on outside curve), 3. Horncore length on upper curve (tip to burr), 4. Horncore length on lower curve (tip to burr), 5. Vertical diameter of horncore (at right angle to longitudinal axis), 6. Transverse diameter of horncore (at right angle to longitudinal axis), 7. Horncore circumference (at right angle to longitudinal axis), 8. Cranial width (between upper centres of horncore burrs), 9. Cranial width (at constriction between horncores and orbits).

[b] Estimated measurement.

FIELD USE OF MACROFEATURES FOR CORRELATING TILLS AND ESTIMATING THEIR AGES: A REVIEW

A. MacS. STALKER

ABSTRACT

Many geological sections in areas of multiple glaciation lack the organic materials necessary for absolute and relative dating of their tills, and for correlation of the tills to deposits of known age elsewhere. However, a till sheet in the field normally exhibits large-scale features, beyond those dependant upon composition, that suffice to distinguish it from most other till sheets in an area, allow its correlation with deposits of known age elsewhere, and may even, at times, indicate its approximate age. In general, the value of these features for such purposes has been underestimated or even disregarded. Examples used here are drawn from the writer's experience with the Laurentide deposits of the southwest Canadian Prairies, but only a few of the many types of features available are described.

These macrofeatures derive from environmental conditions prevailing during deposition and during the post-depositional history of the till, and they remain fairly constant over much of south Alberta. Those arising partly or chiefly from post-depositional factors include oxidization and weathering, compaction, jointing, and style of breakage. Oxidization and weathering have a long record of use for estimating ages and for correlation, and so are not discussed here. Compaction is a valuable tool, for it depends mainly on the number and thicknesses of later glaciers that compressed the till; it becomes worthless, however, where older tills surface beyond the limits of younger glaciers and so are little indurated. Jointing and style of breakage normally are sufficiently distinctive for each till to permit its correlation to deposits of known age elsewhere.

The prominent columnar structures displayed by many of the southern Alberta tills are of combined depositional and post-depositional origin. They are of great value for recognition and correlation of all but surface tills. On the other hand, till colours found in south Alberta are determined by both composition and post-depositional history. Other things being equal, the darker a till the greater its age. The Illinoian tills in that region are very dark brown or grey, depending upon whether they were laid down by southeast of southwest flowing ice; the Pre-classical Wisconsin tills are moderately dark brown or grey, whereas the typical Classical Wisconsin tills are buff or light yellowish brown. The oldest Illinoian till is nearly black, and it maintains this sombre colour, where not weathered, not only over a broad spectrum of bedrock types but also at depth in section, and at the surface in those places where it represents the sole glaciation. The other tills similarly retain their characteristic colours over broad areas. This allows these colour values to be used for estimating till ages.

The effects of environment during deposition are demonstrated by the two "contorted" tills of southern Alberta. The contortion includes strong mixing with underlying material, along with diapirs and flame structures from the till and underlying formation shooting into and piercing each other, and in places completely enveloping segments of the other. The interfingering and mixing may be so intense that no accurate boundary can be drawn between the two units. The contortion, which is especially strong in the younger, distinguishes these two tills from all others, and so enables their correlation from section to section and to deposits of known age elsewhere.

Altogether, the use of the full suite of the macrofeatures found in any area can give a fairly good indication of the relative ages and positions in the stratigraphic section of the tills there, and of their relationship to other tills found elsewhere.

INTRODUCTION

In these days of more refined methods for absolute dating of Quaternary deposits, and of more sophisticated means of till study, it becomes difficult to write about some of the rather mundane, but more traditional, field methods. They remain important, however, and still form the basis for much of the geological work on the Quaternary. Undoubtedly many of them are used routinely and without much thought about their nature. Others, however, have been largely neglected in recent years, and it is to these methods, based on features that are independent of composition, that attention is here directed.

Many of the outstanding field characteristics of a till do depend chiefly on composition. Others, however, are determined to a far greater extent by conditions that prevailed during their deposition or subsequent history. This discussion is based on the premise that these conditions, in most cases, were sufficiently different for each till to endow it with specific, inherent features that distinguish it sufficiently from other regional tills to allow its ready identification in the field, its correlation to tills elsewhere, and often to enable an estimate of its age. They further offer a means for testing the reasonableness of ages obtained by other methods, including absolute dating.

There are numerous types of these large-scale features which, for convenience in this paper, are hereafter called macrofeatures. Only a few of them - more precisely compaction, fracture pattern, colour, deformation, columnar structure - are reviewed here. Some of the others, such as weathering and soil formation, have long been used to establish relative ages of till sheets and as marker horizons, and for estimating absolute ages, even though their use has also decreased of late. Most, however, have received little attention in recent years, and their efficacy for correlation and determination of age has commonly been undervalued or even disregarded. They obviously are of greatest value in the study of complicated, multi-till sections that, as is only too well known, typically lack the fossils or other organic remains necessary for correlation and radiocarbon dating.

Different suites of macrofeatures undoubtedly can be identified and used in various regions. However, the examples given here are drawn from the writer's experience with the Laurentide deposits of the southwest portion of the Great Plains of Canada. That general area, and places mentioned and sections described, are shown in Figure 1. For a number of reasons, that region is probably as close as one can come to an ideal place to study macrofeatures. First, the southwestern plains have seen a long series of glaciers, most of which are represented by at least local remnants of till. Second, the approximate thicknesses, extents, and limits of many of these glaciers are known (Stalker and Harrison, 1977). Third, the successive Laurentide glaciers that invaded the region were of generally decreasing strength. This is most important, for it allows examination of the tills both at the surface beyond the limits of later glaciers and also at depth in section. This successively decreasing strength of the glaciers was also pointed out by Stalker and Harrison (1977) in their study of the glacial history of the southwest Great Plains. In that study they reported that the first glacier known to have invaded the region, the one that laid down the Labuma Till, was by far the thickest and extended well beyond the limits of later ones. The relative sizes of the following two glaciers, which laid down the Maunsell and the Brocket tills in that order, are uncertain, but both were much weaker than the Labuma glacier. After the Brocket advance each successive glacier was smaller than the preceeding one.

Using the macrofeatures, including some of those described in this paper, Stalker and Harrison correlated their Laurentide tills with deposits of better known age found farther east in the province, and especially those exposed in the river sections found along the South Saskatchewan River near Medicine Hat and Lethbridge (Figure 1). These correlations enabled them to propose ages ranging from early Illinoian to late Wisconsin for their tills. The macrofeatures also enabled Stalker (in press) to correlate certain tills of the Cameron Ranch Section (Figure 1, Section 16) with the Medicine Hat sections, and thereby to divide the Laurentide tills of southern and central Alberta into three groups, which he described as Illinoian, Preclassical Wisconsin, and Classical Wisconsin in age.

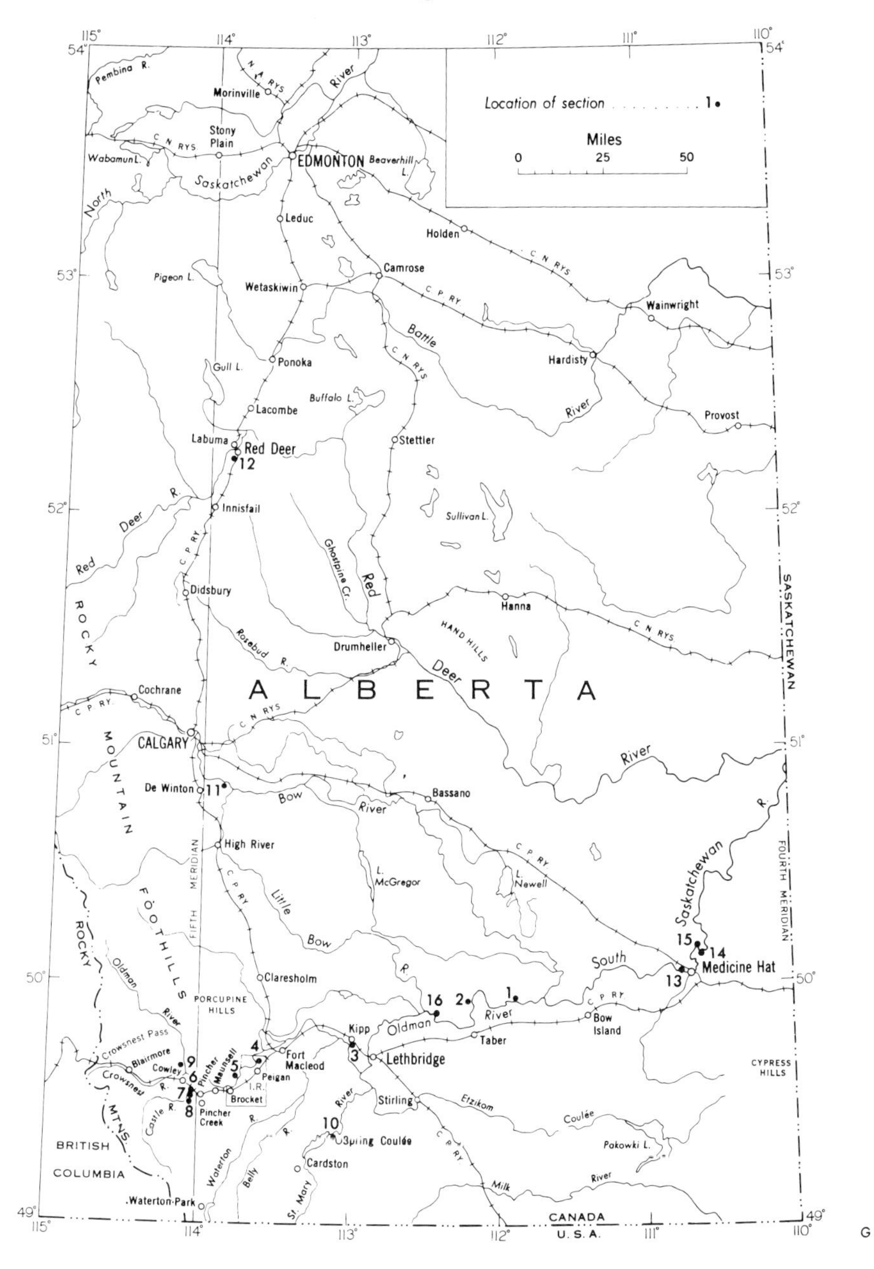

Figure 1 Index map of the southwestern Prairies of Canada. Numbers refer to described sections mentioned in text. Numbers 1 to 12 were described by Stalker, 1963; 13 to 15 by Stalker, 1969; and number 16 (Cameron Ranch Section) by Stalker, in press.

The ages of the deposits near Medicine Hat, to which both Stalker and Harrison (1977) and Stalker (in press) correlated, are based primarily on vertebrate paleontology (Harington, 1978; Stalker, 1976; Stalker and Churcher, 1970, 1972). This is especially the case for those beyond the range of radiocarbon dating. This use of vertebrate chronology causes difficulty in relating the stratigraphy of the southwest Great Plains to stratigraphies proposed for other parts of the continent for, as Stalker and Harrison (1977, p. 883-884) pointed out:

> "Even though the terminology employed is the same, a great deal of the vertebrate chronology developed for North America may be out of phase with the glacial chronology of the mid-continent. In particular, the pre-Wisconsin portion of the vertebrate chronology may embrace more time than the corresponding section of the glacial chronology. For this reason... the terms of the mid-continent chronology are used in only the most general sense to indicate approximate ages of events. This applies particularly to events preceding the Classical Wisconsin, or roughly before 30,000 years ago, for which radiocarbon dating is unavailable."

Stage and stadial names are used here in the manner that they were employed by Stalker and Harrison (1977). As a result, although the terms may be similar to those in the standard, mid-North American chronlogy, they are not necessarily in all cases fully time equivalent. They are used here reluctantly and solely for want of a better alternative. Their use, however, does give an indication of the ages of the various units. Happily, since Stalker and Harrison's 1977 report, a growing realization of the greater length of the glacier record has brought the two chronologies closer to agreement.

PAST WORK

Descriptions of the Laurentide tills found in the southwestern Canadian Prairies have shown remarkable consistency over the years, even though estimates of the number and ages of till sheets present have varied considerably. The first descriptions are found in Dawson (1885, p. 143c) and in scattered references throughout Dawson and McConnell (1895). The 1885 description evidently encompassed all the tills of the Laurentide sequence, for it stated: "colour varies considerably, ranging from dark blackish or bluish-grey to lighter tints of the same, and often becoming grey or fawn-coloured, especially where weathered." These correspond to the colours later described by Stalker (1960) for the Labuma, Maunsell, and Buffalo Lake tills respectively. Next Horberg (1952) divided the till sequence near Lethbridge into three units, which he described in detail. He suggested all were of Wisconsin Age. Later Stalker (1960), while working near Red Deer, Alberta, similarly separated three till units which he named, from older to younger respectively, Labuma, Maunsell, and Buffalo Lake. These tills correspond roughly to the three units of Horberg, but Stalker suggested older ages than Horberg for the two lower tills. Stalker (1963) next described twelve important Quaternary sections (Figure 1, Nos. 1-12), mostly along the South Saskatchewan drainage system, and named the Brocket Till, (the Laurentide till directly overlying the Maunsell). In 1969 he described a further three sections near Medicine Hat (Figure 1, Nos. 13-15). These 1960, 1963 and 1969 papers, along with a description of the important Cameron Ranch Section (Figure 1, No. 16) on Oldman River northeast of Lethbridge (Stalker, in press), provide the basis for the present study. In these papers Stalker suggests till ages ranging from early Illinoian to late Wisconsin. In the meantime Horberg (1954) and Alley (1973) had discussed the past interrelationships of Cordilleran and Laurentide glaciers. The work on these interrelationships, which is based chiefly on interfingering of the two sets of till sheets, was continued by Stalker (1972, 1976) and Stalker and Harrison (1977).

THE FEATURES

Origin

General statement

This paper emphasizes use of the macrofeatures, rather than their origin. For convenience, however, the features are discussed according to whether they formed mainly during deposition of the till, post-depositionally, or both combined, and short sketches of the probable means of development are given. More importantly, however, the features must be shown to be substantially independent of composition, for otherwise they could arise merely from variations in the underlying bedrock and in local direction of ice-movement, and roughly similar characteristics would likely be repeated in several of the till sheets. This would hinder the differentiation of the tills, and so lessen the value of the macrofeatures for correlation and determination of relative ages. The influence of composition is studied first.

Composition

In his early studies of the tills in the Red Deer - Stettler region of Alberta, Stalker (1960) concluded that the dissimilarities between the Labuma, Maunsell, and Buffalo Lake tills of Laurentide provenance were not, in the main, due to differences in composition. On page 22 he pointed out: "Any difference in the material composing the three tills, or in its relative grain size, is generally smaller than lateral variations within each till." On page 23 he further stated despite these lateral variations: "The three tills have been described elsewhere and apparently retain the same characteristics everywhere in the province [Alberta]." He was referring to features such as colour, permeability, compaction, massiveness, hardness, and resistance to water erosion, but the statement applies equally well to the other features described here. He then listed places where those tills had been recorded while retaining the same traits, commonly over vastly contrasting types of bedrock. The number of such records has increased markedly during the ensuing twenty years (*e.g.*, Alley, 1973; Stalker, 1963, 1969, 1972, 1976, in press). The retention of unique traits by these and other till sheets over various bedrock formations, consisting of materials as diverse as dark marine shale, buff siltstone, sandstones of various shades, and also over expanses of coarse gravel, certainly indicates a certain independence of the features from composition.

The general independence of these features from compositional control is further demonstrated in vertical section. The outcrops found along South Saskatchewan River in southern Alberta reveal two basic, overall directions of Laurentide ice flow, one to the southwest and the other to the southeast. Each direction is represented by several till sheets, and even a single exposure may show two or more tills laid down by glaciers that advanced in the same direction and over the same bedrock formations. Although this would cause one to expect rather similar compositions for these tills, the traits they display are, nonetheless, very diverse. This phenomenon is particularly well shown at the Cameron Ranch Section (Figure , No. 16) (Stalker, in press), but is also evident at numerous other sites (Dawson, 1885; Horberg, 1952; Stalker, 1963, 1969, 1977). In these cases it is difficult to ascribe the distinctive characteristics of these tills to compositional controls.

Composition does, indeed, limit the ability of a till to acquire many of the features reviewed in this paper. Extremely bouldery till, for example, could never develop the distinctive columnar structures or breakage patterns found in many of the tills. However, both the retention by a till of its distinctive traits over markedly diverse bedrock formations and the presence in section of tills with vastly dissimilar features, laid down by glaciers travelling in much the same direction over similar bedrock formations, indicate a marked independence of these features from compositional control. Certainly within compositional limits that are not exceeded by the Laurentide tills of the western Prairies, the macrofeatures described here are largely independent of composition.

Other origins

Those macrofeatures that are not primarily controlled by composition can only develop during deposition of the tills and during their post-depositional history; hence the threefold division used here of depositional features, combined depositional and later origin, and post-depositional features. The post-depositional features, which are illustrated by compaction, fracture pattern, and colour, are described first, followed by the depositional features, represented by the deformation shown in two of the till sheets. Lastly those of combined origin, exemplified by the spectacular columnar structure found in most of the tills, are described.

Post-Depositional Features

Compaction

The weight of overriding glaciers appears to be the chief instrument in compacting a till. Other processes are so insignificant they can be ignored for our purposes. For instance, subsequent burial by material such as lake or river deposits could, theoretically, cause compaction, but these deposits would seldom reach thicknesses great enough to have much effect. For certain other processes, such as cementation, time available was too short to produce much induration of our Quaternary tills. However, although the weight of overlying ice is the main cause of compaction, a glacier is unlikely to have much impact on its own till. This is demonstrated by the minor compaction displayed by those surface tills that were subjected solely to the weight of their own glaciers. Certainly an ice-sheet could not compress the ablation till it laid down from its surface during downmelting. As for its basal till, much of that is laid down as a glacier is retreating and thinning, and thereby losing its ability to compress underlying material. Further, while the ice-sheet still persisted locally its freshly deposited drift would tend to be saturated and so less compressible, particularly if escape routes for excess water were blocked. It appears, therefore, that subsequent glaciers are needed to compact a till significantly, and undoubtedly the magnitude of that compaction depends chiefly upon the thickness of those glaciers, and especially that of the thickest one, if other factors, such as height of water-table, are relatively constant. This suggests the possibility of using their relative compactness for recognizing the various tills in the field.

The magnificent river sections of south Alberta display very clearly these differences in compaction. The lowest tills are dense, hard, and resistant to erosion, whereas the top till rarely shows much compaction. Its general weakness is demonstrated on river banks by a marked retreat of the cliff face across it and by the ease with which it becomes overgrown. The intermediate tills are likewise intermediate in compaction and in steepness of outcrop face. The differences in compaction are well developed in the twelve sections described by Stalker (1963). In those twelve sections (Figure 1, Nos. 1-12) the two basal tills are typically described as "indurated" (sections 3, 7) or "well-indurated" (sections 4, 11), "compact" (sections 1 2,3,4,7,10,11), "well-consolidated" (section 2), or "hard" (sections 2, 10). Where not so described, these two tills generally contain sand or silt stringers and inclusions, or else have been deformed by slumping or other processes. The top till, on the other hand, rates terms such as "unconsolidated" (section 1), "Poorly consolidated" (sections 2, 10), "only weakly indurated" (section 2), or "poorly indurated and does not form steep cliffs" (section 7). The inbetween tills are in the intermediate range. For instance, at the Brocket Section (section 5) the two lowest Laurentide tills, the Labuma and Maunsell, are described as "indurated and compact," the overlying Brocket Till is described as "poorly consolidated, though more compact and better indurated than overlying till," which in turn is described as "unconsolidated". The top till, there misidentified as Buffalo Lake Till, is described as "the most poorly consolidated till in this section and it is much slumped." The situation at the comprehensive Cameron Ranch Section is similar (Stalker, in press).

The differences in compaction have obvious merits for separating tills and for correlating them from section to section. They also suggest a method of determining relative ages, for a well-compacted till tends to be older than a noticeably less compacted one, whether at the same or different outcrops. Even where a consolidated till lies at the surface it probably does not represent the last glaciation; more likely it was compressed by a later glacier that failed to leave drift there, or else the drift it did leave was subsequently swept away. Compaction also can be the means of distinguishing tills that are in contact; for example a poorly compacted till and an overlying, better compacted one are unlikely to be from the same glacier, even though they are similar in other ways. An exception could be where the last glacier left ablation till over a somewhat better consolidated basal till. Although in this case neither of these till subunits is likely to be well consolidated, the relative difference might be enough to suggest the presence of two distinct till sheets derived from different glaciers.

However, because tills that escaped later glaciation tend to be unconsolidated no matter whether they are young or old, this criterion is of little value beyond the limits of younger glaciers. For instance, in southern Alberta and even in the west where progressively older tills surface, everywhere the surface till is about equally well consolidated. Thus the oldest (Labuma) till shows little consolidation beyond the limits of the succeeding Maunsell and Brocket glaciers, and they in turn are relatively weak beyond the boundaries of the still younger, Wisconsin glaciers. This has caused confusion, for example in sections 4, 5,7,8 of Figure 1. In those sections the surface till beyond the Classical Wisconsin glacial limits of Stalker (1977, 1978) and of Stalker and Harrison (1977) was incorrectly interpretated by Stalker (1963) as Buffalo Lake Till of Classical Wisconsin Age. Part of this confusion was caused by an overestimation - fairly general at that time - of the strength and extent of the Classical Wisconsin ice advances, but a more important cause of the error was the degree of consolidation of this surface till, which resembled that found in the Buffalo Lake Till farther east. Stalker (in press) has since recognized that the unconsolidated nature of the western till is more a consequence of lack of subsequent, overriding glaciers than of youth, and now assigns a greater age to it.

Fracture pattern

The fracture pattern of a till can be a valuable aid to its recognition and correlation. In southern Alberta an old till tends to break into smaller and sharper pieces than a young one. The oldest (Labuma) till, for instance, breaks into narrow chips 1 to 2 cm long, the Maunsell Till gives rectangular pieces 5 to 10 cm to a side, the Brocket Till, as a rule, gives similar-shaped but somewhat larger fragments, whereas the youngest tills give large, irregular lumps or are too little consolidated to produce any fracture pattern whatsoever. This sequence indicates the relative age of a till and aids in correlation. Similar sequences undoubtedly exist in other regions. However, fracture pattern appears closely linked to compaction and consolidation, and like them is useless where old tills surface beyond the range of younger ones. Also like compaction and consolidation, it appears to be produced during the post-depositional history of a till. Basically, fracture pattern acts as an adjunct to confirm the results obtained from the study of compaction.

Colour

Colour, and especially its value or lightness, is a most intriguing property of a till and can, in some regions, be a valuable tool for correlating till sheets and determining their relative ages. In general, on the southwestern Great Plains, the darker a till the greater its age. Horberg (1952) early drew attention to the different colours and intensity values shown by the different tills of southern Alberta when he described his basal till as "dark brownish gray" (p. 310), and his lower and upper tills as "dark gray" (p. 311, 316). In 1960 Stalker pointed out the marked contrasts in colour of the various Laurentide tills in Alberta, when he stated (p. 23):

"The Labuma till is dark blue or black, perhaps due to a large content of carbonaceous material derived from Cretaceous and Tertiary coal seams and carbonaceous shale, and to long burial below the water-table. However, it seems to retain this colour even when near the surface and above the water-table. The Maunsell till is light to dark blue, also due to its content of carbonaceous material and long burial under the water-table, though for a shorter time than the Labuma till. The Buffalo Lake till is more variable in colour, with varying browns, yellows, greys, and light blues where unoxidized. Upon drying the three tills become light in tint, to various shades of grey."

Later on Stalker (1963, p. 5-8) outlined the relative lightness or darkness of shade of the various till sheets and (in press) gave a general description of their overall change in hue and value from lowest to youngest. Although each of his three groups of tills (Illinoian, Preclassical Wisconsin, Classical Wisconsin) includes both brown and gray tills, depending chiefly upon whether the glacier responsible for each till came from northwest or northeast, the Labuma, Maunsell, and Brocket tills of his lowest, or Illinoian, group are by far the darkest, with the basal (Labuma) one nearly black. The Preclassical Wisconsin tills are intermediate in value, being medium dark brown or bluish to grayish brown, but noticeably lighter in colour than the tills of the underlying Illinoian group. The Classical Wisconsin (Buffalo Lake) tills show a variety of colours, but all are lighter in value, typically being buff, beige, or light brown. Further, colour was one of the characteristics Stalker (1960, p. 23) was referring to when he stated that the tills "apparently retain the same characteristics everywhere in the province."

In a quotation given earlier, Stalker (1960, p. 23) suggested that the colours of his tills were controlled by both composition and post-depositional influences. Whether or not the origins he postulates for the different colours and values are correct is immaterial here; what is important is that the differences are present, they are remarkably consistent, and they can be used to correlate the various tills, determine their relative ages, and even, in south Alberta, obtain an estimate of their true ages. Colour has further value because it apparently is independent of whether the till has been overridden by later glaciers, and so is useful right up to the limits of glaciation, thus avoiding one of the chief limitations of both the compaction and fracture pattern criteria discussed previously. Further, it is largely independent of whether a till is buried or lies at the surface. More attention to colour would have avoided the misidentifications of tills in Stalker (1963, sections 4,5,7,8) mentioned above under consolidation. Certainly the tills there mistaken for the Buffalo Lake Till of Classical Wisconsin Age approach in colouring and value or lightness the Preclassical Wisconsin tills observed in sections farther east (*e.g.*, Island Bluff Section, No. 15, of Stalker, 1969; see Figure 1, No. 15).

Depositional Features

Deformation

Two of the Laurentide till sheets found on the southwest Great Plains are characterized by a strong, distinctive pattern of deformation or contortion. Strong deformation appears restricted to these two tills, though other till sheets may be affected locally and there is always the possibility that additional deformed tills may yet be discovered. The deformation in these tills is greatest near the base and decreases noticeably upward. Further, it is much better developed in the younger of them. Stalker (1969, p. 10) described this deformation as follows:

"This unit is referred to as the 'contorted till' because its strong deformation, best developed towards the base of the unit, is a distinguishing trait practically everywhere. None of the other tills show this deformation to nearly the same extent. It reveals itself in convolutions; in inclusions of sand, silt, or bedrock completely enclosed by till, or of till completely incased in sand, silt, or bedrock; in stringers of till jutting into underlying deposits and lenses of those deposits injected upward into the till; and in strong internal folding and faulting. These features commonly are accentuated by abundant inclusions from the underlying material, whether bedrock or surficial deposits, that contrast vividly with the till itself. In places the deformation is so strong and inclusions so abundant that it is impossible to draw a demarcation between the till and underlying deposits."

These characteristics apply in lesser degree to the other (older) contorted till. The numerous inclusions of sand and silt referred to, which at times have a high water content, weaken both these tills enough to inhibit them from forming either the steep cliff faces or the columnar structures described in the next section.

The writer considers the deformation to be a phenomenon caused by conditions prevailing at time of deposition. If it were post-depositional, the two most likely causes would be permafrost developed during subsequent glaciations or pressures exerted by overriding glaciers. However, both can be rejected. Post-depositional permafrost should have disturbed the upper parts of the units more strongly than the lower, or the reverse of what is seen. Further, permafrost should have acted nearly as strongly upon any other tills or deposits it found overlying these two units, but only these two till sheets, and neither intervening nor overlying tills, nor any other deposits, are significantly affected. Further, in regard to the possible effect of overriding glaciers, Stalker (1977, p. 402) stated, with regard to his "Unit XXIII" at Medicine Hat:

"It is ... strongly contorted and deformed, noticeably more so than any of the other tills, and in places it is strongly intermixed with underlying deposits. This till may form the surface west of Lethbridge, where it has similar traits."

If this suggestion that the traits occur equally well, both where the till is buried in section and exposed at the surface, is correct, they obviously are not caused by subsequent glacier overriding. Further, other till sheets have suffered such overriding with consequent strong compaction without development of the deformation. It appears, therefore, that both subsequent permafrost and glacier overriding can be eliminated as causes, and that the deformation developed near time of till deposition. The writer considers that the deformation formed under permafrost conditions as the glacier advanced over deeply frozen ground, and that the interaction between the glacier and underlying material, along with thawing of the permafrost, churned together and intermixed the lower part of the till the glacier was depositing and the upper part of the underlying unit. Other factors, such as high ground-water pressure due to the glacier blocking the escape of both surface and ground water, could have been important, particularly when one considers the ubiquity of plastic, bentonitic beds on the Prairies. However, no matter what the cause, suitable conditions must have been rare during ice-advances into the region, because well developed deformation appears restricted to these two till sheets. Probably most of the glacier advances in this region were over ground that was not deeply frozen, if frozen at all.

These contorted tills, which the writer considers to represent the earliest glacier advances of both the Preclassical and Classical Wisconsin stages, are of immense value in correlation, for obviously they should correlate to deformed tills found in other sections, rather than to any non-deformed units that might be present. Further, their differing intensity of deformation allows these two tills to be readily distinguished from each other in most outcrops. Altogether, they form excellent marker beds. As with colour, this criterion can be used beyond the limits of younger glaciers.

It would be useful to determine whether similar deformation or other macrofeatures caused by ice-advance under permafrost conditions can be found in other regions. If they can, they might enable long-range correlation from those regions to the tills of south Alberta. This should not, of course, limit the search for other types of features, developed from conditions prevailing at time of deposition, that might prove equally valuable for both correlation and determination of relative ages in those regions, and for long-range correlation to south Alberta.

Features of Combined Depositional and Post-Depositional Origin

Columnar structure

Of features in the southwest Great Plains that owe their origin to combined depositional and post-depositional effects, the most spectacular are the long, vertical columns displayed in several of the till sheets. These are caused by long, vertical joints, whose effect Stalker (1960, p. 22) described as follows:

> "On cliffs and river banks the jointing of the Maunsell till results in distinctive bluffs and columnar forms, as the till slumps or breaks along these fractures and tends to give vertical faces as much as 70 feet high. Dawson (1885, pp. 143c, 144c) described such features along the Oldman River, south of the Red Deer-Stettler area. Similar cliffs along Bow River, also south of this area, are formed by the Labuma and Maunsell tills. The jointing in the Buffalo Lake till is irregular and not very prominent. This till is weak and rarely forms vertical bluffs."

He further noted: "If the fracture planes in these two tills (Labuma and Maunsell) are not horizontal or vertical it can be assumed that the till has slumped or has been contorted by overriding glaciers." Columns that are not near to vertical are rare.

The ability to form columns is a typical attribute of most tills seen on the southwest Prairies, the contorted tills being the chief exceptions. Apparently the contortions, along with the inclusions of sand, silt, clay, and bedrock, weaken the contorted tills sufficiently to inhibit formation of columns. This can also happen in normally columnar tills, where inclusions of non-till material have also checked their local formation. The absence of columns in the tills that were contorted near time of deposition does indicate some control over their formation by conditions prevailing during deposition. Equally, however, the rarity of well developed columnar structures in surface tills that escaped overriding by later glaciers indicates the importance of post-depositional controls. Undoubtedly other post-depositional phenomena, such as shrinkage caused by drying, are often of significance. Altogether, in summary, column formation appears to be a function of both depositional and post-depositional events, and so the columns are used here to illustrate the features of combined origin.

The better development of the columns in some tills, such as the Labuma and Maunsell, than in others, and their absence in the contorted tills, is an invaluable aid to correlation. However, as well developed columns can be found in both old and young tills, except where they are at the surface, the columns are of little value for determination of age.

CONCLUSIONS

Each till has its own individual, largescale features that are independent of composition. These help in field recognition of the till, and commonly enable its correlation from section to section and to sites where its age can be determined. They can also be used in places to determine relative ages of tills and even to indicate true age where other means of dating are unavailable. Even if such other means of dating are available, the macrofeatures can still serve as a check on results obtained by those means, including absolute dates. Use of these macrofeatures has been much neglected of late, and particularly since the improved means of absolute dating.

There are many types of till features available for these purposes, and undoubtedly different suites of them can be identified for use in different regions. The examples for this paper have been drawn from studies made of the Laurentide tills of the southwest part of the Great Plains of Canada, which is almost an ideal spot to study them. Correlation and determination of the relative ages of the many tills present in that region are based largely on field use of such macrofeatures. Luckily, that region also has sections where ages can be obtained by other means - primarily vertebrate paleontology - to which the till units can be correlated. Of the abundant macrofeatures available there, only a few have been chosen to illustrate each of the main types: those formed near time of till deposition, those formed post-depositionally, and those of combined origin. However, those chosen are undoubtedly the most spectacular found in south Alberta, and the most valuable for the purpose.

REFERENCES CITED

Alley, N.F., 1973, Glacial stratigraphy and the limits of the Rocky Mountain and Laurentide ice sheets in southwestern Alberta, Canada: Canadian Petroleum Geol. Bull., v. 21, p. 153-177.

Dawson, G.M. 1885, Report on the region in the vicinity of the Bow and Belly Rivers, Northwest Territory: Geol. Survey Can., Report of Progress 1882-3-4, pt.C., 168 p.

Dawson, G.M. and McConnell, R.G., 1895, Glacial deposits of southwestern Alberta in the vicinity of the Rocky Mountains: Geol. Soc. Amer. Bull., v. 7, p. 31-66.

Harington, C.R., 1978, Quaternary vertebrate faunas of Canada and Alaska and their suggested chronological sequence: National Museums Can., Museum of Natural Sciences, Ottawa: Syllogeus No. 15, 105 p.

Horberg, Leland, 1952, Pleistocene drift sheets in the Lethbridge region, Alberta: Jour. Geol., v. 60 p. 303-330.

________________, 1954, Rocky Mountain and continental Pleistocene deposits in the Waterton region, Alberta, Canada: Geol. Soc. Amer. Bull., v. 65, p. 1093-1150.

Stalker, A. MacS., 1960, Surficial geology of the Red Deer-Stettler Map-area, Alberta: Geol. Survey Can., Memoir 306, 140 p.

________________, 1963, Quaternary stratigraphy in southern Alberta: Geol. Survey Can., Paper 62-34, 52 p.

________________, 1969, Quaternary stratigraphy in southern Alberta, Report II: Sections near Medicine Hat: Geol. Survey Can., Paper 69-26, 28 p.

________________, 1972, Southern Alberta: in Rutter, N.W. and Christiansen, E.A., eds., Quaternary geology and geomorphology between Winnipeg and the Rocky Mountains: Montreal, Twenty-fourth Internat. Geol. Congress, Guidebook, Field Excursion C-22, p. 67-79.

________________, 1976, Quaternary stratigraphy of the southwestern Canadian Prairies: in Mahaney, W.C., ed., Quaternary Stratigraphy of North America: Dowden, Hutchinson & Ross Inc., Stroudsburg, Pa., p. 381-407.

________________, 1977, The probable extent of Classical Wisconsin ice in southern and central Alberta: Canadian Jour. Earth Sci., v. 14, p. 2614-2619.

________________, 1978, The geology of the ice-free corridor: The southern half: Canadian Jour. of Anthropology, v. 1, p. 11-13.

__________________, Quaternary stratigraphy in southern Alberta, Report III: The Cameron Ranch Section: Geol. Survey Can., Paper (in press).

Stalker, A. MacS. and Churcher, C.S., 1970, Deposits near Medicine Hat, Alberta: Geol. Survey Can., Display chart with marginal notes.

___________, 1972, Glacial stratigraphy of the southwestern Canadian Prairies; the Laurentide record: Montreal, Twenty-fourth Internat. Geol. Congress, Sect. 12, p. 110-119.

Stalker, A. MacS. and Harrison, J.E., 1977, Quaternary glaciation of the Waterton-Castle River region of Alberta: Canadian Petroleum Geol., Bull., v. 25, p. 882-906.

EVALUATION OF RELATIVE PEDOSTRATIGRAPHIC DATING METHODS, WITH SPECIAL REFERENCE TO QUATERNARY SUCCESSIONS OVERLYING WEATHERED PLATFORM MATERIALS

CHARLES W. FINKL Jr.

ABSTRACT

Paleosols which lack datable materials may be assigned to relative age sequences by evaluating the nature of soil materials and the degree of pedological organization. The chronological ordering of pedological episodes, which is important to reconstruction of paleoenvironments where sequences of soils and weathering zones occur, is based on stratigraphic principles and practices. Episodic development in the pedosphere may be established by applying concepts of separate identity, lateral continuity, ascendancy and descendancy, weathering differentials, and pedogenic persistence to soil mantles. These concepts are used in the stratigraphic ordering of soils after proofs of existence have been established for independent (buried) soil layers.

Although pedological features and soil fabrics provide a basis for identifying the presence of rock- and soil stratigraphic units in whole soil profiles, they also help establish the relative ages of juxtaposed soil materials. The nature of skeleton grains, especially those which are etched, embayed or inherited from previous cycles, are important to evaluations of soil process and stage of development. Organization of soil plasma into distinct fabrics can, when correlated with specific pedostratigraphic intervals, provide an additional means of establishing chronological sequences based on geographically disjunct paleosols. These methods are particularly useful in deeply weathered terrains, such as those associated with tropical and subtropical cratonic regions, where Quaternary soil-stratigraphic units overlie Tertiary weathering zones. Micromorphological techniques effectively differentiate pedogenic episodes where younger soil-stratigraphic layers contain soil parent materials derived from pre-existing weathering mantles.

INTRODUCTION

Soils are complicated three-phase (solid, liquid, and gas) natural systems that occupy portions of the weathered surface of the earth. Although soils exist as discrete or individual bodies, they co-mingle with larger entities forming continuums which make up soil mantles. These mantles, as defined by Fridland (1974), have wide geographic distribution and are distinguished from one another on the basis of inherent characteristics. Surface soil mantles are comprised of contemporary soils, weathering zones, and exposed paleosols. Buried soils, parts of subsurface mantles, represent pedogenic phases that have been arrested by deposition of overburden. Many exposures exhibit multiple layers of weathered materials successively stacked in vertical sequence (Figure 1).

Subdivision of the soil continuum has taken many tacks but no single method of doing so has yet been agreed upon (Butler, 1980). Part of the classificatory dilemma is undoubtedly related to the fact that soil layering, *i.e.* the stratigraphic succession of unrelated soil materials, is more prevalent than generally appreciated.

Figure 1 Soil development in alluvial terraces along the Stelly River in northwestern Australia. A surface Red Earth (RE) overlies a truncated Red Podzolic (RP) soil which in turn covers an older soil developed in calcareous alluvium (CA). (Photo: William M. McArthur)

Separation of soils on a chronostratigraphic basis requires careful application of absolute and relative dating methods.

Understanding of soil age relationships is critical to comprehension of certain physical and chemical characteristics, duration and intensity of pedogenic processes, and even management of soil resources. Old or senile soils exhibit properties quite different from those associated with soils recently formed. Soils containing datable materials can be assigned absolute time frames. Many soils are placed in relative time sequences because they are not amenable to precise age determinations. Relative age sequences, though not as definitive as absolute methods, often provide a reasonable basis for estimating the timing of pedogenic episodes. Specific soil-forming processes, operating for certain lengths of time, are known to produce the same effects over and over again. Although soil nitrogen and phosphorus levels markedly fluctuate over the short term, they may reach quasi-equilibrium in 10^2-10^3yr (Jenny, 1941; Crocker & Major, 1955; Smeck, 1973). The development of argillic horizons or humus-iron pans of podzols developed in sands may require intervals on the order of 10^3-10^4 yr. The formation of thick mottled and pallid zones of deep lateritic profiles may require 10^6-10^7 yr (Whitehouse, 1940; Hanlon, 1950; Idurm & Senior, 1978; Kronberg *et al.*, 1979). Some resistant epidiagenetic features such as ferruginous duricrusts, of the type described by Maignien (1959), formed as long ago as the Oligocene (Quilty, 1977), but may have taken only 10^2-10^3 yr to harden under desiccating conditions (Mörner, 1978). Leneuf and Aubert (1960) estimate that it takes something on the order of 22,000-77,000 yr for 1 m of gneiss to become completely ferralitized in a humid tropical environment. Deep weathering in Nigeria is thought to have taken 660,000 to 2,310,000 yr to reach 30 m (Jejè, 1970). Such features thus serve as proxies where soil age is inferred.

Soils associated with stable platforms, especially those not subjected to the direct affects of Quaternary glaciations and which presently lie in tropical and subtropical regions, have rather different histories from soil mantles in Northern Hemisphere cratons. The morphotectonic stability of the Australian, Brazilian, and parts of the African cratons permitted deep weathering mantles to develop on suitable surfaces over vast expanses of territory. Intense chemical weathering continued on and off for milleniums on the West Australian craton until largely inhibited by Oligocene induration (Finkl & Fairbridge, 1979). The great extent of deeply weathered cratonic regions (the West Australian Shield alone occupies some 718,000 km^2) points to the need for greater understanding of soil age relationships in these "unglaciated" regions.

Examples of relative age determinations discussed here are derived from studies on the Australian craton. The concepts and methods are, however, relevant to other landscapes where Quaternary soil mantles overlie deeply weathered substrates.

CHRONOLOGICAL ORDERING OF PEDOLOGICAL EPISODES

Concepts of soil age revolve around precise measurements of time or estimates based on comparative soil anatomies which give a relative time reference. The so-called absolute dating methods focus on isotopic age data, namely zircon-fission track ages, radiocarbon residence times, uranium series dating, thermoluminescence (see discussion in Moorbath, 1960; Yaalon, 1971) or correlation of land sequences with ^{18}O stages in the oceanic record (Kukla, 1977). Relative dating methodologies, on the other hand, may employ stratigraphic principles (Figure 2) or rely on notions of relative soil age using soil-geomorphic relationships and appraisal of intensity of profile development.

Absolute Dating Methods

Although absolute dating methods themselves do not fall within our purview, it is relevant to emphasize the importance of these datums to relative chronologies. ^{18}O stages in the oceanic record for example, no matter how remote they may seem from land-based sequences, helped establish the timing of Quaternary successions in Europe. Using river terraces as a link, Kukla (1977) was able to construct an improved climato-stratigraphic model of the Pleistocene in Central Europe where there were sequences of loess bands and soils.

Ages of soil mantles can also be estimated when intercalated with layers of volcanic ash. In the American West, for example, useful stratigraphic marker beds are associated with Mount St. Helens, Mount Rainier, Mazama, Glacier Peak, and Pearlette ash falls (see discussions in Wilcox, 1965). Tephrochronologies may indeed bracket pedogenic episodes but they themselves are complicated by ash falls closely spaced in time (Young & Powers, 1960). The Glacier Peak tephra in the northern Cascades of Washington are divisible into at least nine separate layers (Porter, 1978). The Mazama ash may also contain multiple ash falls within apparently well-defined marker horizons. Mack *et al.* (1979) suggest, on the basis of petrographic studies and radiometric dates, that this marker horizon resulted from at least two eruptions where the upper unit falls in the range 6,600-7,000 yr BP and the lower one around 8,300 ± 80 yr BP. Bioturbation, frost activity, and creep also sometimes inhibit tephrostratigraphic investigations.

Relative Age Sequences

Whether keyed to dated deposits in stratigraphic sequences on land or linked to marine sediments offshore, it is still often necessary or advantageous to determine relative ages of independent soil systems. Soil mantles may, for example, be stacked between other types of chronostratigraphic units or they may be associated with a particular marker horizon. Where such relationships do not exist, relative dating methods generally focus on stratigraphic techniques which place soil mantles in sequence from oldest to youngest according to the Law of Superposition. Stratigraphy, in the absence of absolute dates, provides a relative chronology of cyclic regolith layering features.

Figure 2 The soil developed in the lower basalt was in turn covered by a later flow which formed a new ground-surface. The contemporary surface soil, according to the Law of Superposition, is younger than the buried soil. Both basalt flows near Dorrigo, New South Wales, are of Tertiary age. (Photo: William M. McArthur)

Pedo-lithologic discontinuities and weathering status, interpreted, for example, in terms of stratigraphic position (Leslie, 1978), assist in the elucidation of deposition, soil formation, and pedosphere stripping. Soil-geomorphic relationships often provide useful clues as to the age of soil mantles when a nexus can be demonstrated between landforms and soils (*e.g.* Playford, 1954; Ruhe, 1954; Wright, 1963; Walker, 1966; Parsons *et al.*, 1970; Young, 1970; Eden, 1971; Churchward, 1977; Brosh & Gerson, 1978; Daniels & Gamble, 1978; Kumar, 1979). Linkages between soil mantles and erosional and/or depositional surfaces also provide hints as to soil age (Buurman, 1972). The "soil stratolandscape" (van Dijk, 1979), an extension of the general concept embracing soil-landform relationships, is a landscape unit with a specific soil-stratigraphic assemblage. Such units contain remnants of *in situ* and depositional paleosol layers and relict weathering zones which display a certain surface morphology. Soil stratolandscapes are normally delineated as geomorphic subdivisions which offer advantage by incorporating results of detailed appraisal of cyclic landscape development in a single unit. Pedo-morpholith studies (*e.g.* van Dijk & Rowe, 1980) in an analogous way define relationships between soil-stratigraphic patterns and landforms. Interpretation of relative soil age on this basis is thus very much tied to studies of soil periodicity, following the K-cycle concept formulated by Butler (1959), landsurface morphology, and stage of landscape development in the broadest Davisian sense. Concepts of effective age and stage of development have been applied with varying degrees of success but may have applicability where the properties of buried soils occur in surface analogues (Simonson, 1954). Field morphology rating systems (Bilzi & Ciolkosz, 1971; Meixner & Singer, 1981) have a useful potential for assisting in the evaluation of soil formation and discontinuities. The method, based on the distinctness of morphological features in adjacent horizons and comparison of discrete horizons in the solum with C horizons within a pedon, may have general use as an indicator of relative soil age. Micromorphological analyses offer scope for increased resolution of many soil age problems, especially in cases where field techniques are too coarse for detailed study. Soil fabric analysis and the investigation of skeleton grain morphologies provide additional means of establishing relative ages of paleosols.

STRATIGRAPHIC PRINCIPLES AND PRACTICES

The occurrence of buried weathering zones in Quaternary successions in the American Midwest first suggested to some workers (*e.g.* Worthen, 1866; McGee, 1891; Leverett, 1898) that the layers or bands could be used as stratigraphic marker horizons. It was later formally proposed to employ paleosols and weathering profiles as stratigraphic units (Richmond & Frye, 1957; ACSN, 1961). The concept proved to be applicable to Quaternary investigations not only in glaciated landscapes of the Midwest (Follmer, 1978) but in western mountain regions as well (Morrison, 1964, 1967). Inter-regional correlations of weathered marker horizons were also attempted (Morrison & Frye, 1965).

Recognition of stratigraphic sequences in deeply weathered tropical and subtropical terraines brought soil stratigraphic studies from mid-latitudes to equatorial realms (*e.g.* Erhart, 1948; Ruhe, 1954). Studies of layered soil materials in the Australian region greatly expanded the scope of the field in the 50' and 60's by including a diverse array of weathered materials ranging from ancient deep weathering mantles and duricrusts (ferricrete and silcrete) to relatively younger soils (*e.g.* Fairbridge & Teichert, 1953; Van Dijk, 1958; Butler, 1959; Jessup, 1960, Churchward, 1961; Walker, 1962; Van Dijk, Riddler & Rowe, 1968; Brewer & Walker, 1969).

The principles of soil stratigraphy, as summarized by Morrison (1967, 1978) and recently reviewed by Finkl (1980), utilize the Law of Superposition and the Principles of Separate Identity (Walker, 1958; Butler, 1959), Lateral Continuity, Ascendancy and Descendancy (Walker, 1966), and Pedogenic Persistence. These concepts have been applied to weathered platform materials and related deposits in efforts to establish a relative soil-stratigraphic sequence for the West Australian craton. Detailed information was mainly derived from the southwestern portion of the craton in a dissected upland region of the Darling Plateau, referred to here as the Blackwood Tableland (Figure 3).

Mantle of Deep Chemical Weathering on the Blackwood Tableland

The lateritic deep weathering mantle is a pancratonic feature which occurs on all major geomorphic surfaces where slopes are less than 10 degrees (Mulcahy, 1960). The depth of weathering averages at least 30 m with mottled and pallid zones developed in crystalline basement and sedimentary rocks alike (Butt and Smith, 1980). Referred to as the Warraniani layer or mantle (Finkl & Churchward, 1976), the weathering zone contains mottled and pallid zones (Ballijup materials) so characteristic of lateritic profiles (Stephens, 1971) and deep regolith (Nannup saprolite) on steep valley side slopes. Lateritic facies in the Warraniani mantle indiscriminantly transgress lithologic boundaries (Figure 4) establishing the pervasive nature of the lateritization process. Some parent materials are more susceptible to intense biogeochemical weathering but most are similarly affected. Because both weathering units, Ballijup materials and Nannup saprolite, occupy the same stratigraphic interval, are coterminous, and grade into one another depending on slope and drainage conditions, they are regarded as facies of the Warraniani layer and are thus coeval in age (Finkl & Churchward, 1976; Finkl & Gilkes, 1976). The saprolite facies, like the lateritic deep weathering, rots most sorts of country rock (Figure 5) to depths of 10 m or more.

The deep weathering zone bears evidence of widespread stripping to produce a variety of surfaces interpreted as etchplains (Mabbutt, 1961; Finkl & Churchward, 1973; Fairbridge & Finkl, 1979; Finkl, 1979). These surfaces are capped by ferruginous duricrust (ferricrete) or buried by overburden containing younger soil mantles.

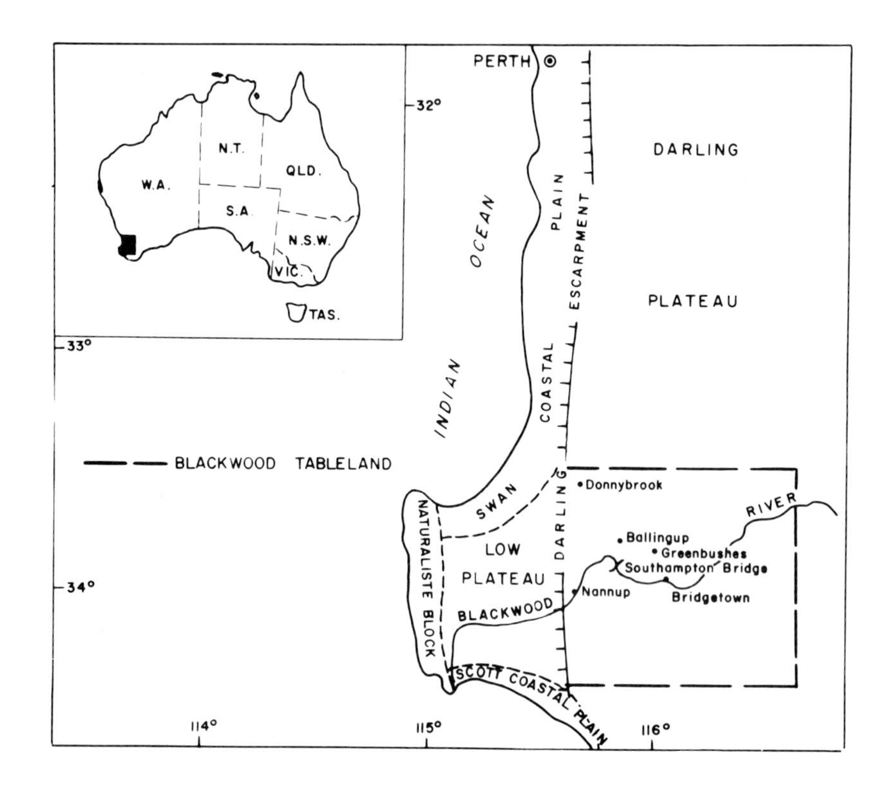

Figure 3 The western margin of the West Australian craton is bounded by the Darling fault scarp. The Blackwood Tableland is a dissected sector of the Darling Plateau with numerous duricrusted plateau remnants. Soil-stratigraphic relationships described for this region were observed throughout the humid southwest with analogues occurring in the arid cratonic interior.

The age of the deep weathering in the Australian region has intrigued geoscientists for the last half century or so. Associated with the great antiquity and stability of the West Australian craton, development of a deep layer of chemical weathering seems to have taken place from late Mesozoic to possibly late Tertiary times (Dury *et al.*, 1968, Geol. Surv. West. Aust., 1975). The chemistry of lateritic soils and ground waters suggested to Kronberg *et al.* (1979) that the deep lateritic profiles on the Brazilian Shield required tens of millions of years to develop. A very stable continental setting free from major erosional events was postulated as being essential to such intensive chemical weathering. The lack of datable materials in the soft lateritic and saprolitic soils required researchers to look for clues elsewhere, notably in Cenozoic marine sediments offshore and in the duricrust itself. Induration of lateritic and saprolitic soils seems to have occurred in the Oligocene (Schmidt & Embleton, 1973; Quilty, 1977) associated in part with the loss of a protective forest cover during a cool semi-arid interval (Kemp, 1978) as well as a probable drop in the ground water table following a fall in sea level at about this time (Galloway, 1970; Leonty'ev, 1970). Subsequent differential stripping (truncation) of regolith materials and degradation of lateritic and saprolitic profiles thus sets the stage for relative age determinations on the craton. The impact of preweathered materials on younger soils has been appreciated in many studies (see summaries in Mulcahy, 1967; Churchward, 1970). The intensely weathered condition of younger soils is thus clearly related to degradation products of the Warraniani deep weathering mantle and only maximum ages of successive soil covers can be suggested. Mulcahy *et al.* (1972) estimated that many differentiated soils on younger surfaces of the craton may have originated in the Last Interglacial or soon afterwards, even though the soils contain materials associated with pre-Quaternary environments.

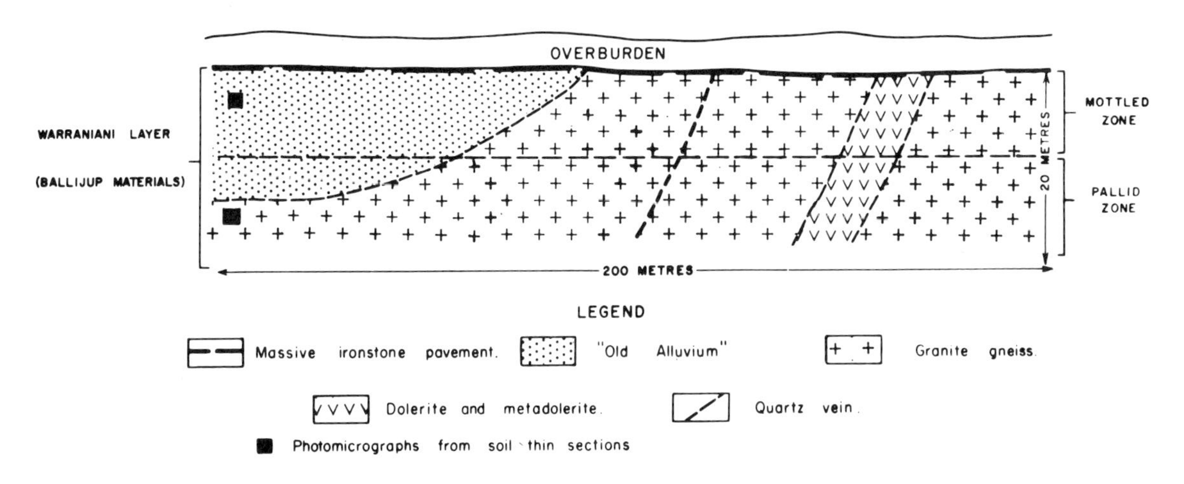

Figure 4 Lateritic deep weathering transgresses lithologic boundaries rotting a variety of country rocks to great depths. The position of the seasonally fluctuating water table is approximated by the mottled zone which overlies bleached rock fabrics in the waterlogged pallid zone.

Clastic Materials

The stratigraphic relationships of superficial clastic materials, established on the basis of field observations, are summarized in Figures 6, 7, and 8. The sequence on the Blackwood Tableland (Figure 6) comprises pisolitic accumulations in valley side slopes and fine-grained alluvium on valley floors. The gravels decrease in particle-size downslope and eventually merge with earthy alluvium. Both units, respectively identified as the Minnijup gravels and Bilbup alluvium (Finkl & Churchward, 1976), are derived from portions of the deep weathering mantle which they now overlie farther downslope. A similar stacking of surficial materials occurs in the shallow Balingup dissection. The ironstone gravels, however, give way to an earthy facies (Cundinup earths) on lower valley side slopes and spurs (Figure 7). Stratigraphic sequences in the deeper Bridgetown dissection are complicated by the interposition of transported Tittibinup clays between the Minnijup and Warraniani layers (Figure 8).

Surficial Materials as Stratigraphic Elements

The stratigraphic positions of *in situ* and clastic materials establish a relative age sequence. The Warraniani layer is the oldest substratum and also the source of pre-weathered materials occurring in the successively younger Tittibinup and Minnijup layers. Application of the Principles of Separate Identity, Lateral Continuity, and Ascendancy and Descendancy lead to the development of this age sequence as detailed by Finkl & Chruchward (1976). Interpretations of *in situ* and depositional bodies as soil layers were justified on the basis of properties normally attributed to pedogenesis, *i.e.* soil structure, soil fabric, illuviated clay, and pedological features. The sedimentary layers thus become the Minnijup and Tittibinup soil layers. They were formalized in terms of the pedoderm concept (Brewer *et al.*, 1970) and ranked stratigraphically, from oldest to youngest: Warraniani Pedoderm, Tittibinup Pedoderm, and Minnijup Pedoderm.

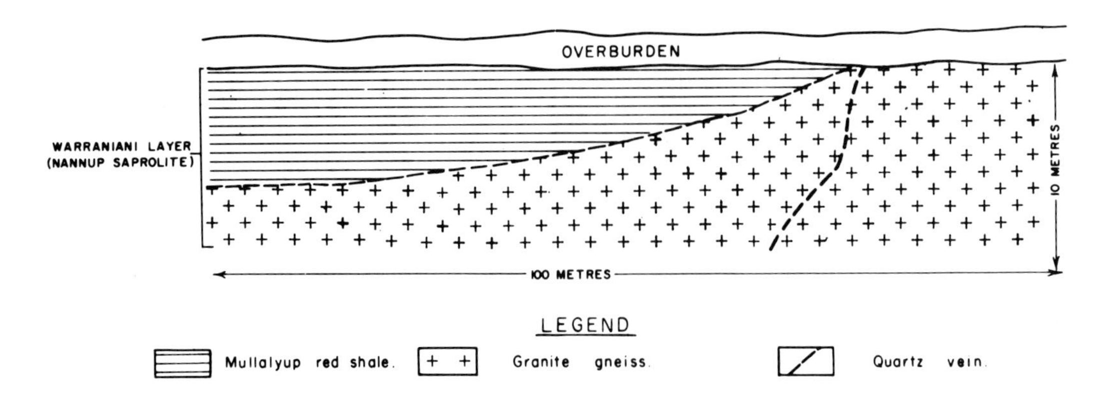

Figure 5 Saprolites, occurring on dissected valley side slopes, are coterminous with laterite profiles. The residual nature of saprolitic weathering, which crosses geological unconformities, is indicated by intact quartz veins and weathered rock fabrics.

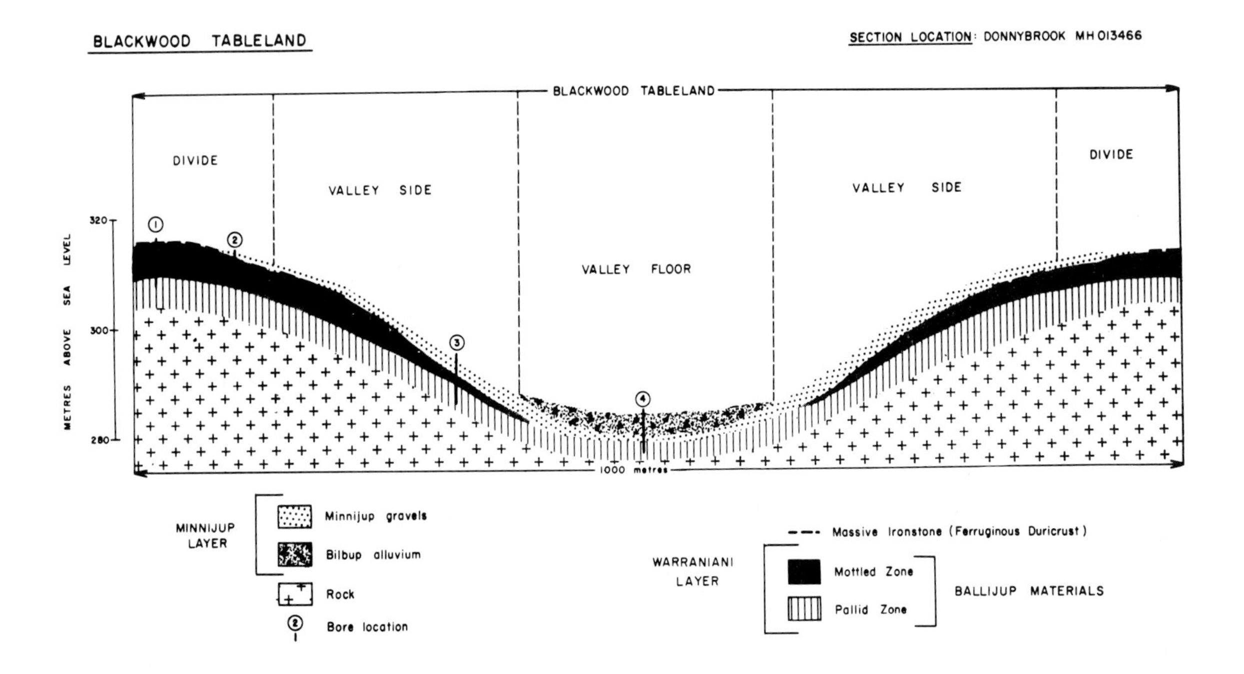

Figure 6 Pisolitic (ironstone) gravels blanket upland valley side slopes but become more finely graded downslope. These colluvial deposits merge with valley floor alluvium, and overlie ferruginous duricrust, soft lateritic materials, and bedrock.

SOIL-GEOMORPHIC RELATIONSHIPS

Considerable controversy has centered on soil-surface age relationships in this region because laterite profiles are found at various levels in the landscape. The designation of high and low level, primary and secondary, and older and younger laterites (Campbell, 1917; Woolnough, 1918) inhibited resolution of contentious issues for decades. Based on investigations along western cratonic margins, Playford (1954) suggested that laterites of major upland and lowland surfaces were of the same age having formed after uplift and partial dissection of uplands. Prider (1966) and Finkl (1971), working in different areas, concurred with Playford's findings. Buried under sand plain deposits and broad floodways containing wash materials, the deep weathering mantle in the cratonic hinterland went largely unnoticed. Playford's

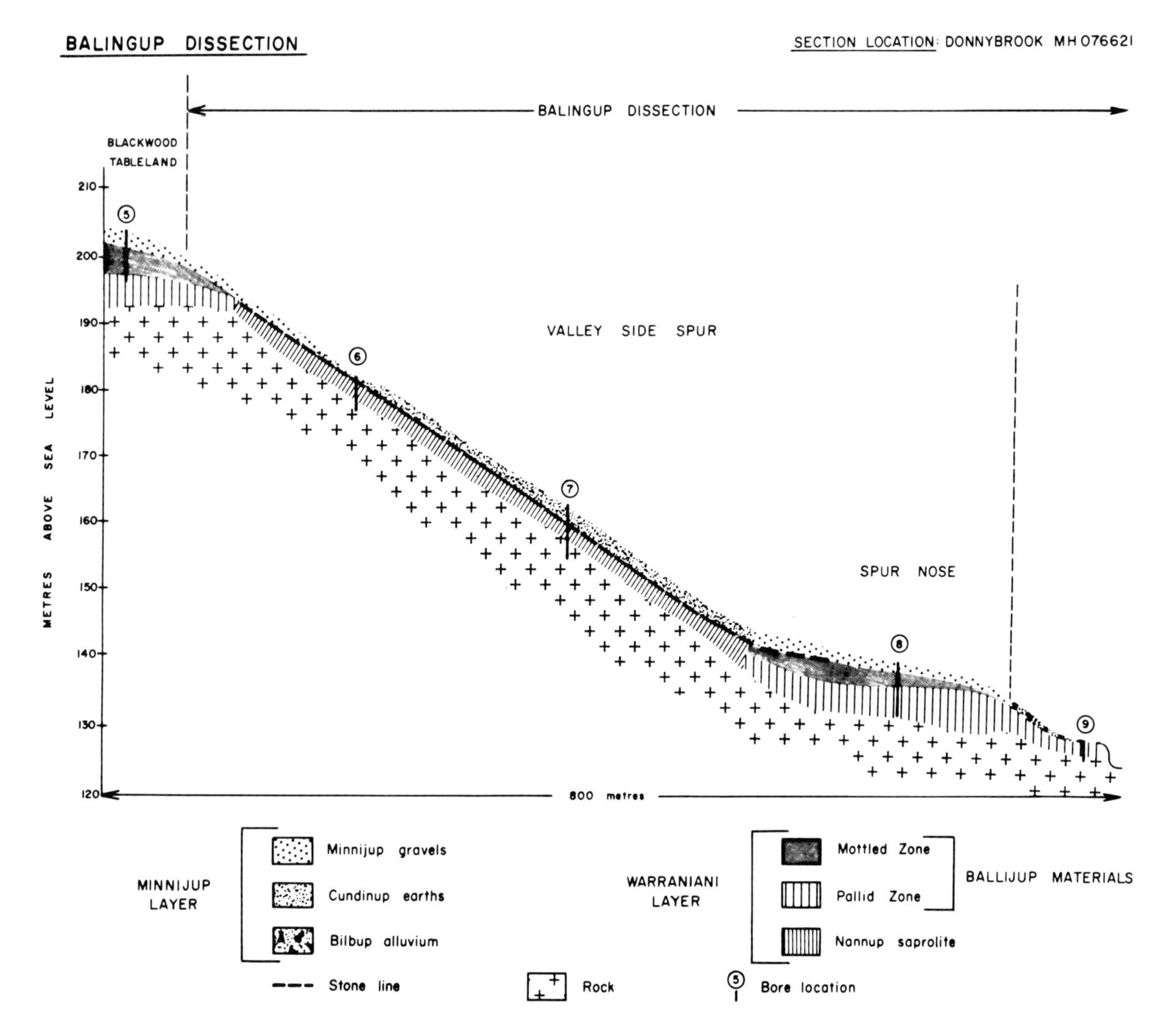

Figure 7 The succession of clastic and *in situ* materials in shallow dissections is similar to upland valleys except for the occurrence of earths and saprolites on side slopes, the former being separated from substrates by a stone line.

deduction was essentially confirmed for the cratonic interior when Churchward (1977) found mottled and pallid zones, under great thicknesses of surficial drift, continuously linking major geomorphic surfaces. Young (1970), working in eastern Australia, also suggested a post-uplift age for the lateritic duricrust.

The deep weathering mantle thus seems to be post-uplift in age. Eocene sediments overlapping the downwarped southern margin of the craton are lateritized and consequently the Warraniani Pedoderm can be no older. Oligocene induration of exposed mottled and pallid zones no doubt inhibited subsequent lateritization. Degradation of deeply weathered substrates and duricrusted residuals, particularly as it relates products of soil and crust formation to stripped surfaces of various ages, has been the subject of detailed investigations as summarized by Stephens (1946, 1971); Mulcahy (1960, 1967); Mabbutt (1961); Churchward (1970); Finkl & Churchward (1973); Mulcahy & Churchward (1973); and Finkl (1979).

STONE LINES, SURFACES, AND SOILS

Stone lines occur as dominant subsurface features in dissected landscapes, their presence being verified by observations in road cuts and old quarries. They appear as lenses of coarse clastic material

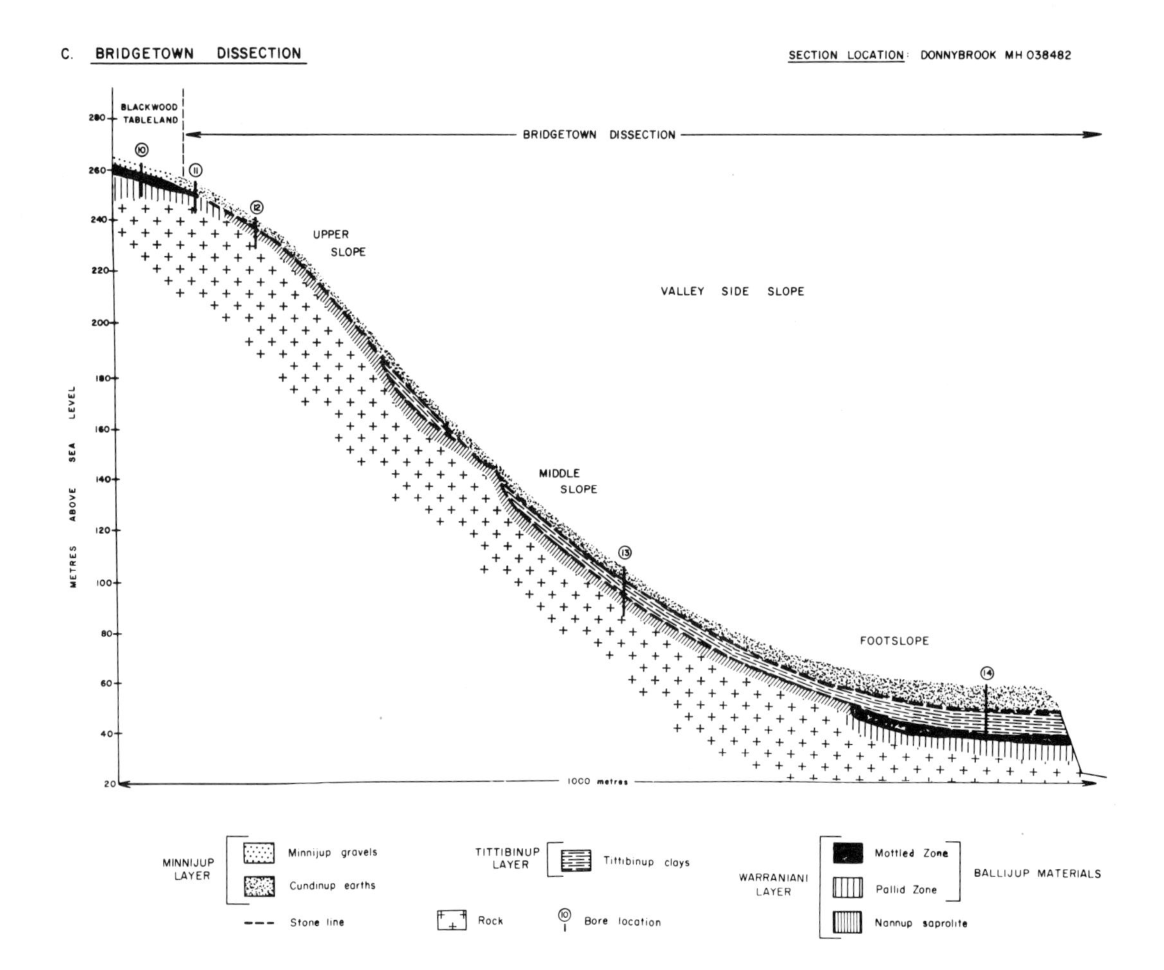

Figure 8 Earths on valley side slopes in deep dissections are separated by stone lines from dark-colored clayey lenses which in turn overlie stone lines set in saprolite or mottled zones. Valley laterites can be traced to upland occurrences by following longitudinal valley profiles upstream. The stratigraphic succession of *in situ* weathering zones and co-alluvial bodies establishes a relative age sequence. Soils developed in each layer are coeval in age and all facies, which collectively make up pedoderms, are ranked according to relative ages, from oldest to youngest, from the Warraniani Pedoderm through the Tittibinup Pedoderm to the Minnijup Pedoderm.

(quartz, dolerite, gneiss, and amphibolite fragments) and are commonly emplaced on truncated soil profiles (Chruchward, 1970). Buried soils usually lack A horizons and much of the B horizon frequently has been stripped away. Transported materials, typically finer-textured stratified colluvium from upslope, similar to the grèzes litées and éboulis ardonées of periglacial regions as described by Guillien (1954) and Tricart (1970), overlie truncated saprolite. Downslope the stone lines distally decrease in particle-size pinching out to finer-textured colluvial toe slope or alluvial deposits.

In the shallow dissections, one near surface stone line is usually buried by a single colluvial unit which in turn is affected by contemporary soil-forming processes producing a still active Holocene pedon (Figure 9). Although multiple lines are evident in deeply dissected terrains, the most widespread example incorporates two stone lines where fine-grained clastics are interposed between the upper and lower stone lines.

Figure 9 The Holocene pedon, developed in a colluvial slope deposit near Southampton Bridge, is separated from an underlying weathering profile by a stone line. The *in situ* nature of the weathered substrate is indicated by the intact quartz veins shown on the lower left hand side of the photo.

In vertical section the stones appear lined-up in single file as more or less horizontal layers giving rise to their name (Kerr, 1881; Sharpe, 1938). In three dimensions they coalesce to sheets or pavements (Parizek & Woodruff, 1957; Ruhe, 1954). Their occurrence in tropical rainforests and semi-arid regions to mid-latitude and subpolar regions has prompted a variety of explanations. Attempts to explain the evolution of stone lines (see summaries in Young, 1976 and Jejè, 1980) generally incorporate pathways associated with (1) soil creep, a steady-state concept where stone lines indicate the lower limit of colluvial soil layers, (2) residual weathering where resistant quartzose materials are concentrated as a coarse lag deposit, (3) faunal pedoturbation, the selective removal of fine soil fractions from the subsurface to the surface leaving coarse fragments to form a distinct horizon, and (4) climatic change where coarse pedisediments are deposited during arid unstable phases and later covered by finer-textured materials during more stable wetter phases. Examples of the biological hypothesis, advocated by proponents of faunal pedoturbation, commonly refer to the industrious activities of termites (Lee & Wood, 1971). It has been suggested, for example, that termites and ants selectively remove fine-textured materials to the ground surface eventually burying the stone pavements creating stone lines (Nye, 1955; De Villiers, 1965; Williams, 1968). The selective sorting process, as described by Holt *et al.* (1980), involves the incorporation of finely graded subsoil materials, brought up from 1-2 m depth, into termite mounds. They further suggest that 20 cm thick A horizons of red and yellow earths in northern Queensland may have accumulated from erosional degradation of termiteria in 8 x 10^3 yrs in a landscape where subsoils may be on the order of 2 x 10^6 yrs old. Springer (1958), on the otherhand, identifies an upward movement of stones in self-mulching desert soils containing clayey, montmorillonitic B horizons. Repeated swelling and shrinking, due to wet and dry phases, apparently pushes stones upward. Essential features of the process have been demonstrated in the laboratory.

Although different processes have been ascribed to the formation of stone lines in soils (see summary in Jejè, 1980), most explanations, recognizing that stone lines represent ecologically disruptive features in the landscape, refer to the colluvial nature of the deposits which are allochthonous with respect to underlying materials (Bigarella *et al.*, 1965; AbSaber, 1967; Ojanuga & Wirth, 1977). Stone lines are commonly regarded as buried examples of what were once eluvial deposits, stone pavement or lag gravels formed at the ground surface, and as such are evidence of stratification in surficial materials. Buried stone lines on the West Australian craton appear to be analagous to the types observed by Ruhe *et al.* (1967) elsewhere because they contain a variety of unrelated lithorelicts, deflect intact quartz veins and small dolerite dikes downslope entraining the rock debris, decrease in particle-size towards distal sections, follow a detailed micro-relief, cap truncated deep weathering profiles, and themselves are overlain by finer-textured transported materials. These features thus seem to signal unstable erosive phases in landscape history where stripping events are followed by deposition of colluvial materials and soil development (Finkl, 1982). The stone lines help establish a relative age sequence by delineating stratification in soil parent materials. The fact that stone lines and materials above them overlie a variety of substrates (Finkl & Churchward, 1976) is further proof of colluvial origin and that they mark prior erosion surfaces which constitute time-stratigraphic boundaries. Micromorphological investigations corroborate many of these deductions (see later discussions).

GRAVEL/SAND-CLAY INTERFACES

Lateritic gravels (Fe-pisolites) are widely distributed on upland valley side slopes. These deposits can be traced to a provenance on duricrusted residuals and interfluves where mottled zone materials have become indurated (Mulcahy *et al.*, 1972). Degradation of duricrusted residuals (breakaways) and massive lateritic pavements releases gravel-sized materials for transport downslope. The colluvial nature of the gravels has been established for widely separated regions on the craton (Mulcahy, 1967; Churchward, 1970; Finkl, 1979) where they overlie mottled and pallid zones, saprolites, and rock. Abrupt wavy boundaries characterize the gravel-clay interface where pisolitic accumulations overlie portions of the deep weathering zone. The relationship has been identified as lithological incompatability between two soil-stratigraphic units (gravel facies of the Minnijup Pedoderm overlying Ballijup materials of the Warraniani Pedoderm) and demonstrated using micromorphological techniques (Finkl & Gilkes, 1976). Similar relationships were observed by Mulcahy (1964) and Brewer & Bettenay (1973) in their investigations of Western Australian spillways and some plains in the arid cratonic interior.

The superposition of pisolitic gravels or sand splays over deeply weathered materials is thus recognized as a lithological discontinuity. Pedogenesis was nowhere observed to transgress these boundaries (see subsequent discussion) and as such they also represent pedological discontinuities. Intervening sediment separates the vertical succession of soil profiles, as is commonly observed in loess or alluvial sections (Fink, 1969; Pećsi, 1975; Morrison, 1978). In other cases where stratigraphic units are moderately thin and soil development comparatively intense, pedogenic phases may be partially imposed one on top of the other. The mergence of a groundsurface solum with a buried soil has been referred to as pedogenic overprinting (Finkl, 1980) and by Ruhe & Olson (1980) as soil welding. Pedo-lithological boundaries thus help delineate parent materials and phases of soil development providing an additional basis for estimating relative soil ages.

MACROMORPHOLOGICAL DEPTH FUNCTIONS

Many soil sections in the region show massive surface earths overlying structured subsoils. The earths, fine-grained distal end members of colluvial sequences on valley side slopes, are regarded as a facies of the Minnijup Pedoderm. The nature of the discontinuity separating the earths from underlying subsoils was partly established by analyzing macromorphological depth functions.

Because the spatial distribution of soil characteristics is depth-dependent, it is possible to graph the vectorial properties of selected features downprofile. Substances which easily migrate in the soil, *e.g.* organic matter, soluble salts, colloidal particles, sesquioxides, as well as changes in macromorphological properties, are amenable to analyses where depth is plotted along the vertical axis and the proportion, relative frequency, or degree of development is shown along the horizontal axis where the scale or order of magnitude is relative to each feature. Plots of specific "depth functions," as Jenny (1941) referred to them, show similar trends with depth in monogenetic soils. Abrupt changes or multiple peaks in the distribution patterns require additional explanation. Churchward's (1961) analysis of macromorphological depth functions (waxing and waning trends in pedality, clay skins, coherence, and porosity) for Victorian soils at Swan Hill showed that such features can be used to identify pedogenic discontinuities in soil sections.

The term earth, as applied pedogenically in Australia, refers to porous soil materials that contain stable crumbs which resist clod formation. The plasma is dominantly floc in materials elsewhere described by Kubiena (1953) as roterde. Boundaries separating the earths from underlying weathered materials are diffuse compared to those separating gravels and sands from subclays. The marked increase in pedality and coherence and decrease in porosity in the vicinity of the interface, which spans a vertical distance of several centimeters, clearly separates unrelated soil materials. Macromorphological depth functions, as displayed in Figure 10, are seen as another parameter for differentiating soil layers which are separated in time and space.

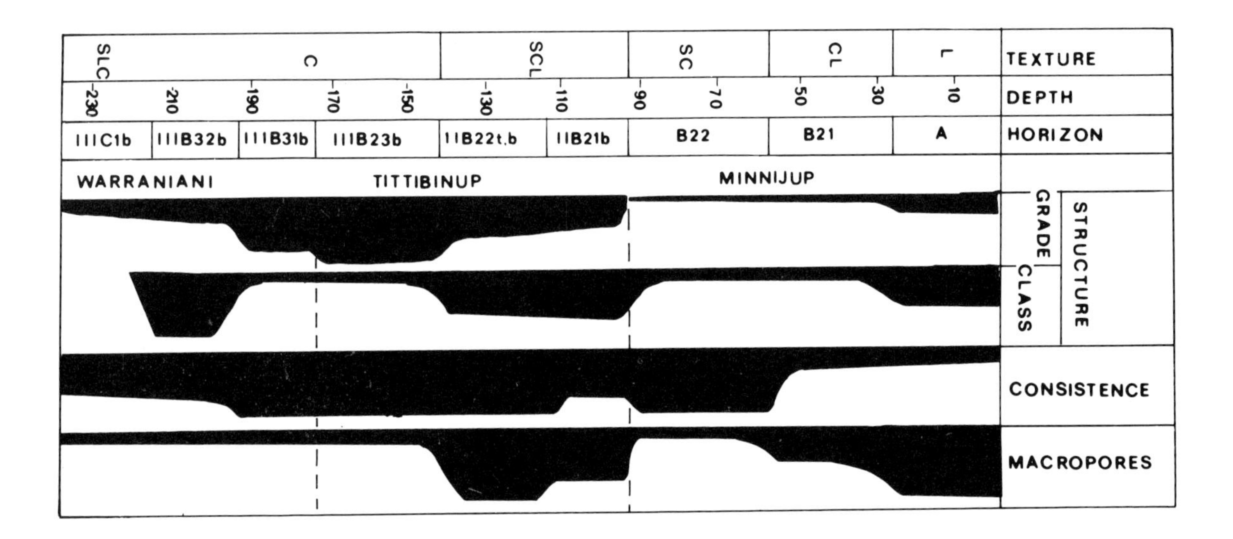

Figure 10 Macromorphological depth functions for the Redshaw soil profile. Abrupt changes or breaks in trends for soil structure, consistence, and porosity at stone lines suggest the presence of unrelated soil materials in stacked profiles.

REVERSE WEATHERING DIFFERENTIALS

The progressive decrease in weathering intensity with depth, through the solum to the base of the profile at the weathering front, typifies "normal" soil development (Jenny, 1941; Nikiforoff, 1959). Exceptions include, for example, (1) weakly developed soils with little change in mineralogy or pedochemical composition with depth and (2) laterite deep weathering profiles where pedological weathering gradually gives way to geochemical and other epidiagenetic processes which rot the parent rocks to depths of 30, 50, or 100 m or more (Holzer & Weaver, 1965). Particularly troublesome is weathered material which has been transported and deposited again, forming a new but entirely weathered parent material, the roche fille of French workers. The widespread occurrence of preweathered parent materials in tropical and subtropical regions (Buringh, 1970) complicates the interpretation of many soil profile sections as polymorphic or polygenetic. Cumulative soils and those where less weathered materials at the soil surface truncate more strongly weathered materials at depth are also inconsistent with this model. Continuous additions of fresh grus from tors to surface soil layers in unglaciated regions of New Brunswick (Canada), for example, produce pronounced weathering differentials. Colluvial grus with inherited gibbsite forms a surface C horizon which overlies truncated B horizons of buried soils (Wang *et al.*, 1981). In tropical regions, layers of pisolitic and/or fresh rock fragments overlying thick clay horizons, or brown (5YR 4/3) and red (2.5YR 4/6) clays overlying mottled/pallid zones or saprolites suggest reversals of normal weathering trends. These "reverse weathering differentials" (Finkl & Churchward, 1976) have been explained in terms of soil layering, soil profiles, or stratigraphic succession of soil materials, where younger soil materials are deposited on older weathered substrates. Such relationships are widespread in the Australian region (Stephens, 1946; Churchward, 1970) and are particularly evident on the weathered cratons (Brewer & Bettenay, 1973).

SOIL FABRIC ANALYSIS

The spatial arrangement of discrete mineral grains, compound particles, and associated voids is described as soil fabric according to Brewer's (1964) terminology. Differences in the size, shape, and arrangement of skeleton grains, soil plasma, pedological features, and voids characterize the nature of individual soil materials. The analysis of soil fabric changes in a spatial context, *i.e.* in both vertical and horizontal directions, provides an independent means for subdividing the soil continuum. Micromorphological techniques are often capable of detecting subtle changes not always apparent at macro- and mesoscales, and in some cases are the only means of differentiating soil materials which appear deceptively similar (Verstraten, 1980). Detailed studies of complex granitic regoliths in northwest Spain, for example, showed characteristic profile fabric zonations which helped differentiate younger surface colluvium from older saprolitic substrates (Bisdom, 1967). Skelsepic plasmic fabrics were common in colluvium whereas the saprolites were characterized by spheroidal weathering conditions by micro-crack networks. In deeply weathered cratonic regions where similar soil materials are juxtaposed, micromorphological techniques are combined with macromorphological methods in efforts to unravel relative soil-age relationships.

Soil Fabric Zones

The micromorphological investigation of some sixty soil-stratigraphic sections on the West Australian craton seemed at first to indicate a bewildering array of soil fabrics. More detailed considerations, however, showed that abrupt changes in soil fabric coincided with other sorts of stratigraphic discontinuities occurring, for example, at stone lines or gravel-clay interfaces. When trends in soil fabric zonation were studied in relation to the positioning of soil mantles, it became apparent that the vertical sequences of soil fabric zones were repetative. Lateral changes in soil fabric were found to reflect facies within pedoderms (Finkl & Gilkes, 1976).

Soil fabric zonation in the Kuringia (P579) soil profile is typical of many tableland soil sections where pisolitic gravels overlie mottled and pallid zones (bore holes 2 and 3 in Figure 6). Micromorphological properties of the type section are summarized in the Kuringia Soil Profile Description (Appendix). Micromorphological depth functions for argillans, nodules, and papules are detailed in Figure 11. The coincidence of abrupt changes in the occurrence and distribution patterns of pedological features at the gravel-clay interface supports the contention that separate episodes of pedogenic development are evident in the profile. The S-matrix of the Minnijup Pedoderm (gravel facies) is characterized by a dark brown undulic porphyroskelic fabric with agglomerated plasma. Consideration of soil fabrics at the level of related distribution of f-matrix (finer material) to f-members (coarser materials) (Brewer, 1979) discloses a gradation from mullgranic to mullgranoidic fabrics in these surface horizons. Abundant organic fragments as well as faecal pellets occur to a depth of about 45 cm. The second fabric zone, corresponding more or less with the B2cn horizon, is an isepic-inundulic porphyroskelic intergrade which becomes redder in color, denser, and more sepic with depth. This fabric zone abruptly terminates at about 72 cm, which marks the upper portion of a truncated mottled zone. The mottled yellow to buff colored soil materials show a strongly insepic porphyroskelic fabric with broad zones of masepic and smaller patches of undulic porphyroskelic fabrics. Argillans are more abundant than in fabric zone 2 where R_2O_3 (sesquioxidic) nodules show a marked decrease in frequency. The complexity of fabric zone 3 increases with depth as the intensified mottling produces ma-mosepic patches. At about 160 cm this zone gives way to a pale-colored strongly weathered rock fabric with included pedological features such as void argillans and papules (Fabric zone 4).

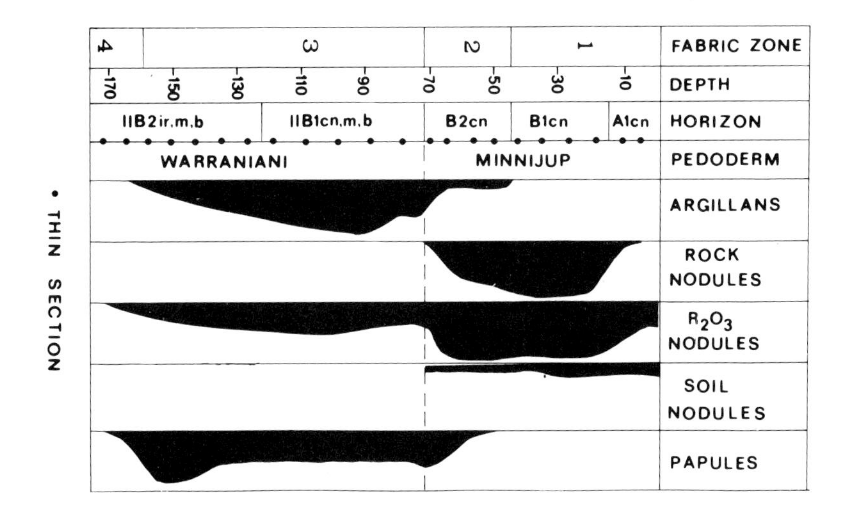

Figure 11 Micromorphological depth functions for the Kuringia soil profile on a truncated margin of the Blackwood Tableland show marked changes in distribution patterns at the gravel-clay interface (72 cm). Rock and soil nodules are confined to the upper soil profile while R_2O_3 nodules, which occur in the younger surface soil and older buried laterite, are most frequent in surface horizons but show a marked decrease at the discontinuity. Void argillans show two frequency maxima, one in each profile. Dark-brown undulic porphyroskelic fabrics with aggregated plasma grade to denser, more birefringent, red colored soil materials in fabric zone 2 (B2cn horizon). Strongly insepic yellow-colored porphyroskelic fabrics in the truncated laterite grade to weathered rock fabrics in zone 4.

The Redshaw (P577) soil profile (type section 13 in Figure 8) offers greater complexity with facies of the Minnijup, Tittibinup, and Warraniani Pedoderms occurring in the section. Macromorphological descriptions of each pedoderm, separated by stone lines, are given in the Redshaw Soil Profile Description (Appendix). The Minnijup Pedoderm contains two fabric zones (Figure 12) which correlate with the pedogenic A and B horizons. The former is characterized by a porous, dark-colored inundulic porphyroskelic fabric with agglomerated plasma. Organic fragments, faecal pellets, and opal phytoliths occur throughout the zone. The plasma becomes somewhat more birefringent with depth tending toward argillisepic elements. Fabric zone 2 is essentially insepic porphyroskelic but contains weak skel-ma-mosepic separations and broad masepic zones in the B2. It is separated from fabric zone 3 by a massive accumulation of rock nodules. The lithorelicts are predominantly weathered amphibolite and contain included argillans. Fabric zone 3 is denser, redder in color, and lacks opal phytoliths and other organic remains. The zone is dominated by insepic porphyroskelic fabrics but contains weak mosepic inclusions. A massive buildup of nodules occurs at about the 170 cm level forming a very dense and compact layer of lithorelicts. Weathered rock fragments and R_2O_3 nodules occur at the base of zone 3. Basal fabrics grade from weak ma-insepic porphyroskelic in zone 4 to red and yellow mottled weathered rock fabrics in zone 5. Multiple peaks in the distribution patterns of pedological features or abrupt terminations of individual trends at stone lines suggest the presence of unrelated soil materials.

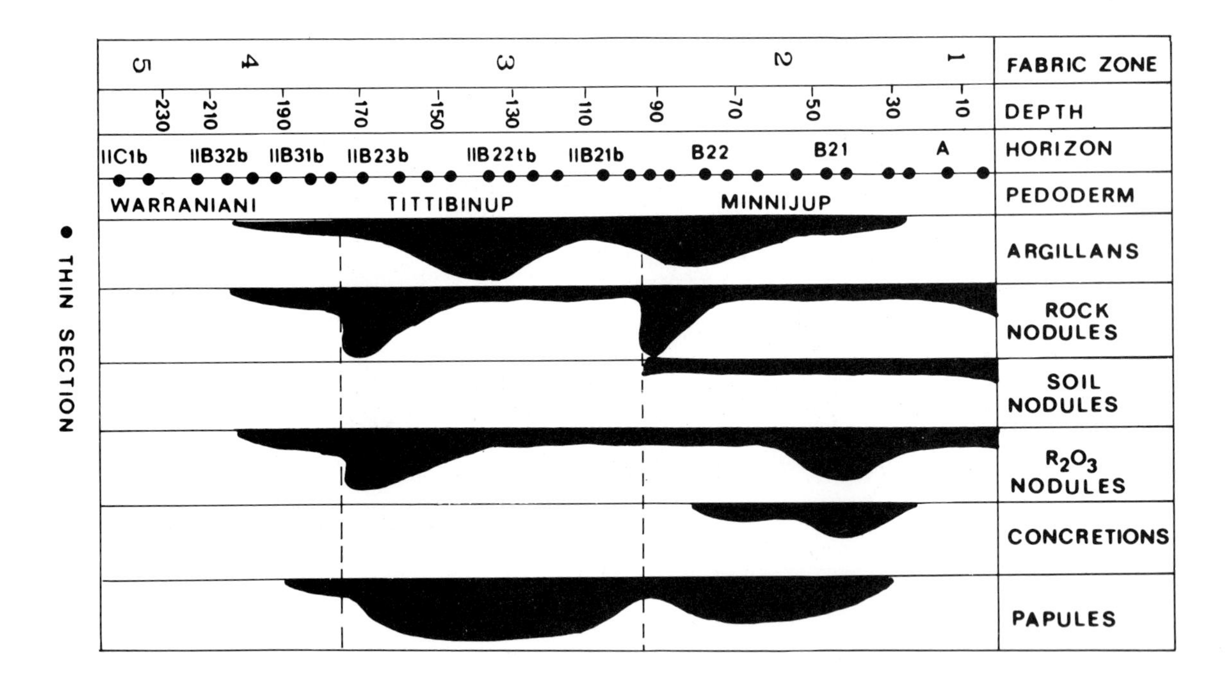

Figure 12 Micromorphological depth functions for the Redshaw soil profile. Stone lines separate the three pedoderms. Fabric zonation is partly related to intensity of profile development within pedoderms and to the presence of unrelated soil materials. Prominent breaks in fabric trends occur at the discontinuities as do marked changes in frequency distribution patterns for pedological features.

Soil Materials, Facies, and Relative Age Sequences

Although the Kuringia and Redshaw profiles are but two of many examples depicting vertical changes in soil fabric, it should be appreciated that fabric zonation coincides with discontinuities observed in the field. Additional sections from different landscape positions (see Figures 6, 7, and 8) produced a range of fabrics associated with each pedoderm. The lateral changes in soil fabric thus represent facies changes by reflecting differences in the nature of preweathered parent materials, accumulation of organic matter, and drainage conditions.

Consideration of soil as a material (Hall, 1945; Paton, 1978), as opposed to soil as an individual, facilitates the organization of soil fabrics in descriptive terms. The term "soma," derived from soil material as suggested by R. Brewer (pers. comm.), could be combined with a geographic name to describe a given range of soil materials. Grisoma (after Grimwade township), for example, identifies soil material associated with the Minnijup Pedoderm (Figure 13). Six varieties of Grisoma commonly occur on the West Australian craton, namely: normal, pelleted, argillasepic, organic, gravelly, and nodular. Characteristics of these materials are summarized in Table 1. Properties associated with Balisoma, summarized in Table 2, characterize soil fabrics in the Tittibinup Pedoderm (Figure 14).

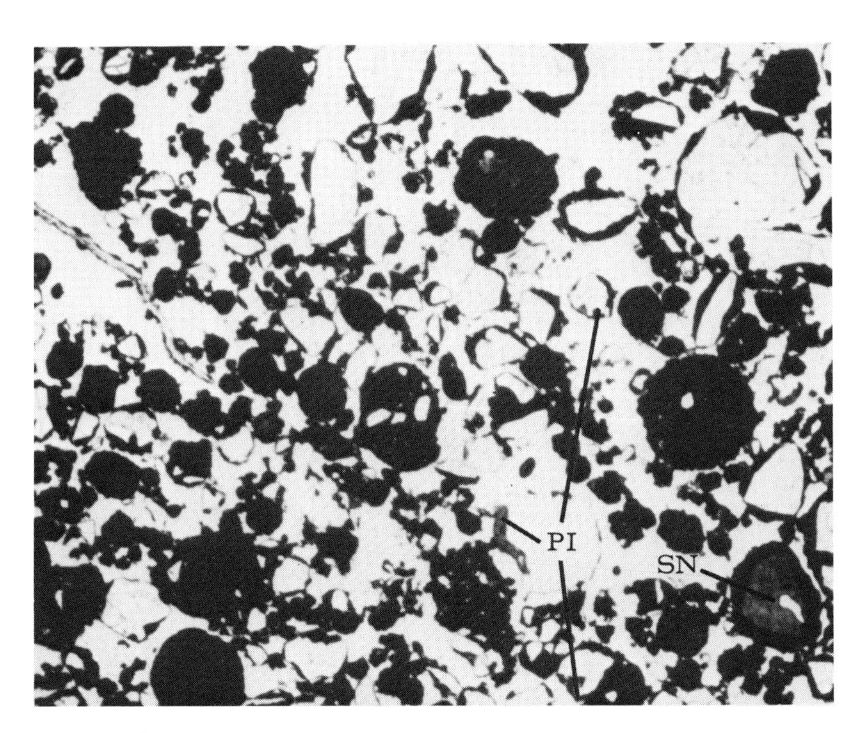

Figure 13 Normal Grisoma in the Minnijup Pedoderm. The organic-rich agglomerated plasma characterizes surface soil materials. Note the soil nodule (SN) derived from pallid zone materials farther upslope and plasma infusion (PI) in embayed quartz grains. (Plane light 2.5X)

Mottled zone materials are yellow to buff colored in reflected light. Insepic porphyroskelic fabrics commonly surround opaque mottles which contain included argillans. Weak ma-mosepic fabrics are interspersed with other complex fabric zones. Pale-colored soil materials in the lower portions of the laterite profiles have predominantly insepic porphyroskelic fabrics with weak mosepic zones. Bleached mineral grains, especially exfoliated micas, and localization of sesquioxides are also common features. Saprolitic facies contain weathered rock fabrics sometimes with included soil fabrics and pedological features.

After the succession of soil materials has been correlated with pedoderms, soil fabrics themselves can be used to indicate soil layering which in turn implies age differentials. Every attempt has been made to avoid a *circulus vitiosus* but once the correlation of macro- and micro-morphological features has been made, there is no reason to ignore it. Agglomerated, organic-rich, undulic-inundulic porphyroskelic

Figure 14 Normal Balisoma from the buried Tittibinup Pedoderm showing ma-mosepic porphyroskelic fabric with void argillans (VA). The dense red-colored plasma is partially masked by sesquioxidic accumulations (dark patches). (Plane light 10X)

fabrics are always associated with the Minnijup Pedoderm. Gravelly or nodular varieties of Grisoma are associated with a lateritic provenance on upland valley side slopes and colluvial toeslopes in dissections.

Normal, pelleted, or organic varieties occur on dissected slopes whereas argillasepic varieties indicate a thinning of the pedoderm with more sepic materials lying not far below.

The succession of soil materials is summarized in Figure 15 for key landscape positions. Varieties of Grisoma occur everywhere at the ground-surface and represent the youngest soil materials. Denser, more birefringent soil fabrics showing greater pedological organization always occur as a subsurface layer below the upper stone lines in dissections. Intensely weathered substrates mantle all major landscape positions and are overlain by weak to moderately well developed soil fabrics.

Age of soil materials is thus inferred from micromorphological data which delineate soil fabric zones and the nature of soil materials. Use of soil fabric analysis on an *ad hoc* basis is not only feasible but an appropriate means for estimating relative soil age.

SKELETON GRAIN MORPHOLOGY AND AGE OF SOILS

Skeleton grains of a soil material, as defined by Brewer and Sleeman (1960), include individual mineral grains and resistant siliceous or organic bodies larger than colloidal size. Although relatively stable, most grains in humid tropical environments show the effects of biochemical weathering processes, sometimes passing through a papular stage as in the case of some feldspars (Figure 16), to form soil plasma. Surface textures and grain shape thus seem to offer clues to mineral provenance, the geochronology of strongly weathered terrains, and the nature of weathering phenomena in tropical pedochemical microenvironments (Cleary and Conolly, 1972; *e.g.* Powers *et al.*, 1979). Features such as striations, etch patterns, dissolution pits, quartz overgrowths,

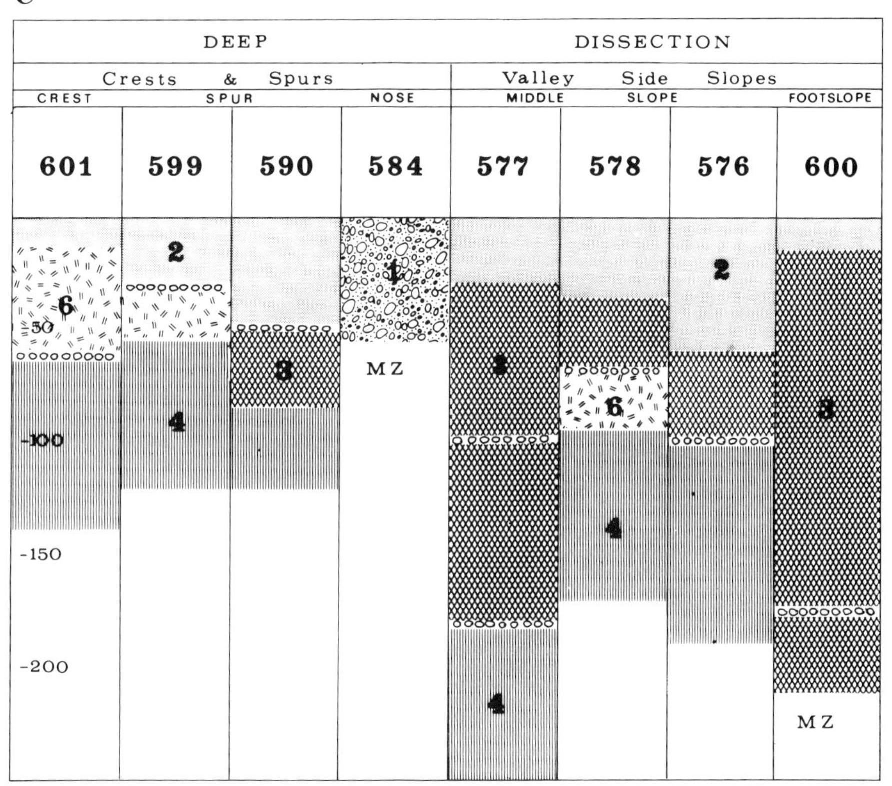

Figure 15 The succession of soil materials in key landscape positions (A,B,C) on the West Australian craton. Discontinuities in the numbered soil-stratigraphic sections are marked by stone lines (small circles) and abrupt changes in patterns of soil fabric zonation. Pisolitic gravels (1) and organic-rich agglomerated plasmas, referred to here as Grisoma (2), are always associated with surface soils in the Minnijup Pedoderm. Denser, more strongly developed fabrics occur in buried soils. Sepic plasmic fabrics, characterized by Balisoma (3), and "mica fabrics" (6) dominate the Tittibinup Pedoderm whereas substrates in the Warraniani Pedoderm contain mottled/pallid zone materials (MZ/PZ), pseudogranoblastic fabrics (5), and weathered rock fabrics (4). Macro- and micromorphological descriptions are given in Finkl (1971) and are filed at the C.S.I.R.O. Perth Lab.

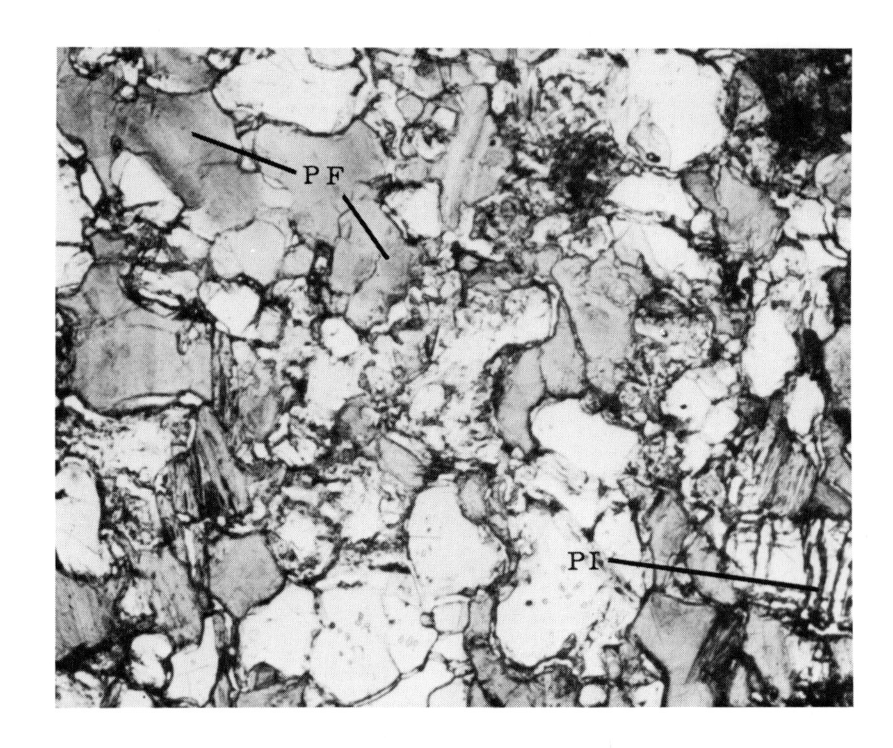

Figure 16 Papule flood (PF) showing feldspars and hornblende grains weathering to clay. Note plasma infusion (PI) in mineral grains but general lack of a plasma groundmass in this weathered rock fabric. (Plane light 40X)

scaling, silica flowers, pellicles, exfoliation, and plasma infusion are interpreted as indicators not only of provenance but also of weathering regimes and relative soil age (Walker, 1969; Wilson, 1970; Ojanuga, 1973; Douglas & Platt, 1977; Eswaran & Stoops, 1979).

In spite of long term interest in features contended to be diagnostic of the environmental history of the deposit (*e.g.* Babington, 1821), relatively few studies have focused attention on the morphology of weathered skeleton grains prior to erosion and transport. The influence of pedochemical weathering is, however, recognized in many studies which attribute highly etched and rounded solution-weathered quartz grains to long periods of leaching (Raeside, 1959; Crook, 1968; Little *et al.*, 1978). Microbial degradation or weathering by complexing organic acids, as commonly occurs in deep regoliths under tropical rain-forest conditions, produces a different sequence of mineral stability than traditionally quoted for pure water (Huang & Keller, 1972; Wilson, 1970, 1975; Berthelin & Belgy, 1979) and is particularly important to feldspar stability in soils due to potassium chelation by plants (Evans, 1964). Van der Waals (1967) further warns that corrosion of quartz grains and deposition of silica on them may be more the result of the micromilieu rather than age. Feldspar dissolution in natural weathering is controlled by pH (Nixon, 1979) but grains with solution pits appear to represent near end members of weathering environments. Berner *et al.* (1980), on the other hand, report that the nature of etch features on soil grains is primarily controlled by crystallography and not the type of etchant. Etch pits were observed to occur on the grains where dislocations outcrop and, as might be expected, rates of selective etching differed for coexisting grains in the same soil.

Several etch mechanisms are thus apparently capable of producing similar surface grain texture, and even though different dissolution rate processes may give rise to different equilibrium shapes, several generalizations still appear valid: (1) paleosols may be the source of super-mature sands, (2) texturally mature grains tend to be produced by chemical weathering in physiographic settings which are tectonically quiescent, (3) strongly weathered skeleton grains are products of extended biochemical leaching regimes, and (4) etched or solution-rounded grains are polycyclic and distinct from "first cycle" grains (Ingersoll, 1974).

Several of the features associated with the chemical weathering of mineral grains in a tropical environment, such as described by Crook (1968), Doornkamp & Krinsley (1971), Eswaran & Sousa (1975), Eswaran & Heng (1976), Eswaran & Bin (1978), and Eswaran & Stoops (1979), were observed in regoliths on the West Australian craton. Grains showing the effects of epidiagenetic alteration were most abundant near the soil surface in the Minnijup Pedoderm. Intensely weathered grains were less frequent in the Tittibinup Pedoderm and rare in the underlying autochthonous zone. Intensity of weathering (etching), as well as the formation of solution pits and embayments. decreased with depth. Embayed skeleton grains and "runiquartz," similar to that described by Eswaran and Stoops (1979), often occurred within large R_2O_3 nodules, derived from petroplinthite, but were most commonly free-floating and invaded by plasma material. Other embayed or stressed quartz grains showed undulose extinctions with neoformed quartz in expanded solution pits.

It is interesting to note in passing that many examples of the etch features observed in soil skeleton grains also commonly occur in achrondites (meteorite fragments lacking chrondules). Does the amusing possibility exist that some of the runiquartz observed in the Minnijup Pedoderm is an inherited feature of interstellar origin, or have terrestrial (pedo-biogeochemical) processes and extraterrestrial etch processes produced similar forms?

In sum, the oldest mineral grains, *i.e.* strongly weathered skeleton grains with modified surface textures and embayments filled with plasma, were predominantly associated with the youngest soil mantle (Minnijup Pedoderm). The occurrence of these stressed grains in R_2O_3 nodules, soil nodules, and lithorelicts suggests diverse modes of formation and polycyclic origins. Weathered free-floating grains (runiquartz) near the surface are probably younger than those set in fragments of petroplinthite and may be products of soil-water relationships, analagous to those described by Douglas and Platt (1977) in a glacial terrain where A horizons dry out faster than subsoils. Embayed grains in lithorelicts may represent the remains of small core stones which survived intense lateritic weathering. They may have recently entered the contemporary weathering environment after becoming entrained in the surface colluvial soil mantle. These features are rare in the *in situ* weathering zone below stone lines because most of the grains have weathered to secondary minerals or occur as papules or just bleached grains.

DISCUSSION AND CONCLUSION

Modern soils in the deeply weathered West Australian craton are variously developed in reworked colluvial and alluvial deposits of different ages. Quaternary geomorphic activity has involved repeated erosion and deposition of preweathered materials that were and *post facto* remain descriptively similar. Superpositioning of soil profiles to form soil-stratigraphic sequences establishes a relative soil chronology for the region. Breaks or discontinuities in soil-stratigraphic profiles separate pedogenic events which are characterized by a distinct array of soil materials. Reliable relative age sequences depend on recognition of cyclic pedogenic developments in the soil mantle but proofs of existence and separate identity, both often taken for granted, are still basic to studies of soil age. Incomplete sequences where parts of sections (usually upper portions of soil profiles) have been eroded away, pedogenic overprinting (soil welding), and preweathered parent materials are built-in problems which must be comprehended in relative soil dating methodologies.

Stage of Soil Development

Intensity of soil profile development within and between pedoderms offers additional scope for assessing soil age. Application of Muir's (1969) Principle of Developmental Sequences, which orders dynamic soil processes (losses from horizons and gains to horizons) in terms of those passing from least to most developed, or Simonson's (1959) general theory of pedogenesis leads to consideration of soil age relationships based on soil properties. The occurrence of sparingly soluble salts, iron-humus complexes, Fe, Al-sesquioxides, neoformed quartz, and illuviated clay are, in a general way, related to increasing profile development.

Soil development in Holocene pedons of the Minnijup Pedoderm involves accumulation of organic matter at the soil surface, a moderate degree of structural development, aggregation of plasma, mobilization and minor segregation of iron and aluminum, and a moderate degree of clay illuviation. Most of the argillans are kaolinans and goethans, although some montmorillans and vermiculans also occur (Finkl, 1971). As a matter of consensus, it has been observed that with increasing age or proportion of plasma, soil materials tend to have porphyroskelic fabrics (Brewer & Sleeman, 1969). Soil materials in the Tittibinup Pedoderm thus contrast with the undulic-inundulic agglomeroplasmic fabrics in the surface Minnijup Pedoderm. Amorphous sesquioxidic organoargillans and secondary silica in the form of grass opal phytoliths are also commonly associated with upper horizons of these younger surface soils. Sesquioxidic glaebules, concretions (concentric fabric) and nodules (undifferentiated fabric), can occur in any kind of profile as they are inherited from pre-existing soils and sediments. Derived from very old degraded lateritic profiles, Fe-Mn glaebules now form quite large surface accumulations on upland valley side slopes and on sput noses as part of the Minnijup Pedoderm. Soils developed in these gravelly deposits show minimal profile development.

The moderate grades of fine and medium angular blocky structure, degree of soil compaction, clear horizonations, strongly sepic soil fabrics, and well developed pedological features in the Tittibinup Pedoderm suggest a moderately strong phase of soil development compared to that associated with surface Holocene pedons. The extensive occurrence of complex void argillans seems to support a prolonged phase of weathering suggesting in turn that the Tittibinup soil materials, expressed in terms of Balisoma fabrics, are more strongly developed (older) than surface Grisoma.

Studies of lateritic substrates (Finkl, 1971; Finkl & Gilkes, 1976) show that the deeper pallid zones are mainly bleached weathered rock fabrics where there is a high proportion of rock nodules with patches of gray-colored sepic soil fabrics. Mottled zones in contrast, contain an abundance of well developed pedological features as well as strongly segregated sesquioxides and clay accumulations indicating that this zone is the more strongly pedogenically organized portion of soft laterite profiles. These soil materials are thus, by far, the oldest in the region and all other weathered materials, including those in pallid zones, are relatively younger.

Climate, Groundsurfaces, and Soil Age Sequences

Relative pedological chronologies on the West Australian craton depend, in the first instance, on interpretations on the nature of deep weathering events. Prolonged subaerial weathering in a stable morphotectonic setting which produced laterites coeval in age has been contrasted with interpretations of so-called older (primary, plateau) and younger (detrital, valley) laterites (see discussions in Woolnough, 1918 and Prider, 1966). Although an extended phase of deep weathering, possibly initiated in the later Mesozoic, is now preferred (*e.g.* Prider, 1966; Fairbridge & Finkl, 1979; Finkl, 1971, 1979), it is the well-defined Oligocene induration event (Quilty, 1977) which sets the clock for the pedochronological sequence. Variable stripping of the deep weathering mantle preceeded induration and in the Quaternary there were other unstable phases, mostly associated with increasing aridity which affected much of the Australian continent (Bowler, 1976). Quaternary climatic history in Western Australia, based in part on biotic distributions (Wyrwoll, 1979), is characterized by contrasting wet and dry regimes. Relatively dry conditions punctuated by high precipitation events of short duration seem to have prevailed until the end of the Pleistocene (Wyrwoll & Milton, 1976). Alternation of Quaternary paleoclimates contributed to unstable geomorphic conditions which favored stripping of weathered materials forming new ground-surfaces with lag gravel or coarse slump deposits which in turn became stone lines when buried by finer-grained colluvium. These deposits, which must be regarded as post-duricrust in age because they contain fragments of the ferricrete, were then pedogenically altered to form pedoderms. A significant increase in precipitation at the onset of the Holocene, associated with a rising sea level and warmer sea surface temperatures, is reflected in surface pedons of the relatively youngest soils on the craton.

REFERENCES CITED

Ab'Saber, A.N., 1967, Problemas geomorfologicas da Amazonia brasiliera: Atlas do Simposio sobre a biota amazonica, v. 1, p. 35-67.

American Commission on Stratigraphic Nomenclature, 1961, Code of stratigraphic nomenclature: Amer. Assoc. Pet. Geol. Bull., v. 45, p. 645-665.

Babington, B., 1821, Decomposition of hornblende and feldspars in laterite formation: Trans. Geol. Soc., v. 5, p. 328-329.

Berner, R.A., Sjöberg, E.L., Velbel, M.A. and Krom, M.D., 1980, Dissolution of pyroxenes and amphiboles during weathering: Science, v. 207, p. 1205-1206.

Berthelin, J. and Belgy, G., 1979, Microbial degradation during simulated podzolization: Geoderma, v. 214, p. 297-310.

Bigarella, J.J., Mousinho, M.R. and da Silva, J.X., 1965, Processes and environments of the Brazilian Quaternary: VII INQUA Congress (Symposium on Cold Climate Processes and Environments, Fairbanks, Alaska), p. 3-71.

Bilzi, A.F. and Ciolkosz, E.J., 1971, A field morphology-rating scale for evaluating pedological development: Soil Science, v. 124, p. 45-49.

Bisdom, E.B.A., 1967, Micromorphology of a weathered granite near the Ria de Arosa (N.W. Spain): Leidse Geologische Mededelingen, v. 37, p. 33-67.

Bowler, J.M., 1976, Aridity in Australia: age, origins and expression in aeolian landforms and sediments: Earth Science Reviews, v. 12, p. 279-310.

Brewer, R., 1964, Fabric and Mineral Analysis of Soils, New York, Wiley, 470 p.

__________, 1979, Relationship between particle size, fabric and other factors in some Australian soils: Australian Jour. Soil Res., v. 17, p. 29-41.

Brewer, R. and Sleeman, J.R., 1960, Soil structure and fabric: their definition and description: Jour. Soil Science, v. 11, p. 172-185.

__________, 1969, The arrangement of constituents in Quaternary soils: Soil Science, v. 107, p. 435-441.

Brewer, R. and Walker, P.H., 1969, Weathering and soil development on a sequence of river terraces: Australian Jour. Soil Res., v. 7, p. 293-305.

Brewer, R., Crook, K.A.W. and Speight, J.G., 1970, Proposal for soil-stratigraphic units in the Australian Stratigraphic Code: Jour. Geol. Soc. Australia, v. 17, p. 103-109.

Brewer, R. and Bettenay, E., 1973, Further evidence concerning the origin of the Western Australian sand plains: Jour. Geol. Soc. Australia, v. 19, p. 538-541.

Brosh, A. and Gerson, R., 1978, Properties of some tropical soils on a sequence of landsurfaces in Uganda: Geografisker Annaler, v. 60A, p. 129-141.

Burgess, A. and Drove, D.P., 1952, The rate of podzol development in sands of the Woy Woy District, N.S.W.: Aust. Jour. Botany, v. 1 (1), p. 83-95.

Buringh, P., 1970, Introduction to the Study of Soils in Tropical and Subtropical Regions: Wageningen, The Netherlands, Centre for Agricultural Publishing and Documentation, 99 p.

Butler, B.E., 1959, Periodic phenomena in landscapes as a basis for soil studies: C.S.I.R.O. (Australia) Soil Publ. No. 14.

____________, 1980, Soil Classification for Soil Survey, Oxford: Clarendon Press, 129 p.

Butt, C.R.M. and Smith, R.E. (eds.), 1980, Conceptual models in geochemistry: Australia: Journal Geochemical Exploration, v. 12, p. 91-137, esp. Regional Features, p. 96-111; Ironstones, p. 118-119; Weathered Bedrock, p. 122-125; Soils, p. 126-127; Transported Overburden, p. 129-130.

Buurman, P., 1972, Paleopedology and stratigraphy on the Condrusian peneplain (Belgium): Agricultural Research Report No. 766. (Wageningen, Centre for Agricultural Publishing and Documentation).

Campbell, J.M., 1917, Laterite: its origin, structure and minerals: Mineralogical Magazine, v. 17, p. 67-77 *et seq.*

Churchward, H.M., 1961, Soil studies at Swan Hill, Victoria. I. Soil layering: Jour. Soil Science, v. 12, p. 73-86.

________________, 1970, Erosional modifications of a lateritized landscape over sedimentary rocks. Its effect on soil distribution: Australian Jour. Soil Res., v. 8, p. 1-19.

________________, 1977, Landforms, regoliths and soils of the Sandstone-Mt. Keith area, Western Australia: C.S.I.R.O. (Australia) Land Resources Management Series No. 2.

Cleary, W.J. and Conolly, J.R., 1972, Embayed quartz grains in soils and their significance, Jour. Sedimentary Petrology, v. 42, p. 899-904.

Crocker, R.L. and Major, J., 1955, Soil developed in relation to vegetation and surface age at Glacier Bay, Alaska: Jour. Ecology, v. 43, p. 427-448.

Crook, K.A.W., 1968, Weathering and roundness of quartz sand grains: Sedimentology, v. 11, p. 171-182.

Daniels, R.B. and Gamble, E.E., 1978, Relations between stratigraphy, geomorphology and soils in coastal plain areas of southeastern U.S.A.: Geoderma, v. 21, p. 41-65.

DeVilliers, J.M., 1965, Present soilforming factors and processes in tropical and subtropical regions: Soil Science, v. 99, p. 50-57.

Doornkamp, J.C. and Krinsley, D., 1971, Electron microscopy applied to quartz grains from a tropical environment: Sedimentology, v. 17, p. 89-101.

Douglas, L.A. and Platt, D.W., 1977, Surface morphology of quartz and age of soils: Soil Sci. Soc. Amer. Jour., v. 41, p. 641-645.

Dury, G.H., Langford-Smith, T. and McDougall, I., 1968, A minimum age for the duricrust: Australian Jour. Sci., v. 31(10), p. 362-363.

Eden, M.J., 1971, Some aspects of weathering and landforms in Guyana (formerly British Guiana): Zeitschrift für Geomorphologie, N.F., v. 15, p. 182-198.

Erhart, H., 1948, Sur la genèse des sols de Lubilash de l'Afrique: C.R. Academie Science Paris, v. 227, p. 598-600.

Eswaran, H., Sys, C. and Sousa, E.C., 1975, Plasma infusion - A pedological process of significance in the humid tropics: Annals Edafologia Agrobiologie, v. 34, p. 66-673.

Eswaran, H. and Stoops, G., 1979, Surface textures of quartz in tropical soils: Soil Sci. Soc. Amer. Jour., v. 43, p. 420-424.

Eswaran, H. and Wong Chow Bin, 1978, A study of a deep weathering profile on granite in penisular Malaysia: III. Alteration of feldspar: Soil Sci. Soc. Amer. Jour., v. 42, p. 154-158.

Eswaran, H. and Yeow Yew Heng, 1976, The weathering of biotite in a profile on gneiss in Malaysia: Geoderma, v. 16, p. 9-20.

Evans, W.D., 1964, The organic solubilization of minerals in sediments: Advances Organic Geochemistry, p. 263-270.

Fairbridge, R.W. and Teichert, C., 1953, Soil horizons and marine bands in the coastal limestones of Western Australia: Jour. Proc. Royal Soc. New South Wales, v. 86, p. 68-87.

Fairbridge, R.W. and Finkl, C.W., Jr., 1979, Cratonic erosional unconformities and peneplains: Jour. Geol., v. 88, p. 69-86.

Fink, J., (ed., 1969, La stratigraphic des loess d'Europe: Bull. Assoc. Francaise Etude Quaternaria, suppl. p. 3-12.

Finkl, C.W., Jr., 1971, Soils and Geomorphology in the Middle Blackwood River Catchment, Western Australia (Ph.D. Thesis): Univ. of Western Australia.

______________, 1971, Levels and laterites in southwestern Australia: Search, v. 2 (10), p. 382-383.

______________, 1979, Stripped (etched) landsurfaces in southern Western Australia: Australian Geographical Studies, v. 17(1), p. 33-52.

______________, 1980, Stratigraphic principles and practices as related to soil mantels: CATENA, v. 7, p. 169-194.

______________, 1982, On the geomorphic stability of cratonic planation surfaces: Zeitschr. f. Geomorph., N.F., v. 26, no. 2, p. 137-150.

Finkl, C.W., Jr. and Churchward, H.M., 1973, The etched landsurfaces of southwestern Australia: Jour. Geol. Soc., v. 20, p. 295-307.

______________, 1976, Soil stratigraphy in a deeply weathered shield landscape in south-western Australia: Australian Jour. Soil Res., v. 14, p. 109-120.

Finkl, C.W., Jr. and Gilkes, R.J., 1976, Relationships between micromorphological soil features and known stratigraphic layers in Western Australia: Geoderma, v. 15, p. 179-208.

Finkl, C.W., Jr. and Fairbridge, R.W., 1979, Paleogeographic evolution of a rifted cratonic margin: S.W. Australia: Palaeography, Palaeoclimatology, Palaeoecology, v. 26, p. 221-252.

Follmer, L., 1978, The Sangamon Soil in its type area - A review, in Mahaney, W.C., ed., Quaternary Soils: Norwich, England, Geo. Abstracts, p. 125-166.

Fridland, V.M., 1974, Structure of the soil mantle: Geoderma, v. 12, p. 35-41.

Galloway, R.W., 1970, Coastal and shelf geomorphology and late Cenozoic sea levels: Jour. Geol., v. 78, p. 603-610.

Geological Survey of Western Australia Staff, 1975, Geology of Western Australia: Western Australia Geological Survey Memoir 2, 514 p.

Guillien, Y., 1954, Interprétations général des grézes litées: Bull. Assoc. Geographie Francaise, 6e serie, p. 111.

Hall, A.D., 1945, The Soil: An Introduction to the Scientific Study of the Growth of Crops: London, Murray.

Hanlon, F.N., 1950, The bauxites of New South Wales: their distribution, composition, and probable origin: Royal Soc. New South Wales Jour., v. 78, p. 94-112.

Holt, J.A., Coventry, R.J. and Sinclair, D.F., 1980, Some aspects of the biology and pedological significance of mound-building termites in a red and yellow earth landscape near Charters Towers, North Queensland: Australian Jour. Soil Res., v. 18(1), p. 97-109.

Holzer, L. and Weaver, G.D., 1965, Geomorphic evaluation of climatic and climatogenetic geomorphology: Annals Assoc. Amer. Geographers, v. 62(4), p. 592-602.

Huang, W.H. and Keller, W.D., 1972, Organic acids as agents of chemical weathering of silicate minerals: Nature Physical Sciences, v. 239 (96), p. 149-151.

Idnurm, M. and Senior, B.R., 1978, Paleomagnetic ages of Late Cretaceous and Tertiary weathered profiles in the Eaomanga Basin, Queensland: Palaeography, Palaeoclimatology, Palaeoecology, v. 24(4), p. 263-277.

Ingersoll, R.V., 1974, Surface textures of first cycle quartz sand grains: Jour. Sedimentary Petrology, v. 44, p. 151-157.

Jejè, L.K., 1970, Some aspects of the Geomorphology of South Western Nigeria (Ph.D. Thesis): University of Edinburgh, Edinburgh.

__________, 1980, A review of geomorphic evidence for climatic change since the Late Pleistocene in the rain-forest areas of southern Nigeria: Palaeogeography, Palaeoclimatology, Palaeoecology, v. 31, p. 63-86.

Jenny, H., 1941, Factors of Soil Formation, N.Y., McGraw-Hill, 781 p.

Jessup, R.W., 1960, Identification and significance of buried soils of Quaternary age in the south-eastern portion of the Australian arid zone: Jour. Soil Sci., v. 11, p. 197-205.

Kemp, E.M., 1978, Tertiary climatic evolution and vegetation history in the southeast Indian Ocean region: Palaeogeography, Palaeoclimatology, Palaeoecology, v. 211, p. 169-208.

Kerr, W.C., 1981, On the action of frost in the arrangement of superficial earthy material: Amer. Jour. Sci., v. 21, p. 345-358.

Kronberg, B.I., Fyfe, W.S., Leonards, O.H. and Santos, A.M., 1979, The chemistry of some Brazilian soils: element mobility during intense weathering: Chemical Geology, v. 24, p. 211-230.

Kubiëna, W.L., 1953, The Soils of Europe: London: Thomas Murby, 317 p.

Kukla, G.S., 1977, Pleistocene land - sea correlations. I. Europe: Earth Science Reviews, v. 13, p. 307-314.

Kumar, A., 1979, Geomorphology of Simdega and its adjoining area, Bihar: National Geographical Society India Research Bulletin No. 22.

Lee, K.E. and Wood, T.G., 1971, Physical and chemical effects on soils of some Australian termites, and their pedological significance: Pedobiologia, v. 11, p. 376-409.

Leneuf, N. and Aubert, G., 1960, Essai d'evolution de la vitesse de ferralitisation: Proc. 7th Internatl. Cong. Soil Sci., p. 225-228.

Leontyev, D.K., 1970, Changes in the level of the world ocean in the Mesozoic-Cenozoic: Oceanology, v. 10, p. 210-217.

Leslie, D.M., 1978, An interpretation of a section through Quaternary deposits and paleosols at Taiaroa Head, Otago Peninsula, New Zealand: New Zealand Soil Bureau Science Report No. 34.

Leverett, F., 1898, The weathered zone (Sangamon) between the Iowan loess and the Illinoian till sheet: Jour. Geol., v. 6, p. 171-181, *et seq*. 238-243 and 244-249.

Little, I.P., Armitage, T.M. and Gilkes, R.J., 1978, Weathering of quartz in dune sands under subtropical conditions in eastern Australia: Geoderma, v. 20, p. 225-237.

Mabbutt, J.A., 1961, A stripped landsurface in Western Australia: Trans. Inst. British Geographers, v. 29, p. 101-114.

Mack, R.N., Okazaki, R. and Valastro, S., 1979, Bracketing dates for two ash falls from Mount Mazama: Nature, v. 279, p. 228-229.

Maignien, R., 1959, Soil cuirasses in tropical West Africa: African Soils, v. 4, p. 5-42.

McGee, W.J., 1891, The Pleistocene history of northeastern Iowa: U.S. Geological Survey Annual Report, v. 11, p. 189-577.

Meixner, R.E. and Singer, M.J., 1981, Use of a field morphology rating system to evaluate soil formation and discontinuities: Soil Sci., v. 131, p. 114-123.

Moorbath, S., 1960, Radiochemical methods, in Smales, A.A. and Wagner, L.R., eds., Methods of Geochemistry: N.Y., Interscience, p. 247-296.

Morrison, R.B., 1964, Lake Lahontan: geology of the northern Carson desert, Nevada: U.S. Geological Survey Professional Paper 401.

______________, 1967, Principles of Quaternary soil stratigraphy, in Morrison, R.B. and Wright, H.E., eds., Quaternary Soils, Reno, Nevada, Center for Water Resources Research, Desert Research Institute, p. 1-69.

______________, 1978, Quaternary soil stratigraphy - concepts, methods and problems, in Mahaney, W.C., ed., Quaternary Soils: Norwich, England, Geo Abstracts, p. 77-108.

Morrison, R.B. and Frye, J.C., 1965, Correlation of the middle and late Quaternary successions of the Lake Lahontan, Lake Bonneville, Rocky Mountain (Wasatch Range), southern Great Plains and eastern midwest areas: Nevada Bureau Mines Report No. 9.

Mörner, N.A., 1978, Low sea levels, droughts and mammalian extinctions: Nature, v. 271, p. 738-739.

Muir, J.W., 1969, A natural system of soil classification: Jour. Soil Sci., v. 20, p. 153-166.

Mulcahy, M.S., 1960, Laterites and lateritic soils in south western Australia: Jour. Soil Sci., v. 11, p. 206-226.

____________, 1964, Lateritic residuals and sandplains: Australian Jour. Sci., v. 27, p. 54-55.

____________, 1967, Landscapes, laterites and soils in southwestern Australia, in Jennings, J.N. and Mabbutt, J.A., eds., Landform Studies from Australia and New Guinea, Canberra: Australia National Univ. Press, p. 211-230.

Mulcahy, M.S., Churchward, H.M. and Dimmock, G.M., 1972, Landforms and soils on an uplifted peneplain in the Darling Range, Western Australia, Australian Jour. Soil Res., v. 10, p. 1-14.

Nikiforoff, C.C., 1959, Reappraisal of the soil: Science, v. 129, p. 186-196.

Nixon, R.A., 1979, Differences in incongruent weathering of plagioclase and microcline - cation leaching versus precipitates: Geology, v. 7, p. 221-224.

Nye, P.H., 1955, Some soil forming processes in the humid tropics, II. The development of the upper slope member of the catena: Jour. Soil Sci., v. 6, p. 51-62.

Ojanuga, A.G., 1973, Weathering of biotite in soils of a humid tropical climate: Soil Sci. Soc. Amer. Proc., v. 37, p. 644-646.

Ojanuga, A.G. and Wirth, K., 1977, Threefold stonelines in southwestern Nigeria: evidence of cyclic soil and landscape development: Soil Science, v. 123, p. 249-257.

Parsons, R.B., Balster, C.A. and Ness, A.O., 1970, Soil development and geomorphic surfaces, Willamette Valley, Oregon: Soil Sci. Soc. Amer. Proc., v. 34, p. 485-491.

Parizek, E.J. and Woodruff, J.F., 1957, Description and origin of stone layers in soils of the southern states: Jour. Geol., v. 65, p. 4-34.

Paton, T.R., 1978, The Formation of Soil Material: London, Allen & Unwin, 143 p.

Pećsi, M., 1975, A Magyarországi löszselvények lithostratigrafiai tagalósa (Lithostratigraphic subdividison of the loess sequences in Hungary): Foldrajzi Közlemények, v. 3-4, p. 217-230. (extended English Summary).

Playford, P.E., 1954, Observations on laterite in Western Australia: Australian Jour. Sci., v. 17, p. 11-13.

Porter, S.C., 1978, Glacier Peak tephra in the North Cascade Range, Washington: stratigraphy, distribution and relationship to late - glacial events: Quaternary Research, v. 10, p. 30-41.

Powers, L.S., Brueckner, H.K. and Krinsley, D.H., 1979, Rb-Sr provenance ages from weathered and stream deposited quartz grains from the Harney Peak Granite, Black Hills, South Dakota: Geochimica et Cosmochimica Acta, v. 43, p. 137-146.

Prider, R.T., 1966, The lateritized surface of Western Australia: Australian Jour. Sci., v. 28, p. 443-451.

Quilty, P.G., 1977, Cenozoic sedimentation cycles in Western Australia: Geology, v. 5, p. 336-340.

Raeside, J.D., 1959, Stability of index minerals in soils with particular reference to quartz, zircon and garnet: Jour. Sedimentary Petrology, v. 29, p. 493-502.

Richmond, G.M. and Frye, J.C., 1957, Status of soils in stratigraphic nomenclature: American Assoc. Pet. Geol. Bull., v. 41, p. 758-763.

Ruhe, R.V., 1954, Geology of the soils of the Nioka Ituri, Belgian Congo: I.N.E.A.C. (Bruxelles), Science Series No. 65.

Ruhe, R.V., Daniels, R.B. and Cady, J.G., 1967, Landscape evolution and soil formation in southwest Iowa: U.S. Department Agriculture, Soil Conservation Service Technical Bulletin 1349.

Ruhe, R.V. and Olson, C.G., 1980, Soil welding: Soil Science, v. 130, p. 132-139.

Sharpe, C.F.S., 1938, Landslides and Related Phenomena: New York, Columbia Univ. Press, 137 p.

Schmidt, P.W. and Embleton, B.J.J., 1976, Paleomagnetic results from sediments of the Perth Basin, Western Australia, and their meaning on the timing of regional lateritization: Palaeogeography, Palaeoclimatology, Palaeoecology, v. 19, p. 257-273.

Simonson, R.W., 1954, Identification and interpretation of buried soils: Amer. Jour. Science, v. 52, p. 152-156.

______________, 1959, Outline of a generalized theory of soil genesis: Soil Sci. Soc. Amer. Proc., v. 23, p. 152-156.

Smeck, N.E., 1973, Phosphorus: an indicator of pedogenic weathering processes: Soil Science, v. 115, p. 199-206.

Springer, M.E., 1958, Desert pavement and vesicular layer of some soils of the desert of the Lahontan Basin, Nevada, Soil Science Society America Proceedings, v. 22, p. 63-66.

Stephens, C.G., 1946, Pedogenesis following the dissection of lateritic regions in southern Australia: Bulletin Council Scientific Industrial Resources (Australia) No. 206.

______________, 1971, Laterite and silcrete in Australia: A study of the genetic relationships of laterite and silcrete and their companion materials and their significance in the formation of the weathered mantle, soils, relief and drainage of the Australian continent: Geoderma, v. 5, p. 3-52.

Tricart, J., 1970, Geomorphology of Cold Climates: London, Macmillan, 324 p.

Van der Waals, L., 1967, Morphological phenomena on quartz grains in unconsolidated sands, due to migration of quartz near the earth's surface: Mededlingen Netherlands, Geologische Stichting N.S., v. 18, p. 47-5].

van Dijk, D.C., 1958, Principles of soil distribution in the Griffith-Yenda district, N.S.W.: C.S.I.R.O. (Australia) Soil Publication No. 11.

__________, 1979, Developing a geographic-geomorphic approach to soil-land classification: 10th New Zealand Geography Conference, p. 264-266.

van Dijk, D.C., Riddler, A.M.H. and Rowe, R.K., 1968, Criteria and problems of groundsurface correlations with reference to a regional correlation in south-eastern Australia: 9th Internatl. Cong. Soil Science Trans., v. 4, p. 131-138.

van Dijk, D.C. and Rowe, R.K., 1980, Soil stratigraphy in the Murray Valley at Albury-Wodonga: A new approach to surficial stratigraphy: Proc. Royal Society Victoria, v. 91, p. 109-125.

Verstraten, J.M., 1980, Water-Rock interactions: British Geomorphological Research, Group Research Monograph No. 2, (Norwich, Geo. Books).

Walker, G.F., 1969, The decomposition of biotite in the soil: Mineralogical Magazine, v. 28, p. 693-703.

Walker, P.H., 1958, A study of Cyclic Soils, their Relation to Landscapes and their Development on the South Coast of New South Wales (Ph.D. Thesis): University of Sydney.

__________, 1962, Soil Layers on hillslopes: A study at Nowra, N.S.W.: Jour. Soil Science, v. 3, p. 167-177.

__________, 1966, Postglacial environments in relation to landscapes and soils on the Carey drift, Iowa: Iowa Agricultural Experiment Station Research Bulletin, v. 549, p. 837-875.

Wang, C., Ross, G.J. and Rees, H.W., 1981, Characteristics of residual and colluvial soils developed on granite and of the associated pre-Wisconsin landforms in north-central New Brunswick: Canadian Jour. Earth Science, v. 18, p. 487-494.

Whitehouse, F.W., 1940, The lateritic soils of western Queensland: University of Queensland Papers in Geology, v. 2(NS), p. 2-22.

Wilcox, R.E., 1965, Volcanic ash chronology, in Wright, H.E. and Frey, D.G., eds., The Quaternary of the United States: Princeton, N.J., Princeton Univ. Press, p. 807-816.

Williams, M.A.J., 1968, Termites and soil development near Brooks Creek, Northern Territory: Australian Jour. Sci., v. 31, p. 153-154.

Wilson, N.J., 1970, A study of weathering in a soil derived from biotite-hornblende rock: Clay Minerals, v. 8, p. 291-303.

__________, 1975, Chemical weathering of some primary rock-forming minerals: Soil Science, v. 119, p. 349-355.

Woolnough, W.G., 1918, The physiographic significance of laterite in Western Australia: Geological Magazine, v. 6, p. 385-393.

Worthen, A.H., 1866, Geology of Illinois: Vibona, Geological Surv. Illinois, 523 p.

Wright, R.L., 1963, Deep weathering and erosion surfaces in the Daly River Basin, Northern Territory: Journal Geol. Soc. Australia, v. 10, p. 151-164.

Wyrwoll, K.H., 1979, Late Quaternary climates of Western Australia: evidence and mechanisms: Jour. Royal Soc. Western Australia, v. 62, p. 129-142.

Wyroll, K.H. and Milton, D., 1976, Widespread late Quaternary aridity in Western Australia: Nature, v. 264, p. 429-430.

Yaalon, D.H., 1971, Soil-forming processes in time and space, in Yaalon, D.H., ed., Paleopedology: Jerusalem, International Society of Soil Science and Universities Press, p. 29-39.

Young, R.W., 1970, A probably post-uplift age for the duricrust on the south coast of New South Wales: Search, v. 1, p. 163-164.

Young, A., 1976, Tropical Soils and Soil Survey. Cambridge, Cambridge University Press, p. 134.

Young, E.J. and Powers, H.A., 1960, Chevkinite in Volcanic ash: American Mineralogist, v. 45, p. 875-881.

EVALUATION OF DATING METHODS USED TO ASSIGN AGES IN THE WIND RIVER AND TETON RANGES, WESTERN WYOMING

W.C. MAHANEY, D.L. HALVORSON, JAMES PIEGAT and K. SANMUGADAS

ABSTRACT

Glacial and nonglacial deposits in the Wind River and Teton ranges of western Wyoming have been dated by radiocarbon, topographic position, surface morphology, weathering characteristics, lichenometry and soil stratigraphy. Of these many methods radiocarbon, lichenometry, weathering features, and soil stratigraphy are the most useful in differentiating deposits which formed mainly during periods of glaciation. No lichen growth-rate curve is available for the mountains of western Wyoming. However, numerous lichen transects across Neoglacial deposits yield relative data on lichen size and percent cover that correlate with the Front Range, Colorado, where growth rates for *Rhizocarpon geographicum* are known with precision. Surface weathering features that assist in deposit differentiation and correlation include: boulder frequency ratio of fresh to weathered stones, weathering rind thickness, and depth of pitting on stone surfaces.

Soil profiles from well-drained sites in deposits containing granitic, gneissic, and granodioritic clasts are used to establish relative age. Soil morphology, particle size, organic constituents, selected soil-chemical parameters, primary mineral alteration and clay mineral composition are all used to establish a soil chronosequence. Soils in the sequence are, from youngest to oldest, post-Gannett Peak, post-Audubon, post-Indian Basin, post-Pinedale, post-Bull Lake, and pre-Bull Lake. In particular, changes in clay mineral composition with depth and changes in the ratio of oxalate-extractable to dithionite-extractable iron oxide are important age indicators. Petrographic analysis of fine and very fine sand separates (250-63μm) reveals changes in the ratio of quartz to plagioclase feldspar, which assist in age differentiation.

INTRODUCTION

The use of multiple geomorphic and pedologic criteria to differentiate deposits in geologic successions provides an important tool for Quaternary stratigraphers. Relative dating (RD) techniques that have been used in the Teton and Wind River ranges of western Wyoming include: changes in deposit morphology, topographic position, weathering characteristics, lichenometry and soil stratigraphy. Absolute dating (AD) is confined to radiocarbon methods because datable pyroclastic layers are absent, and Neoglacial successions are above timberline where dendrochronology cannot be applied. Uncontaminated organic materials are scarce and the few dates available document only the close of Pinedale Glaciation and termination of the Indian Basin advance.

AD and RD methods outlined above have been used at many field localities to differentiate deposits in the geologic succession. Of these many methods radiocarbon, weathering features, lichenometry, and soil stratigraphy have the greatest utility. Used individually each suffers from limitations that hinder refinement of the stratigraphic record. The use of several methods at each locality increases dating precision and helps strengthen the chronology. In this paper the advantages and disadvantages of each method are discussed and evaluated.

FIELD AREAS

The Teton Range (Figure 1), a Tertiary fault block tilted to the west, makes up part of the Middle Rocky Mountain Province as defined by Fenneman (1931). Extending from the Pitchstone Plateau in Yellowstone Park to the Teton Pass Mountains in the south, the range varies from 15 to 23 km in width. Upthrusting and erosion at the east side of the fault block has exposed a complex of Precambrian rocks in which granites, gneisses, schists and pegmatites predominate (J. Reed, pers. comm., 1980). Along the backslope of the Range Paleozoic and Mesozoic strata prevail. Jackson Hole, to the east, is a downfaulted block floored with Cretaceous and Tertiary rocks, covered to the north end by glacial till and outwash. Representative basins with nearly complete glacial sequences include Moran, Paintbrush, Cascade, Garnet, and Avalanche canyons. Along both flanks of the range glaciers are generally found above 3,000 m.

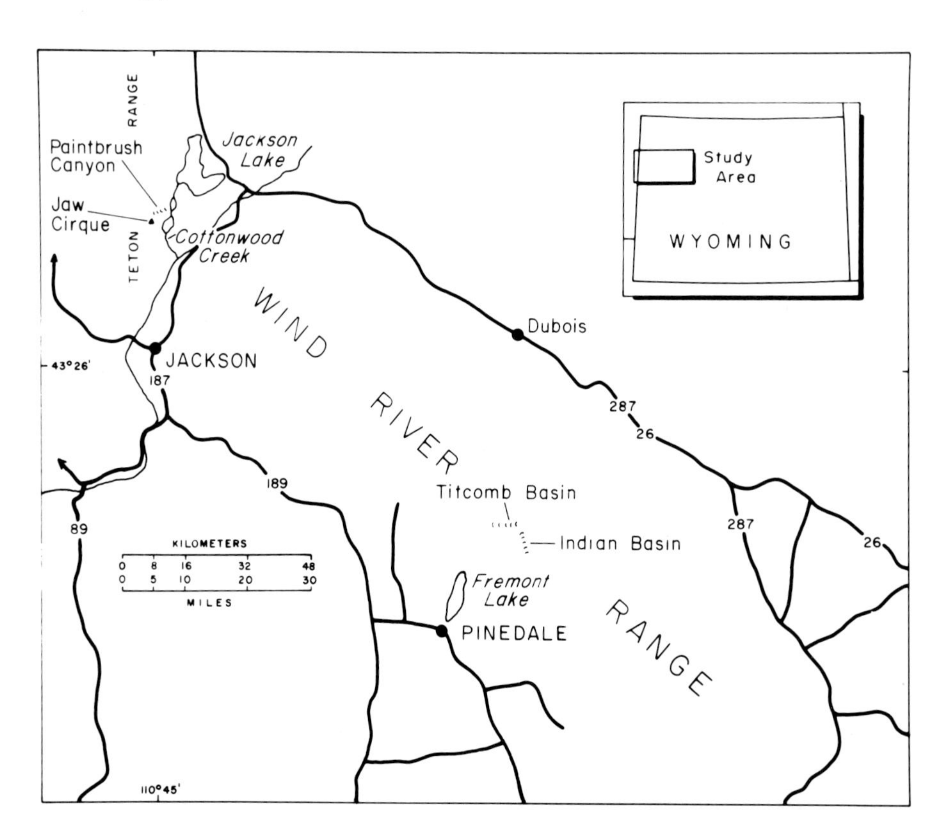

Figure 1 Location map showing (a) the location of Cottonwood Creek, Paintbrush Canyon and Jaw Cirque in the Teton Range, Wyoming, and (b) the location of Titcomb and Indian Basins, and Fremont Lake, Wind River Range, Wyoming.

The Wind River Range (Figure 1) forms a broad anticlinal uplift, trending northwest-southeast. Extending from South Pass to the Gros Ventre Mountains, the range is 190 km long and 60 km wide. The uplift of Tertiary age exposes a core of Precambrian granite, gneiss, and schist (Worl, 1968). Paleozoic and Mesozoic rocks cover the eastern slopes, whereas conglomerates, shales, tuffs and volcanics of Tertiary age crop out on the western slope. Along the crest of the range a complex of igneous and metamorphic rocks with fine, medium, coarse, and porphyroblastic textures prevails. Linearity of major valleys suggests the importance of faults in controlling drainage in the area (Mahaney, 1978). Glacial and periglacial deposits contain clasts of dioritic gneiss, gabbrodioritic gneiss, quartz-monzonitic gneiss, and granitic gneiss. Representative glaciated basins include Green River, New Fork, Titcomb, Indian, Pine Creek, Fremont, Falls Creek, Silver Creek, and Little Popo Agie (Mahaney, 1978). Along both flanks of the range glaciers are generally found above 3,300 m and ice is more common in cirque basins with north or northeast orientations. As in the Tetons, all modern glaciers occupy portions of cirque floors originally formed by Pleistocene and Holocene ice.

Climatic data for alpine and subalpine zones in the Tetons are lacking, but a summary of a 15-year period (1955-70) is available for Moran, Wyoming, located in Jackson Hole (U.S. Dept. Commerce, 1970). Temperature extrapolations for the mountain basins (∿3,000 m) are summarized by Mahaney (1975, p. 143). The average annual temperature approximates -4.7°C at 3,000 m with a maximum of +3.0°C and a minimum of -12.5°C.

Climatic data for stations in intermontane basins adjacent to the Wind River Range are available from the U.S. Dept. Commerce (1965). Temperature extrapolations for the mountain valleys (∿3,000 m) are summarized by Mahaney (1978, p. 231). The average annual temperature in the alpine basins approximates -3.5°C.

Vegetation above timberline in both ranges is dominated by perennial sedges, grasses, and herbaceous plants consisting of Dryas, sedge-grass, and willow-sedge stand types. The timberline contains procumbent whitebark pine *(Pinus albicaulis)*, limber pine *(Pinus flexilis)*, subalpine fir *(Abies lasiocarpa)*, and Engelmann spruce *(Picea engelmannii)* that merges at ∿3,000 m into a subalpine forest of Engelmann spruce *(P. engelmannii)*, and subalpine fir *(A. lasiocarpa)*. The lower limit of the subalpine forest merges with a montane forest of lodgepole pine *(Pinus contorta)*, Douglas fir *(Pseudotsuga menziesii)* and aspen *(Populus tremuloides)* (∿2,500 m in the Tetons; 2,750 m in the Wind River Range). Sagebrush vegetation *(Artemisia tridentata)* is the dominant ground cover below the lower limit of trees.

METHODS

Several dating methods are routinely used to assign relative and absolute ages to deposits. These include various rock-weathering parameters, lichen data, soil morphology and loess thickness, and radiocarbon.

Weathering Characteristics

Several different weathering parameters are measured on samples of 50 or 100 stones at each field locality. Counts are made of boulder frequency; that is, the number of boulders over 0.3 m diameter in unit areas of 10m x 10m (100m^2) or 5m x 5m (25m^2), following the pioneering work of Blackwelder (1915) and Nelson (1954). Usually the area is paced out in the field and total counts are made of all stones without regard to lithologic differences. The results allow differentiation between deposits of glaciation rank, but are not generally useful for differentiation at the substage or lesser level (Mahaney, *et al.*, 1981).

The ratio of fresh to weathered stones can be determined by recording the sound produced by boulders struck with a hammer. A sharp ring and strong rebound of the hammer indicates a fresh boulder, whereas a weathered rock produces a dull sound and weaker hammer recoil. Although this method is admittedly not particularly sensitive, it is quickly done in the field. Alternatively, one could use the fretting ratio of Sharp (1969), where a weathered stone is defined as one where more than 50 percent of the stone surface is weathered to a depth greater than the average grain diameter. Care should be taken to select stones of similar rock type on a single valley side to avoid lithologic and microclimatic effects. Data produced as weathering ratios allow differentiation among deposits at the glacial stage rank (Pinedale *vs.* Bull Lake), but are not generally useful for differentiation at the substage or stadial level (Mahaney, *et al.*, 1981).

Weathering rinds are indicative of the degree to which iron-bearing minerals oxidize and discolor the outer periphery of clasts. The thickness of rinds measured perpendicular to stone surfaces is related to the time since deposition. This method has been described by numerous workers including Birkeland (1973), Mahaney (1973, 1978, 1981) and Kiver (1974). Most workers measure only the maximum thickness of discoloration, neglecting the irregularly developed rind commonly found on many stones. Some investigators undoubtedly measure discoloration along fracture faces which may amount to 2 or even 3 times the rind thickness. Our practice is to ignore fracture faces and to measure both the maximum and minimum rind (Tables 1 and 2). We have also measured the degree of internal discoloration of stones, but find it controlled more by lithology than time, and difficult to quantify.

Weathering pits in boulders, cobbles or rock outcrops can be measured with the aid of a dial or vernier caliper. Depth and width of pits reveal differences between sets of deposits, provided that some care is taken to avoid porphyritic rocks, and in choosing only representative lithologies (*e.g.* granodiorite in the Wind River Range and granite and felsic gneiss in the Teton Range). Accurate width measurements are reasonably easy to obtain, but determining depth is more difficult. Measurements are made by fully extending the caliper depth rod and placing it in the bottom of the pit perpendicular to the pit rim. The main beam is then brought to a position level with the top of the pit and a measurement is made. Fifty measurements are taken at each site and the means calculated.

The data in Tables 1 and 2 show the degree to which deposits can be differentiated using measurements of boulder frequency, weathering ratios, rinds, and pits. The data allow differentiation at the glacial stage level, while differentiation between stades is not always possible. The presence/absence of some minimum rind development is useful in distinguishing Indian Basin deposits from deposits of late-Pinedale age.

Lichenometry

Maximum diameters and percent cover of the four dominant crustose lichens -- *Rhizocarpon geographicum, Lecanora thomsonii, Lecanora aspicilia,* and *Lecidea atrobrunnea* - are used to differentiate deposits in the Neoglacial succession. The method rests on the following assumptions:

1 - the largest lichen thallus represents the oldest and fastest growing lichen on a substrate,

2 - lichen growth (including percent cover) is indicative of elapsed time since deposition,

3 - the substrate falls within the life span of an individual lichen species, and

4 - effects of climatic change are smoothed out over time.

Deposits in the Teton Range (Jaw Cirque and Paintbrush Canyon) (Figure 1) were sampled to determine the largest diameter of each crustose lichen. Percent cover was estimated from percentage diagrams in Oyama and Takehara (1970). Measurements of individual thalli were made to the nearest millimeter. Only the largest elliptical-shaped thalli were measured, and only maximum diameters were used to assign relative ages. Thalli with irregular shapes and those found in depressions were bypassed to avoid problems associated with "interfingering" by individuals of the same species (Mahaney, 1973) and snowkill (Curry, 1969), respectively. Additional sampling restrictions followed by other workers (Beschel, 1957, 1958; Benedict, 1968; and Andrews and Webber, 1964) including standardization of rock type, prevailing wind, exposure to ice crystal blasting, aspect, available moisture, and surface stability were followed in this study.

Table 1. Weathering data[a] for sites in Indian Basin (IB) and Island Lake (IL) areas, Wind River Range, Wyoming

Site	Age	Stone Frequency (stones >0.3m diameter 25m²)	Weathering Ratios[b]			Weathering Rinds			Weathering Pits		
			% Fresh	%Weathered	n	Mean Maximum (mm)	Mean Minimum (mm)	n	Mean Depth (mm)	Mean Width (mm)	n
IB10b	Audubon	60	100	0	50	2.6	0	25	-	-	-
IB11		65	100	0	50	1.6	0	50	-	-	-
IB12		58	100	0	50	2.3	0	50	4.2	23.5	50
IB13b		60	-	-	-	2.6	0	50	-	-	-
IB19		69	100	0	50	2.4	0	50	5.3	29.3	50
IB41		57	100	0	50	1.6	0	50	3.3	26.2	50
IB5[c]	Indian Basin	41	-	-	-	2.1	0	50	6.7	60.6	50
IB9		37	100	0	50	3.6	0	50	-	-	-
IB10a		45	100	0	50	4.1	0	50	-	-	-
IB8		35	-	-	-	3.0	0	50	-	-	-
IB14		32	100	0	50	3.8	0	50	6.2	73.7	50
IB22		31	100	0	50	2.8	0	50	-	-	-
IB43		30	100	0	50	4.2	0.16	50	5.3	38.0	50
IB16	Late Pinedale	20	-	-	-	0	-	-	11.6	94.8	50
IB21		25	98	2	50	6.0	0.04	50	8.0	38.5	50
IL1		15	98	2	50	11.8	0.84	50	12.4	76.3	50

[a] (-) not sampled

[b] Weathered/fresh differentiation is based on the acoustical properties of stones. Fresh stones give a sharp clang, weathered stones a dull thud.

[c] Samples taken from B horizon in soil pit.

Table 2. Weathering data for sites in Paintbrush Canyon (PB), Jaw Cirque (JAW), Leigh Lake (PB) and Cottonwood Creek (MOS) areas, Teton Range, Wyoming

Site	Age	Stone Frequency (Stones >0.3m diameter/25m^2)	Weathering Ratios[a]			Weathering Rinds		
			% Fresh	% Weathered	n	Mean Maximum (mm)	Mean Minimum (mm)	n
JAW2	Audubon	70	100	0	100	1.5	0	50
JAW4		76	100	0	100	1.8	0	50
PB5		61	100	0	100	2.6	0	50
PB9		64	100	0	100	1.1	0	50
PB20b		60	100	0	100	1.2	0	50
PB20c		66	100	0	100	0.5	0	50
PB20d		70	100	0	100	0.6	0	50
JAW3	Indian Basin	45	100	0	100	3.1	0	50
JAW5		41	100	0	100	2.5	0	50
PB1		38	100	0	100	3.5	.03	100
PB20a		46	100	0	100	2.7	0	50
PB21		31	100	0	100	2.0	0	100
PB22		34	100	0	100	2.0	0	100
JAW1	late-Pinedale	18	99	1	100	6.7	0.1	50
PB2		20	96	4	100	12.6	0.06	25
PB13		18	98	2	100	8.9	0.9	100
PB14		17	99	1	100	8.2	0.1	100
PB15		15	98	2	100	8.1	0.4	100
PB19		22	97	3	100	5.6	0.5	50
PB16	Pinedale	17	98	2	100	5.6	0.1	100
PB17		14	97	3	100	5.7	0.08	50
PB18		12	97	3	100	6.9	0.2	50
MOS1	Bull Lake	4	68	32	100	14.9	1.9	50

[a] Weathered/fresh differentiation is based on the acoustical properties of stones. Fresh stones give a sharp clang, weathered stones a dull thud.

The data in Table 3 are used to identify Neoglacial advances; *i.e.* Gannett Peak, Audubon, and Indian Basin.[1] *R. geographicum* and *L. thomsonii* are the two most important crustose lichens because their maximum diameters fall into three discrete groups. *L. aspicilia* is absent on Gannett Peak deposits, but grows rapidly on older substrates. *L. atrobrunnea,* absent on most deposits of Gannett Peak age, appears on Audubon and Indian Basin deposits. Percent lichen cover varies across the Neoglacial deposits, ranging from nil on Gannett Peak and 10-40 percent on Audubon substrates to 40-75 percent on Indian Basin surfaces. Because of the geologically short life span of individual thalli (∿ 3,000 yrs.) the use of lichenometry is restricted to deposits of Neoglacial age.

Lichenometry allows the separation of Gannett Peak and Audubon deposits which are indistinguishable on the basis of surface morphology and only marginally identifiable by weathering characteristics. Lichenometry can also be used as an AD tool, but it cannot be overemphasized that a growth curve must be established (using historic, radiometric and/or dendrochronologic controls) for each geographical area studied. If the data for maximum thalli diameters and percent cover between correlative units in the Teton (this paper, Table 3) and Wind River (Mahaney, 1978, p. 235) ranges are compared some differences emerge, the apparent result of climatic and/or temporal variations. Thus, a separate growth curve must be made for each range if lichenometry is to be used as an AD method.

Soil Stratigraphy

Deposits in geologic successions are separated using soil morphology, particle size distributions, clay mineralogy, and selected soil chemical parameters. Individual pedons are assigned to soil stratigraphic units on the basis of stratigraphic position and morphological features which may be continuously recognized and mapped (Mahaney and Fahey, 1976; Mahaney, 1978, 1982a). Such units form, *in situ*, in deposits by the action of pedological processes working downward on fresh, unaltered materials. Soil stratigraphic units are assigned informal names coinciding with the age of the deposits in which they form (*e.g.* use of the prefix "post" to avoid a proliferation of geologic names). A few examples illustrating the use of soils and soil properties in differentiating Quaternary deposits follows.

Pleistocene Sequences

Moraines enclosing Fremont Lake in the Wind River Range have been described by numerous workers (Blackwelder, 1915; Richmond, 1965, 1974; and Mahaney, 1978). Sites on pre-Bull Lake, Bull Lake, and Pinedale moraines (Figure 2) are located to the east of the lake under montane forest. Soils at these sites were examined to determine specific properties that would prove useful in differentiation. As shown in Table 4 soil depth and loess thickness generally increase with age, as indicated by higher silt content in the sola and upper subsoil of the three profiles. Clay content generally increases over time, with the highest values in the pre-Bull Lake soil. The high value of 28 percent in the post-Pinedale soil may result from airfall influx as this site is located on a high moraine ridge 250 m above Fremont Lake. Soil pH distributions range from slightly acidic in the post-Pinedale soil, to moderately acidic in the post- and pre-Bull Lake soils. Greater variation in pH with depth which occurs in the two older soils, is attributed to variable movement of H^+ ions.

Clay and primary mineral distributions (Table 4) in the fine clay separates (<2 μm) reveal information related to sourcing of the ice and/or subsequent weathering and is useful in age differentiation. Of special significance is the presence of high amounts of kaolinite in

---

[1] This stratigraphic term replaces "Early Neoglacial". It is defined from the type section in Indian Basin, Wind River Range (Mahaney, 1982a).

Table 3. Lichen characteristics[a] for Neoglacial deposits in Jaw Cirque (JAW) and Paintbrush Canyon (PB) areas, Teton Range, Wyoming

Site	Age	*Rhizocarpon geographicum*	*Lecanora thomsonii*	*Lecanora aspicilia*	*Lecidea atrobrunnea*	% cover
PB7	Gannett Peak	9	10	nil	15	nil
PB8		15	16	nil	nil	nil
JAW2	Audubon	46	67	70	65	nil-30
JAW4		55	72	66	61	nil-25
PB5		50	83	55	54	5-40
PB9		45	61	64	55	5-30
JAW3	Indian Basin	86	105	115	109	5-50
JAW5		85	120	97	68	nil-50
PB1		123	155	245	225	50-75
PB20a		110	112	189	136	40-70
PB20b		69	90	146	115	35-60
PB21		85	119	140	110	40-80
PB22		78	132	232	173	35-75
PB23		95	116	129	114	40-80

[a]Lichen measurements of the maximum diameter in mm.

the FL9 profile, which is thought to result from weathering in a moist environment with a greater weathering intensity. This correlates closely with a pre-Bull Lake soil described in the Front Range of Colorado (Mahaney and Fahey, 1980). The absence of chlorite and smectite, and the small amount of mixed-layer clays and vermiculite in the post-Pinedale soil contrast sharply with the post-Bull Lake and pre-Bull Lake soils. The large amounts of orthoclase and plagioclase feldspars in the pre-Bull Lake soil may reflect a different source for the ice. Overall, the upward decrease in feldspar in the profiles is greatest in the pre-Bull Lake soil.

The soil chemistry in Table 5 supports the clay mineralogy described above. The higher K^+, Ca^{+2}, and Mg^{+2} in the pre-Bull Lake soil follow from the greater amounts of illite, smectite, mixed-layer illite-smectite, and vermiculite, respectively. The data indicate that the smectite is calcic, not sodic. Base saturation is higher in the post-Pinedale soil, but declines to 35 percent in the upper solum of the post-Bull Lake soil. Variations in organic matter and nitrogen apparently reflect differences in bulk density and variable leaching in the profiles. The notion that leaching varies over time is supported by the organic matter and nitrogen distributions in the three profiles. Organic matter content decreases slowly with depth in the post-Pinedale soil, decreases abruptly in the post-Bull Lake soil, and tends to increase in the lower solum and subsoil of the pre-Bull Lake pedon.

In the Teton Range Pleistocene deposits can be separated by the degree of soil expression. Representative soils in Pinedale and Bull Lake deposits are shown in Table 6. As in the Fremont Lake sequence, soil depth and loess thickness increase with age. The high silt content in the post-Bull Lake soil is representative of pre-Wisconsinan soils in the Teton Range that have substantial loess caps. Soil reactions (pH), although slightly more acidic in the post-Bull Lake solum, are not considered age-dependent. Whereas the primary mineral distributions provide little information on age, the clay mineral composition yields data that are important in age determination and paleoclimatic reconstruction. Within the 1:1 clay minerals, the increase in kaolinite with age may well represent increased weathering over time. The presence of halloysite in the post-Bull Lake soil may reflect hydration of kaolinite in an environment wetter than today (Mahaney, 1981). The increase in 2:1 clay minerals, especially smectite, illite-smectite, and vermiculite, in the post-Bull Lake soil reflects a similar

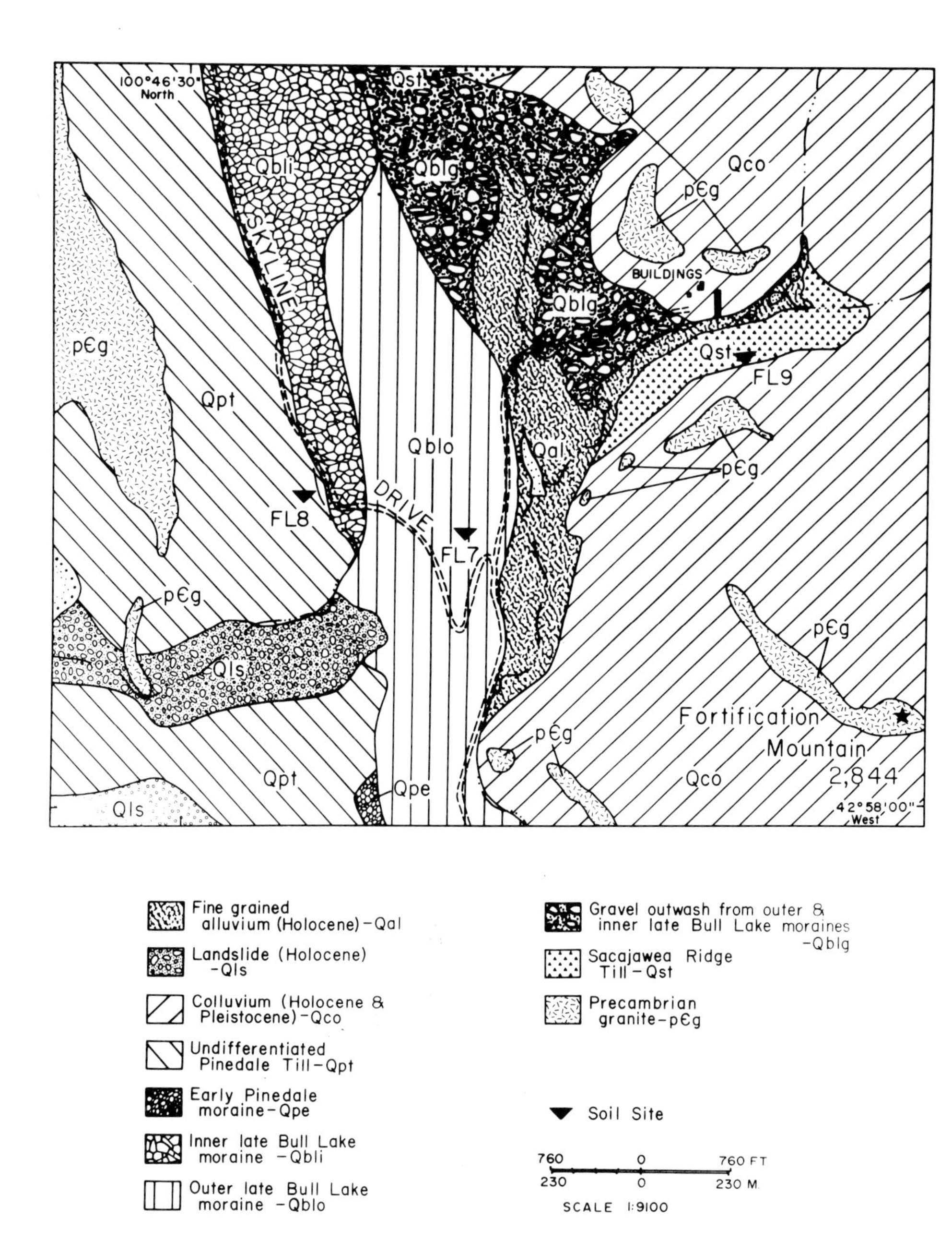

Figure 2 Surficial geology and soil sites in late-Quaternary glacial deposits, east of Fremont Lake in the montane forest zone, Wind River Range, Wyoming.

pattern that occurs in the Pleistocene sequence at the type Pinedale locality (Mahaney, 1978, and this paper). Although it is entirely possible that the clay mineral composition of these paleosols originates by some combination of weathering over time, aeolian input, and/or paleoclimate, the differences appear common and widespread, making them useful in relative age determinations.

Analysis of Cumulative Particle Size Curves

Particle size distributions commonly allow differentiation of soils in geologic successions. If the amount of clay in the parent material of each soil is fairly uniform, then changes in clay content in B horizons of soil-stratigraphic units should reflect relative amounts of weathering over time and/or airfall influx. Post-Bull Lake (MOS1) and Pinedale (PB18) soils in the Teton Range show heavier textures in B horizons because clay content increases with age (Figures 3 and 4). The data show that the tails (<3.9 μm) of the B horizon curves for soils MOS1 and PB18 tend to flatten with increasing age, a result of

Table 4. Physical[a], chemical and mineral[b] properties of post-Pinedale soils in montane and sagebrush steppe vegetation zones, Wind River Mountains, Wyoming.

Site	Age and Parent Material	Elevation (m)	Vegetation	Soil Horizon	Depth (cm)	% Sand 2mm-63μm	% Silt 63-4μm	% Clay <4μm	pH (1:1)	Clay Minerals (<2μm)							Primary Minerals			
										K	H	I	S	Mx	V	C	Q	O	P	B
FL8	Pinedale till	2450	Montane Forest	A11	0-6	43.0	29.0	28.0	6.2	-	-	tr	-	-	-	-	xxx	tr	tr	-
				A12	6-22	52.5	34.0	13.5	6.4	x	-	xx	-	-	-	-	xxx	x	x	-
				B21	22-28	55.9	32.1	12.0	6.4	x	-	xx	-	-	tr	-	xxx	x	x	-
				IIB22	28-38	63.9	27.1	9.0	6.4	tr	-	tr	-	tr	-	-	xxx	x	x	-
				IICox	38-102	65.8	25.2	9.0	6.3	?	-	x	-	-	?	-	xxx	tr	x	-
				IICn	102+	67.9	25.1	7.0	6.3	?	-	tr	-	-	-	-	xxx	tr	x	tr
FL7	Bull Lake till	2445	Montane Forest	A2	0-5	55.9	35.6	8.5	5.3	tr	-	-	tr	-	-	-	xxx	tr	x	-
				B21t	5-23	56.4	32.6	11.0	5.9	tr	-	-	x	x	-	tr	xxx	tr	x	-
				B22	23-54	61.7	28.8	9.5	5.9	-	-	-	-	-	-	-	xxx	tr	x	-
				Clox	54-87	58.4	34.6	7.0	6.5	-	-	-	x	tr	-	-	xxx	tr	x	-
				IIC2ox	87-119	67.2	25.8	7.0	6.6	-	-	-	tr	tr	-	tr	xxx	tr	x	-
				IIC3ox	119-193	64.4	26.6	9.0	7.3	-	-	-	xxx	xx	-	xxx	xxx	x	xx	tr
				IICn	193+	68.8	24.2	7.0	7.3	-	-	-	x	x	-	x	xxx	x	xx	tr
FL9	pre-Bull Lake till	2400	Montane Forest	A1	0-4	44.6	37.9	17.5	5.7	tr	-	x	-	x	x	-	xx	tr	tr	-
				B21t	4-10	44.0	34.0	22.0	5.5	x	-	x	-	x	x	x	xxx	tr	tr	-
				B22t	10-36	43.4	22.1	34.5	5.6	xx	-	xxx	xxx	xxx	-	xx	xxx	-	x	-
				B23t	36-60	48.2	22.8	29.0	5.5	x	-	xxx	xxx	xxx	xxx	-	xxx	tr	x	tr
				B24t	60-95	50.0	25.5	24.5	5.7	xx	x	xxx	xxx	xxx	-	-	xxx	xxx	xx	-
				IIB25t	95-122	63.6	14.4	22.0	6.6	xxx	-	xxx	xxx	xxx	-	-	xxx	xxx	xxx	-
				IIB26t	122-158	61.9	10.1	28.0	5.7	x	-	xxx	xxx	xxx	-	-	xxx	-	xx	-
				IIC1ox	158-175	73.5	10.5	16.0	6.0	x	-	xx	xxx	xx	-	-	xxx	xxx	xxx	x
				IIC2ox	175-200	77.4	13.6	9.0	5.7	-	-	xxx	xxx	xxx	-	-	xxx	xxx	xxx	tr
				IICnm	200+	80.2	10.8	9.0	6.9	tr	-	xxx	xxx	xxx	-	-	xxx	xxx	xxx	-

[a] Data are given in weight-percentages of sand, silt and clay (<2mm). Coarse particle sizes (2000-63μm) determined by sieving; fine particle sizes (63-1.95μm) determined by hydrometer.

[b] Mineral abundance is based on peak height: nil (-); minor amount (tr); small amount (x); moderate amount (xx); abundance (xxx). Clay minerals are kaolinite (K), halloysite (H), illite (I), smectite (S), vermiculite (V), mixed-layer illite-smectite (Mx). Primary minerals are quartz (Q), plagioclase feldspar (P), orthoclase (O), and biotite (B).

Table 5. Selected chemical properties in soils of mid- to late-Pleistocene age, Wind River Range, Wyoming.

Site	Horizon	Depth (cm)	Extractable[a] Cations (meq/100g)				CEC meq/ 100g	Base Saturation (%)	Organic Matter (%)	Nitrogen (%)
			Na^+	K^+	Ca^{+2}	Mg^{+2}				
FL8	A11	0-6	1.2	1.2	27.8	4.8	41.1	85	19.8	0.875
	A12	6-22	0.1	0.9	7.7	1.7	14.8	70	5.1	0.218
	B21	22-28	0.1	0.8	4.3	1.3	10.2	64	3.7	0.148
	B22	28-38	0.1	0.5	2.9	0.8	5.9	73	1.3	0.058
	Cox	38-102	0.1	0.2	2.3	0.6	5.0	64	0.6	0.029
	Cn	102+	0.1	0.1	1.6	0.5	3.9	59	0.3	0.015
FL7	A2	0-5	0.1	0.4	3.0	0.7	12.1	35	6.4	0.178
	B21t	5-23	0.0	0.2	2.8	0.7	5.7	65	1.0	0.024
	B22	23-54	0.0	0.1	1.2	0.5	3.4	53	0.4	0.013
	C1ox	54-87	0.1	0.1	1.3	0.6	3.0	70	0.1	0.008
	C2ox	87-119	0.1	0.1	2.3	0.7	4.1	78	0.1	0.009
	C3ox	119-193	0.2	0.1	4.0	1.1	5.7	95	0.1	0.016
	Cn	193+	0.2	0.2	3.0	1.0	4.6	94	0.2	0.006
FL9	A1	0-4	0.03	0.9	13.8	2.3	46.0	37	14.1	0.240
	B21t	4-10	0.08	0.5	5.3	1.3	27.9	26	3.3	0.064
	B22t	10-36	0.16	0.4	6.5	2.3	20.0	47	0.4	0.030
	B23t	36-60	0.17	0.4	10.7	4.3	28.2	55	0.5	0.039
	B24t	60-95	0.07	0.4	11.5	5.8	23.0	77	0.9	0.006
	IIB25t	95-122	0.09	0.3	11.3	4.4	22.4	72	0.7	0.007
	IIB26t	122-158	0.15	0.5	20.8	7.1	29.7	96	0.7	0.010
	IIC1ox	158-175	0.16	0.4	15.1	6.3	21.6	100	0.7	0.008
	IIC2ox	175-200	0.06	0.2	12.1	4.1	20.4	89	1.2	0.029
	IIC3nm	200+	0.17	0.2	8.9	3.1	14.1	88	0.7	0.009

[a]- in 1N ammonia acetate of pH 7.0

Table 6. Selected physical[a], chemical and mineral[b] properties of late-Pleistocene soils, Teton Range, Wyoming.

Site	Age and Parent Material	Elevation (m)	Vegetation	Soil Horizon	Depth (cm)	% Sand 2mm-63μm	% Silt 63-4μm	% Clay <4μm	pH (1:1)	Clay Minerals (<2μm) K	H	I	S	Mx	V	C	Primary Minerals Q	O	P
PB18	post-Pinedale	2120	Montane Forest	A1	0-9	52.7	35.3	12.0	6.0	x	-	tr	-	-	-	tr	xxx	x	x
				IIB2	9-34	69.8	19.7	10.5	6.2	x	-	tr	-	-	x	tr	xxx	tr	x
				IIC1ox	34-87	92.8	1.2	6.0	6.1	tr	-	tr	-	-	-	-	xxx	tr	xx
				IIC2ox	87-116	92.9	6.1	1.0	6.3	-	-	-	-	-	-	-	xxx	tr	xx
				IICn	116+	75.5	18.5	6.0	6.3	tr	-	tr	-	-	-	-	xxx	tr	xx
MOS1	post-Bull Lake	2060	Montane Forest	A21	0-14	7.0	78.5	14.5	5.4	x	-	xx	-	-	-	-	xxx	x	x
				A22	14-24	5.9	61.1	33.0	6.1	x	-	x	-	-	-	-	xxx	x	x
				B21rh	24 55	2.7	54.8	42.5	5.8	xx	-	xxx	tr	x	x	-	xxx	x	x
				C1ox	55-92	6.7	54.8	38.5	6.3	xx	x	xxx	xxx	xxx	xx	-	xxx	xxx	xxx
				C2ox	92-110	14.4	51.6	34.0	6.7	xx	x	xxx	xxx	xxx	xxx	-	xxx	xx	x
				C3	110-143	19.4	38.6	42.0	8.3	tr	-	-	-	x	-	-	xxx	x	-
				IIC4ox	143-170	55.9	23.1	21.0	8.5	xx	-	-	-	x	-	tr	x	x	tr
				IID	170+														

[a]Data are given in weight-percentages of sand, silt and clay (<2mm). Coarse particle sizes (2000-63μm) determined by sieving; fine particle sizes (63-1.95μm) determined by hydrometer.

[b]Mineral abundance is based on peak height: nil (-); minor amount (tr); small amount (x); moderate amount (xx); abundant (xxx). Clay minerals are: kaolinite (K), halloysite (H), illite (I), smectite (S), mixed layer illite-smectite (Mx), chlorite (C). Primary minerals are: orthoclase (O), plagioclase feldspar (P), and quartz (Q).

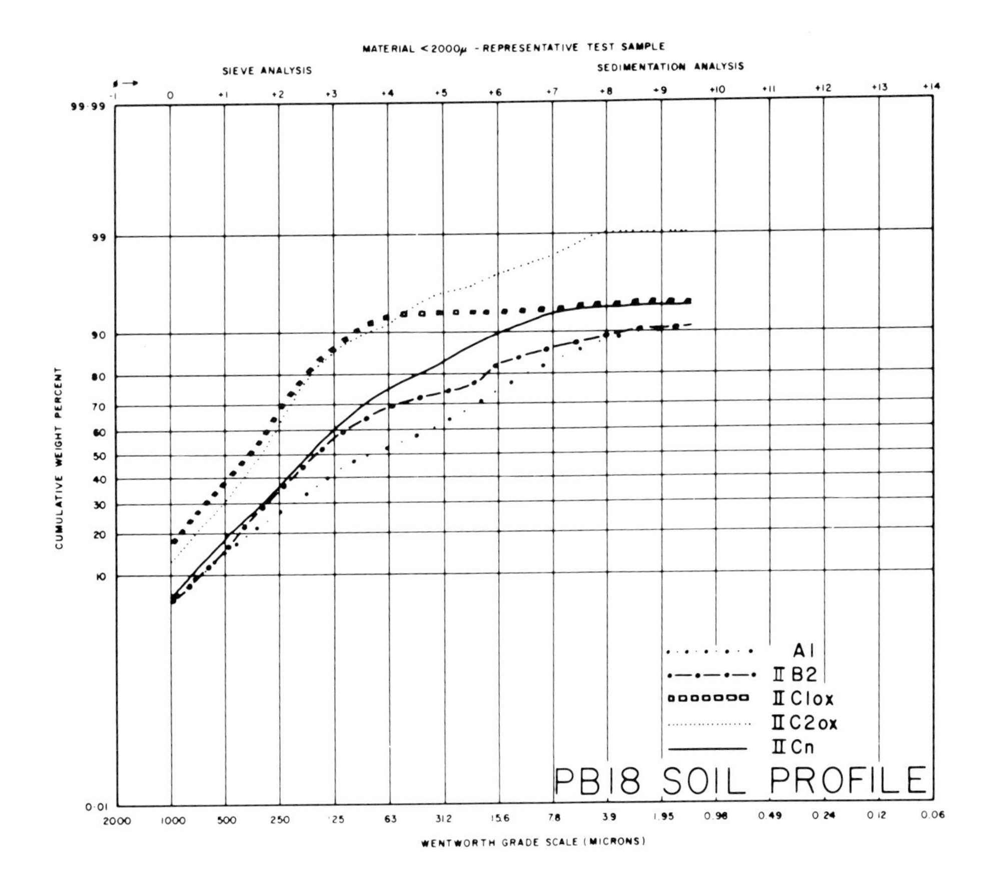

Figure 3 Particle size distribution curve for PB18 soil profile (post-Pinedale soil).

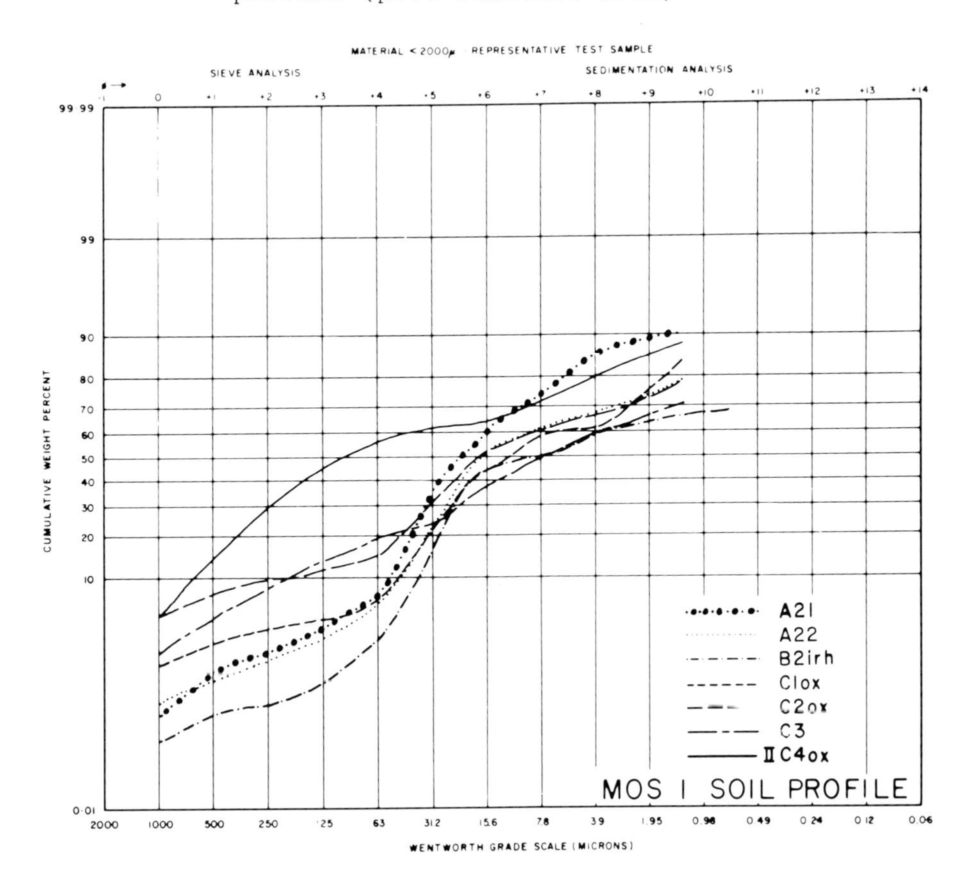

Figure 4 Particle size distribution curve for MOS1 soil profile (post-Bull Lake soil).

clay buildup over time. Not all of the clay accumulates by *in situ* weathering, as post-Bull Lake soils in the Teton Range have thick loess caps suggesting large airfall influxes of silt during the late Quaternary (probably during the Pinedale Glaciation). Both soils display upward fining sequences in the profiles, the result of loess being translocated downward into the solum by percolating water. The older soil has a thicker loess cap (≃143 cm) and an overall silt content nearly four times greater than in the younger soil. This condition is attributed to a longer time for weathering and greater amounts of loess blown in from nearby Pinedale glacial limits. The higher amount of clay in the post-Bull Lake soil is also undoubtedly partly a function of greater time for weathering and airfall influx.

The centers of gravity (mean phi) for the B horizons of four post-Pinedale profiles in lower Paintbrush Canyon and one post-Bull Lake soil in a prominent end moraine near Cottonwood Creek, Teton Range, (Figure 1) are calculated from:

$$\bar{x}\phi = \frac{16\phi + 50\phi + 84\phi}{3} \qquad (1)$$

The data (Table 7) clearly show a considerable difference between B2 horizons in the two soil stratigraphic units, which is useful in differentiating these soils.

Late-Pinedale-Indian Basin soil sequences

In many of the high alpine valleys of the Wind River and Teton ranges, between 2,800 and 3,300 m elevation, moraines of late-Pinedale and Indian Basin age are prominently displayed. Differentiation of these deposits is often difficult on the basis of superficial characteristics, and requires detailed weathering, soil stratigraphic, and isotopic data. The most detailed information can be gleaned from morphological, chemical, and clay mineral analysis of individual pedons (see Mahaney, 1978). A representative sequence can be found in Titcomb Basin in the Wind River Mountains (sites TB14 and TB1; see Mahaney, 1978, p. 242-243; location p. 227-228), and the use of particle size distribution, clay mineralogy, and chemical parameters in differentiating two soil stratigraphic units there have been discussed by Mahaney (1978). Major differences include increased clay content in the post-Pinedale soil, higher amounts of 2:1 clay minerals, such as smectite, in the B and C horizons of the post-Indian Basin soil, and higher percentages of basic cations and salts in the post-Indian Basin soil.

Extractable iron oxides

Data from the soils under montane forest near Fremont Lake were used to determine the degree to which extractable Fe oxides (Coffin, 1963; McKeague and Day, 1966) are useful age indicators. As shown in Table 8, organically-bound Fe (pyrophosphate extractable Fe_p) has similar values in the two youngest soils, increasing in the lower solum of the pre-Bull Lake soil. Organically complexed Fe appears to increase slightly with greater age, and tends to be distributed through the profiles to considerable depth.

Oxalate extractable Fe (Fe_o), representing the organic plus the amorphous Fe, forms slowly reaching the highest values in the post-Bull Lake soil, thereafter decreasing to low amounts in the subsoil of the oldest profile. This trend suggests that more amorphous Fe is converted to crystalline Fe with greater age. Dithionite extractable Fe (Fe_d) decreases from a maximum of 1.36% in the post-Pinedale soil to 1.18% in the post-Bull Lake soil, and then rises to nearly 6.0% in the pre-Bull Lake soil. Lower overall clay values (Table 4) cause the low value in the post-Bull Lake soil.

The Fe ratio (Fe_o/Fe_d) is initially quite variable, increasing in the subsoil as Fe_o increases and Fe_d decreases with depth. In the post-Bull Lake soil the ratio is distributed more uniformly with depth. The ratio drops in the pre-Bull Lake soil, and in the IIB26t horizon falls

Table 7. Mean phi calculations for post-Pinedale and post-Bull Lake soils, Teton Range, Wyoming.

Site	Horizon	Age	Percentile[a]			
			16	50	84	$\bar{x}\phi$
PB19	B2	post-Pinedale	1.1	4.2	7.7	4.3
PB16	B2		1.1	3.8	7.2	4.0
PB17	B21ir		0.8	3.0	6.4	3.4
	B22ir		1.7	2.8	6.5	3.7
PB18	B2		0.9	2.6	6.4	3.3
MOS1	B2irh	post-Bull Lake	4.9	6.8	14.0	8.6

[a]The phi values for the 16th, 50th, and 84th percentiles taken from particle size distribution curves. The sand separates were determined by sieving; fine particle sizes (<63μm) by hydrometer.

to <0.10. Here the percent of Fe_d far outweighs the percent of Fe_o.

The Fe ratio is an important age indicator, especially for Pleistocene soil sequences. The data herein support the findings of Alexander (1974), who analyzed variations in the Fe ratio in a sequence of Quaternary soils formed in stream deposits along the east flank of the Sierra Nevada, California. His overall trend of low to high to low ratios parallels data described in this study.

Quartz/feldspar ratios

Weathering intensity can be determined using weathering ratios of light minerals following procedures set forth by Ruhe (1956). In the light mineral suite quartz is more resistant than feldspar. Counts of resistant and weatherable minerals for each size fraction (250-63μm) in each soil horizon are made under a binocular microscope with an electronic point counter. Larger quotients indicate more resistant than weatherable minerals, whereas smaller quotients reflect more feldspar present in the sample. When ratios for the Cox and Cn horizons are similar, we assume that parent materials have similar lithologic compositions. Higher quotients from A and B horizons suggest that weathering processes have removed some feldspars leaving higher ratios of quartz/feldspar. Applying the quartz/feldspar weathering indices to two soil sequences in the Wind River Range provides data useful in interpreting the relative weathering states of individual profiles. In the Fremont Lake area (for profile morphology and location of sites see Mahaney, 1978) five profiles FL1, 2, 3, 4, and 6 were analyzed. In Figure 5, the post-Pinedale soils have quartz/feldspar ratios that range from 1.0 to 1.5; most profiles are uniform with depth. Changes in the post-Bull Lake soils indicate that the quartz/feldspar ratio increases from 3.0 to 3.5 due mainly to increased weathering over greater time. The data support the hypothesis that the Bull Lake Glaciation is pre-Wisconsinan in age and may be correlative with the Illinoian Glaciation in midwestern and eastern North America. Furthermore, this would make the post-Bull Lake soil coeval with the Sangamon Interglacial Stage (∿100,000 BP). Additional evidence for a pre-Wisconsinan age of the post-Bull Lake soil is found in Mahaney (1981).

In Titcomb Basin (Figure 1) changes in quartz/feldspar ratios in a soil sequence formed in glacial deposits show some interesting variations (Figure 6). The soils range in age from ∿100 yrs BP (TB15; post-Gannett Peak), ∿1000 yrs BP (TB6; post-Audubon), ∿3000 yrs BP (TB14; post-Indian Basin) and ∿8000 yrs BP (TB1 and 3; post-Pinedale). The ratio is close to 1.0 for all parent materials in all profiles. There is little change between the post-Audubon and post-Indian Basin

Table 8. Soil Color and iron in the pyrophosphate, citrate-dithionite, and acid oxalate extracts for sites in the Wind River Range

Site	Age	Horizon	Depth (cm)	Hue	value/chroma		Fe (%) in extract			Fe Ratio (Fe_o/Fe_d)
					moist	dry	Pyrophosphate (Fe_p)	Citrate-Dithionite (Fe_d)	Acid Oxalate (Fe_o)	
FL8	post-	A11	0-6	10YR	2/3	3/2,4/2	0.16	1.05	0.52	0.50
	Pinedale	A12	6-22	10YR	3/2	4/2	0.24	1.35	0.47	0.35
		B21	22-28	10YR	4/3	4/3	0.22	1.36	0.44	0.32
		IIB22	28-38	10YR	5/3	6/3	0.12	1.34	0.63	0.47
		IICox	28-102	10YR	6/3	7/3	0.09	1.07	0.77	0.72
		IICn	102+	2.5Y	5/3	8/3	0.07	1.04	0.72	0.69
FL7	post-	A2	0-5	10YR	5/2	6/2	0.20	1.12	0.82	0.73
	Bull	B21t	5-23	10YR	4/4	7/3	0.16	1.18	0.96	0.81
	Lake	B22	23-54	10YR	6/4,6/3	8/2,7/2	0.09	0.97	0.77	0.79
		C1ox	54-87	10YR	5/3,5/4,6/3	8/2	0.06	0.87	0.69	0.79
		IIC2ox	87-119	10YR	5/3	8/3	0.05	0.83	0.69	0.83
		IIC3ox	119-193	10YR	5/4,5/6,6/4	8/4	0.06	0.91	0.64	0.70
		IICn	193+	2.5Y	6/4	8/3	0.06	0.88	0.66	0.75
FL9	pre-	A1	0-4	10YR	2/2,2/3	4/2	0.42	1.70	0.82	0.48
	Bull	B21t	4-10	10YR	5/4	5/3	0.32	0.81	0.71	0.88
	Lake	B22t	10-36	10YR	6/4	7/3	0.07	1.83	0.43	0.23
		B23t	36-60	10YR	6/4	7/4	0.08	2.69	0.63	0.23
		B24t	60-95	10YR	5/4	6/4	0.18	1.86	0.77	0.41
		IIB25t	95-122	7.5YR	4/4	6/4	0.13	4.58	0.70	0.15
		IIB26t	122-158	10YR	4/4	5/4,6/6	0.12	5.98	0.37	0.06
		IIC1ox	158-175	10YR	5/4	6/4	0.09	3.18	0.58	0.18
		IIC2ox	175-200	7.5YR	5/6,4/4	6/4,7/4	0.10	2.30	0.45	0.20
		IICnm	200+	10YR	8/1		0.08	3.24	0.36	0.11
				2.5Y	7/3	6/4,7/4				

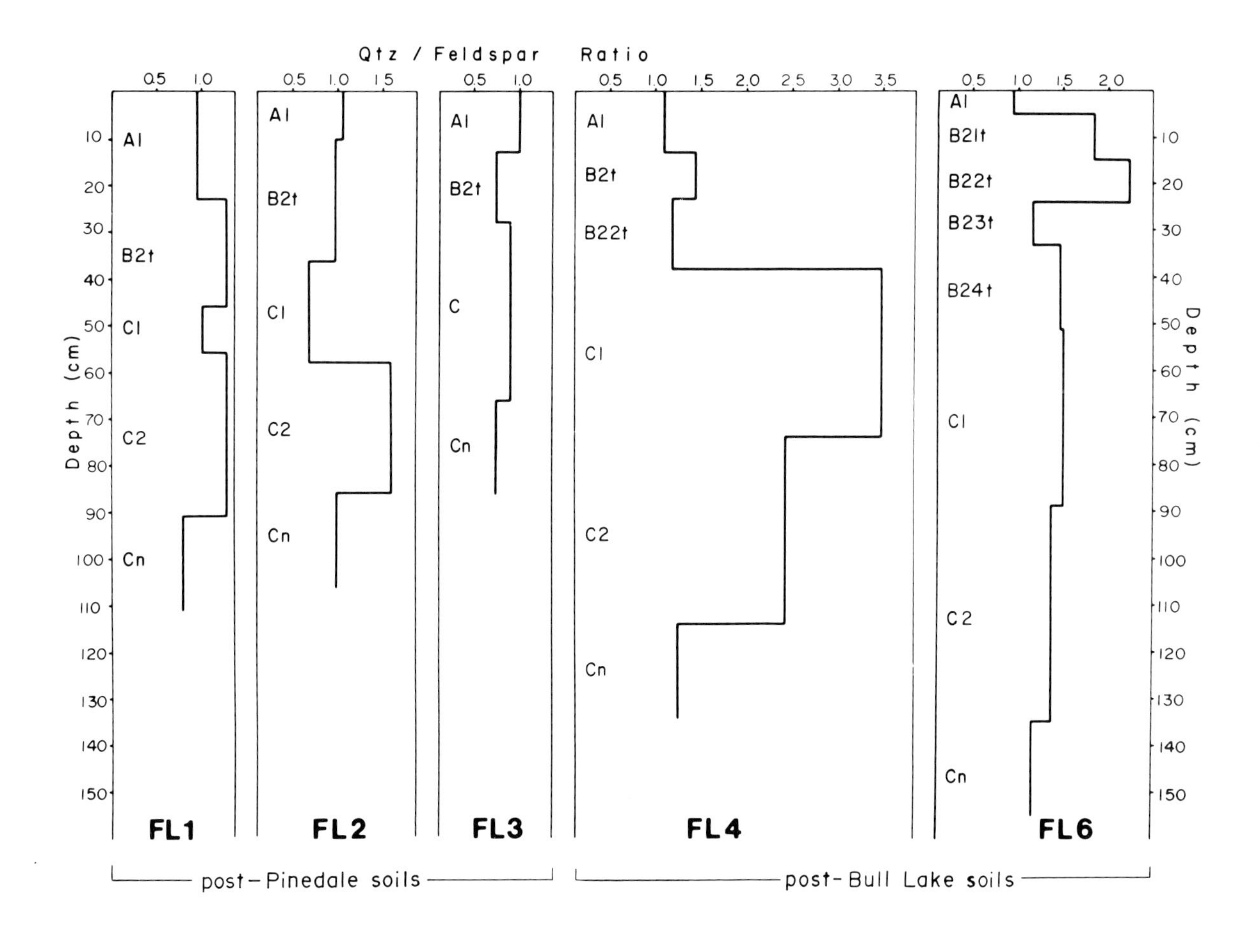

Figure 5 Mineral weathering ratios of post-Pinedale and post-Bull Lake paleosols, south of Fremont Lake in the Wind River Range, Wyoming. See Mahaney, (1978), p. 230 for location; p. 262-64 for soil descriptions.

profiles except that the A1 horizon contains less feldspar in the post-Indian Basin soil. The post-Pinedale soils (TB1 and TB3) contain horizons with considerably lower feldspar counts. Quartz/feldspar ratios appear to have great utility in subdividing soils in alpine glacial successions. The quotients obtained, however, may provide valid differentiation criteria only for individual drainages. Additional research is needed to determine if the magnitude of change in ratios among soils is uniform in different drainages of one range.

Radiocarbon dates

Organic materials from the bottom of two short cores in bogs on the late-Pinedale moraine system in Titcomb Basin yielded radiocarbon ages of 7,380 ± 150 yrs BP (Gak-8361; site TB24) and 7,940 ± 190 yrs BP (Gak-8216; site TB23) (Mahaney, 1982b, in preparation). These dates provide minimum ages for late Pinedale drift in Titcomb Basin and maximum ages for profile TB1. They indicate an approximate time zero of ∿8,000 years BP for the beginning of soil development in late Pinedale substrates. No finite radiocarbon dates exist for the beginning of the Indian Basin advance in the Wind River and Teton ranges, but using a radiocarbon date of 5050 ± 170 yrs BP (Gak-8359; sample NR6-A1b) from a stone-banked lobe on Niwot Ridge in the Front Range, Colorado, we infer an approximate date for the onset of Neoglaciation of 5,000 yrs BP.

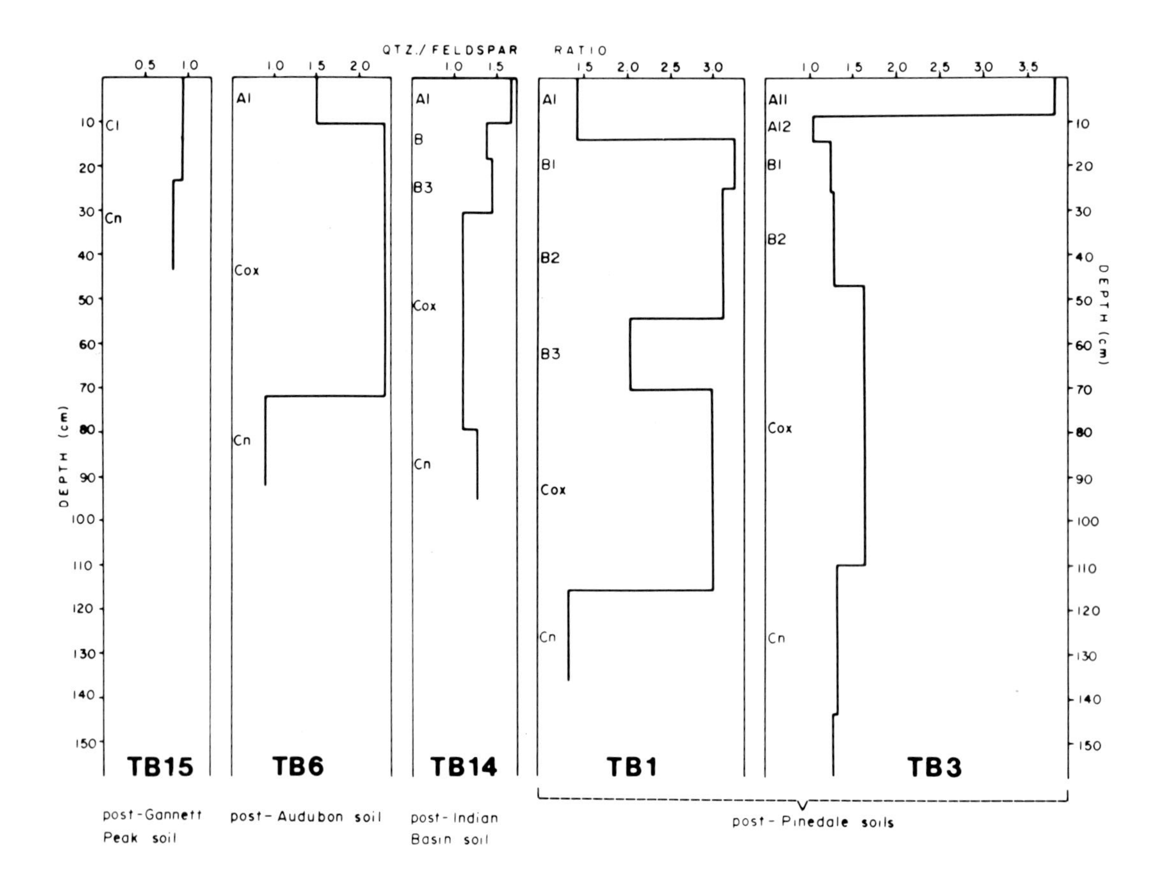

Figure 6 Mineral weathering ratios of Holocene paleosols and soils in Titcomb Basin, Wind River Range, Wyoming. See Mahaney (1978) p. 227-28 for location, p. 256-59 for soil descriptions.

This correlates closely with estimates made by Benedict (1968, 1973). A date from a soil buried beneath lacustrine sand indicates that the Indian Basin advance ended more than 3,050 ± 120 yrs BP (Gak-6024). This sequence documents a rise in the level of Lower Titcomb Lake (for location see site TB9, Mahaney, 1978, p. 228).

CONCLUSIONS

Various absolute (AD) and relative (RD) dating methods useful in establishing ages for deposits in the mountains of western Wyoming have been discussed. Radiocarbon dating is very useful, but because datable organic materials are scarce, most workers must rely on certain relative methods that require a broad range of geologic, geomorphic and pedologic expertise. Every attempt should be made to use a number of different methods in establishing geological chronologies.

Of the RD methods, soil stratigraphy has great utility in deposit differentiation as many individual properties depend on elapsed time since deposition. Because soil properties approach steady state at different rates, we have concentrated on specific characteristics that undergo change over long time periods (*e.g.* 0.5 x 10^6 years). Thus, depth of weathering, particle size changes (especially upward fining sequences and shifts in mean phi), relative clay mineral assemblages, quartz/feldspar ratios (within the fine and very fine sand fractions), and some soil chemical parameters provide a detailed body of data useful in relative age determination. As more absolute dates become available it is imperative to establish lichen growth curves, weathering rind curves, and iron oxide curves for individual ranges.

ACKNOWLEDGEMENTS

We thank B.D. Fahey (Guelph University) and W.N. Melhorn (Purdue University) for critical reviews of this paper. N. Stokes assisted in the field in 1972. Students in the York University Mountain Geomorphology Field Camp (1973-75 and 1978-81) assisted with the field work. L.J. Gowland helped with the excavation and analysis of several profiles discussed in the text. Soil and sediment samples were analyzed in the Geomorphology and Pedology Laboratory at York University with the assistance of G. Berssenbrugge. Rock and sand samples were analyzed at the North Dakota Geological Survey, Grand Forks, N.D., and Geomorphology Research Center, Purdue University, W. Lafayette, Indiana. Research was financed by grants from York University, and carried out with the cooperation of the U.S. Forest Service and National Park Service. We are grateful to Mr. Bob Wood (NPS) for permission to carry out the field work.

REFERENCES CITED

Alexander, E.B., 1974, Extractable iron in relation to soil age on terraces along the Truckee River, Nevada: Soil Sci. Soc. Amer. Proc., v. 38, p. 121-124.

Andrews, J.T. and Webber, P.J., 1964, A lichenometrical study of the northwestern margin of the Barnes Ice Cap: a geomorphological technique: Geographical Bull., Ottawa, no. 22, p. 80-104.

Benedict, J.B., 1968, Recent glacial history of an alpine area in the Colorado Front Range, U.S.A. II Dating the glacial deposits: Jour. of Glaciology, v. 7, no. 49, p. 77-87.

______________, 1973, Chronology of cirque glaciation, Colorado Front Range: Quaternary Research, v. 3, p. 584-599.

Beschel, R.E., 1957, A project to use lichens as indicators of climate and time: Arctic v. 10 p. 60.

______________, 1958, Lichenometrical studies in West Greenland: Arctic, v. 2, no. 4, p. 254.

Birkeland, P.W., 1973, Use of relative age-dating methods in a stratigraphic study of rock glacier deposits, Mt. Sopris, Colorado: Arctic and Alpine Research, v. 4, no. 4, p. 401-416.

Blackwelder, E., 1915, Post Cretaceous History of the Mountains of Central Western Wyoming: Jour. Geol., v. 23, pp. 97-117; 192-217; 307-340.

Coffin, D.E., 1963, A method for the determination of free iron in soils and clays: Can. Jour. Soil Sci., v. 43, p. 7-17.

Curry, R.R., 1969, Holocene climatic and glacial history of the central Sierra Nevada, Calif., in Schumm, S.A., and Bradley, W.C., eds., United States Contributions to Quaternary Research: Geological Society of America Paper 123, Boulder, Colorado, p. 1-47.

Fenneman, N.M., 1931, Physiography of Western U.S., v. 2, N.Y., McGraw-Hill, 534 p.

Kiver, E.P., 1974, Holocene glaciation in the Wallowa Mountains, Oregon, in Mahaney, W.C., ed., Quaternary Environments: Proceedings of a Symposium, Geog. Mongr. No. 5, York Univ., p. 169-195.

Mahaney, W.C., 1973, Neoglacial chronology in the Fourth of July Cirque, Colorado Front Range: Geol. Soc. Amer. Bull., v. 84, p. 161-170.

______________, 1975, Soils of post-Audubon age, Teton Glacier area, Wyoming: Arctic and Alpine Research, v. 7, no. 2, p. 141-153.

______________, 1978, Late-Quaternary stratigraphy and soils in the Wind River Mountains, western Wyoming: in Mahaney, W.C., ed.: Quaternary Soils, Norwich, U.K., Geoabstracts Ltd., p. 223-264.

______________, 1981, Paleoclimate reconstructed from paleosols: evidence from the Rocky Mountains and East Africa: in Mahaney, W.C., ed.: Quaternary Paleoclimate, Norwich, U.K., Geoabstracts Ltd., p. 227-247.

______________, 1982a, Correlation of Quaternary glacial and periglacial deposits on Mount Kenya, East Africa with the Rocky Mountains of western U.S.: in Abstracts Vol. I, XI INQUA Congress, Moscow, U.S.S.R., p. 208.

______________, 1982b, Superposed Neoglacial/late Pinedale tills, Summer Ice Lake - Titcomb Basin, Wind River Mountains, Wyoming: Boreas, submitted.

Mahaney, W.C. and Fahey, B.D., 1976, Quaternary Soil Stratigraphy of the Front Range, Colorado: in Mahaney, W.C., ed.: Quaternary Stratigraphy of North America, Stroudsburg, Pa., Dowden, Hutchinson and Ross, p. 319-352.

______________, 1980, Morphology, composition and age of a buried paleosol on Niwot Ridge, Front Range, Colorado, U.S.A.: Geoderma, v. 23, p. 209-218.

Mahaney, W.C., Fahey, B.D. and Lloyd, D.T., 1981, Late Quaternary glacial deposits, soils and chronology, Hell Roaring Valley, Mount Adams, Cascade Range, Washington: Arctic and Alpine Research, v. 13, no. 3, p. 339-356.

McKeague, J.A. and Day, J., 1966, Dithionite and oxalate extractable Fe and Al as aids in differentiating various classes of soils: Can. Jour. Soil Sci., v. 46, p. 13-22.

Nelson, R.L., 1954, Glacial geology of the Frying Pan River drainage, Colorado: Jour. Geology, v. 62, p. 325-343.

Oyama, M. and Takehara, H., 1970, Standard Soil Color Charts, Japan Research Council for Agriculture, Forestry and Fisheries,

Richmond, G.M., 1965, Type moraines of the Pinedale Glaciation: in Schultz, C.B. and Smith, H.T.U., eds.: Guidebook for Field Conference E., Lincoln, Nebr., Nebr. Acad. Sci. Ser. VII, INQUA Congr., p. 34-36.

______________, 1974, Geologic map of the Fremont Lake South quadrangle, Sublette County, Wyo., U.S. Geol. Survey Geol. Quad. Map GQ-1138.

Ruhe, R.V., 1956, Geomorphic surfaces and the nature of soils: Soil Sci., v. 82, p. 441-455.

Sharp, R.P., 1969, Semiquantitative differentiation of glacial moraines near Convict Lake, Sierra Nevada, Calif.: Jour. Geol., v. 77, p. 68-91.

U.S. Dept. Commerce, 1965, Climatography of the U.S., No. 86-42: Climate Summary of the U.S., (Supplement for 1951-1960), Wyoming, Washington, U.S. Govt. Printing Office, 77 p.

___________, 1970, Climatography of the U.S., No. 20-48, NOAA, Climatological Summary, Moran, Wyo.

Worl, R.G., 1968, Taconite in the Wind River Mountains, Sublette County, Wyo.: Preliminary Rept. No. 10, Laramie, Wyo.

DATING METHODS APPLICABLE TO LATE GLACIAL DEPOSITS OF THE LAKE AGASSIZ BASIN, MANITOBA

R.W. KLASSEN

ABSTRACT

A radiocarbon chronology based on nineteen dates from wood, shells, peat, gyttja and bone from sites within and adjacent to the Lake Agassiz basin in Manitoba suggests this lake began in southern Manitoba before 14,000 years ago and drained into the Tyrrell Sea about 8,000 years ago. Relative dating methods including roundnesses of beach pebbles, average rates of sedimentation and ice retreat each account for about one-half to two-thirds of the 6,000 year span suggested by the radiocarbon dates.

The discrepancies between the various durations inferred from each of the dating methods to some extent reflect the limitations of the methods. However, the relative dating methods do not account for the duration of several low water stages and stillstands of the glacier.

Consideration of the stratigraphic positions of the most reliable radiocarbon dates and minimal time intervals inferred from relative dating suggests the oldest radiocarbon date may be about 1000 years too old and that Lake Agassiz existed between about 13,000 and 8,000 years ago in Manitoba.

INTRODUCTION

The various chronologies proposed for glacial Lake Agassiz during the last several decades are based on radiocarbon dates from a variety of organic materials. The purpose of this paper is to appraise these radiocarbon chronologies of Lake Agassiz in Manitoba on the basis of the stratigraphy of dated sites in the vicinity of Assiniboine delta and by relative dating methods including sedimentation rates based on varve counts, rate of ice retreat and the degree of roundness of beach pebbles.

Relative dating methods provide a measure of the approximate duration of certain intervals of lake history and a means of appraising the radiocarbon chronologies. This approach offers a way of evaluating the reliability of the oldest finite dates obtained from organic detritus.

BACKGROUND

This paper focuses on published and unpublished studies dealing primarily with the chronology of glacial Lake Agassiz in Manitoba. Johnston (1946, p. 17) assigned ages to the main beaches in Manitoba by correlating with Antev's (1939) varve chronology for the Great Lakes region. He concluded that the oldest (Herman) beach formed in southern Manitoba about 20,000 years ago and that the youngest beaches in the northern part of the basin formed about 4,000 years ago. According to this chronology, the various stages of Lake Agassiz in Manitoba occurred over some 16,000 years.

The advent of radiocarbon dating in the late 1950's resulted in a substantial revision of Johnston's chronology. Elson (1957) cited five radiocarbon dates between about 12,400 and 8,000 years ago that spanned the so-called "Lake Agassiz 1-11" interval. These radiocarbon dates were included in a subsequent list of twenty-one radiocarbon dates that formed the basis of the most detailed chronologic history of Lake Agassiz in Manitoba to date (Elson, 1967, p. 88-94). Elson's chronology placed the duration of Lake Agassiz in Manitoba between "probably more than 12,400 years ago" and about 7,500 years ago. Two additional radiocarbon dates of about 12,800 years (Table 1, I-1682) and 8,500 years (Table 1, GSC-896) led Elson (1971, p. 289) to revise the previous chronology somewhat.

Later revisions to the chronology of the beginning of Lake Agassiz in Manitoba are included in reports by Klassen (1972, p. 553), Christiansen (1979, p. 926) and Teller and Fenton (1980, p. 32). Additional radiocarbon dates from the western margin of Lake Agassiz and the prairies to the west led Klassen (1972, p. 553) and Christiansen (1979, p. 926) to suggest that Assiniboine delta began forming between about 15,000 and 14,000 years ago. According to the latest proposal by Teller and Fenton (1980, p. 32) delta building occurred considerably later. They suggest the last advance of a glacier into North Dakota occurred about 12,000 years ago, followed by retreat to a position north of the delta by about 11,000 years ago. This time frame accommodates the established chronology of late Wisconsin events south of the International Boundary but requires the rejection of the oldest postglacial radiocarbon dates from southern Manitoba (Teller and Fenton, 1980, p. 30).

CURRENT STATE OF RADIOCARBON CHRONOLOGY

Radiocarbon dates on a variety of organic materials from postglacial sediments within and adjacent to the Lake Agassiz basin in Manitoba suggest Lake Agassiz began before 14,000 years ago and ended about 8,000 years ago. These radiocarbon dates included a new date of about 13,900 years (Ritchie, 1976, p. 1799) from the surface of Assiniboine delta, a date of about 8300 years (Lowdon *et al.*, 1977, p. 12) from a beach north of Lake Winnipeg formed during a final stage of Lake Agassiz and 5 new radiocarbon dates from postglacial sediments within or adjacent to the northern end of the Lake Agassiz basin (Figure 1, Table 1).

Several recent publications have raised the question of the validity of certain radiocarbon dates on organic detritus from limnic sediments (Karrow and Anderson, 1975; Nambudiri *et al.*, 1980; Mathewes and Westgate, 1980). The so-called "old carbon effect" (Mathewes and Westgate, 1980, p. 1460) is believed to have resulted in an age some 4000 years too old for a date in the 14,000 year range in eastern Canada (Karrow and Anderson, 1975, p. 1809) and in ages some 1400 years too old for dates in the 4000 year range in southern British Columbia. Teller and Fenton (1980, p. 31) rejected the oldest non-wood dates (Table 1) on the basis of conclusions drawn by Nambudiri *et al.* (1980) regarding radiocarbon dating of postglacial sediments from Lake Manitoba. The oldest material dated was thought to be contaminated by dead-carbon from "pre-Quaternary organic matter derived from lignites and shales" (Teller and Fenton, 1980, p. 31).

The stratigraphic positions and relative elevations of a number of radiocarbon dated sites in post-Assiniboine delta sediments are shown in Figure 2. The relative elevations of the dated sites shown in Figure 2 are chronologically significant for they relate to a sequence of events within a single fluvio-lacustrine system. The main events were the initial excavation of Assiniboine valley across the delta and the subsequent infilling and re-excavation of the valley. The post-depositional changes in relative elevations of the various sites must be minimal as they are within or adjacent to Assiniboine Valley which roughly parallels Johnston's (1946, p. 2) isobase 5 drawn from Lake Agassiz beaches.

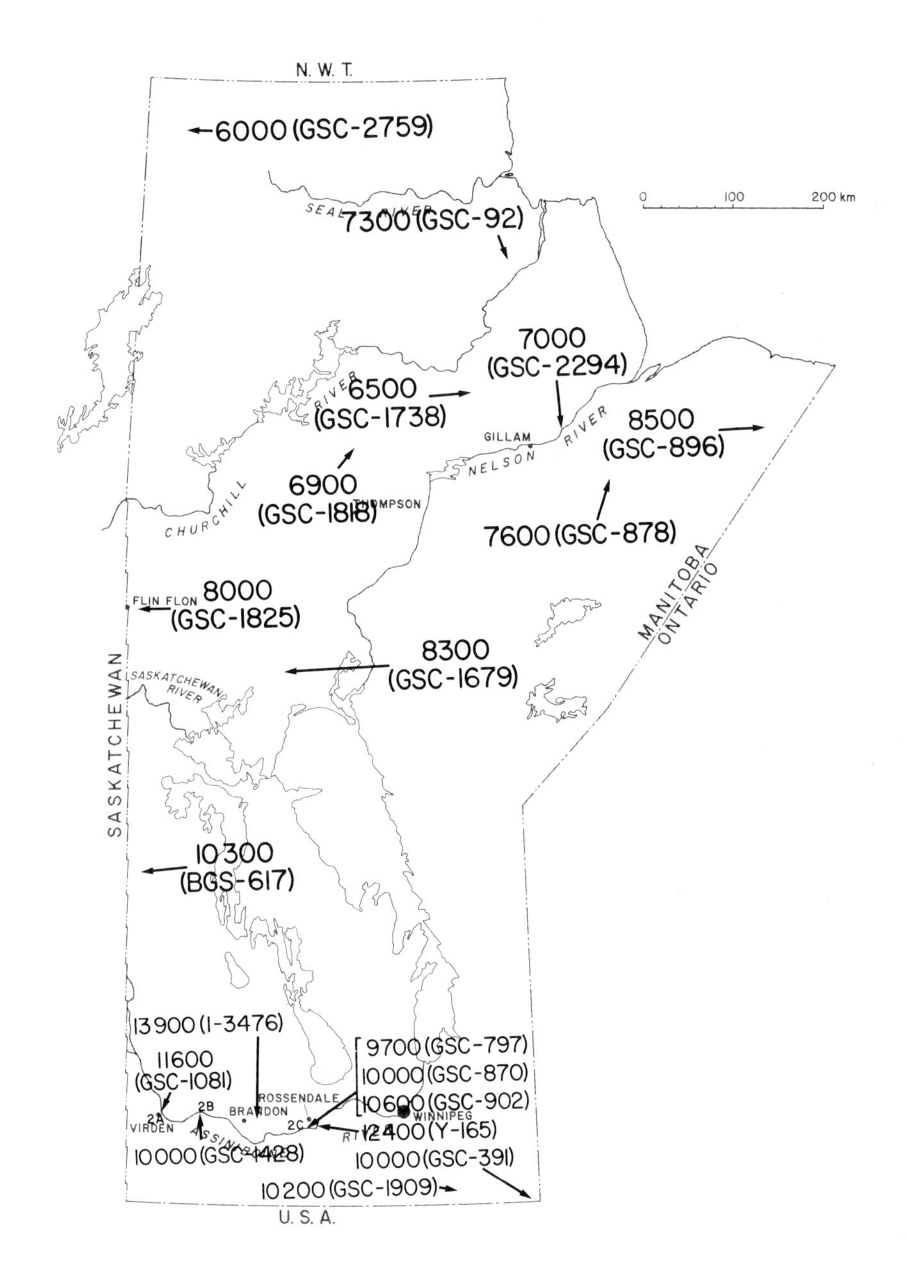

Figure 1 Locations of selected radiocarbon dates of late glacial and postglacial ages in Manitoba, and cross sections Figures 2a, 2b and 2c.

The oldest date of about 13,900 years BP (Table 1, No. 1) came from the highest site directly overlying delta sediments. This age is some 2000 years greater than the approximate ages of 12,000 years from peat and wood (Table 1, Nos. 2, 3 and 4) in post-delta, alluvial fill at sites at 18 m or more lower in elevation. The oldest date may well be too old because of contamination by dead carbon from the underlying sand that has a carbonate content of about 25 percent (Klassen, 1971, p. 253). Contamination from lignites and shale, as suggested by Teller and Fenton (1980, p. 31), appears unlikely as there is no evidence of the pre-Quaternary pollen (Ritchie, 1980, pers. comm.) that was found so abundantly in contaminated samples studied by Nambudiri *et al.* (1980, p. 124).

The two dates of about 12,000 years BP (Table 1, Nos. 2 and 3; Figure 1) were on peat within alluvial fill in a gully excavated into delta sediments. Contamination of the dated material is possible, though only to a lesser extent that in the material that yielded the

Table 1. Selected Radiocarbon Dates Relevant to the History of Lake Agassiz in Manitoba.

Years BP	Lab No.	Material	Lat.	Long.	References	Comments
Dates from deposits younger than main part of Assiniboine delta but older than Campbell beach						
1) 13,900±240	I-3476	organic detritus	*49°50'N	99°35'W	Ritchie, 1976, p. 1799	Brandon; dated sediment was from a postglacial lake within a swampy flat on the upper part of Assiniboine delta at *ca.* 375 m a.s.l.
2) 12,400±420	Y-165	peat	49°47'N	98°35'W	Preston *et al.*, 1955	Rossendale; alluvial fill *ca.* 4 m below surface of abandoned channel on delta at *ca.* 328 m a.s.l.
3) 12,100±160	BSC-1319	peat	49°47'N	98°35'W	Lowdon *et al.*, 1971, p. 282	Duplicate of Rossendale sample Y-165; *id.* by M. Kuc as peatmoss *(Scorpidium scorpioides)*
4) 11,600±430	GSC-1081	wood detritus	49°51'N	100°50'W	Lowdon *et al.*, 1971, p. 286	Virden; core from basal zone of Assiniboine alluvium (Klassen, 1975, p. 18 and 58) at *ca.* 18 m below surface at *ca.* 375 m a.s.l.
**********						
Dates from deposits related to Campbell beach						
5) 10,600±150	GSC-902	plant detritus	49°45'N	98°39'W	Lowdon and Blake, 1970, p. 65	Rossendale; fluvio-lacustrine sediment *ca.* 18 m below Campbell terrace at 320 m a.s.l.
6) 10,300±200	BGS-617	bone	52°11'N	101°26'W	Teller, 1980, p.6	Swan River, water worn bison bone fragment between upper and lower Campbell beach at *ca.* 344 m a.s.l.
7) 10,200±80	GSC-1909	organic detritus	49°06'N	96°14'W	Lowdon and Blake, 1976, p. 7	Sundown; *ca.* 8 m below surface of Campbell strandline at *ca.* 326 m a.s.l.
8) 10,000±150	GSC-870	wood	49°45'N	98°45'W	Lowdon and Blake, 1970, p. 65	Rossendale; (SW side basin) alluvium, *ca.* 8 m below surface of Campbell terrace in Assiniboine valley across delta at *ca.* 320 m a.s.l.
9) 10,000±280	GSC-1428	wood	49°54'N	100°18'W	Lowdon and Blake, 1973, p. 22	Alexander; Assiniboine alluvium *ca.* 9 m depth below surface at *ca.* 354 m (Klassen, 1975, p. 58)
10) 9,900±160	GSC-391	wood	49°00'N	95°14'W	Lowdon *et al.*, 1967, p. 10	Buffalo Point; SE *ca.* 2 m below Campbell terrace surface at *ca.* 323 m a.s.l. lower Campbell beach
11) 9,700±140	GSC-797	wood	49°45'N	98°39'W	Lowdon and Blake, 1970, p. 65	Rossendale; alluvium, *ca.* 4 m below surface of Campbell terrace at *ca.*

							320 m a.s.l. in Assiniboine valley across delta
				**********			
			Date from deposits related to late stage in northern Manitoba				
12)	8,310±180	GSC-1679	freshwater shell	54°10'N	98°50'W	Lowdon *et al.*, 1977, p. 12	beach ridge *ca.* 245 m a.s.l., north of Lake Winnipeg
				**********			
			Dates from post Lake Agassiz deposits in northern Manitoba				
13)	8,530±220	GSC-896	marine shells	54°29'N	90°24'W	Lowdon and Blake, 1970, p. 64	marine beach at *ca.* 125 m a.s.l., *ca.* 10 m below marine limit
14)	7,970±150	GSC-1825	gyttja	54°45'N	101°41'W	Lowdon and Blake, 1975, p. 14	Flin Flon; basal layer in small lake at *ca.* 305 m a.s.l.
15)	7,570±140	GSC-878	marine shell	56°02'N	93°17'W	Lowdon and Blake, 1970, p. 64	Hayes River; shells in living position in silty clay at *ca.* 114 m a.s.l.
16)	7,030±170	GSC-2294	marine shell	56°31'N	94°05'W	unpublished	Gillam; marine clay at 90 m a.s.l. *ca.* 30 m below the highest Tyrrell Sea beaches
17)	6,920±150	GSC-1818	organic debris	56°21'N	97°58'W	Lowdon and Blake, 1975, p. 14	Thompson; Settee moraine, basal organic layer in Kettle Lake *ca.* 305 m a.s.l.
18)	6,490±170	GSC-1738	peat	56°52'N	95°47'W	Lowdon *et al.*, 1977, p. 12	Recluse Lake; basal peat in permanently frozen bog over till
19)	5,990±80	GSC-2759	peat	59°28'N	101°13'W	Lowdon and Blake, 1979, p. 15	Moorby Lake; basal peat from permanently frozen bog

*Location is corrected from 49°35'N 99°15'W given originally by Ritchie (1976, p. 1795)

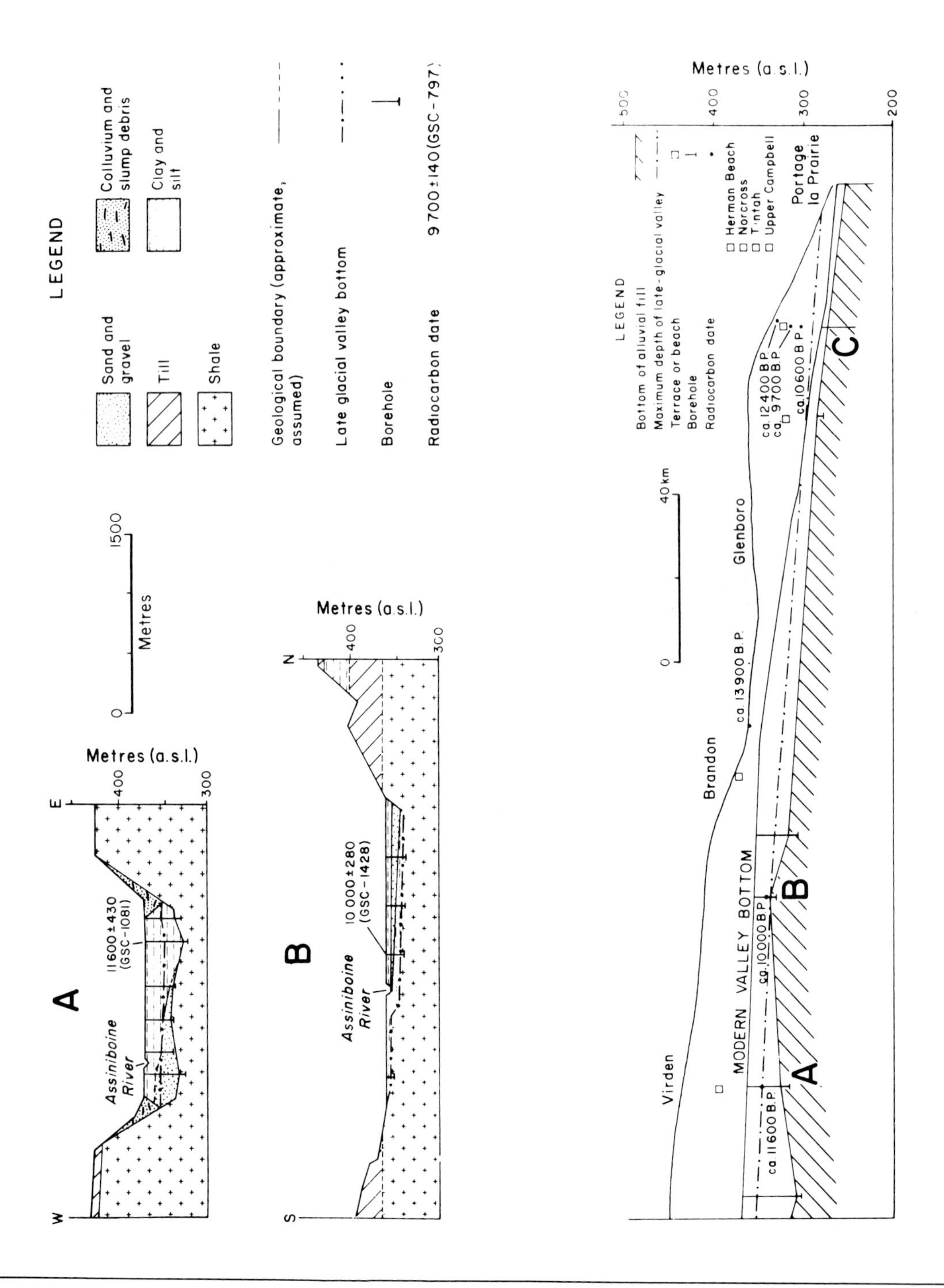

oldest date. However, the wood date of 11,600 years (Table 1, No. 4; Figure 1) from within Assiniboine alluvium some 40 km upstream from the delta, is another minimal post-delta date as shown by its stratigraphic position and elevation relative to the delta surface (Figures 2 and 2A). This date is corroborated by the 10,000 year wood date (Table 1, No. 9; Figure 1) at a stratigraphically higher position within Assiniboine alluvium further downstream (Figure 2 and 2B).

A series of radiocarbon dates from wood and organic detritus (Figure 2C) date the fluctuations of Lake Agassiz associated with a rising phase between about 10,800 and 9,700 years BP. Stream and lake deposits of an earlier valley fill form a terrace within Assiniboine valley where it crosses the outer part of the delta. The fossils

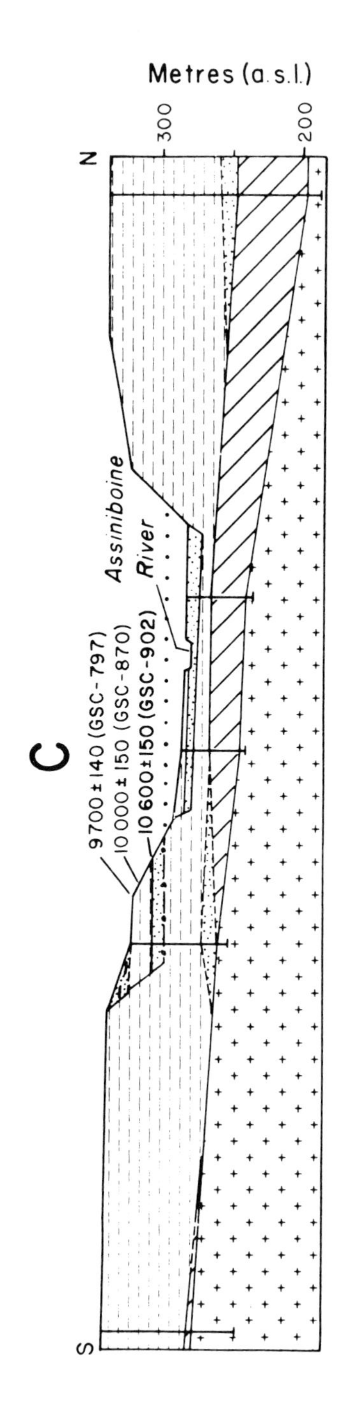

Figure 2. Long profile (at top) and cross sections (A, B, C) of Assiniboine valley between Virden and Rossendale, showing elevations and stratigraphy of radiocarbon dated sites (modified after Klassen, 1972).

(ostracods and pollen) and dated material within the fill led Klassen and Elson (1972, p. 10 and 11) to propose the following sequence of events:

> "The succession seen in this section records three intervals of inundation of the Assiniboine valley by Lake Agassiz. During the earliest interval (Unit 1) the lake was at least at the Campbell level 1,050 feet (320 m) a.s.l. and a relatively cool climate (max. air temp. 26°C) prevailed. Shortly before about 10,600 years BP the lake fell below 980 feet (300 m) a.s.l. and the climate became warmer (Unit 2, max. air temp. *ca.* 31°C). This low water phase probably corresponds to an unconformity radiocarbon dated in North Dakota and Minnesota between 9,900 and 11,000 years

BP. Before 10,000 years BP the climate again became cooler (Units 3 and 4, max. air temp. *ca.* 26°C) and the lake rose to about the Campbell level. About 10,000 years BP the climate was warmer (max. air temp. 35°C) and the lake was below the 1,010 foot (310 m) level (Unit 5). A variety of plants *(Picea, Pinus, Betula, Artemesia, Myriophylum;* identified by R.J. Mott, Geological Survey of Canada) grew adjacent to the stream that occupied the valley. A warmer and drier interval (Unit 6) and a moister interval (Unit 7) preceded the final inundation of the valley by Lake Agassiz (Unit 6). After *ca.* 9,700 years BP the climate became cooler (max. air temp. 26°C) and Lake Agassiz rose to at least 1,070 feet (325 m) a.s.l."

The Campbell beach that can be traced along much of the western margin of the basin in southern Manitoba (Figure 3) formed during those intervals. Some 200 km north and northeast of the northernmost part of the Campbell beach in Manitoba (Figure 3) radiocarbon dates from bog bottoms and shells (Table 1, Figure 1) are about 8,000 years old and 7,000 to 6,500 years old near the Hudson Bay Lowland. With the exception of a marine shell date of about 8,500 years (Table 1, No. 13) from the Lowland, the progressively younger dates to the north reflect the direction of deglaciation, and suggest that contamination by old carbon is not a problem here, even though the substrata in places has a significant carbonate content. For example, the fine fraction of the till beneath the 6500 year old peat samples at Recluse Lake (Table 1, No. 18) has 20 percent carbonate (Klassen, in press).

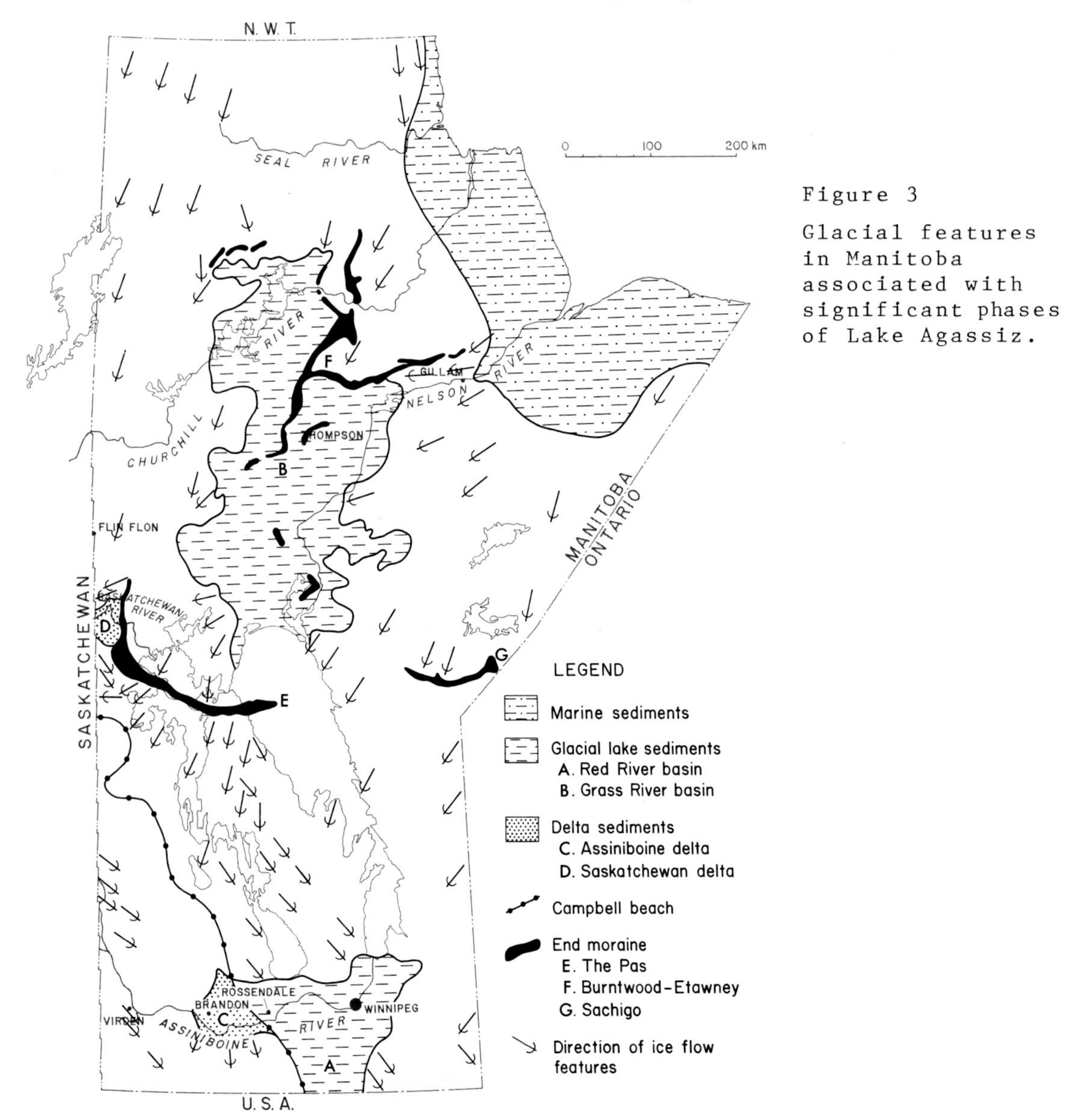

Figure 3

Glacial features in Manitoba associated with significant phases of Lake Agassiz.

GLACIAL LAKE SEDIMENTS AND CHRONOLOGY

Glaciolacustrine silt and clay are widely distributed in the northern (Grass River basin) and southern (Red River basin) parts of the region covered by Lake Agassiz in Manitoba (Figure 3). Elsewhere within the region, glaciolacustrine deposits are thin and patchy and extensive areas consist of eroded till or bedrock. Antevs (1931, p. 47-50) calculated the rate of ice recession in the Grass River basin from studies of varved clay. However, studies of the glaciolacustrine sediments in the Red River basin have focussed primarily on distribution and stratigraphy (Elson, 1967, p. 45-50; Fenton and Anderson, 1971; McPherson *et al.*, 1971; Teller, 1976; Teller and Fenton, 1980), and their chronologic implications have not been considered. There are various reasons for this, such as the general abandonment of varve counting as a dating method with the advent of radiocarbon dating, the lack of good exposures, and the discontinuity of sediments north of the Red River basin.

The thick, "laminated" to massive clay and silt (Teller, 1976, p. 31-36) that constitutes the bulk of the offshore sediment in the Red River basin apparently reflects a considerable span of time. A rough estimate of this time can be made on the basis of the average sedimentation rate indicated by Antev's (1931) varve counts in the Grass River basin. The average thickness of varves measured across some 225 km of the Grass River basin between Wekusko Lake and Split Lake (Antevs, 1931, p. 57-67) is 1.5 cm. Thicknesses average 1 cm in the distal (western) part of the basin and 2 cm in the proximal (eastern) part. The thickness of offshore sediment in the Red River basin averages about 40 m thick (Teller, 1976, p. 34-35) suggesting 2600 years for deposition. Furthermore, a significant hiatus occurred between the lower predominantly clayey sediments averaging some 30 m in thickness and the upper 10 m of predominantly silty sediments (Elson, 1967, p. 45-50; McPherson *et al.*, 1971, p. 280). The latter likely correlate with the post-delta sediments within Assiniboine valley (Figure 2C).

The average thickness of varved clay at 17 sites measured by Antevs (1931, p. 51) across the Grass River basin was 2.5 m. Borings in the vicinity of Thompson in the western part of the basin penetrated clays up to 9 m thick, but much ground ice was included (Klassen, 1976, p. 36). According to the average thicknesses obtained from Antevs' measurements, the sediment represents about 160 years of deposition.

Rate of Ice Recession

The study of the varved sediments of the Grass River basin by Antevs (1931, p. 47-70) provided basic data for calculating the average rate of ice retreat in this region. The average rate of retreat, based on Antevs (1931, p. 50) figures for the annual retreat rate between 40 sites covering some 225 km, was roughly 300 m per year. This figure compares well with the rate of 200 to 300 m per year proposed by Ritchie (1976, p. 1809) for northward migration of the boreal forest, following deglaciation of the Canadian prairies, and also with the rate of 275 m per year, during final stages of deglaciation of this region, proposed by Christiansen (1979, p. 934).

An average retreat of 250 m per year between the northern margin of Assiniboine delta and the western margin of the Hudson Bay Lowland is likely a maximum figure, because the rate during the early stages of deglaciation was probably somewhat slower than it was during the final stages. Christiansen (1979, p. 934) suggested an initial rate of 60 m per year and final rate of 275 m per year during deglaciation of southern Saskatchewan. Continuous deglaciation, not considering halts in retreat, or ice readvances, across the Lake Agassiz basin, at a rate of 250 m per year would require some 3000 years.

Roundness of Beach Pebbles

Some 20 abandoned beaches have been identified in Manitoba and related to the broad chronologic framework of Lake Agassiz history (Johnston, 1946; Elson, 1967 and 1971). Elson's (1971) study of beaches north and south of the International Boundary demonstrated that the degree of roundness of the beach pebbles reflected the duration of various water planes and provided a means of establishing the relative ages of the beaches.

The Campbell beach (Figure 3) represents the longest period of stability of Lake Agassiz (Elson, 1971, p. 288). It is interesting to note that the pebbles from this beach are rounder than those from present day Lake Manitoba beaches. The pebbles from the other beaches are significantly less rounded, particularly those from the youngest beaches which exhibit essentially the same degree of roundness as pebbles from the till source.

Elson (1971, p. 290) concluded that roundness studies "can aid in apportioning the time spanned by radiocarbon dates". He calculated that the beaches below and including the lower Campbell formed during a 1400 year interval. The results of the roundness studies along with radiocarbon dates were the basis for the Lake Agassiz chronology shown in Figure 4.

CHRONOLOGIC IMPLICATIONS OF REGIONAL ICE FLOW PATTERNS AND ICE-MARGINAL MORAINES

The manner in which ice lobes were deployed and the nature of end moraines within and adjacent to the Lake Agassiz basin have implications concerning the history of Lake Agassiz. Ice-flow markings usually provide a means of determining the relative sequence of at least the last two flow directions. The marginal moraines, in turn may provide a means of determining whether or not the fluctuations of certain ice lobes were synchronous and indicate the manner in which the lobes retreated.

A prominent system of interlobate kame moraines in the northern part of the Lake Agassiz basin (Figure 3) shows synchronous deployment of two distinct ice lobes in this region (Klassen, in press). They reflect flows from ice centers east (Labradorian) and north (Keewatin) of this region as proposed by Tyrrell (1913). Shilts *et al.* (1979, p. 537) proposed that outflow from these centers during the last glacial maximum was synchronous and that they met somewhere between Nelson River and Churchill.

The pattern of ice flow markings in the southern part of the Lake Agassiz basin (Figure 3) reflects the coalescence of Labradorian and Keewatin ice. However, within the west-central part of the basin, the southeast markings (Keewatin) are modified or masked by the southwest (Labradorian) ones that also cross The Pas moraine. The superimposed features reflect a readvance of Labradorian ice following an interval of general retreat of both Labradorian and Keewatin ice from the southern part of the basin. Additional evidence for a major readvance of Labradorian ice, following that general retreat, is a clayey till overlying the regional silty till in the area between The Pas moraine and Assiniboine delta (Klassen, 1965, p. 5).

This readvance predates the Campbell beach for that beach is continuous across the area overridden by the readvance (Figure 3). Most likely the rising phase of Lake Agassiz, as recorded in the Assiniboine valley fill, reflects this readvance. A recently reported (Nielson *et al.*, 1981, p. A-43) readvance of Labradorian ice to the Hartman moraine in western Ontario (Hughes, 1965) after 11,400 years ago (fresh-water shell date GSC-3114, unpublished) probably correlates with the readvance in Manitoba.

IMPLICATIONS OF ABSOLUTE AND RELATIVE DATING METHODS FOR LAKE AGASSIZ CHRONOLOGY

Radiocarbon chronology suggests that the earliest stage of Lake Agassiz in Manitoba began about 14,000 years ago and that its final

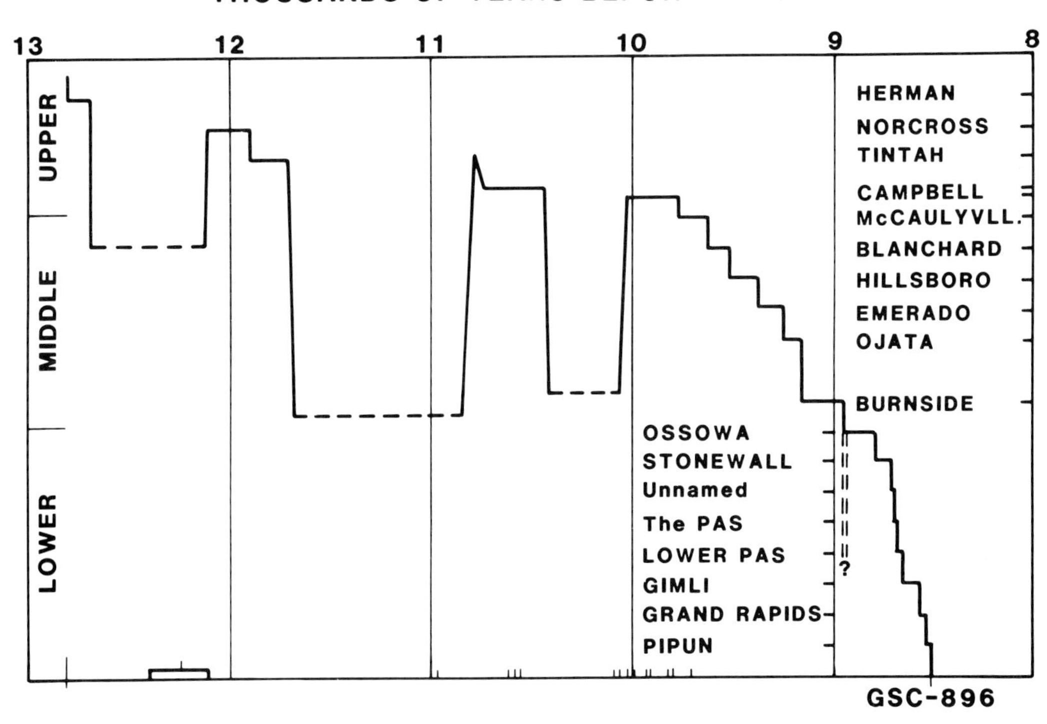

Figure 4 Duration of various levels of Lake Agassiz inferred from radiocarbon dates, the positions of which are indicated by short lines along the abscissa, and beach pebble roundnesses (after Elson, 1971).

drainage into the Tyrrell sea occurred about 8,000 years ago. Uncertainties concerning the validity of the oldest dates, discussed in preceding sections, suggest the lake may have begun later but there is little reason to question the time inferred for its final drainage. Figure 5 gives a comparison of the time intervals deduced by means of various dating methods. It shows the duration of the lake based on radiocarbon dating as a continuous line representing 6,000 years, and time intervals deduced by the relative dating methods are shown at inferred positions within that absolute time frame. Sums of the total years obtained by each dating method are shown along the right ordinate. The 4,500 year sum deduced from ice retreat includes the 3000 years calculated for continuous retreat across the basin north of Assiniboine delta and the 1500 years allocated to a major readvance from just north of Assiniboine delta.

The differences in the sums obtained for each of the dating methods in part reflect the degrees of inaccuracies of the methods. A significant consideration however, is that the time recorded by the relative dating methods does not include the duration of low water levels and the stillstand of the ice that formed The Pas moraine prior to a major readvance.

Conclusions from this appraisal of absolute and relative methods of dating Lake Agassiz deposits in Manitoba are as follows:

1. The stratigraphic positions and distribution of the most reliable radiocarbon dated sites establish that Lake Agassiz ended about 8,000 years ago, and began at least 12,000 years ago in Manitoba. The lake lasted at least 4,000 years.

2. The sedimentation rate and pebble roundness methods that account for some 3000 years of lake history fail to include

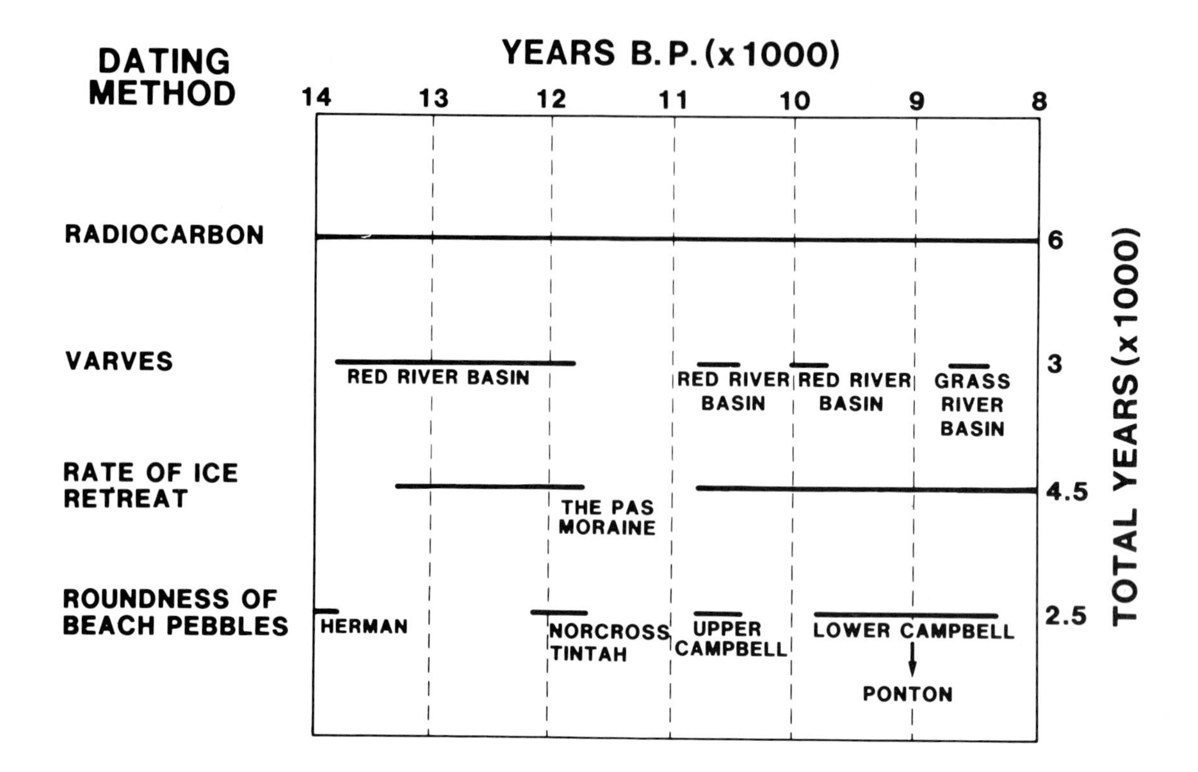

Figure 5 Duration of Lake Agassiz in Manitoba according to radiocarbon dates, compared to the duration of various stages deduced by relative dating methods.

the length of the low water stages. The 4,500 years deduced from the estimated rate of ice retreat suggests the low water stages may represent some 1500 years of lake history.

3. The radiocarbon chronology based on available dates suggesting the duration of Lake Agassiz was from roughly 14,000 to 8,000 years ago may require some modification based on the possible contamination of the oldest dates and the somewhat shorter duration inferred from relative dating methods. A time frame between about 13,000 and 8,000 years BP is proposed as a realistic accommodation of the time spans arrived at by the dating methods discussed in this paper.

REFERENCES CITED

Antevs, Ernst, 1931, Late-glacial correlations and ice recession in Manitoba: Geological Survey of Canada, Memoir 168, 76 p.

______________, 1939, Late Quaternary upwarpings of north-eastern North America: Journal of Geology, v. 47, no. 7, p. 707-720.

Christiansen, E.A., 1979, The Wisconsinan deglaciation of southern Saskatchewan and adjacent areas: Canadian Journal of Earth Sciences, v. 16, no. 4, p. 913-938.

Elson, J.A., 1957, Lake Agassiz and the Mankato-Valders problem: Science, v. 126, no. 3281, p. 999-1002.

____________, 1967, Geology of glacial Lake Agassiz, in Mayer Oakes, W.J., ed., Life, Land and Water, Proceedings of the 1965 Conference on Environmental Studies of the Glacial Lake Agassiz region: Winnipeg, University of Manitoba Press, 414 p.

____________, 1971, Roundness of glacial Lake Agassiz beach pebbles, in Turnock, A.D., ed., Geoscience Studies in Manitoba: Geological Association of Canada, Special Paper no. 9, p. 285-291.

Fenton, M.M. and Anderson, D.T., 1971, Pleistocene stratigraphy of the Portage la Prairie area, Manitoba, in Turnock, A.D., ed., Geoscience Studies in Manitoba: Geological Association of Canada, Special Paper no. 9, p. 271-276.

Johnston, W.A., 1946, Glacial Lake Agassiz, with special reference to the mode of deformation of the beaches: Geological Survey of Canada, Bulletin 7, 20 p.

Karrow, P.F. and Anderson, T.W., 1975, Palynological study of lake sediment profiles from southwestern New Brunswick: Discussion: Canadian Journal of Earth Sciences, v. 12, p. 1808-1812.

Klassen, R.W., 1965, Surficial geology of the Waterhen-Grand Rapids area, Manitoba: Geological Survey of Canada, Paper 66-36, 6 p.

______________, 1971, Nature, thickness and subsurface stratigraphy of the drift in southwestern Manitoba: in Turnock, A.C., ed., Geoscience Studies in Manitoba, Geological Association of Canada, Special Paper no. 9, p. 235-261.

______________, 1972, Wisconsin events and the Assiniboine and Qu'Appelle valleys of Manitoba and Saskatchewan: Canadian Journal of Earth Sciences, v. 9, no. 5, p. 544-560.

______________, 1976, Landforms and surface materials at selected sites in a part of the Shield north-central Manitoba, with appendix of testhole data and soil index properties by Jean Veillette: Geological Survey of Canada, Paper 75-19, 41 p.

______________, in press, Surficial geology of north-central Manitoba: Geological Survey of Canada, Memoir.

Klassen, R.W. and Elson, J.A., 1972, Southwestern Manitoba: in Quaternary geology and geomorphology between Winnipeg and the Rocky Mountains, Guidebook Field Excursion C-22, International Geologic Congress, p. 3-23.

Lowden, J.A. and Blake, W. Jr., 1970, Geological Survey of Canada radiocarbon dates IX: Geological Survey of Canada, Paper 70-2, Part B, p. 46-86.

___________, 1973, Geological Survey of Canada radiocarbon dates XIII: Geological Survey of Canada, Paper 73-7, 61 p.

___________, 1975, Geological Survey of Canada radiocarbon dates XV: Geological Survey of Canada Paper 75-7, 32 p.

___________, 1976, Geological Survey of Canada radiocarbon dates XVI: Geological Survey of Canada, Paper 76-7, 21 p.

__________, 1979, Geological Survey of Canada radiocarbon dates XIX: Geological Survey of Canada, Paper 79-7, 57 p.

Lowden, J.A., Fyles, J.G., Blake, W., Jr., 1967, Geological Survey of Canada radiocarbon dates: Geological Survey of Canada, Paper 67-2, Part B, 45 p.

Lowden, J.A., Robertson, I.M. and Blake, W., Jr., 1971, Geological Survey of Canada radiocarbon dates XI: Geological Survey of Canada, Paper 71-7, p. 255-324.

__________, 1977, Geological Survey of Canada radiocarbon dates XVII: Geological Survey of Canada, Paper 77-7, 25 p.

Mathewes, R.W. and Westgate, J.A., 1980, Bridge River Tephra: revised distribution and significance for detecting old carbon errors in radiocarbon dates of limnic sediments in southern British Columbia: Canadian Journal of Earth Sciences, v. 17, p. 1454-1461.

McPherson, R.A., Leith, E.I. and Anderson, D.T., 1971, Pleistocene stratigraphy of a portion of southeastern Manitoba: in Geoscience Studies in Manitoba, Geological Association of Canada, Special Paper, No. 9, p. 277-283.

Nambudiri, E.M.V., Teller, J.T. and Last, W.M., 1980, Pre-Quaternary micro-fossils - A guide to errors in radiocarbon dating: Geology, v. 8, p. 123-126.

Nielson, E., McKillop, W.B. and McCoy, J.P., 1981, Age of the Hartman moraine and the Campbell level of Lake Agassiz, northwestern Ontario: Abstracts, Geological Association of Canada, Annual Meeting, the University of Calgary.

Preston, R.S., Person, E. and Deevey, E.J., 1955, Yale natural radiocarbon measurements 11: Science, v. 122, p. 954-960.

Ritchie, J.C., 1976, The late-Quaternary vegetational history of the Western Interior of Canada: Canadian Journal of Botany, v. 54, no. 15, p. 1793-1818.

Shilts, W.W., Cunningham, C.M. and Kaszycki, C.A., 1979, Keewatin ice sheet - Re-evaluation of the traditional concept of the Laurentide ice sheet: Geology, v. 7, p. 537-541.

Teller, J.T., 1976, Lake Agassiz deposits in the main offshore basin in southern Manitoba: Canadian Journal of Earth Sciences, v. 13, p. 27-43.

__________, 1980, Radiocarbon dates in Manitoba: Mineral Resources Division, Geological Report GR80-4, Manitoba Department of Energy and Mines, 61 p.

Teller, J.T. and Fenton, M.M., 1980, Late Wisconsinan glacial stratigraphy and history of southeastern Manitoba: Canadian Journal of Earth Sciences, v. 17, no. 1, p. 19-35.

Tyrrell, J.B., 1913, Hudson Bay exploring expedition, 1912: Ontario Bureau of Mines, 22nd Annual Report, p. 161-209.

THE QUATERNARY SUCCESSION IN THE RIO BLANCO BASIN, CORDON DEL PLATA, MENDOZA PROVINCE, ARGENTINA: AN APPLICATION OF MULTIPLE RELATIVE-DATING TECHNIQUES

W.J. WAYNE

ABSTRACT

In the basin of Río Blanco, which flows eastward from the Cordón del Plata west of Mendoza, Argentina, moraines and fossil rock glaciers record several periods of activity during and since the last major glaciation, the Vallectios (=Wisconsinan). Farther down the valley than the moraines of Vallecitos age are three older tills, one of which is exposed almost as far downstream as the junction with Río Mendoza, more than 17 km beyond the outer Vallecitos moraines.

No vegetation borders the modern glaciers and rock glaciers in this dry part of the Central Andes, and efforts to obtain datable organic material proved fruitless. A lense of coarse volcanic ash that overlies one of the tills was dated by zircon fission-track methods as between 100,000 and 200,000 years old.

Relative dating (RD) techniques were used, therefore, in order to try to develop a stratigraphic order for the deposits. Granites are rare in the Río Blanco basin and, although the quartzites and rhyolites that dominate the area weather, few effects of weathering on these boulders could be quantified. For deposits of Vallecitos age and younger, the most consistent data came from 1) morainal and rock glacier position and overlap, 2) sharpness of topography, 3) loess thickness, 4) soil profile characteristics, 5) vegetation cover, and 6) lichens. For the older (pre-last glaciation) deposits, data for comparison was provided by: 1) topographic position, 2) stratigraphic sequence, 3) soil profile characteristics and development, and 4) boulder weathering.

INTRODUCTION

The Central Andes between Santiago, Chile and Mendoza, Argentina, consist of 3 major parts, the Cordillera Principal, Cordillera Frontal and Precordillera and includes Mt. Aconcagua, at approximately 7000 m, the highest peak in the Western Hemisphere. The piedmont plains on each side of the Andean Chain are arid, and the well developed agricultural economy depends heavily on snow melt and glacier meltwater runoff. Glaciers are the Andean type (Polanski, 1954) which have snowfields feeding small areas of open glacier above debris-covered ice; in many, the terminal part is a long ice-cored rock glacier tongue. Farther down the valleys are inactive and fossil rock glacier deposits. For the past century, most of the glaciers of the dry Central Andes have been shrinking. Thermokarst pits are common in the debris-covered segments of many, and some rock glacier termini have become inactive.

The Cordón del Plata is one of the ranges of the Cordillera Frontal in the northwestern part of the province of Mendoza, Argentina (Figure 1). From its highest peak, Cerro del Plata (6100 m), it extends northward through Cerro Blanco (5490 m) and southwestward to the lower Cordón de Santa Clara. The crest of the range forms the divide between Río Tupungato, which flows northward to enter Río Mendoza at Punta de

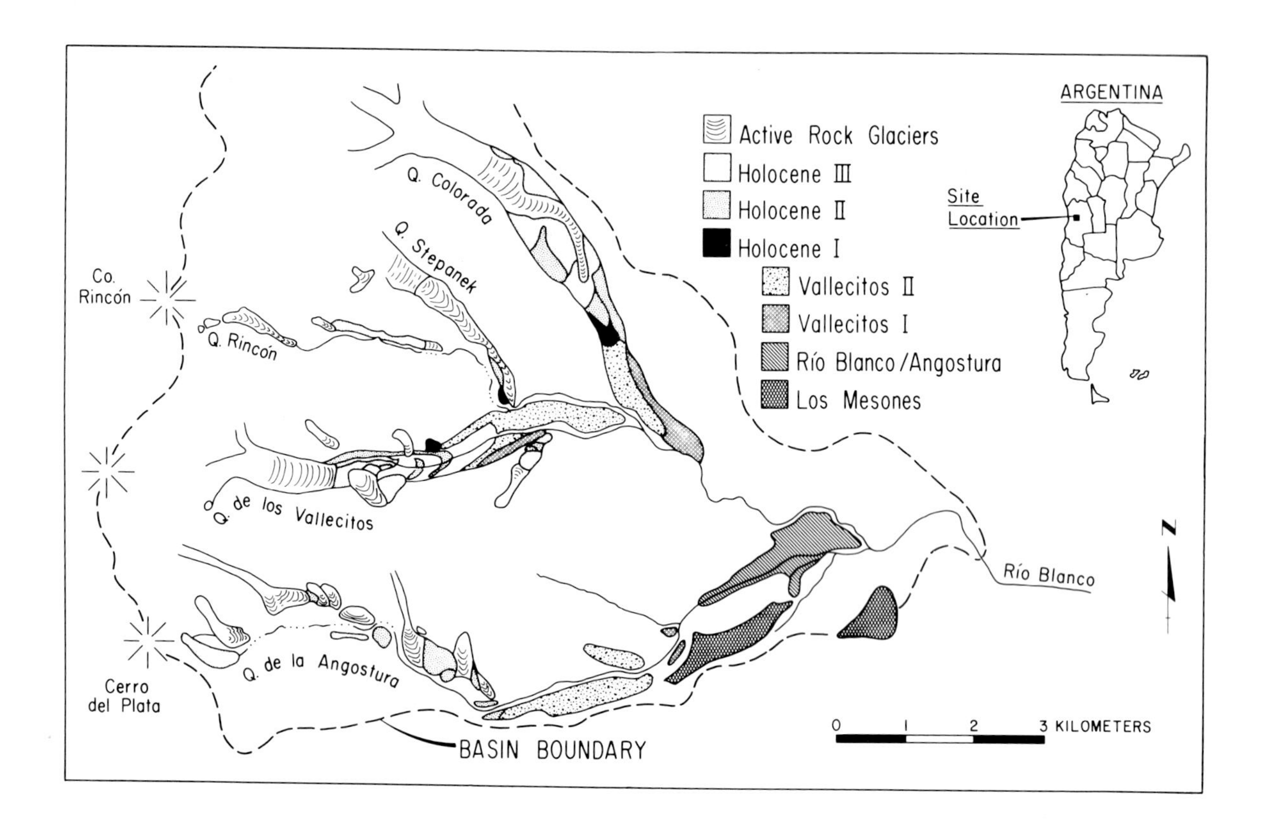

Figure 1 Map showing rock glaciers and glacial deposits of the upper Río Blanco basin, Cordón del Plata, Mendoza Province, Argentina.

Vacas and the eastward-flowing tributaries of both Río Mendoza and Río Tunuyán. Two Río Blancos drain the northern part of the Cordón del Plata; one flows northward and the other eastward.

The area selected for this investigation was Quebrada de los Vallecitos, a relatively accessible area of the Cordón del Plata, where glacial moraines as well as a large number of active, inactive, and fossil rock glaciers could be studied. The study was continued into Quebrada de la Angostura drainage which joins that from Q. de los Vallecitos to form Río Blanco (Figure 1), a major tributary of Río Mendoza. An area of 85 km^2 is drained by these two streams above their junction, which is the start of Río Blanco. These sub-basins lie between 33°02' and 32°52'S lat. and run from 67°25' to 67°26'E long.

The climate of this part of Argentina is continental, with relatively low precipitation (400-600 mm) and with large diurnal temperature changes. Incomplete records for a little more than 2 years from the meteorological station at Vallecitos (2470 m) suggest a mean annual temperature of approximately 5°C and 450 mm precipitation. An 8-year record at a station near the south edge of the area, Estancion Las Aguaditas (2225 m), indicates a mean annual temperature of 7.6°C and a precipitation of 294 mm (Estrella *et al.*, 1980). The altitude at which 0°C is the mean annual temperature should be 3400 m in the basin of Río Blanco, based on a lapse rate of 0.6°C per 100 m.

This study was undertaken in an effort to determine whether the fossil rock glaciers in the Central Andes had been active continuously or only spasmodically since disappearance of the last glaciers about 12,000 years ago. If their activity had been spasmodic it might be possible to determine the frequency and timing of advance. A second purpose was to work out criteria by which active and inactive rock glaciers and debris-covered glaciers, all of which contain ice that represents water in storage for the irrigated Mendozan plains, can be distinguished reliably from fossil rock glaciers by airphoto study. The project was part of a cooperative program between the National Science Foundation (U.S.) and the Instituto Argentino de Nivología y Glaciología (IANIGLA), a part of the Consejo Nacional de Investigaciones Científicas y Técnicas (CONICET).

GENERAL GEOLOGY

Between Co. del Plata and Co. Blanco, the Cordón del Plata is composed of quartzites and related metasediments; rocks of volcanic origin, primarily rhyolites and andesites; and a few plutonic rocks. Within these two sub-basins, the southern part is composed almost wholly of dark-colored quartzites, but the northern part is dominated by reddish-brown rhyolites. Caminos (1965) and Polanski (1972) mapped the spine of the Cordón del Plata as the El Plata Formation, a Paleozoic unit more than 7000 m thick. Quartzitic rocks of the El Plata Formation dominate the slopes of Quebrada de la Angostura, Q. de los Vallecitos, and Q. Rincón, and make up some of the southwest side of Q. Stepanek. Along the southwest side of Q. Stepanek and to the northeast in Q. Colorada, the metasedimentary rocks of the El Plata Formation give way to rhyolitic tuffs and andesitic breccias, which are part of the Variscan Volcanic Association (Caminos, 1965, p. 370-378).

Plutonic rocks are not abundant in the basin. Coarse-grained biotite granite of the stock of Cuchilla de las Minas (Caminos, 1965) crops out between 3400 and 3500 m in Q. de la Angostura, and the distribution of erratic boulders in one moraine suggests that it is buried beneath Pleistocene sediments in Q. Stepanek. Deeply weathered granitic rocks also crop out on the slopes near the junction of Vallecitos and Angostura Creeks. The only other plutonic rocks recognized are medium- and coarse-grained diabasic gabbros, which occur as dikes.

GEOMORPHIC FRAMEWORK

From Co. del Plata northward, most of the crest of Cordón del Plata is an arete punctuated by a few horns. The cordon widens at Co. del Plata, though, and includes broad sloping surfaces of cryoplanation below which glaciers have gouged the valleys. All the valleys show clear evidence of having been enlarged and scoured by glacial ice, but frost shattering and avalanching have cut channels through the once-smooth valley walls, and the U-shape characteristic of the troughs carved by alpine glaciers is becoming less obvious. Nevertheless, polished and striated rock surfaces can be seen in some places between avalanche chutes and cones and on bedrock surfaces in the valleys above 3500 m. The cirques that head along the main ridge of the range contain small uncovered ice glaciers. The lower parts of the main valleys of Angostura and of Vallecitos are largely filled with moraines deposited by moving ice, although small rock glaciers have formed along the north wall in the upper valleys since the disappearance of the glacier. In the other two valleys, Stepanek and Colorada, the bulk of the sediments consists of active, inactive, and fossil rock-glacier debris. Below about 2700 m, the topography is dominantly produced by running water, but both glacial and fluvial sediments are exposed along the valley and channel walls.

Both Vallecitos and Angostura creeks carry meltwater from winter precipitation and the small glaciers and rock glacier in the glacially widened upper parts of their valleys. Angostura Creek flows through Holocene outwash bordered and overlain by rock glaciers on the north side and loose talus on the south to 3500 m, where it falls through a granite outcrop. Between 3500 and 2700 m it flows down a steep, V-shaped trench between a massive moraine and the bedrock valley side. The glacially scoured bedrock walls through this reach of the valley have been modified by erosion and mass wasting. About 3 km downstream from the granite falls the stream turns northward through a narrow gorge about 1000 m long. The valley is little more than 100 m wide where the stream passes through this "narrows". At the end of the constrictions, it turns northeastward again and the valley widens from 75 m to 300 m in a distance of 1 km. It widens to 550 m at the junction with Vallecitos Creek, but constricts abruptly to a width of only 200 m where the two creeks join to become Río Blanco. For a little more than 100 m the stream passes through a second "narrows", then the valley opens abruptly where it leaves the east edge of the Cordón del Plata. From this point to its junction with Río Mendoza, about 11 km downstream, Río Blanco becomes entrenched beneath a wide alluvial slope.

GLACIAL GEOLOGY

The Cordón del Plata, along with other ranges in the dry Central Andes, supported considerably larger glaciers during the last major glaciation which terminated, according to Mercer (1976, p. 155), by 12,000 BP in Patagonia. In the Cordón del Plata, ice of this glaciation, named the Vallecitos Stage (Corte, 1957, p. 14), extended down-valley to about 2600 m, where it deposited end moraines across the valleys of both Vallecitos and Angostura Creeks.

Little disagreement exists among geologists who have studied the area that glaciers expanded extensively in the valleys of the Mendoza Andes during the last glacial age. Whether or not glaciers were as large during earlier cold periods of the Pleistocene has been questioned however. Desante (1946), Groeber (1954) and others described diamictons that they regarded to be evidence of an earlier glacial advance onto the piedmont in the valleys of Ríos Diamante and Atuel in the southern part of Mendoza Province. Corte (1957) discussed briefly the moraines of the last glaciation and described several exposures of a diamicton that is exposed in the steep valley walls of Río Blanco and can be traced from the junction of Q. de la Angostura and Q. de los Vallecitos to Potrerillos, where Río Blanco enters the valley of Río Mendoza. He considered this diamicton to be till deposited during an earlier glaciation.

Polanski (1963, 1965) disagreed with the identifications made by Groeber and by Corte of these diamictons as till. Instead, he regarded them to be volcanic mudflows that, along with the bouldery gravels accompanying them, represent the stratigraphic record of uplift of the Central Andes. Polanski argued that the region of Lat. 32°S is and always has been too dry to have supported Pleistocene glaciers sufficiently large to have reached the piedmont and he agreed with Stappenbeck (1917) that the most extensive glaciation of the Andes was the last one.

In this report the relative dating methods that proved useful in reconstructing the Pleistocene history of the Río Blanco valley are described. Also additional data on the diamictons in the basin of Río Blanco that I believe will firmly establish their glacial origin are discussed.

Little vegetation borders the modern glaciers and rock glaciers in this dry part of the Central Andes, and efforts to dig through one rock glacier margin to obtain datable organic material proved fruitless. Lenses of coarse volcanic ash that overlie a till exposed along Río Blanco were dated by zircon fission-track methods (Glen Izett and

Charles Naeser, letter 12, Dec. 1980) as between 100,000 and 200,000 years old, but are too young to be dated more precisely. Relative dating (RD) techniques were used, therefore, in order to try to provide some stratigraphic order for the sequences of deposits mapped.

RELATIVE DATING METHODS

Since 1931, when Blackwelder demonstrated the value of comparing erosional modification of moraines and the weathering of granitic boulders at and just below the surface to distinguish glacial deposits of different stages in the mountains of western U.S., North American glacial geologists have relied on these techniques to work our morainal sequences. Blackwelder's criteria were refined by Sharp (1939), Richmond (1962), and other students of alpine glacial deposits during the decades that followed. Most recently, a more comprehensive set of criteria has been worked out so that measurements of several features can be compared to establish the probable relative age of a particular drift and to correlate it from one valley to another (Birkeland, 1973; Carroll, 1974; Burke and Birkeland, 1979; Birkeland *et al.*, 1979, 1980; Burke and Birkeland, 1980).

Weathering processes take place continuously, and the degree to which weathering will affect rock decomposition and soil profile development is dependent on time, as well as on the factors of climate, rock type, slope, and biota. Landform morphology also changes with time, and may be affected by both erosion and deposition. Those changes are subject to measurement and aspects of the rate of change determined. Some of the parameters to be measured are more useful for short spans of time, others for longer periods; thus a few features should be quantifiable for most deposits of Pleistocene age. Some subjectivity is present in the method, and the specific measurements of two or more workers may not be identical. Nevertheless, most geologists will pick major breaks in a sequence at the same place.

Use of RD techniques in southern South America has been limited. Flint and Fidalgo (1964, 1969) examined the degree of weathering of granitic clasts in the soil profile to distinguish three tills in the region around Bariloche. They also used the height of outwash above present drainage and subjective evaluations of the degree of erosional modification of the landforms. They examined soil profiles but were unable to detect sufficient differences in the weakly developed profiles to use them. Caviedes and Paskoff (1975) recognized differences in the degree of valley erosion since deglaciation; they used these morphologic changes along with alteration of heavy minerals and soil profile thickness to distinguish deposits of three glaciations in the Aconcagua and Elqui valleys of Chile. Mercer (1976) mentioned the much greater weathering and erosion of three pre-Llanquihue tills, the Casma, Colegual, and Río Frío, of the Lago Llanquihue area in Chile west of Bariloche, Argentina. He noted the progressive thickness of the weathering rinds on andesitic clasts in these tills. Wayne (1981) used many of the RD parameters used currently in North America alpine studies in working out a late glacial-Holocene glacial geomorphic history for the upper part of the Río Blanco basin.

The following discussion covers the RD methods that I have found useful in efforts to identify the deposits of different glacial advances in the Río Blanco basin. The principal parameters that were measurable and in varying degress useful in this study included morphostratigraphic relationships, multiple till exposures, morphologic changes, soil profile development, surface and subsurface boulder weathering, diagenetic changes on surface of quartz sand grains, vascular plant cover and lichen cover.

Morphostratigraphy

Advancing ice builds moraines that may overlap parts of older moraines; likewise active rock glaciers may bury older sediments. Such morphostratigraphic relationships are most useful among younger deposits that have undergone little erosional change but are of decreasing value in older materials. Some of the Holocene sediments show clearly their relationships by morphostratigraphic overlap (Figure 2).

Figure 2 Quebrada de los Vallecitos, successive morpho-stratigraphic overlap: Holocene III in background and rock glacier on right overriding Holocene II lateral moraine overriding toe of Holocene I rock glacier, which lies across Vallecitos II lateral moraine.

Multiple Till Exposures

Holocene deposits are only slightly dissected, but the meltwater streams flowing from the glacier remnants have cut away the sides of moraines in a few places. In addition, downstream from the outermost end moraines of the last glaciation, stream entrenchment along Río Blanco exposes, two tills in several places. With such exposures a till can be recognized by its physical characteristics and traced from one exposure to the next (Figure 3).

Figure 3 Exposure along Río Blanco of two tills, which can be distinguished in the field by color (the lower till is brown, probably from oxidation, and the upper one is greenish gray) and by the relative abundance of boulders.

Changes in Morphology

Active rock glaciers are angular and freshly deposited moraines have sharp topography, but in time their surfaces become rounded through erosion, mass wasting, and deposition. A progressive reduction in slope and increase in roundness of surface features characterizes the increases in age of deposits (Figures 4 and 5).

Figure 4 Surface of rock glacier deposits in Quebrada Colorado, showing progressive softening of topographic form, reduction of slope angles, and increase in vegetation cover with increasing age. Active rock glacier front in background; foreground is Vallecitos II surface.

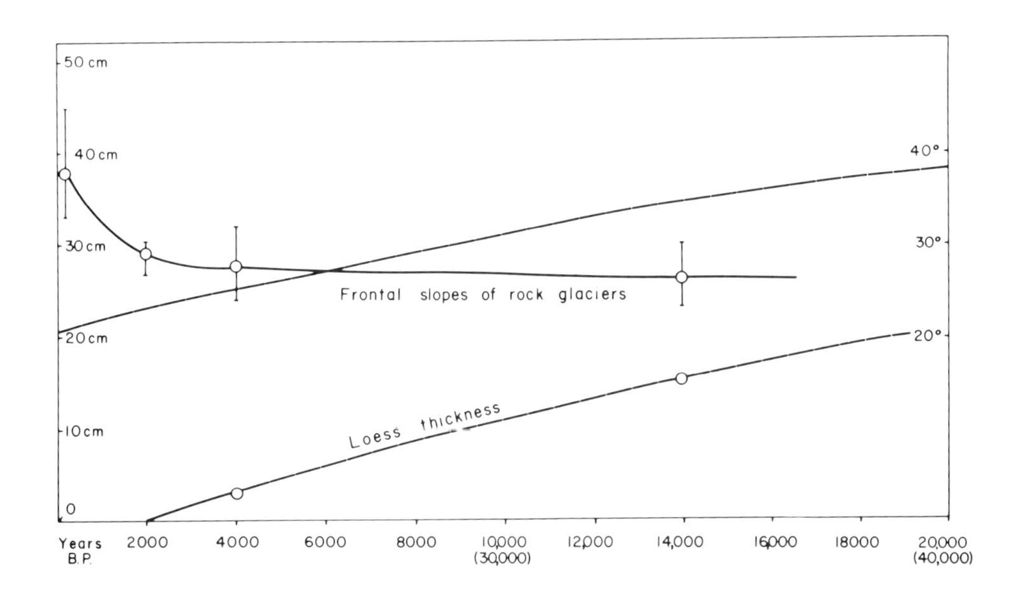

Figure 5 Chart showing increase in loess thickness and decrease in slope angle of rock glacier fronts with increasing age.

Dust accumulates on upland surfaces in dry alpine areas, and the thickness slowly increases. The source in many regions is the occasional dust storms on the adjacent arid piedmont region. Very young moraines and rock glaciers have little or none, but a measurable layer is present on older surfaces. Dust accretion surely does not take place continuously and uniformly on morainal surfaces, nevertheless, a plot of average loess thickness on Holocene and Vallecitos surfaces in the Río Blanco basin against time is nearly linear (Figure 5). The relationship does not hold for older deposits, though.

Surface Weathering

Chemical and physical weathering processes begin to alter boulders at the surface of moraines and rock glaciers as soon as they are exposed. Granites show the most readily measurable changes and are the rock type that has been used most commonly in measurements of boulder weathering. Unfortunately, some regions that have extensive moraines and rock glacier deposits contain few or no outcrops that could have supplied granitic clasts to the glaciers eroding them. In the Río Blanco basin only two small sites were source areas for granite clasts, so measurements of the degree of weathering of granites was only minimally useful. One biotite granite outcrop though, had been covered by glacial ice during the last major glacial advance (Vallecitos), and part of it was again covered during a Holocene glaciation. The recently glaciated surface is polished and striated; the parts covered several millenia earlier are pitted and the surface is grusified. All granite boulders observed, totalling fewer than 100, were rounded, had surfaces etched through weathering, and many were pitted.

Rhyolite porphyries, which shatter readily in the present frost climate above 3500 m, were the most abundant rock in two of the rock-glacier-filled valleys studied; quartzites made up the bulk of the boulders in the other two. When first broken from the outcrop, the edges of most quartzite boulders are sharply angular. The edges lose their sharpness with exposure to weathering processes, though, so boulder-edge rounding provided one measurement of some value. I looked for clasts that were 30 cm or greater in intermediate diameter and counted at least 25 at each site, if that many were available, listing each in one of 4 categories: angular, subangular, subrounded, and rounded. Loss of angularity seemed to be slowly progressive in the Andean tundra, but rounding was much more noticeable where boulders lay in woody shrub-covered parts of the moraines. Fire spalling may have accelerated the rounding on surfaces where sufficient vegetation existed to have supported brush fires.

A split-to-non-split ratio of boulders was determined wherever split boulders were found. Boulder splitting however, seemed to be more strongly affected by nearness to a later glacier front and the periglacial climate conditions that accompanied it than to time alone. The areas of most highly split boulders also contain abundant sorted circles.

Where rocks that develop weathering rinds are abundant, rind-to-non-rind ratios and thickness of rinds are useful measurements. In the Río Blanco basin measurable rind development seems to have taken place only on fine grained dioritic rocks, which are rare in the deposits. Only a few measurements were made, but the data seemed to be consistent for Late Pleistocene and Holocene deposits.

Subsurface Weathering

Boulders that are buried and thus kept moist are subjected to hydrolysis, which causes them to disintegrate. Plutonic rocks in the zone of weathering are most readily affected and their soundness has been used in the Sierra Nevada and Rocky Mountains as a guide to the length of time they have undergone weathering in place (Birkeland, 1973; Burke and Birkeland, 1979; Carroll, 1974). This technique has little value in the separation of the deposits of and since the last major glaciation in the Río Blanco basin because of the paucity of plutonic rocks. It has some value, however, for comparing older tills, in which the few plutonic rocks were completely grusified and quartzite clasts had become fragmented.

Soil profiles develop slowly in the alpine environment; nevertheless, the process is sufficiently rapid that distinctive changes will take place within a few thousand years. Even the slight development on young materials can be recognized and measured. Although certainly not a linear function (Colman, 1981), soil-profile development is progressive with time, so study of the profiles is an important method to help distinguish deposits from the very young to those of early Pleistocene age. Accurate and detailed field descriptions, using Munsell color designations, permit recognition of many glacial deposits. The data from laboratory analysis, particularly the kind and amount of clay minerals present through the profile and the degree of alteration of minerals such as hornblende, aids in the characterization. Difficulties were encountered in obtaining adequate laboratory analyses beyond grain size distribution, so soil profile studies used in this project are based largely on field descriptions. All but a small number of the soil profiles examined required the excavation of a pit from 30 to 90 or more cm in depth.

Quartz Grain Surface Textures

Some sediment transport processes import distinctive sets of features to the surfaces of detrital quartz grains (Krinsley and Doornkamp, 1973). Grains carried by moving ice commonly are fractured so that they have a high relief and unique patterns unlike those imparted to quartz grains in other environments. Because surface textures of glacially transported grains are so distinctive, the presence of a relatively high proportion of grains with these textures in a sediment means that the grains were carried in glacial ice, and, if no other pattern is overprinted the sediment from which they came, probably is glacially deposited.

Sand grains were separated from a small sample of each of the tills or diamictons thought to be a till in the Río Blanco valley and in two major headwater streams. After separation, the grains were washed in distilled water to remove clay-sized particles. Monocrystalline quartz grains were picked from each sample and 15 of them mounted on stubs and coated with gold for examination under a scanning electron microscope (SEM).

Nearly all grains from the samples taken from the tills and rock glacier deposits of the last (Vallecitos) glaciations were completely and sharply fractured (Figure 6a, b). Most grains from the older tills showed comparable surface textures, but some had one or more surfaces that had not been broken in transport. In addition, grains from the oldest sediment thought to be a till showed an etching overprint on the glacially formed fracture pattern (Figure 6). Similar etching has begun but is only slightly developed on grains from the tills of intermediate ages (Figure 6). These diagenetic changes, though subtle, evidently are progressive, so that it may be possible to use them as one additional means of ordering tills within a sequence relative to each other, provided the differences are of at least stage magnitude. The rate of diagenetic etching surely is too slow for it to have value in separating deposits of substages of a glaciation.

Biotic Cover

Lichens require light and moisture in order to become established on rocks. Thus they are rare on the surfaces of active rock glaciers and do not begin to grow on morainal boulders until glacial ice has melted from the area. The rate of growth of *Rhizocarpon geographicum sensu lato* has been used in many parts of the world as a tool to determine an absolute age for late Holocene glacial deposits (Benedict, 1967, 1968; Beschel, 1973). Because climate, and particularly available moisture, directly affect the rate of growth of this and other lichens, a thallus growth-rate curve must be developed for each area

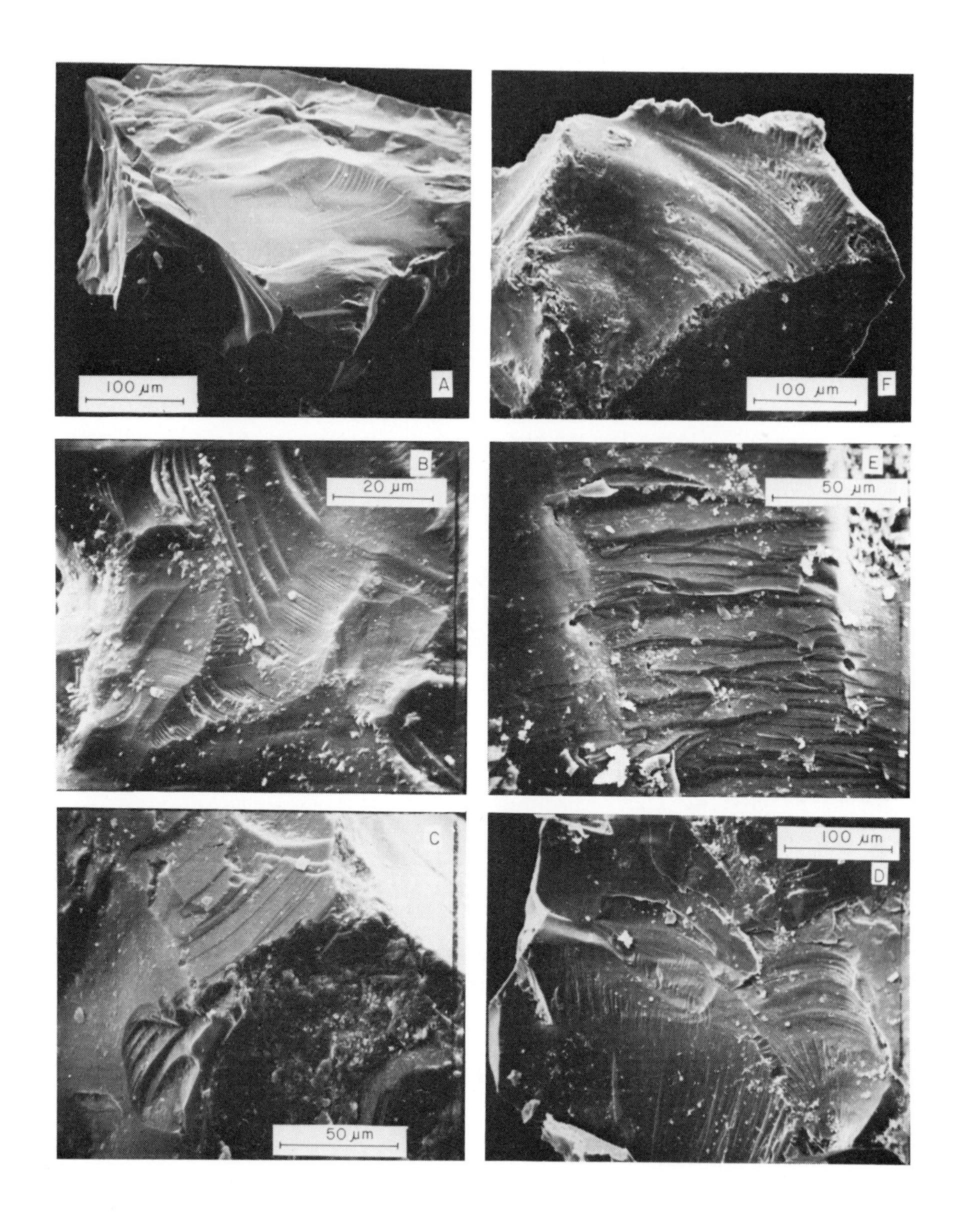

Figure 6 Scanning Electron Microscope images of quartz grain surfaces.
a. and b. Q. Colorado, Vallecitos II Stage rock glacier
c. Q. de los Vallecitos, Vallecitos Stage, till
d. Q. de la Angostura, Río Blanco Stage, till
e. Q. de la Angostura, Angostura Stage, till
f. Mesón del Plata, Los Mesones Stage, till

where lichenometry is to be used to calculate the age in years of a particular deposit. Even though a growth rate curve has not--or can not--be worked out for a particular area, two measurements - maximum thallus diameter and percent of lichen cover - are useful in relative dating (see Mahaney *et al.*, this volume).

In alpine regions, vascular plants are more sparse on younger deposits than they are on older (and lower) surfaces. Part of the difference is age, which may be related to both loess thickness and soil development. Part may also be climatic. Nevertheless a distinct difference in plant cover was noted on groups of moraines and on fossil rock glaciers in the Río Blanco basin. These estimates of cover were one additional parameter that seemed to have value in the Holocene deposits ranging from a few hundred to a few thousand years old.

APPLICATION OF RELATIVE DATING DATA

Throughout all deposits, maturity and character of soil profile development provided effective information for use in classifying by age. Although the soil profile development was only one of several kinds of data used to determine the number of Holocene and Late Pleistocene glacial advances, it was one that seemed to be particularly reliable. For the older deposits, it and till superposition were the two used with the greatest degree of confidence. A description of a soil profile that is characteristic of each of the deposits mapped is in the appendix.

Holocene Deposits

Several kinds of RD measurements are sufficiently distinctive to permit separation of the Holocene glacial deposits into three groups (Table 1). Perhaps the most important is that the moraines and rock glaciers of the most recent ice advance, which are the equivalent of the recently ended "Little Ice Age", remain ice cored, and many of the rock glaciers still are active. All soil profiles are weak, but show increasing thickness and intensity of color with age. Both lichen cover on boulders and total vegetation cover of the surface increases with the age of these deposits. Only the oldest of the Holocene rock glaciers has a loess cover, and the frontal angles of rock glaciers become less steep with increasing age (Figure 5). In some valleys, successive partial overlap makes recognition of relative position simple, so that measurable data can be related readily and used to identify deposits of correlative age in other valleys.

Vallecitos/Wisconsinan/Würm Deposits

The oldest and outermost deposits that have undisputed glacial morphology are those of the last Pleistocene glaciation, named by Corte (1957) the Vallecitos Stage. Each of the measurable parameters used to distinguish and identify the moraines and rock glaciers of Holocene glacial phases are applicable to those of Vallecitos age as well (Table 1). Loess thickness is greater, and rock glacier frontal slope angles are lower (Figure 2). Lichen cover on boulders and vegetation cover of morainal surfaces are significantly greater than are those on the younger deposits. Both soil profile development (Table 1, Appendix) and loess cover clearly show that the Vallecitos moraines fall into two groups. The younger surely are the same age as those dated in Patagonia (Mercer, 1976) as 13,000 to 19,000 or so years old; the older group may match those Mercer found to have been deposited more than 40,000 years ago.

Pre-Vallecitos Deposits: Tills vs Mudflows

Diamictons greater in age than those in the moraines of Vallecitos age are present at lower altitudes in Río Blanco valley. Although Corte (1957) described a diamicton that can be traced to the valley of Río Mendoza, 17 km downstream from the outermost Vallecitos Stage moraine, and called it the Río Blanco till, Polanski (1953, 1963, 1965) argued that the last glaciation was the most extensive in the dry Mendoza Andes, and that the diamictons farther down valley are mudflows that record uplift of the Cordillera.

Before these extra-Vallecitos diamictons can be considered evidence of earlier glaciations, it will be necessary to determine whether they are in fact more likely tills or mudflow sediments. They fill the lower part of the Río Blanco valley, where they are overlain by bouldery gravel and sand and, in one place, by a bed of clean, volcanic ash. The source of the ash has not been determined, but it may have come from one of the recently active volcanic cones farther south in Mendoza Province.

Table 1. Relative Dating (RD) characteristics of the Late Pleistocene and Holocene glacier and rock glacier deposits of the upper part of the Río Blanco drainage basin

CHARACTERISTICS	VALLECITOS I	VALLECITOS II	HOLOCENE I	HOLOCENE II	HOLOCENE III ("Little Ice Age")
Position of moraines or rock glaciers	End moraines low in valley (below 2600m) Outer lateral moraine smoothly rounded	Moraines and fossil rock glaciers fill valleys between 2600 and 3400 m	Rock glaciers at 3200 m	Clustered moraines extend to 3450 m Rock glacier tongue to 3100 m	Ice-cored moraines beyond present ice margin Active or recently inactive rock glaciers to 3250m, fronts 33°-45°, contain ice
Modification of constructional landforms	Loess variable, 35-80 cm; Moraines partly destroyed by stream	Loess averages 20 cm; Moraines trenched but little valley widening Rock glacier fronts 23°-30°	Loess averages 5cm Rock glacier front 25° - 32° Smoothly rounded	No loess Trenched only where meltwater stream passes through. Rock glacier fronts 27°-30°	Ice-cored moraines starting to collapse Some rock glaciers inactive, rounded crests, still contain ice
Boulder weathering	Granites etched, rounded, pitted to 3 cm Quartzites subrounded: (A)0-(SA)57-(SR)43-(R)0	Granites etched, rounded, pitted to 3 cm Non-granitic rocks: (A)10-(SA)37-(SR)32-(R)21 Rinds 0.5 to 1.0 mm	No granites: quartzites: (A)29-(SA)-62(SR)9-(R)0 Rinds 0.5mm on diorites	No granites; quartzites: (A)20-(SA)73-(SR)7-(R)0 Rinds 0.2mm on diorites	All angular
Soil profile development	Moderate, Mollisols B-horizon massive to weak blocky structure 5YR3/3-5/4(B)/10YR 4/4(C)	Moderate, Mollisols B-horizon massive 7.5YR4/4(B)/10YR4/4 (C)	Weak, Inceptisols Cox horizon loess/debris 7.5YR4/4 to 5YR4/4(B)/ 7.5YR4/4(C)	Weak, Inceptisols Cox 0-5 cm (till) 10YR4/3(B)/2.5Y 5/2(C)	None, Entisols
Lichens	10% to 90% covered Diameters meaningless; multigenerations	50% to 70% covered Diameters meaningless multigenerations	30% to 40% covered *R. geographicum* 70mm	10% to 40% covered *R. geographicum* to 60mm	Cover less than 2% No *R. geographicum* observed
Vegetation	Fully covered (60%-90%) Grasses, forbs, woody shrubs	Fully covered (60%-90%) Grasses, forbs, woody shrubs	About 40% covered Grasses, forbs, prostrate *Adesmia*	5% to 20% covered Grasses, forbs	Less than 2% covered Succulent forbs
Other features		Large, massive moraines Sorted stripes, lobes on slopes above moraines Avalanche cones beside moraines in valleys	Thick loess accumulation beneath boulders at toe of slope Sorted circles 1-1.5m diam. to 50 m downslope	Sorted circles 1-1.5m diam. on adjacent surfaces 750 m downslope	Many rock glaciers expanding now Ice-cored moraine rock-glacierized
Estimated age	>40,000 BP	19,000-13,000 BP	4600-4000 BP	2700-2000 BP	300 BP-0

Mudflows commonly originate from the frontal regions of active glaciers, but mudflows in the environment of small valley glaciers do not result in massive, thick, several-km-long, valley-filling diamictons such as the one in Río Blanco valley. Alpine mudflows of volcanic origin, such as those that were caused by the recent eruption of Mt. St. Helens in the Cascade Range of the United States, do leave sediments comparable in volume to those of Río Blanco. The Cordón del Plata contains no volcanoes, though, so such a source would be most unlikely. Also, the sediments of volcanic mudflows invariably contain some extrusive volcanic materials; the sand fraction of the Rio Blanco diamictons contains no volcanic debris and is dominated by lithic clasts of fine grained quartzites. Scanning electron microscope (SEM) images of quartz grains from the sand fractions of these diamictons exhibit surface textures characteristic of those produced by glacier transport, as do, the grains from unquestioned tills higher in the valley (Figure 6a). For these reasons, I believe the diamictons of the Río Blanco valley are more likely tills than mudflow sediments, although the possibility can not yet be ruled out that some of the sediment could have originated as mudflow from the front of a glacier. They do represent the deposits of pre-Vallecitos glaciations.

Downstream from the outermost and oldest Vallecitos moraine, Angostura Creek flows in a trench eroded below a boulder-strewn, fan-shaped surface. Till underlies this surface, which contains a veneer of loessal silt about 30 cm thick. A soil profile comparable to that on some of the lower Vallecitos Stage moraines has developed through the loess and into the top of the till. The surface is Vallecitos in age, but it was cut by currents of Vallecitos meltwater across pre-Vallecitos tills. Two tills, the lower one of which has a well developed buried paleosol in it, are exposed in the 8-m-high banks along the creek. At one place, the creek has swept free the valley side and exposed a much thicker section, with a mature carbonate rich soil profile on the upper till and a buried paleosol with a thick argillic B horizon on the lower one (Table 2; Appendix). Clearly, exposures of the two tills in superposition, a standard stratigraphic approach, has clarified the ages of these two relative to each other. The well developed soil profile at the top of the younger till shows it not only to be older than the Vallecitos tills, which have less maturely developed profiles, but also to have formed under more arid climatic conditions.

Both of these tills can be seen in superposition in another exposure about 3 km downstream (Figure 3), but only one of the tills seems to be present in the lower part of Río Blanco valley. The soil profile on the upper one clearly shows the effect of the warm dry climate that evidently existed during the interglacial stage preceeding the Vallecitos Glaciation. As mentioned earlier, an effort to date the volcanic ash bed that overlies the till Corte named the Río Blanco was not wholly successful, but a zircon fission-track study showed it to be between 100,000 and 200,000 years old (Glen Izett, Charles Naeser, letters, December 1980). The two tills in the upper valley exposures have distinctly different boulder/matrix ratios; the lower till is very bouldery, and the upper one is similar to the Río Blanco till downstream. For these reasons, I have identified the upper till in these multiple exposures as the Río Blanco till of Corte (1957) and named the lower one the Angostura till (Wayne, in press).

Still one additional diamicton is present in the Río Blanco valley. It caps a sloping flat-surfaced ridge that stands nearly 200 m above present drainage and is aligned with the upper part of Q. de la Angostura (Figure 1). It has a thick, carbonate-rich soil profile, quartzite boulders exposed are extremely rounded, and many in the soil are extensively weathered. It was included in Polanski's (1963) Los Mesones Formation because it caps Mesón del Plata in the Río Blanco valley. Its topographic position, soil profile development, and boulder weathering characteristics show it to be much older than any of the other glaciogenic sediments (Table 2) in the area.

Table 2. Relative Dating (RD) characteristics of the Pleistocene glacial deposits, Río Blanco drainage basin

Position of constructional landforms	None left. Till present along valley to 1800 m	None recognized; Surface of till buried to about 1950 m	Morainal remnants present about 2200m, 1900m and 1700m. Till present in valley to 1400 m	Moraines and fossil rock glaciers in valley between 2400 m and 3400 m
Modification	Fragments of once sloping plain cap mesetas 75 to 200 m above present streams Some loess in soil profile	Buried beneath younger till between 2500 m and about 1950 m.	Pediment cut across till surface from 2500 to 2100 m: Moraine at 1700 m deeply trenched Coarse gravel over till downstream Loess cap 15 to 30 cm	Moraines fresh to rounded, outer ones moderately breached Inner moraines trenched Loess 20-35+ cm
Boulder weathering	Few boulders visible on surface Only quartzites, all well rounded, many split Quartzites in soil completely weathered. Few granites observed, all completely grusified. Boulders exposed only along edges	No boulders at surface Granites in buried paleosol grusified	Surface boulders not common Only quartzite fragments to 3 m in soil profile at outermost moraine recognized Boulders on higher moraine fragment (A)15-(SA) 21-(SR25-(R)39% with 70% broken	Granites rounded, etched, pitted to 3 cm Quartzites subround to subangular Rinds 0.5 to 1.0 mm V.I:(A)0-(SA)57-(SR)43-(R)0 V.II:(A)10-(SA)37-(SR)32-(R)21
Soil Profile Development	Very strong: at 3000 m argillic B-horizon with strong blocky structure (27cmthick) Aridisol below 2350 m, with thick chalky Bca-Cca horizon Locally strongly cemented (12-80+cm)	Buried paleosol with strongly developed argillic B_2, weak Bca and Cca horizons	Moraine fragments at 1900 & 2300 m: Aridisol with Bca 25-35 cm and Cca(K) horizons 35-60 cm, platy Outer moraine at 1700 m: silty aridisol, Bca 35-60 cm, weakly cemented	Moderate, Mollisols B-horizon massive to weak blocky Acid
Other Features	Grassland, with scattered woody shrubs Rests on weathered andesite in Q. de la Angostura Till calcareous when exposed	Fills bottom of bedrock valley	Capped by ash bed in a few exposures Till not calcareous under pediment but contains secondary $CaCO_3$ in other places	Grass with woody shrubs Till not calcareous
Estimated Age	1.2-1.0 x 10^6 yrs	450,000 yrs BP (?)	200,000-100,000 BP (?)	Wisconsinan/Würm

CONCLUSIONS

The relative dating (RD) techniques that have been developed by investigators of glacial sediments to subdivide and classify the glacial deposits of North America have served well in the study of glaciogenic deposits of the Central Andes of Mendoza Province, Argentina. Standard stratigraphic practices of superposition of both strata and morphologic units, supplemented by weathering and erosional measurements, have led to the recognition of the deposits of four glacial stages and of three phases of glaciation during Holocene time.

ACKNOWLEDGEMENTS

The field work for this report was done between January and June 1980 while I was attached to the Instituto Argentino de Nivología y Glaciología (IANIGLA) in Mendoza, Argentina, through mutual agreement between the National Science Foundation of the U.S.A. and the Consejo National de Investigaciones Cientificas y Técnicas (CONICET) of Argentina. It was supported in part by NSF Grant No. INT - 7920798. I am greatly indebted to Dr. Arturo E. Corte of IANIGLA, who suggested the problems related to the sediments in the Río Blanco drainage basin, investigated in this project. My understanding of the deposits studied was helped greatly by discussions with him both in and out of the field while the field work was in progress. I would also like to thank the staff of IANIGLA, for providing access to an institute vehicle for field work; and to my wife, Naomi Wayne, who served as field and laboratory assistant throughout the project.

REFERENCES CITED

Benedict, J.B., 1967, Recent glacial history of an alpine area in the Colorado Front Range, U.S.A. I. Establishing a lichen-growth curve: Journal of Glaciology, v. 6, no. 48, p. 817-832.

______________, 1968, Recent glacial history of an alpine area in the Colorado Front Range, II. Dating the glacial deposits: Journal of Glaciology, v. 7, no. 49, p. 77-87.

Beschel, Roland (tr. by W. Barr), 1973, Lichens as a measure of the age of recent moraines: Arctic and Alpine Research, v. 5, no. 4, p. 303-309.

Birkeland, P.W., 1973, Use of relative age-dating methods in a stratigraphic study of rock glacier deposits, Mt. Sopris, Colorado: Arctic and Alpine Research, v. 5, no. 4, p. 401-416.

Birkeland, P.W., Colman, S.M., Burke, R.M. Shroba, R.R. and Meierding, T.C., 1979, Nomenclature of alpine glacial deposits, or, What's in a name?: Geology, v. 7, p. 532-536.

Birkeland, P.W., Burke, R.M. and Walker, A.L., 1980, Soils and subsurface rock-weathering features of Sherwin and pre-Sherwin glacial deposits, eastern Sierra Nevada, California: Geol. Soc. Amer. Bull.: v. 91, p. 238-244.

Blackwelder, E.B., 1931, Pleistocene glaciation in the Sierra Nevada and basin ranges: Geol. Soc. Amer. Bull.: v. 42, p. 865-922.

Burke, R.M. and Birkeland, P.W., 1979, Reevaluation of multiparameter relative dating techniques and their application to the glacial sequence along the eastern escarpment of the Sierra Nevada, California: Quaternary Research, v. 11, p. 21-51.

Caminos, Roberto, 1965, Geología de la vertiente oriental del Cordón del Plata, Cordillera Frontal de Mendoza, Asociación Geológica Argentina, Revista, Tomo 20, No. 3, p. 351-392.

Carroll, T., 1974, Relative age dating techniques and a late Quaternary chronology, Arikaree Cirque, Colorado: Geology, v. 2, no. 7, p. 321-325.

Caviedes, C.N. and Paskoff, R., 1975, Quaternary glaciations in the Andes of north-central Chile: Journal of Glaciology, v. 14, no. 70, p. 155-170.

Colman, S.M. 1981, Rock-weathering rates as functions of time: Quaternary Research, v. 15, no. 3, p. 250-264.

Corte, A.E., 1957, Sobre Geología glacial Pleistocenica de Mendoza: Anales del Departamento de Investigaciones Clientificas, Universidad Nacional de Cuyo, Tomo 2, no. 2, p. 1-27.

Dessanti, R.N., 1946, Hallazgo de depósitos glaciales en las huayquerías de San Carlos: Revista Sociedad Geológica Argentina, Tomo 4, p. 270-284.

Estrella, H.A., Heras, V.A. and Guzzetta, V.A., 1980, Registro de elementos climáticos en areas críticas de la Provincia de Mendoza: Instituto Argentino de Investigaciones de las Zonas Aridas, Cuaderno Técnico 1-79, p. 49-71.

Flint, R.F. and Fidalgo, F., 1964, Glacial geology of the east flank of the Argentine Andes between latitude 39°10'S. and latitude 41°20'S: Geol. Soc. Amer. Bull., v. 75, p. 335-352.

__________, 1969, Glacial drift in the eastern Argentine Andes between latitude 41°10S. and latitude 43°10'S.: Geol. Soc. Amer. Bull., v. 80, p. 1043-1052.

Groeber, P., 1954, Bosquejo paleogeografico de los glaciares del Diamante y Atuel: Revista de la Asociación Geológica Argentina, v. 9, no. 2, p. 89-108.

Krinsley, D.H. and Doornkamp, J.C., 1973, Atlas of Quartz Sand Surface Textures: Cambridge University Press, 91 p.

Mercer, J.H., 1976, Glacial history of southernmost South America: Quaternary Research, v. 6, no. 2, p. 125-166.

Polanski, J., 1953, Supuestos englazamientos en la llanura pedemontana de Mendoza: Rev. Asociación Geológica Argentina, T. 8, no. 4, p. 195-213.

__________, 1954, Contribución al conocimiento y la systemática del englazamiento actual de la Alta Cordillera de Mendoza: Rev. Asociación Geológica Argentina, T. 9, no. 4, p. 232-245.

__________, 1963, Estratigrafía, neotectónica, y geomorfología del Pleistoceno pedemontano entre Ríos Diamante y Mendoza: Rev. Asociación Geológica Argentina, T. 17, no. 3-4, p. 127-349.

__________, 1965, The maximum glaciation in the Argentine Cordillera: Geol. Soc. Amer. Bull., Spec. Paper 84, p. 453-472.

__________, 1972, Descripción geológica de la Hoja 24a-b, Cerro Tupungato, Provincia de Mendoza, Dirección Nacional de Geología y Minería, Boletín, No. 128, 114 p.

Richmond, G.M., 1962, Quaternary Stratigraphy of the La Sal Mountains, Utah, U.S. Geological Survey Prof. Paper 324, 135 p.

Sharp, R.P., 1939, Basin-range structure of the Ruby-East Humboldt Range, northeastern Nevada: Geol. Soc. Amer. Bull., v. 50, p. 881-920.

Stappenbeck, R., 1917, Geología de la falda oriental de la cordillera del Plata: Anales del Ministerio de Agricultura de la Nación, T. 12, no. 1.

Wayne, W.J., 1981, La evolucion de glaciares de escombros y morrenas en la cuenca del Río Blanco, Mendoza, VIII Congresco Geológico Argentino.

__________, Multiple glaciation of the Cordon del Plata, Mendoza, Argentina: Palaeogeography, Palaeoclimatology, Palaeoecology, in press.

APPENDIX

Characteristic soil profiles on rock glacier debris and till (field descriptions)

1. Stable ridge along left margin of rock glacier in Q. Colorada, about 675 m upstream from active front Holocene III

0 - 5 cm	Rock chips, base abrupt
5 - 20 cm	Loam with 50% rock fragments >2 mm, 7.5YR4/4, pH 3.8

2. Collapsed rock glacier, Q. Colorada, Holocene II, 50 m downvalley from active face, surface level

0 - 6 cm	Rock chips (1 to 10 cm diam) base abrupt
6 - 17 cm	Sand, silty, loose, 5YR3/4, pH 5.2, base clear
17 - 30 cm	Sand and rock fragments, loose, 7.5YR4/4, pH 5.6, hole stopped at boulder

3. Long rock glacier tongue in Q. Colorada, 200 m downstream from active face, Holocene II, vegetation 5%

0 - 5 cm	Rock chips, base abrupt
5 - 9 cm	Sandy loam, 40% rock fragments >2 mm, 7.5YR4/4, pH 5.2, loose
9 - 25 cm	Silty sand, 60% rock fragments >2 mm, 7.5YR4/4, pH 5.0, loose

4. Crest of left lateral moraine, Q. de los Vallecitos, Holocene II, vegetation 1%

0 - 5 cm	Loam, 10% rock fragments >2 mm, 10YR 4/3, pH 4.2, base clear
5 - 25 cm	Sandy loam, 50% rock fragments >2 mm, 2.5YR4/2, pH 3.8

5. Crest of fossil glacier tongue that overrode an older rock glacier. Holocene I. Vegetation cover 50%, flat.

0 - 2 cm	Rock chips, angular, diam 2 - 10 cm
2 - 8 cm	Gritty loam, 15% clasts >2 mm, 5YR3/2, massive, pH 4.0, base clear
8 - 30 cm	Silty sand, 50% clasts >2 mm, 5YR4/4, loose, pH 5.6

6. Crown of very bouldery, small, fossil rock glacier on left valley wall of Q. de los Vallecitos, overlaps older left lateral moraine and is overridden by younger left lateral moraine. Holocene I or Vallecitos II. Vegetation cover 40 - 50%

0 - 2 cm	Rock chips
2 - 8 cm	Loamy sand, 20% rock fragments >2 mm, 7.5YR3/2, non plastic, non sticky, pH 6.4, base abrupt
8 - 12 cm	Loam, 20% clasts >2 mm, 5YR4/2, sl. plastic, pH 5.0, base clear
12 - 30 cm	Loam, sandy, 50% clasts >2 mm, 7.5YR4/4, loose, pH 4.6

7. Crest of left lateral moraine about 50 m upvalley from morainal loop. Vallecitos II. Vegetation cover 70%; flat: stabilized sorted circles 1 m in diameter on surface.

0 - 8 cm	Loam, 15% clasts >2mm, 10 YR 3/3, sl. plastic, pH 3.8,

base clear

8 - 18 cm Loam, sandy, sl. clayey, 30% clasts >2 mm, 2.5Y4/4, sl. plastic, pH 3.8, base clear

18 - 35 cm Sandy loam, 50% clasts >2 mm, 7.5YR5/4, loose, pH 3.8

8. Left lateral moraine of large morainal mass, Q. de los Vallecitos. At junction with Q. Colorada. Vegetation cover 100%. Vallecitos II

0 - 10 cm Loam, 10% clasts >2 mm, 7.5YR3/2, pH 6.4, base clear

10 - 22 cm Loam, pebbly, 50% >2 mm, 7.5YR3/4, pH 6.0, base clear

22 - 40 cm Silt, sandy, pebbly, 50% clasts > 2 mm, 10YR4/4, loose, pH 6.0

9. Flat area between two lateral morainal ridges, Q. de la Angostura, Vallecitos II. Vegetation cover 100%, 10% slope to E.

0 - 5 cm Loam, 7.5YR3/2, pH 5.6-5.8, base gradual

5 - 12 cm Loam, 10% clasts >2 mm, 7.5YR3/2, sl. plastic, pH 5.2

12 - 44 cm Loam, 25% clasts >2 mm, 10YR4/3, loose, sl. plastic, pH 4.0

10. Crest of end moraine in offset gorge of Q. de la Angostura. Vallecitos I (?). Vegetation cover 90%, slope 5% to E, altitude 2570 m.

0 - 11 cm Loam, sandy, 5YR3/2, sl. plastic, pH 5.0, base clear

11 - 26 cm Loam, 5YR5/3, sl. plastic, massive, pH 5.0-5.2, base clear

26 - 46 cm Loam, sandy, pebbly, 50% clasts >2 mm, some large, 10YR4/4, loose, pH 5.8-6.0

11. Section exposed in stream bank along Q. de la Angostura, 1000 m above junction with Q. de los Vallecitos, Río Blanco till over Angostura till

0 - 7 cm Loam, 5YR3/2, loose, roots abundant, pH 6.4, base diffuse

7 - 13 cm Loam, 5YR4/2, loose, pH 7.0, base abrupt

13 - 35 cm Silt, 7.5YR7/2, chalky, cementation weak to strong, strongly effervescent with HCl, base clear

35 - 6.00 m Till, pebbly, sandy, silty. 5Y5/3, calcareous throughout

6.00-6.15 m Loam, 10YR4/4 mottled with 7.5YR4/4, very compact, base clear

6.15-6.40 m Loam, 7.5YR4/6, very compact, base clear

6.40-6.83 m Loam, 5YR4/4, mottled with 5YR3/3, clayey, weak blocky structure, pH 5.6, base clear

6.83-7.58 m Loam, 7.5YR5/4, with veinlets of 7.5YR7/2. Contains ghost cobbles and grusified granitic boulders

7.58-8.20 m Covered

8.20-11.20 m Loam, silty, pebbly (till), 2.5Y5/4, pH 6.2

PANEL DISCUSSION

The following remarks were aired in the discussion session following the Symposium on Saturday evening, May 23. The discussion panel included Dr. C.S. Churcher (Royal Ontario Museum), Dr. D. Coates (S.U.N.Y., Binghamton, N.Y.), Dr. H.B.S. Cooke (Vancouver, B.C.) and Dr. J. Terasmae (Brock University, St. Catharines, Ontario).

W.J. Wayne (University of Nebraska):

This conference has been very informative to me. I think we made a good quick review of the dating methods that are being used and are working. I learned of some absolute dating techniques that I hadn't been aware of before. After listening to reports on various methods, and particularly in the amino acid techniques, I suspect that I may have been a little too willing to put faith in radiocarbon dates; perhaps fewer of them than I thought are valid. Under some circumstances, we may be able to do as well with estimates on time from relative dating methods. We looked at the relative dating methods and results on glacial sediments in the Brooks Range, central Canada, the Wind Rivers, and Argentina. I'm interested in seeing what might be done further on terrace dating. What do fluvial terraces in Louisiana, Nebraska, the Rockies, for example, mean with respect to the continental and alpine glaciations on the North American continent?

H.B.S. Cooke:

The thing that strikes me most about any kind of conference considering dating and the Quaternary is the fact that Quaternary studies are one of the most involved interdisciplinary studies that can be imagined. It is almost impossible to solve Quaternary problems by an attack of any single conventional discipline. We need to be detectives; we need to interweave the lines of evidence we achieve from different sources. Some years ago Kenneth Oakley of the British Museum made a very important contribution to the question of dating in the archaeological field which, after all, is part of the Quaternary problem. In trying to recognize various levels of dating he distinguished in particular the difference between chronometric (or absolute) dating and relative dating. He also tried to lay down some ground rules for the criteria which represent different levels of reliability of absolute and relative dating so as to form a hierarchy. In this conference it seems to me that we have dealt with both aspects - absolute dating and relative dating. In terms of absolute dating there are essentially only three techniques which Kenneth Oakley would put as first order dating techniques, namely the radiometric methods, (radiocarbon, potassium argon). Amino acid dating belongs to a lower order as it depends on having a calibration derived from a radiometric scale, and it is not of the same order of absoluteness. Even though the radiometric dates may be wrong, they are physically accurate determinations of the sample that has been presented, but one still has to evaluate the meanings of the dates. I will immediately get into conflict with the paleomagnetists by saying that paleomagnetism does not represent a direct method of absolute dating. The reason for this is that paleomagnetism only gives you a signal - it's a yes or a no, normal or reversed. You are therefore fingerprinting your sample and have to use other criteria to identify which normal or which reverse period you take to match up with part of the dated paleomagnetic time scale. In the ocean sediments, paleomagnetism works very well because you can assume that you start at the top and work down so that you begin the scale from the present day. This has caused occasional trouble. We know that Glomar Challenger, for example, published some results reporting that they had reached very interesting deposits in the lower Pleistocene which turned out to be Oligocene! There was a hiatus in between. So paleomagnetic dating is not a first order method. On the other hand, if one allies paleomagnetic dating with other methods, so that one can get a ball park figure in terms of biochronology or absolute chronology or something else to plug in a starting point for the magnetic scale, then the paleomagnetic method provides an extremely important tool. It is the only tool that we have at the present time which will interlink oceanic and terrestrial environments of all kinds and this, I think, is something that was perhaps underplayed in the symposium. So I would like just to reinforce this idea of Kenneth Oakley's of differing orders or ranks for scales of dating and to throw in these thoughts at this stage.

C. Kolb (Louisiana State University):

Dr. Cooke you have just summarized three major facets of absolute dating. Would you do that for relative dating, or do you have this kind of a summary in your mind for relative dating?

H.B.S. Cooke:

In relative dating we start with the fact that the key to any kind of interpretation of geological horizons must be based on the normal geological principles of stratigraphy and superposition. About the only thing that I believe absolutely is if in a relatively undisturbed area, I see one bed overlying another; then I believe that the one on top is younger than the one underneath. Relative dating is obviously the fundamental mechanism by which one derives a series into which one wishes to plug "absolute" ages. So whether we are taking cores from a bog or looking at successions of geomorphological features, one of which cuts another, we are dating these things relatively by the normal geological and geomorphological criteria. So I think that I would call stratigraphy a first order method of relative dating. The geomorphological criteria I think would have to be degraded to second level dating because, while the relative ages are right, the fact that we do not have direct superposition does mean that there are often possible alternative interpretations. One has to choose between these alternatives, so one has to make second order inferences, rather than first order conclusions. "Bed A is underneath B" is a positive statement. But saying that a waterfall has retreated as a result of some geomorphological operation is a second order inference and is not quite of the same value. In relative dating we employ fundamental things like the "law of faunal succession", where we look at sequences of fossil material through known stratigraphic sequences or correlative stratigraphic sequences. Then we use our knowledge of the development of changing life forms through a sequence (as Harington was doing with the Mammoths or the Bisons) to try to recognize fragments of a continuous time series and link them by correlation into other areas. Now, as soon as you involve correlation you are dealing with a totally different problem in dating, and I think this is something that was not really brought out. There are also mechanisms open to the paleobiologist which are in a way dating techniques that we did not discuss. These involve the study of the development of life forms through essentially continuous sequences of strata. If one does that, it is often possible to find metrical criteria on variation in a biological species through the sequence; for example, we can see the change in the number of plates and the thickness of the enamel in elephant teeth. We can actually put numbers on these changes and we can use absolute or other forms of dating to put time into the sequence. We can then put in values for rates of change in the biological series and this enables us to take and analyze them statistically so as to plug in a date which is almost as reliable as some of the dates that we have been getting from other criteria. So this is another tool. I don't think I have really answered your question.

C. Kolb:

You have in a way. I assume that pedological evidence would provide a breakdown in a relative dating sequence. Is this correct?

H.B.S. Cooke:

Sure. What we need in the soil studies area is some kind of a breakthrough to get age values out of soil forming processes. I think in the long run that ^{10}Be may solve this problem, but it is still a long way off.

C. Kolb:

Why did you make a distinction between what you call 'correlation' and 'relative dating'? To me it may be different, but you end up with the same result. For instance, in my area of Louisiana we are interested in trying to find a method to correlate terrace levels. And, of course we are leaning toward a pedological or chemical key, or signature if you will, that will hopefully permit us to say, "well, we've got three terraces over in this valley and we have three terraces in another valley, and the top terrace of this one over here is indeed the same age as another one, or it was laid down at roughly the same time and correlates with it."

H.B.S. Cooke:

I think you have answered your own question in that you have already drawn a distinction between the stratigraphic succession that you see in one place, and the stratigraphic succession you see in another place. Then you must bring in correlation for the purpose of interrelating these two different sequences. So correlation is a step that interrelates two or more sets of stratigraphic observation. New data may alter the correlation, but not the individual sequences. If you have a terrace sequence in which you can actually walk out the terrace along the banks of the river for many tens of kilometres, this is not correlation; this is geological mapping and consists of tracing an actual stratum continuously. But, if you have a break of ten miles or ten kilometres, or 100 kilometres, then you are making a stratigraphic set of observations in more than one location, and you are drawing upon all kinds of criteria that you can use to show that there is a unique relationship between these sets of data. The importance of this is that there are many things in geology which produce, in fact, the same effects. For example, if you have a cycle going from wetter to drier, and back to wet in a river system, you will produce a certain pattern of terraces. If you repeat that pattern in another area, but at another time, you will get a replica of that set of terraces. So, if you repeat the condition you will get a replica of the geomorphic set of evidence. Accordingly, mere similarity of geological, geographical or geomorphic setting does not provide you with the background for making a reliable correlation. What you have to do is to say, "I believe this unit is the same as that one because it has a soil which is sandwiched between units B and C in one section and a similar soil is sandwiched between units P and Q in my other section. They both have the same chemical features and pollen spectra". Actually, I have picked a bad example for with pollen spectra we can be in trouble. It is very difficult to identify particular interglacials in absolute terms. It is possible to recognize that you have an interglacial pollen assemblage, but to say with assurance which interglacial is not so easy. So, this is why I say that we have a very complex discipline in which, in order to make correlations, we need to draw upon all the tools we have been hearing about. It is only by combining a substantial number of the tools that we can finally arrive at good and firm correlations, because we are operating within the framework of our relative time scale.

C. Beaty (Lethbridge):

One brief comment: I think the most telling piece of evidence supporting the proposition that Hutton-Playfair type geology works is represented by the fact that the geologic column was put together by: (A) formulation and then; (B) application of the big rules: original horizontality superposition, faunal succession and assemblies, and crosscutting relationships, and it hasn't been changed. The relative position of the units in the column has not been altered with the advent and availability of the sophisticated techniques of the so-called new geology. It seems to me that that, in itself, is all the argument you need to defend the idea that there still, even in the latter part of the 20th century, is a lot to be said for old-fashioned geology based upon field observations properly comprehended.

H.B.S. Cooke:

This is where it all starts. But I would not underrate the great importance of some of the new techniques. For example, in the field of tephrochronology it is one thing to say we have an ash in the sequence, but it is another thing to say, "this is the same ash as the one that I have 100 kilometres away". The only answer to this has been the development of very sophisticated techniques of trace element analysis which have enabled one to assign fingerprints to these ashes.

R. Klassen (Geological Survey of Canada):

I would like to say something about basic stratigraphy, its interpretation, and what a stratigraphic unit means within a glacial lake basin. On a field trip last fall near the south end of Glacial Lake Agassiz, I noted that geologists looking at lake basin sediment may have quite different interpretations on the genesis of glacial lake basin deposits. What to one geologist is stony lake clay to another is till. Certainly this difference in interpretation is going to make quite a difference in deciphered glacial history. At the north end of the basin a typical succession consists of varved clay beds several inches thick separated by similar thicknesses of stony clay. This sequence grades downward into stony sand similar to sandy till beds seen at the south end of the basin. I would interpret this type of sequence as entirely lacustrine in origin.

C. Kolb:

Could you expect thermoluminescence to work on material transported by streams?

H.B.S. Cooke:

I think it's inconceivable that the material could be carried by a stream and deposited in a terrace without being exposed to solar radiation.

C. Kolb:

Even though it may have been deposited in say a band a foot and a half thick?

H.B.S. Cooke:

I think that's of no consequence. It has to go through the process of slope erosion from the source and transport into the stream system and what you are dating by thermoluminescence is when it ceased to be exposed to sunlight. So once it is buried you get the date of burial. This means that you might also take samples from different levels in your deposit and get some idea of relative ages. Another possibility is paleomagnetism. I was concerned about it because I believe Don Coates and I were talking about this and the problem was that this material is water laid. Currents affect the orientation of magnetic particles but if the currents are too strong the orientation of the little magnets becomes disturbed, whereas if they settle down in the water column the magnets align themselves in the Earth's field. Diagenesis also may deposit magnetic material and you then have the date of the fixing of the chemical remnants, which may have been considerably later than the original depositions. I think you have more of a problem than you imagine. The last major polarity change was 700,000 years ago, so the chances of your getting a reverse to normal change are not very great. What about the little excursions in between? Well the problem there is that excursions are probably a great deal more common than most paleomagnetists presently admit. This means that there are frequent blips of which only a relatively small number have actually been studied; so if you do find such a blip, you need some other dating method to know whereabouts you are within the paleomagnetic time scale.

C. Kolb:

Thermoluminescence?

H.B.S. Cooke:

Yes, thermoluminescence would do as a control. It might not provide you with an exact date, but it might provide you with a ballpark figure. The great hope would be that in a continuous sequence like that you could study the declination variation and you would get curves which could ultimately be matched with other sequences, as has been done in the Great Lakes area. But the sequences for comparison are not yet available.

C. Kolb:

Drifting of the pole, is what you mean?

H.B.S. Cooke:

Yes, but while we can do this for the last 20,000 years we can't do it for 300,000 years ago because we don't have sequences which have been studied in the right way. And if you don't have the fingerprint to match you cannot connect to the time scale. We have relatively good fingerprints from the nearby area, the Gulf of Mexico, over the last 20-25,000 years, but beyond that, no. It's possible that the top level of your sequence is 700,000 or more years old and of course if you find a reversal in it then you do have something positive.

W.J. Wayne:

In the past, many of the geologists who worked on Pleistocene deposits had never seen a glacier or an active permafrost region, so they have had to rely on imagination rather than observed experience to try to understand glacier or frozen ground processes and interpret the deposits they studied. Fortunately, for many of us today, funding and modern transportation methods have permitted many more geologists to visit and work in the environments that existed around the ice sheets during the Pleistocene. We still have problems that are sedimentologic, though; for example, the differences between ice-laid till and debris from a landslide or mudflow sometimes seem to defy identification. I'm trying to find criteria to use with confidence. Fabric works with some tills; other sediments deposited along and beyond the ice margin really are glacier-generated mudflows, yet I'm sure I have called some of them tills.

C.S. Churcher:

I'd like to say a thing or two about presence and absence of fossils. You know when you are dealing with one particular aspect you think you know what you're doing and rely on other people because you think they know something better than you do. But I have found as a paleontologist that I will perhaps say, "I have got a mammoth here." (as Dick Harington said earlier today), and then someone else will say, "You are in the Irvingtonian," when before I thought it was the Blancan. But what does it really tell you? It tells you only you have an elephant in that place and level. It doesn't tell you that there weren't elephants earlier on. And later on you say "Mammoth dies out 10,500 BP, or whenever." But that doesn't tell you that the elephants did die out then, it tells you that's when you last had the last elephant, or the last bit of evidence that elephants were extant. So there is this binary schism between information that you have and information you don't have. All the land mammal faunas that have been identified (to which Dick Harington again referred and Don Savage put together originally), are useful as concepts. But they do make for troubles because somebody will say, "It is old enough, it has to be Ranchalabrean." However, if you've got this certain animal in it, perhaps it can't be Ranchalabrean, and therefore it has to be some other faunal age. We also know that we've found living fossils in different parts of the world and one has to watch for this sort of "ringer" when it turns up. A very good "ringer" that I like to consider is one that is time transgressive in a really broad sense; it is a zebra called *Plesippus* which lived in North America at the end of the Pliocene and into the early Pleistocene. *Plesippus* has various species, *P. shoshonensis* is one of them, from the Shoshone River in Wyoming, which is obviously a good North American species: the genus was isolated paleontologically for years. Meanwhile in Europe Milne-Edwards described a modern African animal as Grévy's zebra (named after the then President of France) when it first became known to modern Europeans, that is, after the Romans who did know it. This modern zebra has now been put into the sub-genus *Dolichohippus*, as has *Plesippus*. And taxonomists say that effectively we've got the same zebra in the Pliocene of North America and in the African Pleistocene and living in the horn of Africa today. Specifically different perhaps, but horses are very hard to tell apart. Now we known nothing about Africa's relations with Asia at this point when it comes to Grévy's zebra. There are some things about which you can speculate. But somewhere across that Asiatic land mass, perhaps in Beringia, you're going to find Grévy's zebras, or *Plesippus* akin to *P. shoshonensis*, whatever we may call them. And somewhere you're going to have to say when it died out and somewhere you're going to say when it came in. What are we going to put as dates on this history? We're going to have to rely - I'm going to have to rely - on all of you to tell me how old my animal is. And then some of you will come around and say, "I've got your animal, how old is it?" There is then this lovely circular reasoning! And I don't see how we avoid it. We've got the same problem with mammoths. Dick Harington gave you a very nice sequence of mammoth teeth and Basil Cooke has just referred to them. You can do various things with the tooth plates. You can make nice little sequences, but you've got parallel lineages, or co-evolution between lineages if you like. One lot of North American mammoths *Elephas columbi* (the Columbian mammoth) starts off with wide plates and, as it became more "advanced", it becomes an animal with narrower plates able to deal with tougher and presumably more fibrous vegetation, as we like to think. Slightly later, during the Sangamon, *Elephas primigenius* (the Siberian or woolly mammoth) entered North America and has nicely compressed teeth, but the earliest North American *E. primigenius* has teeth that are less compressed than the last of the Columbian mammoths in the Pleistocene-Holocene boundary. Thus we have these two groups evolving in parallel but one lags behind the other. And I'm sure that if we could check the record, (I can't point the finger of scorn at anybody particularly) we're going to find animals identified as *Elephas columbi* 'advanced form' which are actually *E. primigenius* 'primitive form'. Thus the dating of faunas or deposits by the relative compression of the teeth is unsure taxonomically. There are lots of pitfalls in palaeontology like this and as a palaeontologist I need your help to overcome them. I think it was Basil Cooke who was saying that we really need to take care over our circular arguments, and I think a lot of our arguments <u>are</u> circular. If they keep on as you might say, revolving without grating, and without too much of a wobble, then we've probably got a concept that is reasonable to go along with and one that's applicable that we can use as a working model.

<u>R. Barendregt</u> (University of Lethbridge):

I was discussing this point earlier with Dr. Churcher and I think it was he who actually made the point that we have genetic drift where animals coming out of a wooded area onto the plains, for example, may develop slightly different characteristics. These variations are not necessarily a criterion for differentiating the two. As animals disperse, divergent characteristics develop and these are not necessarily indicative of different ages. I think you find this in all plant and animal kingdoms. Similarly in the case of the paleomagnetic record, we see normally and reversely magnetized sediments, but a detailed record may show a secular variation which can be a very good correlative tool. This secular variation may be the result of very localized phenomena, nevertheless the record can be useful for correlation of nearby sections. Maybe Dr. Churcher would like to comment further on this in regards to the bison.

C.S. Churcher:

With the bison, yes, there is the matter of the loss of large horns. You heard Dick Harington say how large the early bison were, no matter what name you use for them, and how the modern bison have small horns. But the trouble in this is that the early samples usually were obtained by people who were concerned with only picking up the well-preserved material. Fair enough, it was the easier thing to do and the better preserved material was usually the better ossified material and usually the bigger the bone is the better it is preserved. It is one of the standard rules of thumb of palaeontology that the little ones don't get preserved as well as the big ones and, anyhow, a mammoth leaves more bits and pieces around than does a mouse. So the big bison horn cores and the big back ends of the skull were picked up, and you've got the huge bison. At the same time it seems now there were smaller bison, there were younger bison, or female bison, whatever the bison were that had smaller horns. Really there has been selectivity on the part of the palaeontologists who took up the bigger specimens only, and the whole variation wasn't known. One of the little suspicions I have at the moment, founded on talking to Dr. Michael Wilson in Calgary, is that the earliest bison populations also had small-horned individuals. These may have been females, but we don't know. So what has happened is that the bigger animals became less useful in some specific sense, and they weren't required to have big horns to be successful males, or for whatever the reason was, and there's a physiological economy if you can be successful and have smaller rather than big horns. So, whatever the reason, let's say they came out onto the plains and could be seen more easily. Thus, instead of having large horns, which cost a lot to make, it would be easier to build a beard and a cape, or whatever you call the things the North American bison have and European bison don't. This is a possible adaptation to being a plains animal as hair is insulation, and hair is much cheaper to produce than great big horns. Now it might be said that this could be genetic drift at some rather low level and wasn't important to the animal except in breeding display. However, we've set up a whole chronology on the horn cores and will have to undo this chronology. The chronology of the large bison goes to about 15,000 BP, I think, followed by a sudden reduction in size, and by about 10,000 BP there are basically small horned bison, which is a very quick evolutionary change. Yet bison first appear in the Illinoian of North America, and hundreds of thousands of years passed during which the bison didn't change the size of their horns. So there must have been a change in habit, and probably a change in habitat as well, to go along with the impulse to change horn-size.

R. Barendregt:

Yes, just as they possibly collected only the best and largest specimens of these horn cores so we only build our magnetic chronology on the best records and we've avoided the poorer or uninteresting records. Nevertheless these records may contain information that should have gone into the construction of the detailed palaeomagnetic time scale.

C.S. Churcher:

We've used only the easy ones, I suspect.

R. Barendregt:

Yes. We start out with the easy ones, and I guess that's natural, but we should also continue to look at the complex records. They're telling us more.

<u>C. Stearns</u> (Tufts University):

Isn't one of the illustrations of your collecting bias that it's really - I won't be able to put this as a number, but isn't it say the last 25 to 40 years that rodents have come into their own?

<u>C.S. Churcher</u>:

Yes. I wouldn't like to go along with 25 to 40, whatever it is, but recently, "yes".

<u>C. Stearns</u>:

It isn't because there were no rodents in the fossil record before, but they were harder to collect.

<u>C.S. Churcher</u>:

That's very true. I have horrible experiences occasionally as when I read an author in the 1920's who said in effect, "I collected all these specimens from Kom Ombo in Egypt. There were many rodent bones, but I didn't bother with them." If I only had had the rodent bones I could tell more than having had all the big animals' bones. The rodents and the smaller animals are considered to be better climatic indicators. Of course, they can't get away from the climate; they've got to adapt to local climate. You know a mouse that goes 100 yards from home has gone a long way, but an elephant can walk all the way from, let's say, Saskatoon down to the Texas Gulf Coast and back again without straining itself, in about 4 months. That gives it 8 months, 3 months up north in the summer and 5 months on the coast, with no effort at all. You get problems from this because people don't consider the possibility of large animals' migrations. They don't realize that if you had dozens of mammals that are going south, you might have *Elephas primigenius* coming south in the winter too. There would be this migratory shift going on annually. And so you might find *E. primigenius* in winter and *E. columbi* in summer turning up in the same local deposit. Then of course if you don't consider that this could happen, what are you going to call that mixed population. And who can say what the mixed sample tells you about the age of things. A mixed mammoth population has to be markedly divergent or you get the age wrong on the basis of the tooth plates.

<u>C. Stearns</u>:

In talking about dating a stratigraphic succession we are talking about recognizing events in geologic history, arranging them in the proper order, and then seeing if we can attach something like a time in years or on somebody's calendar. We can still disagree about the <u>nature</u> of the events that we are placing in order and retain complete <u>agreement</u> about the order in which those events occurred. Disagreement about the nature of events may be particularly embarrassing in Quaternary studies because we try to make so many correlations in relation to our notions about sequences of climatic change. The nature of the event therefore seems important. But, in fact, relative age and even absolute age can be established independently of the nature of the event.

<u>D. Coates</u>:

Well there are three questions that I think emerge at an affair like this which are - what did we accomplish, what should we have accomplished, and where do we go from here. We can't solve all the problems of the universe. Is there a next step or a next symposium or will we lose whatever momentum has been gained here. Where does this lead us in the future? What have you learned or what would you say has been accomplished.

Let me give a little preamble concerning the symposium. The one item missing is environmental relevance. I purposely kept this out of my talk to see whether I could give a presentation that did not involve environmental and economic applications of different dating methods and how such methods can be used in the "real" world. For example, I purposely didn't say anything in my paper about nuclear sites. In siting such plants dating of Quaternary events is very important. Nuclear Regulatory Commission guidelines require that plants should not be constructed where there is a fault that has moved within the last 35,000 years, or where there have been multiple occurrences of a fault in the last 500,000 years, or where there is a fault nearby which is of high magnitude. I've been involved in several of these investigations and we have used many of the types of dating that have been mentioned. We've also used others that didn't emerge from these meetings, such as temperature differentials and how these can give a pattern of events through different kinds of fluid inclusion work which is very applicable to the things that we're talking about. The engineers, as Dr. Kolb well knows, when they design dams or flood walls they must know whether they're going to design for a 100-year flood or some other recurrence interval. The people at St. Helen's are now wondering what is going to be the recurrence interval of earthquakes and volcanic activity for that region. So we have a lot to say in terms of the real world, in land use planning and on economic resources and development. All of these are in large measure dependent upon either absolute or relative dating in some way. And this is perhaps one topic that could be addressed at some future meeting.

QUATERNARY STRATIGRAPHY OF THE COASTAL BLUFFS OF LAKE ONTARIO EAST OF OSHAWA

I.P. MARTINI, M.E. BROOKFIELD and Q.H.J. GWYN

ABSTRACT

The shore bluffs between Oshawa and Port Hope have good, albeit discontinuous, exposures of Wisconsin glacial and interstadial deposits. The stratigraphic units show strong lateral variations in thickness and lithology. Most of them have undergone considerable local erosion. Some are found as isolated remnants.

The reconstructed generalized stratigraphic sequence can be correlated on the basis of gross lithology and stratigraphic position to the classical sequence of Scarborough (to date no sufficient organic material has been found in the eastern sections to be dated). Variations that occur between the two areas relate to different environments and to the closer proximity of the eastern sequences to the Precambrian source and to a calcareous shale substratum.

To facilitate communication, tentative informal names have been given to the major till layers. Accordingly, the Port Hope till is correlated with the Sunnybrook till of Scarborough, the Bond Head till with the Seminary (or Meadowcliffe (?)), and the Bowmanville with the former Lower Leaside. Local occurrences of a blocky, silty upper till can be tentatively correlated lithologically with the Halton till.

Detailed analysis of the waterlaid sequences has revealed well developed interstratifications of waterlaid tills (pebbly sandstones) within other lacustrine deposits, and, most importantly, large valleys probably cut during the Plum Point interstadial and filled by cross-bedded, sorted sands, and in part, by Bowmanville till.

INTRODUCTION

Thick interlayered deposits of till, glacio-fluvial and glacio-lacustrine sediments are exposed along the bluffs of the north shore of Lake Ontario between Toronto and Port Hope. They are eroded locally at a very fast rate (30 cm/year, Terasmae *et al.*, 1972). Considerable research has been done on the erosional processes and gross stratigraphy of the deposits (Karrow, 1967, 1974; Singer, 1974). The objectives of this paper and field guide are to establish the stratigraphic framework for those materials, extending eastward the scheme Karrow (1967, 1974) established in the Scarborough area, and to reconstruct sedimentologically the events that led to the emplacement and disruption of great lakes during Wisconsin and Holocene times (Figure 1). One major drawback to this work is the absence of organic material for absolute dating from the sections east of Oshawa. Within this area, and between this area and the Scarborough bluffs. only lithostratigraphic correlations are possible. Because of the regional variability in pebbles and boulders and in the calcite/dolomite ratio, locally influenced by the bedrock, some of the correlations may not stand the test of close scrutiny. However, the resultant stratigraphic scheme is good enough to allow a sedimentological analysis of the various environments to be made (Figure 2).

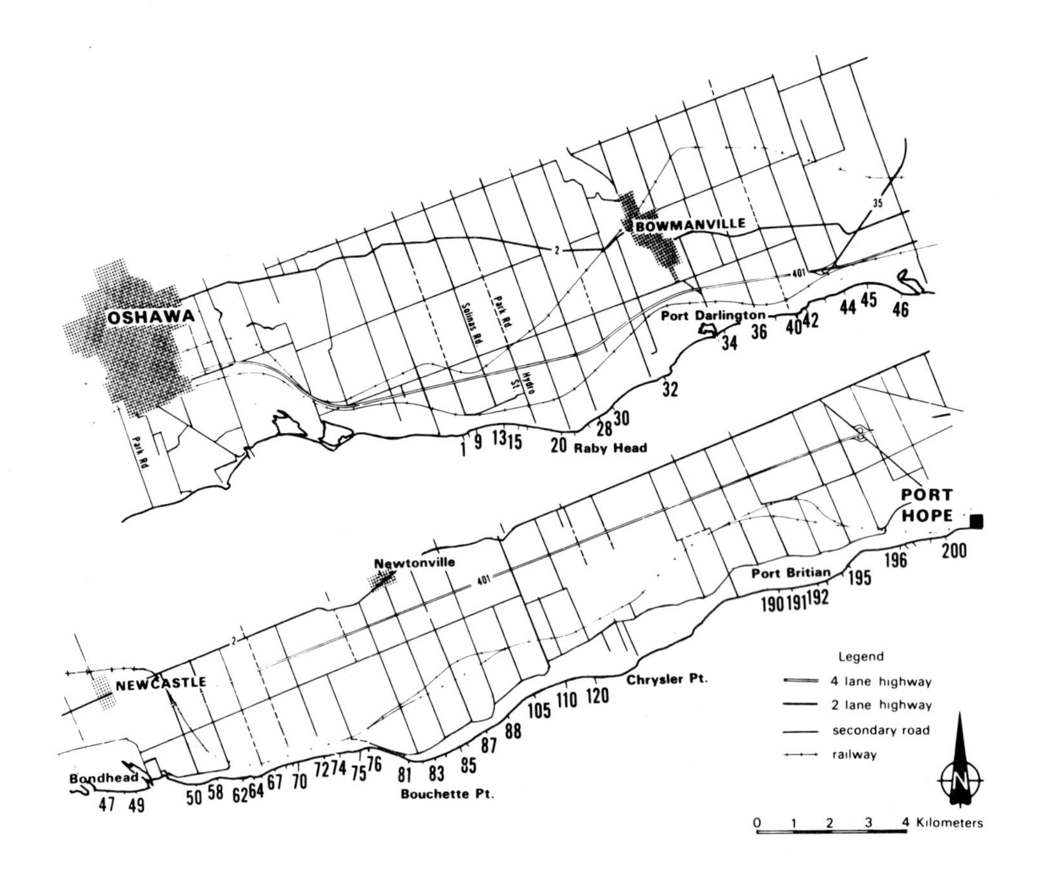

Figure 1 Location map of sections of the coastal bluffs. The numbering sequence of Singer (1974) was followed extending it eastward to Port Hope. Sections 121-189 have not been measured.

Of the several sedimentological facies present, some tills and varved lacustrine sequences show consistent characteristics over a wide distance and they can be utilized as key beds. Four major tills are found in the Oshawa-Port Hope area having characteristics similar to layers of the Scarborough sections and are exposed at the correct stratigraphic position to be tentatively correlated. Other tills are present but they are lenses and do not serve correlative purposes. The four best developed tills, and their correlations are as follows.

The lower till [proposed name: Port Hope Till (Brookfield *et al.*); Lower Glacial Till Unit of Singer, 1974] is a moderately compact, dark grey, silty till with approximately 5% pebbles most of which (more than 90%) are of local carbonate origin. The distribution of these pebbles is patchy, from local virtual absence to clusters of well imbricated stones. The matrix is calcareous (less than 20%) and shows a variable calcite/dolomite ratio averaging approximately 2.9 to 3.0 (Table 1; Singer, 1974; Brookfield *et al.*, in preparation). The concentration of heavy minerals is low (0.25%). This till is best developed in the Port Hope area where it reaches a thickness of 8m. It is present at other localities, but in many other areas it lies below the present day lake level (Figure 2). In the Raby Head area Singer shows that this till and the whole Pleistocene sequence disappear either because of non deposition or most probably because of erosion, approximately 0.5 km off-shore at the edge of the shelf of Lake Ontario (Singer, 1974: Figure 5). This till had been correlated by previous authors with the Sunnybrook Till of the Scarborough Bluffs. Main lithological differences between the two tills are that the Port Hope till has less clay (approximately 30%) and a higher carbonate content (Table 1).

The middle till (proposed name: Bond Head Till (Brookfield *et al.*); probably Middle Glacial Unit of Singer, 1974) is a compact, dark grey,

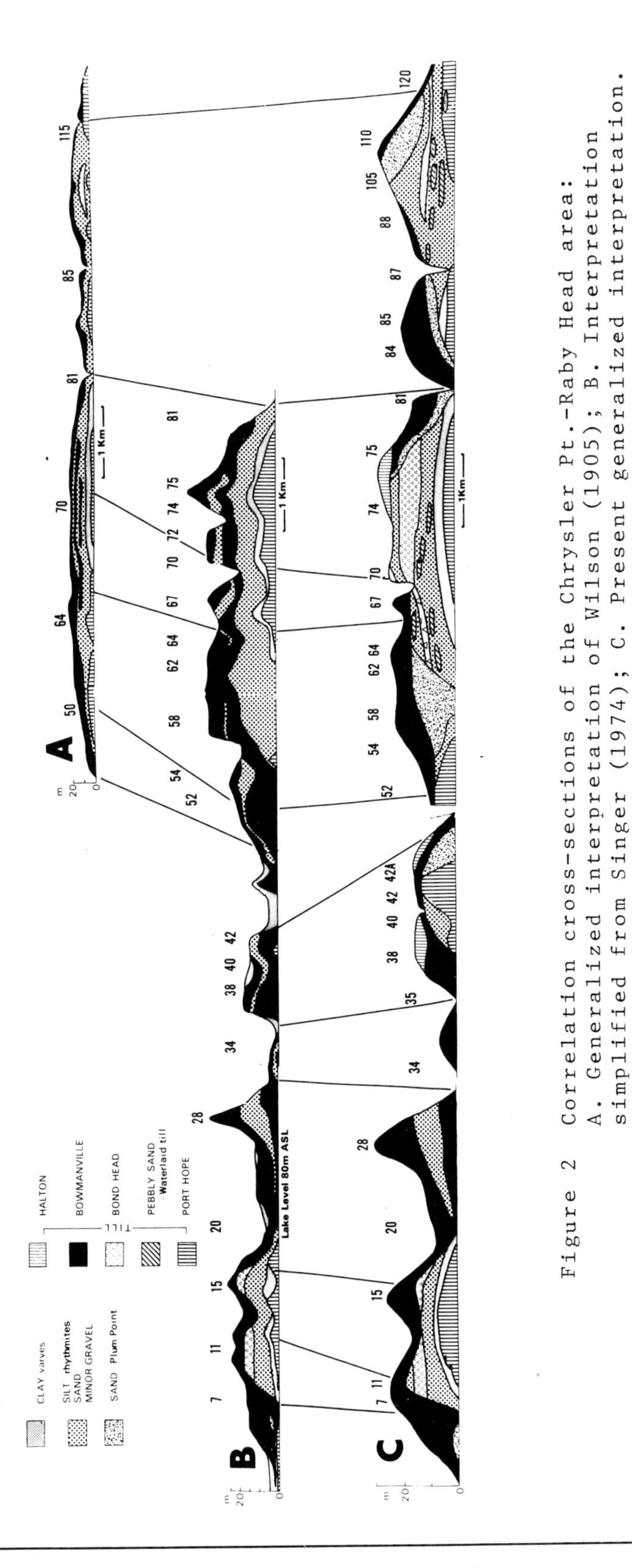

Figure 2 Correlation cross-sections of the Chrysler Pt.-Raby Head area: A. Generalized interpretation of Wilson (1905); B. Interpretation simplified from Singer (1974); C. Present generalized interpretation.

silty (40 to 70%) till with less than 5% pebbles. It has an average high calcite content (35%) and a relatively high average calcite/dolomite ratio (5.5; Singer, 1974). Its stratigraphic position led Singer to correlate it with the Meadowcliffe Till of the Scarborough section. However, its texture is much more similar to that of the Seminary Till (Table 1). It is a distinct cliff-forming till reaching a maximum thickness of 4 m between Bond Head and Bouchette Point, but on most other sections along the bluffs it was not deposited or was eroded off.

Table 1. Till analyses. Data for Scarborough are from Karrow (1967). Data for Newcastle are from Singer (1974), and (in italics) from Brookfield *et al.* (in preparation).

AREA	UNIT	Thick.	Grain Size %			Pebbles %					Carbonate %			
		m	Clay	Silt	Sand	LS	Dol.	SH	SS	PG	Calc.	Dol.	Tot. Carb.	Cal/ Dol.
SCARBOROUGH	Halton Till (Leaside Till) Sandy Till	3-17	20	33	47	71	-	14	-	14	25	6	31	4.0
SCARBOROUGH	Meadowcliffe	0-14	54	35	11	58	5	18	-	19	24	6	30	4.0
SCARBOROUGH	Seminary	0-8.5	33	30	37	55	5	30	-	10	18	7	25	2.7
SCARBOROUGH	Sunnybrook	7-10	45	37	18	45	12	26	-	7	6	6.5	12.5	0.9
NEWCASTLE	*Halton Till*	*1-2*	*10*	*57*	*33*						*25*	*5*	*30*	*6*
NEWCASTLE	Upper Glacial Unit (upper part)	*2-15*	12	37	51						39	6	45	6.5
NEWCASTLE	*Bowmanville Till*		*7*	*42*	*51*						*27*	*7*	*34*	*4*
NEWCASTLE	Upper Glacial Unit (lower part)		*9*	*45*	*46*						*27*	*4*	*31*	*6*
NEWCASTLE		*1-10*	12	39	49						36	7	43	5.2
NEWCASTLE	*"Pebbly Sand"*													
NEWCASTLE	*Bond Head Till*	*1-4*	*14*	*62*	*24*						*19*	*6*	*25*	*3.5*
NEWCASTLE	Middle Glacial Unit		32	54	14						34	7	41	5.5
NEWCASTLE	*Port Hope Till*	*1-13*	*15*	*56*	*27*						*10*	*7*	*17*	*1.3*
NEWCASTLE	Lower Glacial Unit		32	51	17	90	-	-	-	10	21	7	28	2.9

The upper till(s) (proposed name: Bowmanville Till(s) (Brookfield *et al.*); Upper Glacial Unit of Singer, 1974) is a stony (up to 15% pebbles) sandy (40-60%) till. In most places it is split into an upper and a lower sheet by interlayered sand, gravel or by a stone pavement. Although highly variable from location to location, lithologically the two sheets are very similar, except for a looser structure and perhaps a greater amount of Precambrian boulders in the upper unit. Generally these tills are oxidized and are gray brown in color. Occasional thin lenses of sand and apparent stratifications occur in the uppermost unit. The tills are rather calcareous (in excess of 40%; Singer 1974), have a high calcite/dolomite ratio (in excess of 5) and have a low heavy-mineral content (0.54%, Table 1). Locally, they reach a combined thickness of up to 15m. They have a strongly undulating lower and upper surface (Figures 1, 2). Gwyn (Pers. comm.) has traced the Bowmanville Till(s) northward to the Oak Ridge Moraine and westward to Whitby. This till forms the core of the northwest-southeast trending drumlins south of the Oak Ridge Moraine (Figure 3; Chapman and Putnam, 1966, Gravenor, 1957). This till was correlated by Singer with the Leaside Till of Scarborough (Karrow, 1967). More recently, the Leaside Till has been subdivided into two units (Karrow, 1974; Morgan, 1979). On the basis of its lithology and stratigraphic position, the Bowmanville Till(s) is correlated with the 'sandy unit', which was the former lower Leaside Till.

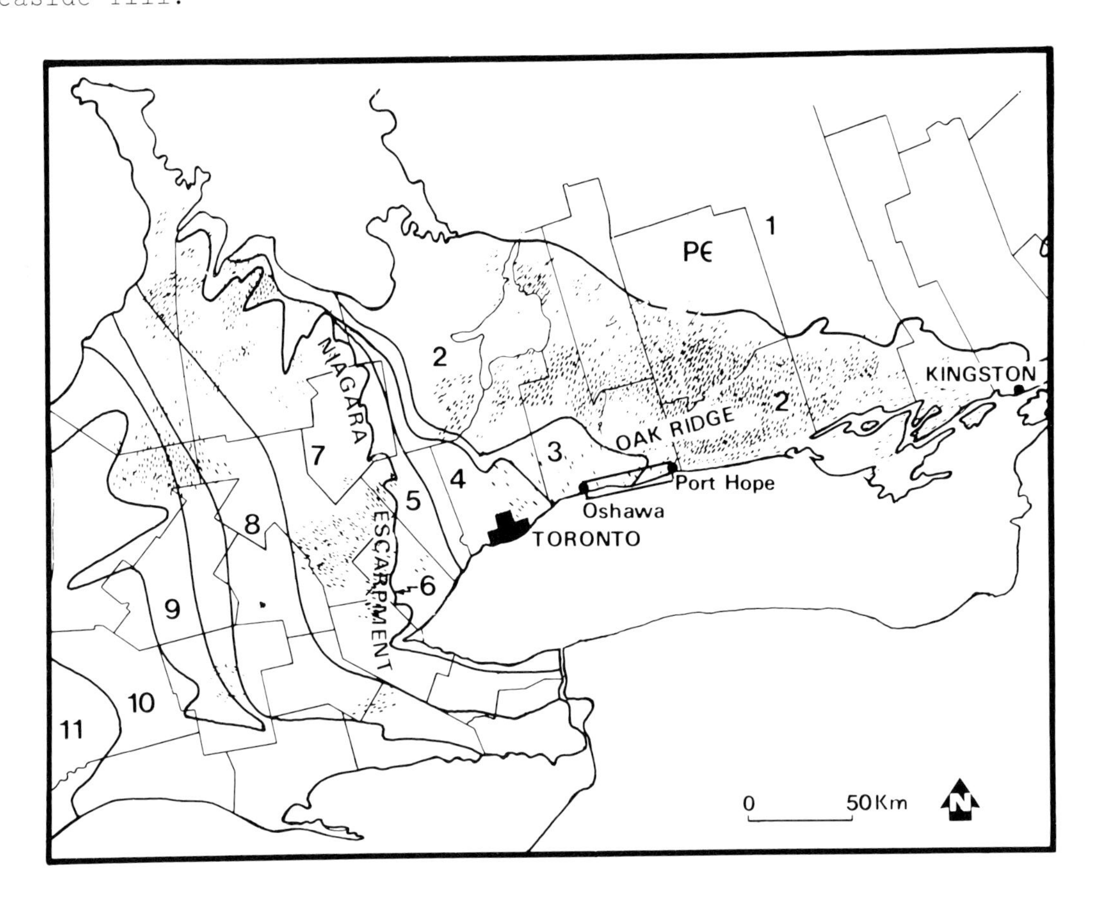

Figure 3 Paleozoic geology and Pleistocene drumlins of parts of Southern Ontario (after Chapman and Putnam, 1966). 1. Precambrian; 2. Trenton-Black River; 3. Collingwood; 4. Meaford-Dundas, Blue Mtn.; 5. Queenston; 6. Medina-Clinton; 7. Lockport-Guelph; 8. Salina and Bass Island; 9. Bois Blanc; 10. Delaware; 11. Hamilton.

The uppermost till was not previously recognized in the bluffs between Oshawa and Port Hope. It is thin (1-2 m) and present only in a few sections. However, it is sufficiently distinct to be correlated lithologically with the Halton Till of the Scarborough and Oak Ridge areas (Gwyn, pers. comm.). It is a grey brown, clayey, silty, essentially pebble free till, showing well developed blocky structure. It has an average carbonate content of approximately 30% and a calcite/dolomite ratio of 6. Where present, it may either directly overlie the Bowmanville Till(s) or it may cap sandy units (Figure 2).

MAJOR INTERSTADIAL UNITS

Four tills form the stratigraphic framework of the Oshawa-Port Hope bluffs (Figure 2). They record major advances of the glacial lobes during Middle and Late Wisconsin times. Interstadial sediments are characterized by recurring sequences of lacustrine and glacio-fluvial deposits, locally interstratified with flow tills and waterlaid, unsorted, lensing massive to weakly stratified sandy tills best described by the field term of 'pebbly sands'. Three major lacustrine events have occurred, with possible other smaller pondings occurring at one time or another in different parts of the section. The Lower Lacustrine event (part of the Clarke unit of Wilson, 1905; Coleman, 1909; Singer, 1974) starts with well developed distal varves (clayey), locally intensely deformed and slumped, grading upwards into increasingly siltier and sandier rhythmites, into proximal varves (sandy) with well defined ripple-drift cross-laminations, and finally into massive to slightly rippled and crossbedded sands (Figures 4,5). Locally these upper sands are cut by deep channels filled with gravel and gravelly sands. In the upper part of this sequence, particularly in the area between Bond Head and Bouchette Point, the lacustrine sequence is irregularly interstratified with subaqueous debris (till like units) and 'pebbly sands' (waterlaid flow till). Generally these tills occur and are best developed in association with the Bond Head Till. The silty nature of the Bond Head Till reflects its cannibalistic behaviour in englobating the fine materials of the underlying lacustrine sequence.

After the retreat of the ice responsible for the sedimentation of the Bond Head Till, a second important lake stage (Upper Lacustrine) is recorded directly on top of the till, or it overlies the sandy units of the Lower Lacustrine event (Figure 2, 4, 5). This second major lacustrine unit shows more proximal and shallower characteristics as the siltier component of the varved couplets prevails. Irregular stratifications of well sorted sands record subaqueous channels very wide (5m) and shallow (10 to 30 cm), possibly scoured and filled by turbidite events. Similarly to the Lower Lacustrine deposits, this Upper Lacustrine unit shows a well defined regressive sequence, the uppermost part of which is not present, however, because it is truncated by a major unconformity.

The absolute time of this unconformity cannot be established through direct evidence from the bluffs. Circumstantial considerations would suggest a major lowering of the lake level during Plum Point times (Dreimanis and Karrow, 1972). The unconformity is characterized by deep valleys cut into the lacustrine sequence of the post-Bond Head till times, locally throughout the Pleistocene sequence down to the present day lake level (Figure 2). These valleys were formed at a particularly low stage (lower than the modern Lake Ontario) of the ancient glacial lakes, when the opening of some southward flowing outlets occurred during the Plum Point interstadial (Mörner, 1971). The valleys were subsequently filled by well sorted, crossbedded nearshore sands locally reworked into well defined channel fills. As the glaciers covered this area, parts of the deposits were eroded to form the sandy overlying tills, and some valleys were filled with the Bowmanville Till. The irregular surfaces of the Bowmanville Till(s) have been recognized by previous authors, however the importance of the preceding unconformity was not, nor were the existence of valleys filled with sand, noted. Indeed, these sands have been previously considered as part of the Clarke Unit (Figure 2). The slopes of the valley sides and the lithology and structural characteristics of the sands of the valleys and their paleocurrent directions, variable but predominantly to the west, differentiate them from the massive and rippled sands of the Lower and Upper Lacustrine sequences. Whereas the lacustrine units together with

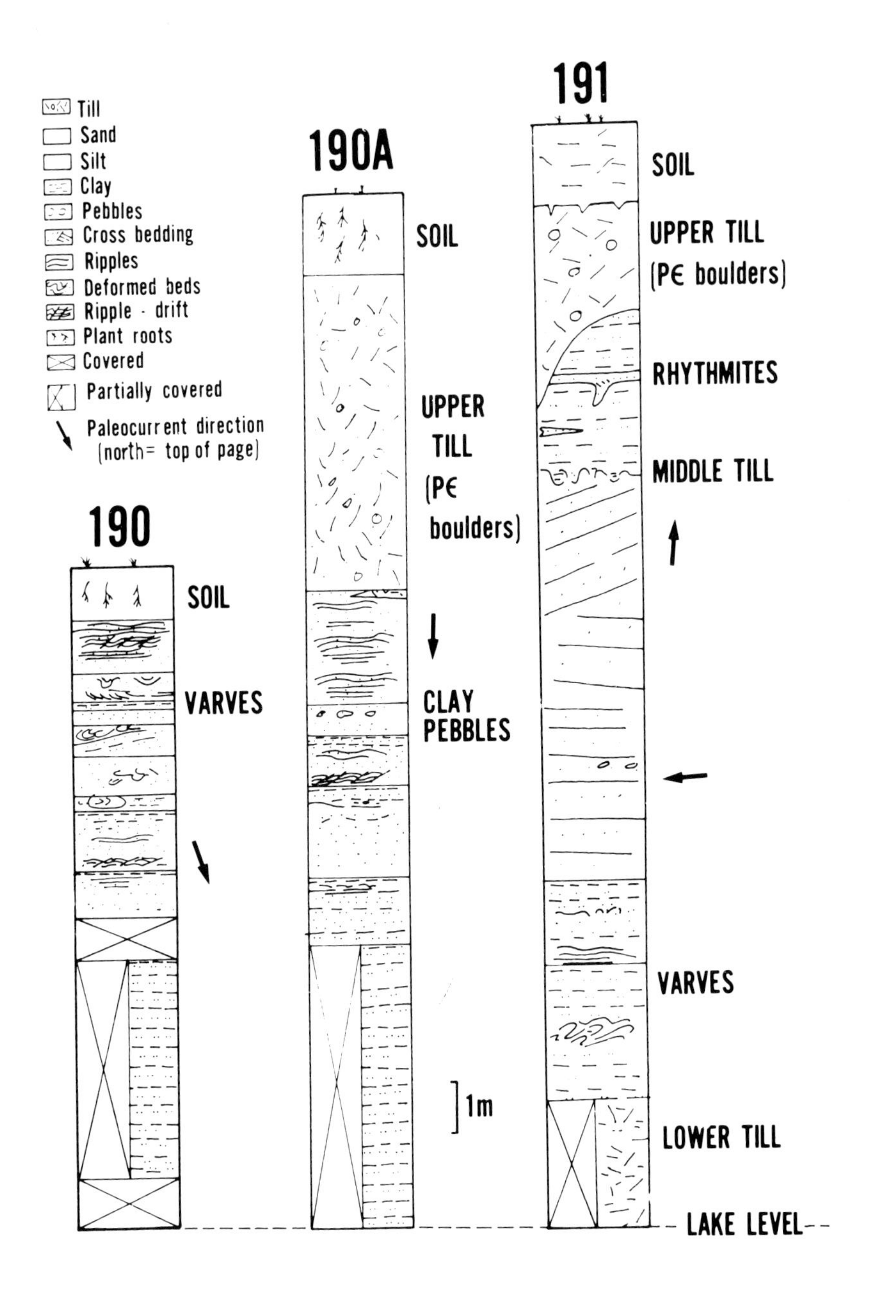

Figure 4 Detailed stratigraphic sections from the Port Britain area.

the Bond Head Till and the sandy waterlaid till interlayers may be correlated to the complex Thorncliffe Formation of Scarborough subdivided locally by the Seminary and Meadowcliffe (Middle Complex of Coleman, 1932) tills, the sandy valley fills and the large unconformity have not been reported in the Scarborough area (Mörner, 1971; Karrow, 1974). If they exist there at all, they should be found in the Upper Thorncliffe sandy units where it may be difficult to recognize the valley fills from the other regular deltaic sands.

The last lacustrine events recorded in the eastern bluffs of the north shore of Lake Ontario have developed over the Bowmanville Till, or where present, over the Halton Till. They relate to the deposits of the early Lake Iroquois. Subsequent changes that have occurred to the distribution of sediments and to the morphology of the area are due to the lowering of the lake level to Lake Ontario (Chapman and Putnam, 1966; Karrow, 1967). The Lake Iroquois deposits are discontinuous in

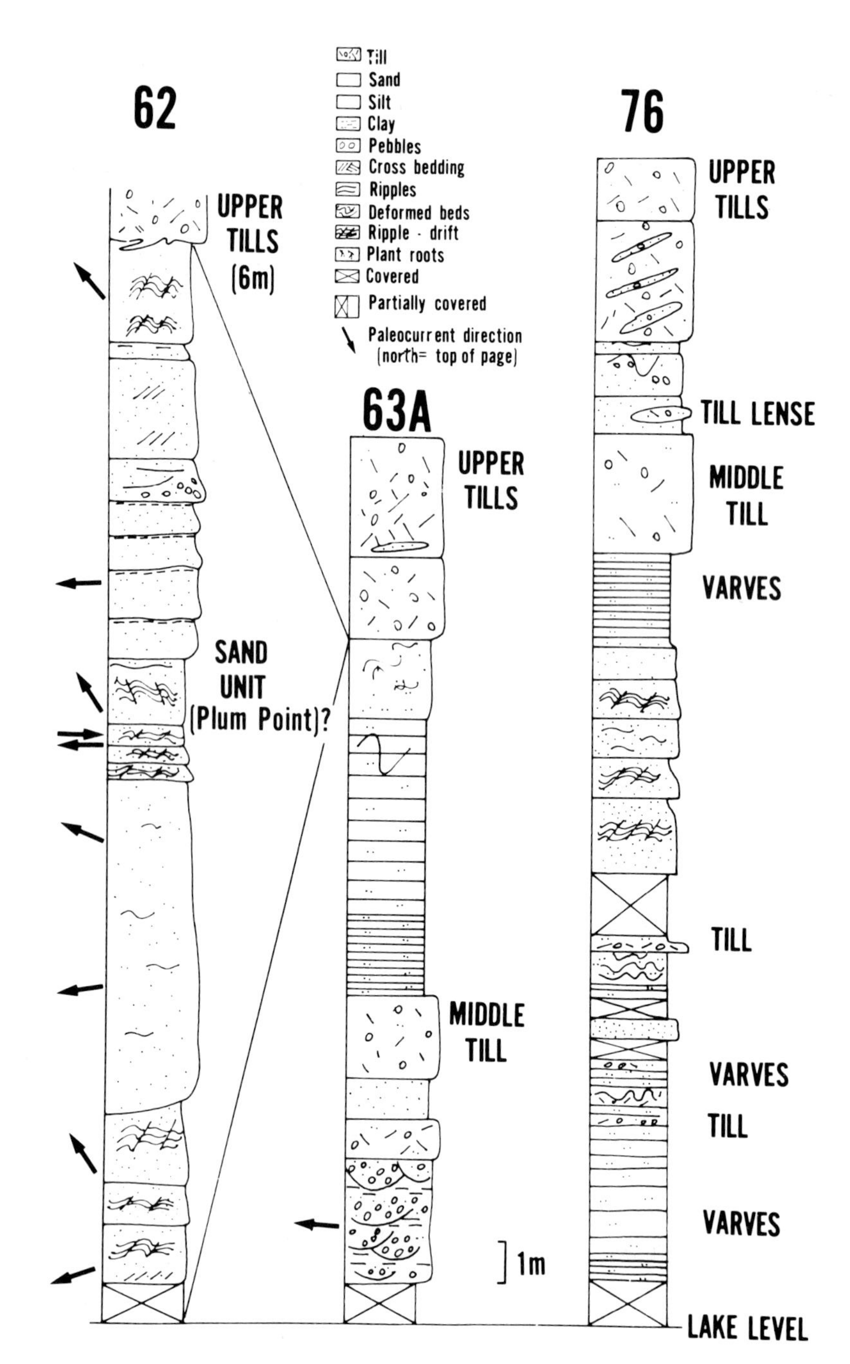

Figure 5 Detailed stratigraphic sections from the Bouchette Pt. - Bond Head area.

distribution as they have been intensely eroded during the lowering of the lake level. Locally in small valleys, good diamictic distal varves are preserved. These grade laterally and particularly landward into sands of nearshore bar and beach deposits. During lowering of the lake level, valleys have formed, and local fluvial sands and gravels were deposited. At the foot of the bluffs, the present day beach is characterized primarily by thin, narrow pebbly deposits.

SUMMARY AND CONCLUSIONS

The series of events recorded in the bluffs between Oshawa and Port Hope is as follows:

- Advance of a glacial lobe which cannibalizes pre-existent fine lacustrine deposits and forms the silty clay, basal Port Hope Till. Locally reworked blocks of varves are recognizable in the upper part of this till.

- Retreat of the glacier and formation of a relatively deep lake where distal varves could form and locally slump off the submerged toe of the glacier or local till highs. The varves become increasingly regular upward and, as the glacier retreats farther eastward, a drop in lake level led to the formation of siltier and sandier proximal varves of shallow lacustrine deposits. Locally, subaerial braided stream conditions were established.

- A rapid readvance of the glacier lobe led to the formation of tills flowing off the front of the glacier into the glacial lake, and interstratification occurred between the waterlaid tills and varved silts and sandy units, some of which were possibly formed as subaqueous outwash.

- This glacial advance culminated in the formation of the clayey silty Bond Head Till which had a discontinuous distribution, mostly because it has been subsequently eroded, but also because of irregular depositional thickness. In places, the Lower Lacustrine unit is separated from an Upper Lacustrine event by a few cm (5-10 cm) of till representing all that was deposited at that locality of the Bond Head Till.

- After the deposition of the Bond Head Till a major retreat of the glacier occurred. This led to the regressive sequence of the Upper Lacustrine unit whose basal layers are made up of well developed but siltier varves than those of the Lower Lacustrine, capped by rippled sands.

- During lower stages of the lake, possibly during Plum Point times, deep valleys were entrenched and partially filled by well sorted crossbedded sands.

- A readvance of the glacial lobe in this area led to reworking of some of the sands of the lacustrine and subaerial events. The sandy tills of the Bowmanville Till(s) were formed under fluctuating lobe conditions. The Bowmanville Till is separated into two units by locally well defined shallow ponds filled by sand. Finally the Halton Till, possibly reflecting a further fluctuation of the glacier or simply a local till facies, was deposited in some areas.

- After this event the glacier retreated, not to return. The Lake Iroquois sequence, characterized by lower distal varves capped by more proximal nearshore sands, was formed. Later lowering of the lake level below modern Lake Ontario resulted in the erosion of part of the Lake Iroquois sequence, the scouring of valleys and possibly the local formation of braided sandy and gravelly sand deposits. Isostatic uplift of the eastern outlet raised Lake Ontario again to its present level.

- Present activities along the lake consist of very active erosion of the bluffs, reworking of the finer materials westward, and formation of thin, narrow, gravelly beaches at the base of the bluffs.

ROAD LOG

Due to the intense spring slumping of the bluffs which drastically modifies the exposures, and because of the reduced accessibility to the outcrops due to mudflows and water level of Lake Ontario, no specific stop description is given here. Instead, the major features to be observed and discussed at selected locations are indicated.

1. Port Britain (Sections 190-191; Fig. 4)

- Lower till capped by deformed varves. Note orientation of recumbent folds.

- Deformed varves grading upward into regular varves, then into sandy rhythmites, and into massive to slightly crossbedded sands.

- Transition between the Lower Lacustrine Sequence and the Upper Lacustrine Sequence marked by thin clay pebbly layers possibly correlated to the Bond Head Till.

- Upper Lacustrine sequence characterized by silt-rich rhythmites, and turbiditic events.

- Upper Till (Bowmanville Till) cutting through the lacustrine sequences.

- To the east of the sections visited a valley (Plum Point (?)) was partially filled by sandy gravels, later covered by Bowmanville Till.

- To the west of section 191 note the undulated lower part of the Bowmanville Till. Are some of the sandy gravels found in cut and fills (Section 190) of Lower Lacustrine age, or are they the precursors of the Bowmanville Till, thus part of the Plum Point (?) valley fill?

2. Bouchette Point East (Sections 80-81)

- Thick section of Bowmanville Till, locally subdivided into two units and showing some pseudo-stratifications in the upper unit.

- To the west of this area is a gradational rise of the section and a development of the Bond Head Till.

3. Bouchette Point West (Sections 74-72; Figure 5)

- Lower till (Port Hope Till) overlain by regular distal varves, some of which are locally folded.

- Vertical transition of distal varves to more proximal rhythmites with well developed ripple-drift cross-laminations.

- Complex interrelations between "pebbly sands" (waterlaid tills) and sandy gravels of cut and fills. What is the age and significance of these cut and fills? Possible subaqueous outwash?

- Good exposures of Bond Head Till.

- Development of upper lacustrine sequence over the Bond Head Till.

- Vertical gradation of the upper lacustrine sequence from diadactic varves to silty rhythmites, to thick, apparently massive, sandy beds.

- Upper till: Is it Bowmanville Till, Halton Till or both?

4. Bond Head East (Sections 58-60; Figure 5)

- Sandy fill of a Plum Point (?) valley.

5. Port Darlington East (Sections 39-42)

- Valley between two till layers filled with sandy, proximal varves showing well developed ripple-drift cross-laminations. The sand is of uncertain age. Singer (1974) interprets it to be a sandy unit separating the lower from the upper till unit of the Bowmanville Till. Brookfield *et al.* offer an alternative solution considering the till exposed at lake level at section 42 to be the Port Hope Till, thus interpreting the section as a complex one where the Port Hope Till is separated from the Bowmanville Till by a remnant of the Lower Lacustrine Unit to the west of section 42 and sand of a Plum Point (?) valley fill to the east.

6. Raby Head East (Sections 27-30)*

- Large valley fill between the upper and lower units of the Bowmanville Till. The valley is filled with rippled and cross-bedded well sorted sands.

REFERENCES CITED

Brookfield, M.E., Gwyn, Q.H.J. and Martini, I.P., Quaternary sequences along the north shore of Lake Ontario: Oshawa-Port Hope. (In Preparation).

Chapman, L.J. and Putnam, D.F., 1966, The Physiography of Southern Ontario: Univ. of Toronto Press, 386 p.

Coleman, A.P., 1909, Classification and nomenclature of Ontario drift: Ontario Bureau of Mines, v. 18, p. 294-297.

____________, 1932, The Pleistocene of the Toronto region; accompanied by Map 418: Ontario Dept. Mines, v. 41.

Dreimanis, A. and Karrow, P.F., 1972, Glacial history in the Great Lakes-St. Lawrence Region, the classification of the Wisconsin(an) Stage, and its correlatives: 24th Intern. Geol. Congress, Section 12, p. 8.

Gravenor, C.P., 1957, Surficial geology of the Lindsay-Peterborough area. Ontario, Victoria, Peterborough, Durham and Northumberland counties, Ontario: Geol. Surv. Canada, Mem. 288, Ottawa.

Karrow, P.F., 1967, Pleistocene geology of the Scarborough area: Ontario Dept. of Mines, Geol. Report 46.

____________, 1974, Till stratigraphy in parts of southwestern Ontario: Geol. Soc. Amer. Bull., v. 85, p. 761-768.

Karrow, P.F. and Morgan, A.V., 1975, Quaternary stratigraphy of the Toronto area: Geol. Assoc. Can., Min. Assoc. Can., N.C. Geol. Soc. Am., Field Trips Guidebook. Univ. of Waterloo, p. 161-179.

Liberty, B.A., 1969, Paleozoic Geology of the Lake Simcoe area, Ontario: Geol. Surv. Canada, Mem. 355, Dept. of Energy, Mines and Resources, Ottawa.

Morgan, A.V., 1979, A field guide to the Don Valley Brick-Pit and the Scarborough Bluffs, Toronto, Ontario in Mahaney, W.C., ed., Quaternary Climatic Change Symposium, Abstracts with Program and Field Guide, York University, Toronto, p. 77-99.

Mörner, N.A., 1971, The Plum Point interstadial: Age, climate and sub-division: Canadian Jour. of Earth Sciences, v. 8, p. 1423-1431.

Singer, S.N., 1974, A hydrogeological study along the north shore of Lake Ontario in the Bowmanville-Newcastle Area: Ontario Ministry of the Environment, Water Resources Report 5d, 72 p.

Terasmae, J., Karrow, P.F. and Dreimanis, A., 1972, Quaternary stratigraphy and geomorphology of the eastern Great Lakes region of Southern Ontario, Guidebook for Excursion A 42: XXIV Intern. Geol. Cong., Montreal, Quebec.

Wilson, A.W.G., 1905, A forty-mile section of Pleistocene deposits north of Lake Ontario: Canadian Inst. Trans., Toronto, v. 8, p. 11-21.

---

*Subject to permission from Ontario Hydro to enter the locality.

INDEX